# MODERN NMR TECHNIQUES FOR SYNTHETIC CHEMISTRY

# NEW DIRECTIONS in ORGANIC and BIOLOGICAL CHEMISTRY

Series Editor: Philip Page

PUBLISHED TITLES

**Activated Metals in Organic Synthesis**
*Pedro Cintas*

**The Anomeric Effect**
*Eusebio Juaristi and Gabriel Cuevas*

**C-Glycoside Synthesis**
*Maarten Postema*

**Capillary Electrophoresis: Theory and Practice**
*Patrick Camilleri*

**Chemical Approaches to the Synthesis of Peptides and Proteins**
*Paul Lloyd-Williams, Fernando Albericio, and Ernest Giralt*

**Chiral Sulfur Reagents**
*M. Mikolajczyk, J. Drabowicz, and P. Kielbasiński*

**Chirality and the Biological Activity of Drugs**
*Roger J. Crossley*

**Concerted Organic and Bio-organic Mechanisms**
*Andrew Williams*

**Cyclization Reactions**
*C. Thebtaranonth and Y. Thebtaranonth*

**Dianion Chemistry in Organic Synthesis**
*Charles M. Thompson*

**Mannich Bases: Chemistry and Uses**
*Maurilio Tramontini and Luigi Angiolini*

**Modern NMR Techniques for Synthetic Chemistry**
*Julie Fisher*

**Organozinc Reagents in Organic Synthesis**
*Ender Erdik*

# MODERN NMR TECHNIQUES FOR SYNTHETIC CHEMISTRY

Edited by
JULIE FISHER
University of Leeds
School of Chemistry
Leeds, United Kingdom

CRC Press is an imprint of the
Taylor & Francis Group, an **informa** business

CRC Press
Taylor & Francis Group
6000 Broken Sound Parkway NW, Suite 300
Boca Raton, FL 33487-2742

First issued in paperback 2021



Version Date: 20140818

ISBN 13: 978-1-03-209894-4 (pbk)
ISBN 13: 978-1-4665-9224-7 (hbk)



Publisher's Note
The publisher has gone to great lengths to ensure the quality of this reprint but points out that some imperfections in the original copies may be apparent.

**Library of Congress Cataloging-in-Publication Data**

Modern NMR techniques for synthetic chemistry / [edited by] Julie Fisher.
pages cm. -- (New directions in organic & biological chemistry)
Includes bibliographical references and index.
ISBN 978-1-4665-9224-7 (hardback)
1. Magnetic resonance microscopy. 2. Spectrum analysis. I. Fisher, J. (Julie), editor.

QD96.N8M624 2015
543'.66--dc23 2014032308

**Visit the Taylor & Francis Web site at**
**http://www.taylorandfrancis.com**

**and the CRC Press Web site at**
**http://www.crcpress.com**

*To John, and our daughters Elizabeth and Abigail.*

**—J. F.**

# Contents

Series Preface .......... ix
Preface .......... xi
Acknowledgements .......... xiii
Editor .......... xv
Contributors .......... xvii

**Chapter 1** The Basics .......... 1
*Julie Fisher*

**Chapter 2** Dynamic NMR .......... 15
*Alex D. Bain*

**Chapter 3** NMR in Ligand Binding Studies .......... 63
*Michelle L. Rowe, Jane L. Wagstaff and Mark J. Howard*

**Chapter 4** Diffusion: Definition, Description and Measurement .......... 125
*Scott A. Willis, Timothy Stait-Gardner, Amninder S. Virk, Reika Masuda, Mikhail Zubkov, Gang Zheng and William S. Price*

**Chapter 5** Multi-Nuclear Magnetic Resonance Spectroscopy .......... 177
*Jonathan A. Iggo and Konstantin V. Luzyanin*

**Chapter 6** NMR and Complex Mixtures .......... 225
*Cassey McRae*

**Chapter 7** Selected Applications of NMR Spectroscopy .......... 263
*Robert Brkljača, Sylvia Urban, Kristian Hollingsworth, W. Bruce Turnbull and John A. Parkinson*

**Index** .......... 329

# Series Preface

## NEW DIRECTIONS IN ORGANIC AND BIOLOGICAL CHEMISTRY

Organic and biological chemistry forms a key major division of chemistry research and continues to develop rapidly, with new avenues of research opening up in parallel with exciting progress in existing activities.

This book series encompasses cutting-edge research across the entire field, including important developments in both new and fundamental aspects of the discipline. It covers all aspects of organic and biological chemistry. The series is designed to be very wide in scope and provides a vehicle for the publication of edited review volumes, monographs and 'how to' manuals of experimental practice.

Examples of topic areas would include

- Synthetic chemistry
- Natural products chemistry
- Materials chemistry
- Supramolecular chemistry
- Electrochemistry
- Organometallic chemistry
- Green chemistry, including catalysis and biocatalysis
- Polymer chemistry
- Protein–protein, protein–DNA, protein–lipid and protein–carbohydrate interactions
- Enzyme/protein mechanism, including cofactor–protein interactions and mechanisms of action
- Pathways in cofactor (including metal) homeostasis
- Synthetic biology – the engineering of natural systems towards novel functions
- Development of physical methods and tools – e.g. new spectroscopic methods, imaging techniques and tools

The volumes highlight the strengths and weaknesses of each topic and emphasise the latest developments emerging from current and recent research.

The series is of primary interest to academic and industrial chemists involved broadly in organic, materials, biological and medicinal chemistry.

It is also intended to provide research students with clear and accessible books covering the different aspects of the field.

**Philip C. Bulman Page, DSc, FRSC, CChem**
*Series Editor*
*School of Chemistry, University of East Anglia, Norwich, UK*

# Preface

There are many nuclear magnetic resonance (NMR) textbooks available (a number of them are referred to in this volume), and a good number of these are in their third or higher editions. A number of these books deal with the application of NMR alongside other spectroscopic techniques (and mass spectrometry) and include guides for spectral analysis. Some texts deal with the physical principles of NMR spectroscopy with a light mathematical touch, while others include a full mathematical (quantum mechanical) description. A smaller number of books deal with the 'nuts and bolts', that is, the practicalities so that the reader may discover how to perform a particular NMR experiment. The current text has an element of each of these features, the emphasis being dictated by my experience of the aspects of NMR on which that different communities feel inadequately informed. For example, early-year researchers who employ NMR to analyse small molecules (organic and inorganic) and/or their interactions with larger molecules often feel that they are not clear on what 'questions' they can ask with NMR, or NMR researchers who may be developing new techniques but feel that they have limited knowledge of the range of applications to which they may be put.

Since the days, over 30 years ago now, I suppose, when there was excitement about the developments that enabled NMR to be used to investigate biomolecular structure and function, the use of NMR has developed in parallel to, and in support of, much novel science. Areas such as, for example, supramolecular chemistries (including organometallic nanocages and vessels, and organic materials in general); rational drug design and high throughput screening; interrogation of complex mixtures for biomarker discovery or process monitoring, to name but a few, have all benefitted from NMR input. In contrast to a number of applications of NMR in biological sciences, these more recent applications do not depend on the use of extensive isotopic enrichment (or dimensionality >2). Rather, they depend on NMR methods that, in many instances, were around long before the first 2-D NMR spectrum was produced. Their resurgence is in part due to improvements in NMR instrumentation and related software.

This volume begins with a reminder of the fundamental principles of NMR, with the level of detail that might be expected of an undergraduate later year chemistry course. In Chapter 2, the huge topic of the impact of dynamic processes on NMR spectral appearance is described with lots of examples and a widely accessible description of the mathematical basis for the observed NMR properties is provided. The relevance of dynamic processes is further illustrated in Chapter 3. The main focus of this chapter is the description of saturation transfer in detecting chemical exchange and the use that this may be put to in studying molecular interactions. NMR is being increasingly used to investigate the diffusion properties of molecules and ions, with the view of using these data to determine molecular weight/shape or association constants. The data are more reproducible than measurements made by, for example, light scattering techniques. In Chapter 4, the pros and cons and practicalities of measuring diffusion are described in detail. The majority of the examples described in the text employ $^{1}H$ or $^{13}C$ detection (directly or indirectly); however, as many synthetic routes involve an 'inorganic' component, it is helpful to be aware of other useful elements and how their NMR spectra may be informative. In Chapter 5, such guidance is provided. NMR has been used to investigate complex mixtures, such as biological matrices (blood, urine, cell suspensions or extracts, etc.), for many years, but this area has recently developed dramatically. In Chapter 6, current NMR approaches to the analysis of biofluids, or tissues, or batch processes, are described alongside an illustration of the approaches used to extract information from the NMR spectra acquired. Finally, in Chapter 7, three examples are provided of application areas: natural products, carbohydrates, and nucleic acids. Each example highlights the NMR experiments appropriate to the type of molecule under investigation and provides an insight into topics such as sample handling.

# Acknowledgements

I am extremely grateful to all of the authors who have contributed to this text. I have very much enjoyed reading their contributions and feel that they have achieved what I hoped for when this project started. Thanks to all of you.

I would also like to thank Nikki Cookson for preparing the illustrations used in Chapter 1, Kathryn Everett (at Taylor & Francis) for her support with this text and, Hilary Rowe, for her encouragement and patience.

# Editor

**Julie Fisher** is a reader in biological NMR at the School of Chemistry, University of Leeds, United Kingdom. She earned her PhD from the University of Liverpool, United Kingdom in 1986, under the supervision of Professor Raymond Abraham, working on empirical methods for predicting $^{1}H$ NMR chemical shifts. She followed this with postdoctoral placements at the National Biological NMR Centre in Leicester, United Kingdom and in the Department of Chemical Physics (part of the Department of Chemistry), University of Southampton, United Kingdom, working on proteins and peptides, respectively. In 1989, Dr. Fisher took up an academic position at the University of Leeds. She has over 90 publications and has co-authored 3 books. Her research interests include the application of NMR to the analysis of the solution properties of novel and/or chemically modified nucleic acid systems, the composition analysis of complex systems (primarily biological fluids) and supramolecular systems including host–guest interactions. Dr. Fisher is the academic lead for the NMR facilities in the School of Chemistry.

# Contributors

**Alex D. Bain**
Department of Chemistry and Chemical Biology
McMaster University
Ontario, Canada

**Robert Brkljača**
School of Applied Sciences (Discipline of Chemistry)
Health Innovation Research Institute
Royal Melbourne Institute of Technology University
Melbourne, Australia

**Julie Fisher**
School of Chemistry
University of Leeds
Leeds, United Kingdom

**Kristian Hollingsworth**
School of Chemistry
University of Leeds
Leeds, United Kingdom

**Mark J. Howard**
School of Biosciences
University of Kent
Kent, United Kingdom

**Jonathan A. Iggo**
Department of Chemistry
University of Liverpool
Liverpool, United Kingdom

**Konstantin V. Luzyanin**
Saint Petersburg State University
Saint Petersburg, Russian Federation

**Reika Masuda**
School of Science and Health
University of Western Sydney
New South Wales, Australia

**Cassey McRae**
Avacta Analytical
Avacta Group PLC
Wetherby, United Kingdom

**John A. Parkinson**
WestCHEM
Department of Pure and Applied Chemistry
University of Strathclyde
Glasgow, United Kingdom

**William S. Price**
School of Science and Health
University of Western Sydney
New South Wales, Australia

**Michelle L. Rowe**
School of Biosciences
University of Kent
Kent, United Kingdom

**Timothy Stait-Gardner**
School of Science and Health
University of Western Sydney
New South Wales, Australia

**W. Bruce Turnbull**
School of Chemistry
University of Leeds
Leeds, United Kingdom

**Sylvia Urban**
School of Applied Sciences (Discipline of Chemistry)
Health Innovation Research Institute
Royal Melbourne Institute of Technology University
Melbourne, Australia

**Amninder S. Virk**
School of Science and Health
University of Western Sydney
New South Wales, Australia

**Jane L. Wagstaff**
MRC Laboratory of Molecular Biology
University of Cambridge
Cambridge, United Kingdom

**Scott A. Willis**
School of Science and Health
University of Western Sydney
New South Wales, Australia

**Gang Zheng**
School of Science and Health
University of Western Sydney
New South Wales, Australia

**Mikhail Zubkov**
School of Science and Health
University of Western Sydney
New South Wales, Australia

# 1 The Basics

*Julie Fisher*

## CONTENTS

1.1 Introduction ........ 1
1.2 Nuclear Magnetisation ........ 2
  1.2.1 Perturbation of Nuclear Magnetisation: Radio-Frequency Pulses and Vector Representations ........ 3
1.3 Chemical Shift ........ 5
  1.3.1 Factors Influencing Chemical Shift ........ 5
1.4 Spin–Spin Coupling ........ 8
  1.4.1 'AX' Spin System ........ 9
1.5 Nuclear Spin Relaxation ........ 10
  1.5.1 Longitudinal Relaxation Time Constant, $T_1$ ........ 11
  1.5.2 Transverse Relaxation Time Constant, $T_2$ ........ 12
1.6 Summary ........ 12
References ........ 13

## 1.1 INTRODUCTION

In subsequent chapters, discussions regarding a number of nuclear magnetic resonance (NMR) techniques that could not be implemented when nuclear magnetic resonance was first discovered are presented. Their advent required, for example, strong magnetic fields and/or cryoprobes to accommodate limited sample availability. Pulsed field gradients (PFGs) have improved solvent suppression, have enabled efficient selective excitation, and have made accessible a different time range to diffusion coefficient measurement. Such developments have, of course, been made in parallel with increasing access to powerful computers and sophisticated software, permitting speedy processing and analysis of the various types and sizes of acquired data sets. Instrumental and software developments in the past 30 to 40 years have meant that NMR spectroscopy is now used in a wide range of scenarios. Synthetic chemists use NMR to elucidate structures of small molecules. It is employed in pharmaceutical industries for structure elucidation and drug development and screening (Chapter 3, Section 7.1). Biochemistry and biotechnology sectors utilise NMR to probe solution structures and functions of biological polymers (Chapter 7), and it is increasingly used in biomedicine (in particular, biomarker discovery; Chapter 6) for the analysis of complex matrices. Materials science (both soft and hard matters) is another application area in which solution and solid-state NMR has proved extremely valuable. While not an exhaustive list of applications, this is an illustration of the breadth of science that has benefitted from this analytical technique. Irrespective of technical advances, the principles, of course, remain the same. Building on the fundamental requirement for detection (that a nucleus possesses a nuclear magnetic moment), factors influencing key NMR parameters – chemical shift, coupling, and relaxation phenomena – have not changed. By way of an introduction to the 'applications' chapters to follow, in this chapter, a brief reminder of the basic NMR observables is provided.

## 1.2 NUCLEAR MAGNETISATION

Nuclei that have either an odd mass and/or an odd atomic number have a spin ($I$), which has the values, 1/2, 1, 3/2, 2 etc. As a result of this spin, the nucleus displays a magnetic moment, μ, given by:

$$\mu = \frac{\gamma I h}{2\pi} \tag{1.1}$$

where $\gamma$ is the magnetogyric ratio, specific to each nuclear type (Chapter 5, Table 5.3), and $h$ is Planck's constant. In the presence of a magnetic field ($B_o$), these nuclear moments arrange themselves in 'allowed' orientations (this is a quantum mechanical system after all) described by the magnetic quantum number, $m_I$, which has values (hence, orientation labels) $-I$, $-(I + 1)$,... $I$; there are $2I + 1$ of these. The energy ($E$) associated with this orientation interaction (with $B_o$) is given by:

$$E = -\gamma h B_o m_I / 2\pi \tag{1.2}$$

It is the perturbation of these nuclear spin orientations that is detected in the NMR experiment, with the selection rule governing the observation being that $\Delta m_I = +/-1$, and thus, the transition (or re-orientation) energy is given by:

$$\Delta E = \gamma h B_o / 2\pi \tag{1.3}$$

Resonance condition (of nuclear magnetic resonance) results when a nucleus of magnetogyric ratio $\gamma$ is placed in a magnetic field $B_o$ and a radiation of frequency $\nu$ (which equals $\gamma B_o/2\pi$) is applied.

The range of energies associated with the various spin orientations is quite small, and if the system is in thermal equilibrium, a Boltzmann distribution of orientation populations is maintained. This is shown for a spin-1/2 (with orientations +1/2 or α, and −1/2 or β) system in Equation 1.4:

$$N_\alpha / N_\beta = e^{\left(\frac{h\nu}{kT}\right)} \tag{1.4}$$

where $N_\alpha$ and $N_\beta$ are the number of spins in the lower (+1/2) and the higher (−1/2) orientations, respectively. For NMR, the radiation frequency is in the radio frequency (RF) region of the electromagnetic spectrum, and thus, the excess spin population (which will be referred to later as 'bulk magnetisation', $M_z$) is of the order 1 in $10^5$

nuclear spins, and this is the reason for the low sensitivity of NMR, compared, for example, to infrared (IR) spectrophotometry. However, this deficiency is more than offset by the fact that the response detected is directly proportional to the number of nuclei producing it.

### 1.2.1 Perturbation of Nuclear Magnetisation: Radio-Frequency Pulses and Vector Representations

Originally, nuclear spin transitions that occur over a frequency range, depending on the local magnetic environment (Section 1.3), would be observed by slowly sweeping through different RF values covering the required range and a response detected as each resonance condition was satisfied; this is the continuous wave experiment. This has been largely superseded by the pulsed-RF experiment, in which a frequency bandwidth is applied, satisfying all resonance conditions at once (in microseconds); all of the experiments described in this book are pulsed-RF techniques. There are a number of excellent textbooks that explain the effect of an RF pulse on nuclear spin populations, free induction decay (FID) collection, and subsequent Fourier transformation, both pictorially[1] and mathematically[2], and no attempt will be made here to reproduce these. However, in view of the content of later chapters, it is appropriate at this point to introduce the vector representation commonly used to illustrate the consequence of the application of RF pulses on bulk magnetisation ($M_z$).

Using a three-dimensional coordinate system, and a single nuclear spin type, the effect of an RF pulse on $M_z$ is shown in Figure 1.1.

In Figure 1.1 the RF pulse (labelled $B_1$) is applied along the $x$ axis, and depending on how long it is applied for and the power level used, the usual ‘convention’ is to depict this vector rotating away from the $+z$ direction toward the $+y$ axis (note that this should be depicted as a rotation toward the $-y$ axis, as adopted in Chapter 4, but the convention used in almost all basic NMR texts is as shown here; the discussion that will follow applies in either case). When the vector actually lies along the $y$ axis, a 90° or a $\pi/2$ RF pulse is said to have been applied. In this representation, the precessional or Larmor frequency ($\omega_o$) is accommodated by rotating the $xy$ plane at that frequency (i.e. we are working in the rotational frame of reference). The impact of a number of different pulse types on $M_z$ is shown in Figure 1.1.

The RF pulse, thus, induces an excited state, with spin populations inverted (compared to the ground state). Prior to regaining the ground state, further pulse trains may be applied, which have the result of modifying the signal ultimately detected as a function of various properties of the nuclear spins system; many examples illustrating this point are provided in subsequent chapters.

As a result of the insensitivity of the NMR technique, it is generally necessary to repeat the ‘pulse data acquisition’ process many times, co-adding data to improve the signal-to-noise ratio. As described at the start of Chapter 3, care must be taken in this to ensure that the ground state is achieved in advance of each repetition. The regaining of the ground state is referred to as ‘nuclear relaxation’ (Section 1.4); thus, a relaxation delay (RD) is generally introduced in advance of each pulse-acquire step. The simplest experimental setup is shown in Figure 1.2, and the conventions used are adopted throughout this book.

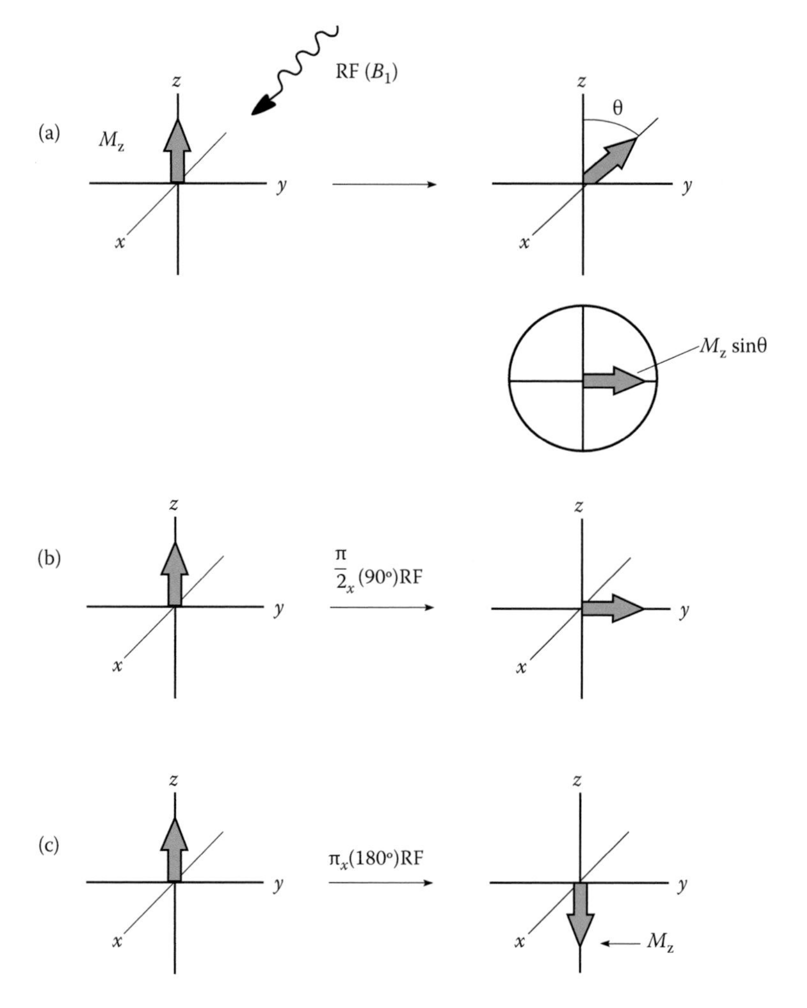


**FIGURE 1.1** Vector representation of the effect of an RF pulse on bulk magnetisation, $M_z$, resulting in rotation by angle θ (a), π/2 or 90° (b), and π or 180° (c).

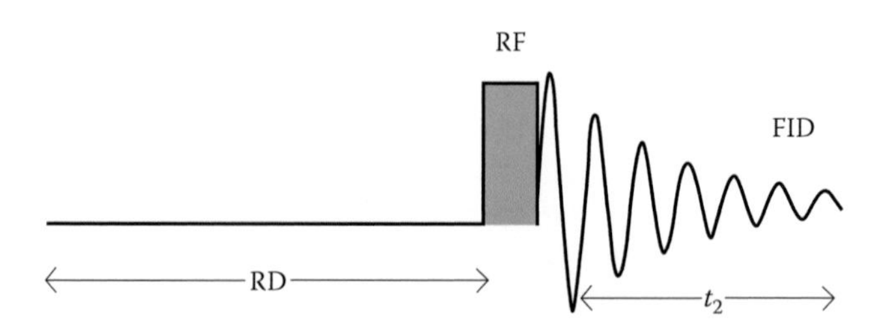


**FIGURE 1.2** Representation of a one-pulse experiment in which RD is the relaxation delay, $t_2$ is the acquisition time, and FID is the free induction decay.

In essentially all descriptions of the effect of RF pulses on bulk magnetization, the assumption is that the applied (external) magnetic field is uniform (homogeneous) across the sample volume. However, some of the most recently developed NMR techniques require the magnetic environment to be varied (along the sample length, by convention, the $z$ axis). This is achieved through the use of a separate set of current-bearing coils (providing an additional magnetic field) surrounding the sample. The current through these coils is varied such that a controllable field gradient may be applied – a PFG. A detailed description of the effect of a PFG on excited nuclear spins is provided in Chapter 4.

## 1.3 CHEMICAL SHIFT

In the foregoing discussion, a 'bare' nucleus was considered, but of course, this does not reflect reality as all nuclei are surrounded by electrons, and these 'shield' the nucleus from the full effect of $B_o$. That is, the magnetic field experienced at the nucleus is smaller than $B_o$; we label the reduced field $B$, and the difference between $B_o$ and $B$ is referred to as 'nuclear shielding', $\sigma$ (Chapter 5, Section 1.1). This parameter is not generally reported; rather, the difference between the nuclear shielding experienced by a given nuclear spin and that experienced by the same nuclear spin type in a 'reference' compound is noted. As nuclear shielding is proportional to the applied field, this must also be taken into account if data are to be readily compared when recorded at different magnetic field strengths. Thus, dividing the nuclear shielding difference (i.e. $B_{reference} - B_{sample}$) by the applied field (by convention, this is taken as $B_{reference}$) leads to the dimensionless parameter referred to as the 'chemical shift' ($\delta$) in units of parts per million (ppm). Substituting for frequency (Section 1.1), the chemical shift may also be given by Equation 1.5:

$$\delta = \frac{(\nu_{sample} - \nu_{reference}) \times 10^6}{\text{oscillator frequency (Hz)}} \tag{1.5}$$

### 1.3.1 FACTORS INFLUENCING CHEMICAL SHIFT

As described above, the magnetic field experienced by the nucleus (and hence, its resonance frequency, $\nu$, or chemical shift, $\delta$) is determined by the electron distribution (circulating) about the nucleus; the surrounding electrons induce their own magnetic field due to their motion as opposed to their intrinsic (electron) spin. This upfield shift (to $B$ rather than $B_o$) is called a 'diamagnetic shift' and is a feature of the spherical $s$ distribution of electrons about a nucleus. When $p$ electrons (and $d$ and so on) are present, the magnetic field experienced at the nucleus is enhanced, and a large low-field, or 'paramagnetic shift', is produced – a nuclear de-shielding (Chapter 5, Section 1.1). (This should not be confused with 'paramagnetism' arising from unpaired electrons. The paramagnetic contribution to nuclear magnetic shielding is a quantum feature.[3]) This electron distribution difference is the reason why the chemical shift range observed for the proton ($^1H$) is small (~10 ppm) compared to all other NMR-detectable nuclei (these can be hundreds of parts per million; Chapter 5).

The majority (although not all) of the techniques described in later chapters utilise $^{1}$H observation; therefore, it is appropriate to consider here factors that influence proton chemical shifts (and coupling and relaxation phenomena).

The electron distribution (and hence, the nuclear shielding) about the hydrogen nucleus will be influenced by the electronic properties of other atoms, substituents, or functional groups. As these vary, so too do the mechanisms by which influence is exerted. It is helpful, therefore, to consider the chemical shift as being determined by a combination of these processes; the relative importance of the different contributions do, however, vary. The different contributors to the (proton) chemical shift are referred to as (1) electric field ($\delta_{EF}$), (2) anisotropy in magnetic susceptibility ($\delta_{MA}$), (3) van der Waals ($\delta_{vdW}$), and (4) solvent factors ($\delta_{solv}$) (Equation 1.6).

$$\delta_H = \delta_{EF} + \delta_{MA} + \delta_{vdW} + \delta_{solv} \quad (1.6)$$

It has long been established that the nuclear shielding experienced by a proton (attached to a carbon) is dependent on the electric field (E) (originating in the functional group) along the C–H bond, with an additional contribution ($E^2$) perpendicular to the bond. The quadratic term becomes insignificant at distances greater than approximately 2 Å between the hydrogen atom and the functional group, and hence, only the direct through-bond electric field need to be considered. Thus, the electronegativity of substituents will have maximal influence on the observed chemical shift, as illustrated in Table 1.1.

The electric field contribution to $\delta_H$ is quite predictable, and it dominates the other factors, such that empirical rules have been developed to enable proton chemical shift prediction for simple small molecules.[3] These rules take no account of the

**TABLE 1.1**
**Chemical Shifts (δ, ppm) of Selected $CH_3X$ Molecules as the Function of Electronegativity of X**

| $CH_3X$ X = | $\delta_H$ | $\delta_C$ | Electronegativity |
|---|---|---|---|
| $SiMe_3$ | 0 | 0 | 1.90 |
| H | 0.13 | −2.3 | 2.20 |
| $CH_3$ | 0.88 | 5.7 | 2.60 |
| $CO{\cdot}CH_3$ | 2.08 | 29.2 | 2.60 |
| $NH_2$ | 2.36 | 28.3 | 3.05 |
| I | 2.16 | −20.7 | 2.65 |
| Br | 2.68 | 10.0 | 2.95 |
| OH | 3.38 | 25.1 | 3.15 |
| F | 4.26 | 75.4 | 3.90 |

*Source:* Abraham, R. J., J. Fisher and P. Loftus: *Introduction to NMR Spectroscopy*. 1988. Copyright Wiley-VCH Verlag GmbH & Co. KGaA. Reproduced with permission.

geometric element of the electric field, which also impacts the substituent-induced chemical shift, and so fall down when the shape of a molecule is such that electric dipole moments (associated with, e.g. the X–C bond, where X is the substituent or functional group) are close in space. They are, however, a good starting point.

In discussing the chemical shift, so far, we have considered the atomic circulation of electrons, but in some instances, notably in aromatic rings, a molecular circulation of electrons needs to be considered. The circulation of π electrons induces a (molecular) magnetic moment (and a secondary magnetic field, $B'$), which opposes (resulting in a high-field chemical shift) or reinforces (resulting in a low-field chemical shift) the external magnetic field, depending on the location within the molecule (Figure 1.3). This is referred to as the 'ring current shift' (RCS; Chapter 3, Section 5.1) and is a special example of anisotropy in magnetic susceptibility.

The RCS is the reason why protons in benzene have a chemical shift of 7.27 ppm, while alkene hydrogens in cyclohexa-1,3-diene display a chemical shift of 5.86 ppm. The large chemical shift of the alkene hydrogens relative to those in the saturated molecule (in which proton chemical shifts are <2 ppm) is in part due to the secondary magnetic field associated with the π electrons (which are anisotropic in their distribution about the Pi bond) of the C=C bond. Similar to the ring current in benzene and other aromatic molecules, this secondary magnetic field is anisotropic, and its influence depends on the geometric relationship between the observed proton and the Pi-bonded functional group.

Electric field and magnetic anisotropy considerations are most important in analysing or attempting to predict proton chemical shifts. van der Waals effects are small, and of course, the influence of solvent is only relevant when solvent changes are required.

The van der Waals contribution to the chemical shift was introduced in an attempt to explain the de-shielding noticed for a nucleus on moving from vapor to liquid state and was suggested as being the result of interactions with the fluctuating electric field (F) of the surrounding medium (in the liquid rather than the solution) whose square is not averaged out. This intermolecular feature may also have an intramolecular counterpart, with the fluctuating electric field being associated with a functional group. The electric field has a $1/r^6$ ($r$ is the distance between the proton and the functional group) dependence, and thus, its importance is very

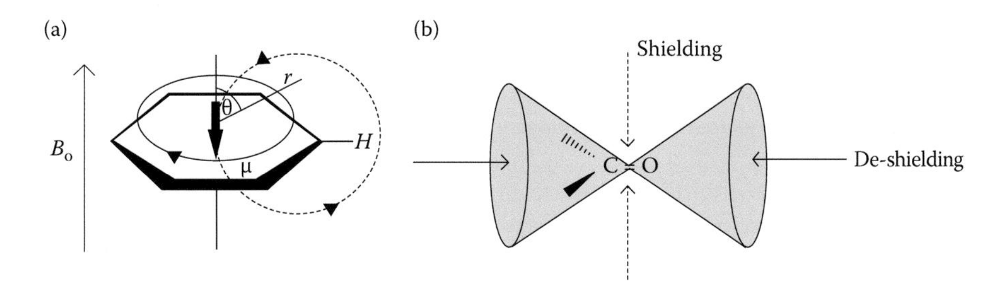


**FIGURE 1.3** Depictions of the additional magnetic field induced by the circulation of π electrons, both examples of magnetic anisotropy, in benzene (in which the secondary magnetic field is referred to as a 'ring current') (a) and around a carbonyl bond (b).

short range. Nevertheless, this term is included in attempts to accurately predict chemical shifts.

All of the factors described above may, of course, be attributed to a solvent, particularly if the solvent is anisotropic in its magnetic susceptibility (as is, e.g. acetone or dimethyl sulfoxide), that is, the solvent contribution to the chemical shift is a combination of electric field, magnetic anisotropy, and van der Waals effects. This is not, however, the total solvent effect as, when a polar molecule is placed in solution, it may, to some extent, polarise the surrounding solvent; this is a reaction field. In attempting to account for solvent-induced shifts, the dielectric constant of the solvent is generally considered, although treating the solvent as a continuum leads to a breakdown in this approach. Thus, the role of solvent in chemical shift determination is complex, and for most purposes, it is sufficient to accept that solvent changes will change chemical shifts, and that may be advantageous in attempts to resolve overlapping resonances.

In later chapters (e.g. Chapter 3 and Section 7.3), the effect of hydrogen bonding on specific (proton) chemical shifts will be referred to. Of course, a hydrogen bond results in a change in the polarization of the H–O or the H–N bond of the hydrogen bond donor, reducing the electron density around the hydrogen nucleus and, as a consequence, a low-field shift. This is simply a short-range example of the electric field contribution to the chemical shift.

## 1.4 SPIN–SPIN COUPLING

In discussing the chemical shift, interactions between nuclear magnetic moments and the external magnetic field, $B_0$, were considered. What was not addressed was the possibility of interactions between nuclear magnets within a molecule; this is what gives rise to the feature referred to as 'spin–spin' or 'scalar' coupling, symbolised by $J$ and measured in hertz.

If we consider a system in which there are two nuclei of the same spin type (for simplicity), both spin-1/2, but different chemical shifts, $\delta_A$ and $\delta_B$, then the magnetic field experienced by A is a result of its nuclear shielding and the magnetic field associated with nucleus B. As the nuclear magnet B can adopt two orientations (almost equally populated), two equal but opposite magnetic fields are imposed on nucleus A, and thus, two peaks are detected equally spaced around the resonance frequency of A. The separation between the two lines is the spin-spin or scalar coupling constant, $J$. The same is true for nucleus B, and therefore, the spectrum for the molecule containing A and B will consist of four lines. If more nuclear magnets are present or if one or more of the nuclear spins is greater than 1/2, then more lines will appear in the spectrum, consistent with the different energies of interactions. The number of lines around each chemical shift is referred to as the 'signal's multiplicity', and the relative intensity of the lines in each multiplet may be determined by considering the energy of each spin–orientation interaction. Descriptions of multiplets and an indication of the size of coupling constants is provided in a number of standard texts and are not included here.[3,4] However, it is appropriate to consider the mechanism of spin–spin coupling in a little more detail and to illustrate it using, once more, the 'AB-type' spin system. It is convention when describing a spin system that the label 'AB' refers to a two-spin system, in which the difference in resonance frequency between nucleus A and nucleus B, $(\nu_A - \nu_B)$, is small compared to $J_{AB}$. If, however, the frequency difference is greater than (or equal to) 4 $J$, then the spin system is labelled 'AX'.

### 1.4.1 'AX' Spin System

Having described the consequences for the NMR spectrum of nuclear spins interacting, it is now worth considering how this interaction is mediated.

If a molecule is held stationary in a magnetic field, the magnetic moments of the nuclei interact directly with each other, and this dipolar interaction is very large. In liquids and solution, rapid molecular motions mean that these dipolar interactions are averaged to zero; thus, these are not the interactions that give rise to the multiplets referred to above. Spin–spin or scalar couplings are transmitted via bonding electrons, that is, they are electron-coupled interactions involving three possible mechanisms: (1) nuclear magnetic moments interacting with electric fields due to orbiting electrons, (2) dipolar interactions between the nuclear and electronic magnetic moments, and (3) interactions between nuclear magnets and electron spins (in *s* orbitals) as a result of the wave function having a finite value at the nucleus (this is the so-called 'contact term'). This latter term dominates couplings involving the hydrogen nucleus; the other terms increase in significance when heavy atom–heavy atom coupling is present (Chapter 5). As a result of this contact interaction, scalar couplings have a strong geometric dependence, and they are also influenced by the number and type of intervening bonds (including the nature of functional groups involved in the bonding network). The reader is referred to textbooks in which values of coupling constants are provided and their structural dependence are explained.[5]

The occurrence of multiple lines around each chemical shift may be explained by considering the simplest coupled system (as referred to above), two spin-1/2 nuclei A and X; the spectrum of such a system will comprise four lines (transitions) just like the AB spectrum, but (as noted earlier) it is implicit in the choice of labels that these nuclear spins have very different resonance frequencies, that is, $\nu_A >> \nu_X$. For such a system, obeying the selection rule (Section 1.2) but extending this to include the principle that, when one nucleus absorbs energy, it does so without affecting any other, then an energy level diagram may be drawn, which illustrates how the four lines (or transitions) arise (Figure 1.4).

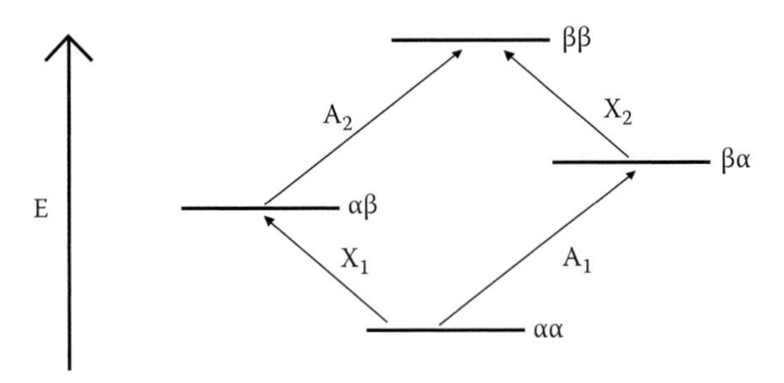


**FIGURE 1.4** Energy level diagram for a two spin-1/2 AX spin system; the arrows indicate allowed transitions, $A_1$, $A_2$, $X_1$ and $X_2$, for the A and X spins.

The quantum mechanical description for an AX system is quite straightforward. The Hamiltonian ($\boldsymbol{H}_1$) for the interaction between the nuclear magnetic moments and the magnetic fields experienced at the nucleus may be written as follows:

$$\boldsymbol{H}_1 = -\frac{\gamma h}{2\pi}(m_A B_A + m_X B_X) \tag{1.7}$$

Substituting for the resonance frequency ($v$) of both A and X, Equation 1.7 may be re-written as follows:

$$\boldsymbol{H}_1 = -(m_A v_A + m_X v_X) \tag{1.8}$$

The interaction between the nuclear magnetic moments of A and X is given by Equation 1.9:

$$\boldsymbol{H}_2 = C \times \bar{\mu}_A \times \bar{\mu}_X \tag{1.9}$$

where $C$ is a constant directly related to the scalar coupling constant, $J$.

The Hamiltonian describing the system as a whole is the sum of $\boldsymbol{H}_1$ and $\boldsymbol{H}_2$. For the simple case of the AX system, the Hamiltonian may be solved explicitly to give the expression in Equation 1.10:

$$E = -(m_A v_A + m_X v_X) + m_A\, m_X\, J_{AX} \tag{1.10}$$

Substituting the values +/−1/2 for $m_A$ and $m_X$ permits the production of the energy level diagram illustrated in Figure 1.4.

When more than a few nuclear spins are involved in mutual spin–spin coupling, this analytical approach is not practical, and computer-assisted spectral simulation is necessary (Chapter 2 and references therein).

Finally, it should be noted that scalar couplings are magnetic field independent and can be both positive and negative, although the sign only becomes apparent when the spectrum is computer simulated.

## 1.5 NUCLEAR SPIN RELAXATION

In Chapters 3 and 6, nuclear spin relaxation is referred to in the contexts of saturation and relaxation-editing, respectively. As an introduction to those topics, it is described in brief here.

As mentioned earlier, 'relaxation' is the term applied to the processes involved in the re-orientation of nuclear spins following the application of an RF pulse. Clearly, energy must be dispersed to regain the ground state (i.e. the state in which there is an excess spin population in the lowest energy orientation, aligned with the applied magnetic field), and there are a number of mechanisms by which this may be achieved. The most significant of these mechanisms are defined by the time constants $T_1$, the spin–lattice or longitudinal relaxation time constant, and $T_2$, the spin–spin or transverse relaxation time constant.

### 1.5.1 Longitudinal Relaxation Time Constant, $T_1$

Following the application of an RF pulse (see Figure 1.1b), the equilibrium magnetisation, $M_z$, is rotated into the $xy$ plane. The new vector $M$ can be represented by its components along the $+y$ and $+z$ axes, the relative size of these being related to the rotation angle θ. Once the RF pulse has ceased the component of magnetisation along the $z$ axis increases in magnitude, longitudinal relaxation is taking place, with energy being lost to the molecular lattice surrounding the nucleus being observed; this is an enthalpic process. The decay of $M$ back to the $z$ axis is an exponential process characterised by the time constant $T_1$ (Section 2.4.4.4) and is expressed mathematically in Equation 1.11:

$$(M_z - M) = M_z(1 - \cos\theta)\exp\left(-\frac{t}{T_1}\right) \tag{1.11}$$

where $t$ is the time following the RF pulse.

Longitudinal relaxation dictates the repetition rate in a multi-pulse experiment, with sufficient time being required for $T_1$ relaxation following each pulse-acquire step to ensure that the ground state is achieved prior to a further excitation pulse (Section 3.1). It is, therefore, wise to measure $T_1$ prior to collecting quantifiable data, and the most common method for doing this is by using the inversion–recovery pulse sequence.[6] This is a 180° – $t$ – 90° – acquired pulse train that is repeated with different values of $t$. The plot of natural log of $(M - M_z)$ against $t$ produces a straight line plot whose gradient is $-1/T_1$. An illustration of the output from an inversion–recovery experiment is shown in Figure 1.5.

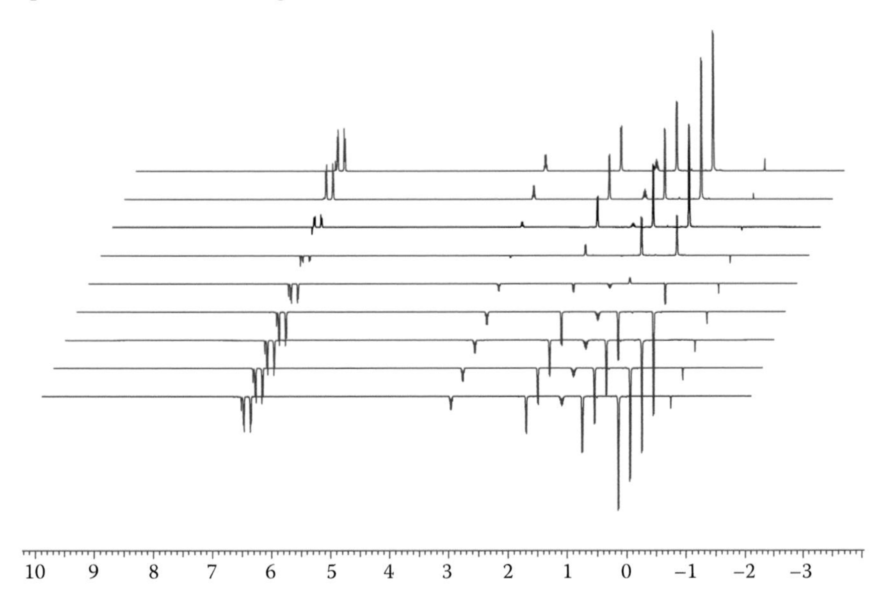


**FIGURE 1.5** Spectra recorded via and inversion–recovery experiment aimed at determining a value for $T_1$ for each of the hydrogen atoms in ibuprofen. The inter-pulse delay, $t$, was increased (from bottom to top) from 0.001 to 10 s.

Considering the exponential factor, it is apparent that, if the RD (or $d_0$ as specified in some pulse sequences) is chosen to be $5T_1$, the amount of magnetisation remaining in the $y$ axis (i.e. not having relaxed) is 0.007 M (i.e. effectively zero).

We have described longitudinal relaxation as a process that involves the dissipation of energy to the molecular lattice; however, we have not discussed the mechanism for this. For efficient relaxation, the lattice must provide local magnetic fields that are fluctuating at an appropriate frequency. These fields are dominated by the magnetic moments of protons within the molecules as they tumble in solution. This is, thus, a dipole–dipole process whose efficiency depends on the effective molecular correlation time, $\tau_c$ (i.e. the average time taken for the vector joining the mutually relaxing spins to rotate through one radian); the nature of the nuclei themselves (i.e. the magnetogyric ratios); and most importantly, the distance between the nuclei involved. This distance dependence is extremely important and is the basis for perhaps the most utilised relaxation phenomenon, the nuclear overhauser effect (NOE). NOE is referred to in almost all of the chapters to follow. Readers are referred to the book by Neuhaus and Williamson[7] for a detailed description of this phenomenon; a more pictorial description is provided in a number of other textbooks.[1,8]

### 1.5.2 Transverse Relaxation Time Constant, $T_2$

Following the application of an RF pulse, nuclear spins interchange energy with one another such that some precess faster than the Larmor (or resonance) frequency and others precess more slowly. The result is a de-focussing of the magnetisation vector in the $xy$ plane. In the extreme, this de-focussing will be such that the sum of the magnetisation vectors in the $xy$ plane is zero. This entropic process is referred to as 'spin–spin relaxation' and is characterised by the time constant, $T_2$. Clearly, when all the magnetisation has returned to the $z$ axis, there is no residual component of magentisation in the $xy$ plane, and therefore, $T_1 \geq T_2$.

Transverse relaxation times may be measured using the spin–echo experiment (Chapters 3 and 6), although such experiments are more frequently employed to spectrum edit on the basis of differences in $T_2$ between different types of molecules in a solution; high–molecular weight molecules have short $T_2$ compared to low–molecular weight molecules. Examples of such applications are provided in Chapters 3 and 6.

## 1.6 SUMMARY

The foregoing discussion was aimed at setting the scene for what is to follow. The basic picture for the impact of an RF pulse on nuclear magnetisation and subsequent relaxation is developed in Chapter 2, with a detailed description of the effect of time-dependent phenomena on the NMR parameters considered here. In Chapter 3, the practicalities of the application of the concepts introduced in Chapter 2 are discussed, with chemical exchange being of fundamental significance to the saturation transfer techniques so elegantly depicted. The manipulation of the nuclear magnetic response by the application of a linear field gradient, following an RF pulse, is the principle subject of Chapter 4. Therein, the impact of time-dependent chemical processes for the NMR-derived diffusion characteristics of a range of moieties is addressed, along with consideration of appropriate experimental setups. The content of Chapters 6 and 7 serves to illustrate the range of applications in which the NMR techniques encountered in earlier chapters may be employed.

## REFERENCES


1. (a) Derome, A. E. 1987. *Modern NMR Techniques for Chemistry Research.* Oxford, United Kingdom: Pergamon Press, 280 pp. (b) Claridge, T. D. W. 2009. *High-Resolution NMR Techniques in Organic Chemistry*. 2nd ed. Oxford, United Kingdom: Elsevier Science, 398 pp. (c) Simpson, J. H. 2012. *Organic Structure Determination: Using 2-D NMR Spectroscopy – A Problem-Based Approach*. 2nd ed. London, United Kingdom: Elsevier, 362 pp.
2. (a) Ernst, R. R., G. Bodenhausen, and A. Wokaun. 1990. *Principles of Nuclear Magnetic Resonance in One and Two Dimensions*. Oxford, United Kingdom: Oxford Science Publications, 610 pp. (b) Levitt, M. H. 2008. *Spin Dynamics*. 2nd ed. Chichester, United Kingdom: Wiley, 686 pp. (c) Keeler, J. 2010. *Understanding NMR Spectroscopy*. 2nd ed. Chichester, United Kingdom: John Wiley & Sons, Ltd, 526 pp. (d) Abragam, A. 1983. Principles of nuclear magnetism. *International Series of Monographs on Physics* 32, 618 pp. (e) Hore, P. J., J. A. Jones and S. Wimperis. 2000. *NMR: The Toolkit*. Oxford, United Kingdom: Oxford University Press, 85 pp.
3. Gunther, H. 2013. *NMR Spectroscopy: Basic Principles, Concepts, and Applications in Chemistry*. 3rd ed. Weinheim, Germany: Wiley-VCH, 734 pp.
4. Zerbe, O. and S. Jurt. 2014. *Applied NMR Spectroscopy for Chemists and Life Scientists*. Weinheim, Germany: Wiley-VCH, 548 pp.
5. (a) Akitt, J. W. and B. E. Mann. 2000. *NMR and Chemistry: An introduction to modern NMR*. 4th ed. Cheltenham, United Kingdom: Nelson-Thornes, 400 pp. (b) Abraham, R. J., J. Fisher, and P. Loftus. 1988. *Introduction to NMR Spectroscopy*. Chichester, United Kingdom: John Wiley & Sons, Ltd, 271 pp.
6. Jacobsen, N. E. 2007. *NMR Spectroscopy Explained: Simplified Theory, Applications, and Examples for Organic Chemistry and Structural Biology*. New Jersey: John Wiley & Sons, Ltd, 668 pp.
7. Neuhaus, D. and M. P. Williamson. 1989. *The Nuclear Overhauser Effect in Structural and Conformational Analysis*. Cambridge, United Kingdom: VCH Publishers Inc, 522 pp.
8. Sanders, J. K. M. and B. K. Hunter. 1994. *Modern NMR Spectroscopy: A Guide for Chemists*. 2nd ed. Oxford, United Kingdom: Oxford University Press, 330 pp.

# 2 Dynamic NMR

*Alex D. Bain*

## CONTENTS

2.1 Introduction: NMR Timescales ..... 15
2.2 Exchange Regimes ..... 18
  2.2.1 Unequal and Equal Populations ..... 21
  2.2.2 Slow Exchange ..... 27
    2.2.2.1 Selective Inversion Relaxation Experiments ..... 28
  2.2.3 Fast Exchange ..... 31
2.3 Kinetics ..... 35
2.4 Theory ..... 37
  2.4.1 Bloch Equations in the Frequency Domain ..... 37
  2.4.2 Bloch Equations in the Time Domain ..... 39
  2.4.3 Bloch Equations with Unequal Populations ..... 45
  2.4.4 Density Matrix Treatment ..... 46
    2.4.4.1 More Complex Systems ..... 48
    2.4.4.2 Coupled Spin Systems ..... 49
    2.4.4.3 Z Magnetisations ..... 51
    2.4.4.4 Spin–Lattice Relaxation Experiments ..... 53
    2.4.4.5 NOESY/EXSY 2-D Experiment ..... 55
2.5 Summary ..... 57
References ..... 58
Further Readings ..... 61

## 2.1 INTRODUCTION: NMR TIMESCALES

Dynamic can mean many things: lively, moving, active, animated, etc. We draw molecules as structures, but we know that there is often some substantial motion associated with them. In the context of nuclear magnetic resonance (NMR), let us start with the assumption that, essentially, all systems are dynamic. In almost all cases, the consequences are not obvious in the spectra, but they may be quite dramatic if the timescales of the motions are appropriate. Using appropriate techniques, a wide range of dynamics can be revealed in many chemical systems. This can then tell us about structure, bonding, and reactions, and may even lead us to structures that we cannot observe directly.

Dynamics has been a part of NMR almost from the beginning. Exchange effects were discovered[1] at about the same time (and in the same laboratory) as the phenomenon of scalar (or *J*) coupling. Thus, part of the thesis of this chapter is that dynamics should almost always be considered, and the examples provided will support this notion. Not only we will cover the classic dynamic NMR line shapes (Figure 2.1, Ref. [2]), but

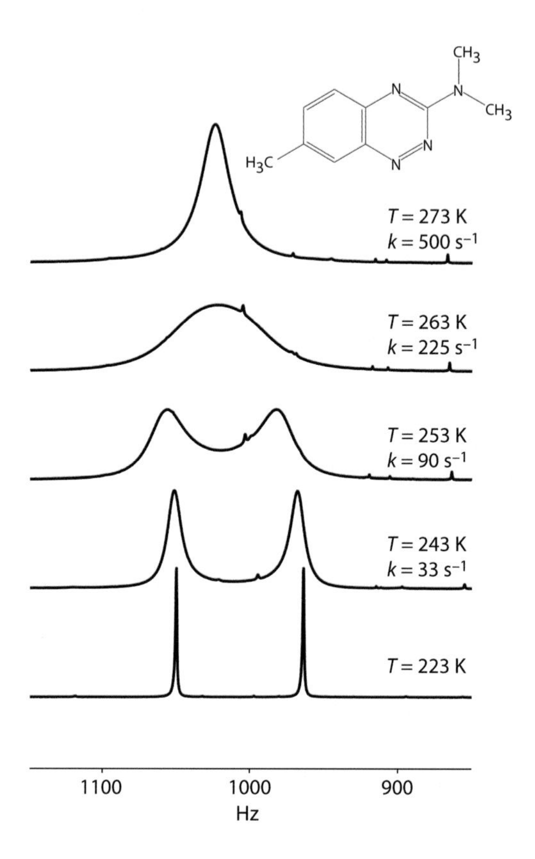


**FIGURE 2.1** Proton spectrum of the methyl region of 3-dimethylamino-7-methyl-1,2,4-benzotriazine as a function of temperature. (Reproduced from Fauconnier, T. et al., *Canadian Journal of Chemistry* 76, pp. 426–430. ©1998 Canadian Science Publishers or its licensors. With permission.)

we will also show how different experiments can reveal molecular motions that are either much faster or slower. In fact, NMR is often the best method to use to explore the dynamics of molecules.

The concept of a timescale is central to the discussion here. 'The NMR timescale' is a rather loose and ill-defined quantity and is probably better called 'an NMR timescale'. What this means is that molecular motion occurs at a rate comparable to one of the frequencies associated with the NMR phenomenon. Often, this is the difference in resonance frequencies between two sites, but the rate can also be compared to spin relaxation times, molecular correlation times, diffusion (Chapter 4), or reaction rates. We show that these effects can all be described theoretically in a straightforward and consistent way. Spin relaxation and dynamic NMR are very closely related, both physically and theoretically, so we will discuss the effect of dynamics on spin relaxation as well as the more obvious changes that it produces in a standard 1-D spectrum.

We begin by illustrating a number of aspects of dynamic NMR with some recent examples then give some background on the kinetics and the theory. The theory is not essential, but it does give an idea of the basic strengths and weaknesses of the methods.

There are many reports of solids, liquids, or gases whose dynamics has been probed by NMR. The following represent just a few illustrative examples. We think of solids as rigid, but many solids have substantial motion that may be detected by NMR.[3] Polymers above their glass transition have significant segmental motion that helps determine their physical properties. Latex rubber is very flexible and behaves almost like a liquid in an NMR spectrometer since it is well above its glass transition at room temperature. Molecules with a roughly spherical shape, such as adamantane or cubane,[4] are often called 'plastic crystals'. They stay in their site in the crystal lattice but can tumble quite freely. In a crystal, small ions can migrate within a larger framework. This is an important phenomenon since mobile hydrogen or lithium ions can carry electricity in fuel cells or batteries. Figure 2.2 shows a 2-D $^6$Li exchange spectroscopy (EXSY) spectrum of a lithium battery material, $Li_3V_2(PO_4)_3$.[5] The lithium ion occupies three different sites, signified by the three signals in the 1-D spectrum. These correspond to the diagonal peaks in the 2-D spectrum, and the cross peaks signify exchange among these sites. The rates of these exchanges can be measured and correlated with the electrical conductivity of the material. Often, one has to go looking for them, but many macroscopic solids show significant dynamic effects.

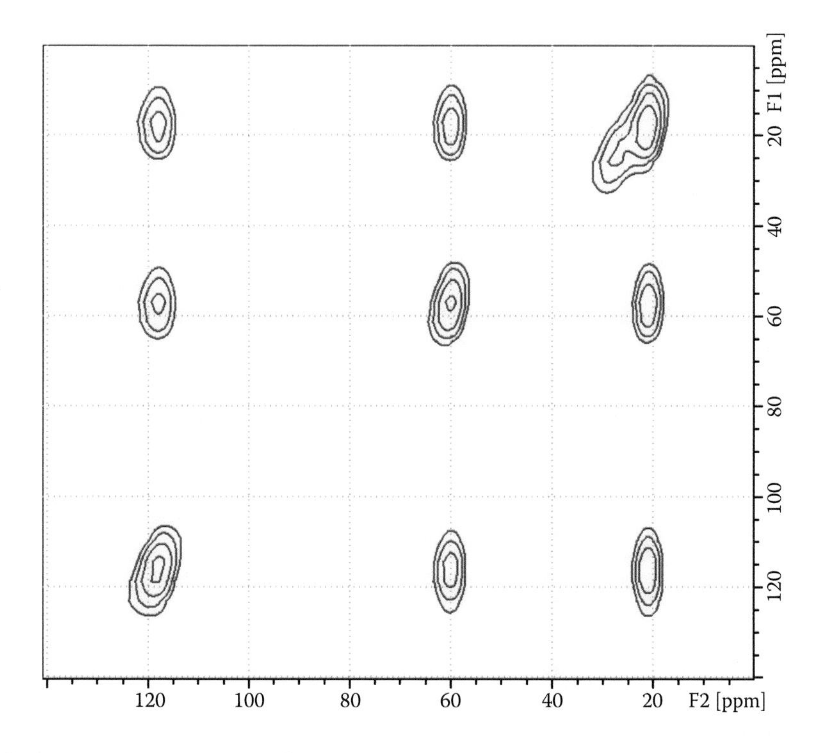


**FIGURE 2.2** 44.17 MHz 2-D $^6$Li EXSY spectrum of $Li_3V_2(PO_4)_3$. The diagonal peaks for the three sites run from bottom left to top right, and symmetrical pairs of cross peaks indicate exchange between the sites. (Data from Danielle Smiley and Dr. Gillian R. Goward.)

The most common studies of dynamics occur in the liquid state. In liquids (and in gases, if the molecule is sufficiently volatile for study[6]), the molecule is no longer constrained by an overall crystal lattice, so many types of dynamics are possible. For small organic molecules, there are numerous examples of restricted rotation about partial double bonds, conformational exchange in rings, etc. Organometallic molecules often depend on ligand mobility and exchange for their chemical properties, so they are also an extremely rich source of examples. The book edited by Jackman and Cotton[7] gives an almost complete review of the NMR exchange literature up to the mid-1970s, covering many of the typical examples, and much of the recent work expands on this. What is missing from this text is the application of multidimensional NMR techniques since they had not yet appeared. Sophisticated pulse experiments extend the classical methods and are essential in the isotope-edited multidimensional studies of biological macromolecules.[8,9] These experiments have revealed in detail the many dynamic processes that are essential to protein function. NMR provides the only way to study this (Chapter 3). These dynamic processes are often classic chemical reactions with breaking and re-forming of bonds, even though the sample is at macroscopic equilibrium. Dynamic NMR is often called 'chemical exchange' since, when a nucleus moves from one magnetic environment to another, it is replaced by the same process. We can think of NMR as following an individual nucleus, so the process is a reaction from the point of view of that nucleus. The spectrum of that nucleus is affected, even though the sample remains macroscopically the same. A classic example is an N,N-dimethyl group of an amide or another Pi-bonding system (Figure 2.1). This adds a double bond character to a formal carbon–nitrogen single bond, and so rotation around that bond is restricted. In the ground state, the nitrogen $p$ orbital overlaps with the $\pi$ system, but in the transition state, they are orthogonal, so the partial double bond character is lost, and the energy is higher.[10] If there is no motion, then there are two distinct resonances for the two different methyl environments (Figure 2.1). As the temperature is raised, the rate of exchange is increased, and the two lines broaden, shift, and eventually, coalesce when the rate is comparable to the frequency difference. In the limit of fast rates, the observed spectrum becomes a single sharp line.

For the amine referred to above, the reaction is within itself: an intramolecular exchange. Intermolecular reactions can also affect an NMR spectrum – for instance, exchange of protons on a hydroxyl group with water in the solvent. In general, even though the sample is at macroscopic equilibrium, there is almost always activity at the molecular level.

## 2.2 EXCHANGE REGIMES

Dynamic NMR can be organised into a number of subclasses: the phenomenon is the same, but its manifestations are different. The main distinctions are slow, intermediate, and fast exchanges. If the exchange is slow, the separate sites are observable, and little perturbation of the line shape is observed (Figure 2.1, bottom). In intermediate exchange, the lines are significantly broadened and may even have started to coalesce. In fast exchange, the signals have coalesced, and only a single line is observed at the average chemical shift. A further classification depends on whether the two sites are equally populated. Sometimes, the initial and final states are chemically identical (mutual exchange), but they may be different molecules as well. In other words, is the equilibrium constant exactly equal to one, by symmetry, or does it have some other value? Finally, is the reaction intramolecular or intermolecular? The former is easier to understand and is the most commonly studied case, but the latter is also very chemically important. Even though the classical kinetics for an intermolecular reaction is different, the description of the NMR spectrum is the same as for an intramolecular reaction. Each of these situations has particular methods that are appropriate, but the dynamics is still the same, and the theoretical approaches are similar. In intermediate exchange, the rates are comparable to the frequency differences between the sites, so we are studying the *xy* magnetisations associated with the transitions. When the rates are slower and more comparable to spin–lattice relaxation rates, we look at the behaviour of the *z* magnetisations, often starting with a nuclear overhauser effect spectroscopy (NOESY)/EXSY–type (Chapter 3) two-dimensional experiment, such as the one in Figure 2.2. Combining these studies of the reaction in both slow and intermediate exchange conditions provides much more reliable data on the thermodynamics of the reaction. To illustrate, there is a class of compounds called 'push–pull ethylenes' (Figure 2.3). These

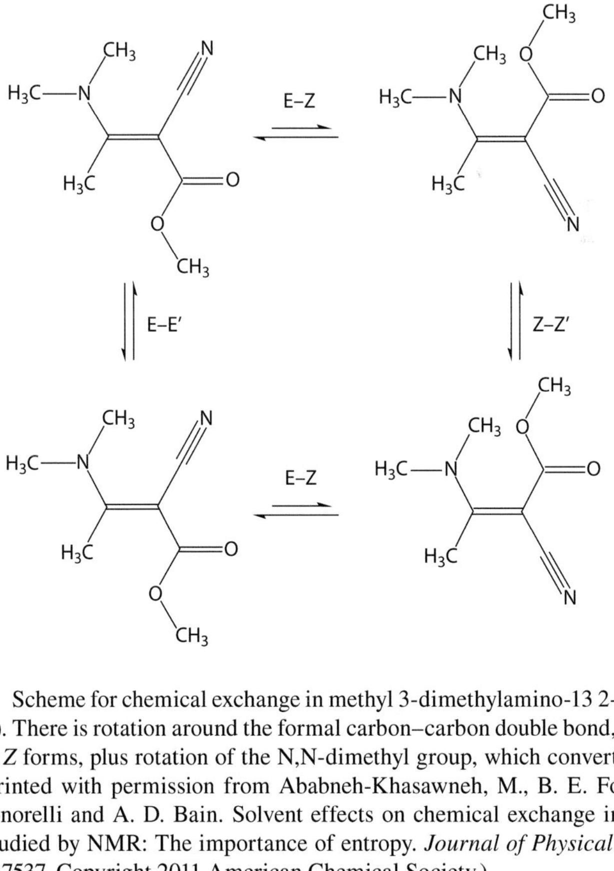

**FIGURE 2.3** Scheme for chemical exchange in methyl 3-dimethylamino-13 2-cyanocrotonate (MDACC). There is rotation around the formal carbon–carbon double bond, interconverting the *E* and *Z* forms, plus rotation of the N,N-dimethyl group, which converts *E* to *E′* and *Z* to *Z′*. (Reprinted with permission from Ababneh-Khasawneh, M., B. E. Fortier-McGill, M. E. Occhionorelli and A. D. Bain. Solvent effects on chemical exchange in a push–pull ethylene as studied by NMR: The importance of entropy. *Journal of Physical Chemistry* A 115, pp. 7531–7537. Copyright 2011 American Chemical Society.)

have electron-withdrawing groups on one end of a carbon–carbon double bond and electron-donating groups on the other. This changes the bonding significantly[11] (the chemical shifts of the two carbons in the double bond differ by ~100 ppm), sometimes to the extent that rotation can occur around the formal double bond at room temperature.[12–15] A combination of slow and intermediate exchange methods has been used to study the solvent effects on such a system,[16] revealing that entropy effects play an important role.

Fast exchange has sometimes been dismissed in the past as not very useful since there is only one line to work with. The important parameter is the spin–spin (or transverse) relaxation time, $T_2$, and that can be tricky to measure reliably. However, new and sophisticated experiments provide a wealth of information, particularly from biological macromolecules.

Just like any other chemical reaction, exchange dynamics follows a reaction coordinate in a potential energy (PE) landscape. This process is often pictured in three dimensions as going up a valley, traversing a pass and descending into another valley, that is, starting with the reagent, passing through a transition state, and going on to a product. The true situation is more complex, of course. There are usually hundreds of dimensions in the PE hypersurface, and the topography in that space may be very convoluted and rugged. The reaction coordinate is a unique direction in the multidimensional space that has the following property. Along the reaction coordinate, the slope may be positive, negative, or zero (at a transition state or intermediate). However, for all perpendicular directions, there must be a local minimum – the slope is zero, and the second derivative is positive. The reaction coordinate (or minimum energy pathway) is the path that maintains these conditions all through the reaction. The study of reaction rates provides a unique way of obtaining some experimental data on these reaction coordinates and the heights of the barriers.

Cyclohexane provides a classic demonstration of dynamic effects in NMR. The chair form of the ring inverts rapidly from one chair form to another chemically identical form in which axial protons have become equatorial, and vice versa. At room temperature, the proton spectrum is observed to be a single line. For the relatively simple case of the dynamics in modified six-membered rings, modern computational methods can explore the PE hypersurface and map out the detailed trajectory to show how the ring wriggles from one chair form to another.[17] The pucker of the ring can be described by three parameters,[18] so we can plot out the geometry at a numbers of these points along the path. Figure 2.4 shows the energy along the exchange path (with a few geometry snapshots) for the case of *trans*-1,4-dibromo-1,4-dicyanocyclohexane, which mimics cyclohexane quite closely. Each point on the path is a set of coordinates in the multidimensional conformational space, and the calculation provides a geometry and energy at each point. It turns out that the conformer with the bromines axial (Figure 2.4a) has roughly the same energy as the one with the bromines equatorial (Figure 2.4b), so the situation is quite similar to cyclohexane itself. In Figure 2.4, the plot starts with the bromines equatorial in the chair form of the molecule. One end flips up to form a boat-like transition state, which then relaxes into a metastable twist-boat intermediate. The other end of the molecule then flips up to form a second transition state, which leads to the final product with the bromines axial. Rings with more than six members may show more complicated spectra.[19]

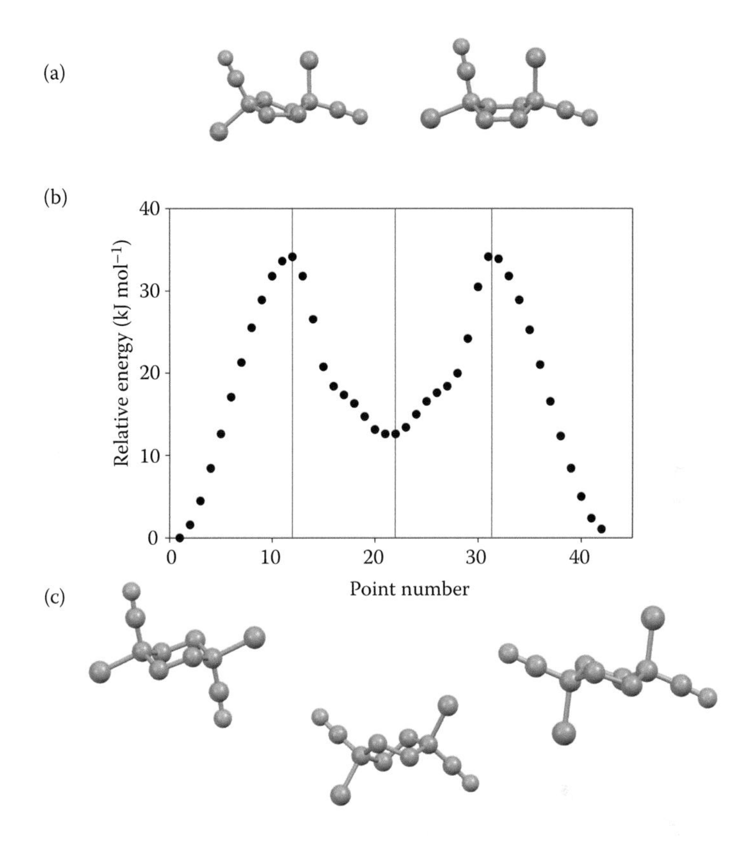


**FIGURE 2.4 (See color insert.)** Schematic of the reaction coordinate for the exchange in trans-1,4-dibromodicyanocyclohexane. Each point represents a geometry along the reaction coordinate, and the vertical line in the middle of the plot shows the locations of the two transition states and the metastable intermediate. (a) Shows the geometries of the two transition states, (b) plot of energy along the reaction coordinate (minimum energy pathway) and (c) shows the geometry of the initial state, the intermediate and final state. (Reprinted with permission from Bain, A. D., M. Baron, S. K. Burger, V. J. Kowalewski and M. B. Rodriguez. Interconversion study in 1,4 substituted six-membered cyclohexane-type rings: Structure and dynamics of trans-1,4-dibromo-1,4-dicyanocyclohexane. *Journal of Physical Chemistry* A 115, pp. 9207–9216. Copyright 2011 American Chemical Society.)

There are much more complex kinetic schemes, but six-membered rings illustrate most of the points we wish to make. Important concepts are the two (or more) stable ground states, the transition states, plus any possible metastable intermediates. All of these have the thermodynamic properties of enthalpy and entropy, and differences between these determine both the equilibrium constants and forward and reverse reaction rates.

### 2.2.1 Unequal and Equal Populations

There are many cases in the literature where the initial and final states are chemically identical – the so-called 'mutual exchange'. For situations as in Figure 2.1, the populations of both states are equal, and the equilibrium constant is 1, by symmetry. This is only a fraction of the chemically interesting systems. In the case of *trans*-1,4-dibromo-1,4-dicyanocyclohexane (Figure 2.4a and c), there are two chemically distinct ground states that happen to have very similar energies. Often, there is a change in conformation, or the unfolding of a protein, where there are two distinct states with unequal populations. Provided that the population ratio is not too extreme (we can deal with some cases where the minor site is a few percent of the major), the systems can be treated in the same way as equally populated ones.

Figure 2.5 shows a beautiful example in the case of the protein T4 lysozyme (T4L), with two conformations referred to as the 'ground' (G) and 'excited' (E) states.[9] Here, different mutations in the protein sequence will change the conformations and their populations, which can be seen in the 2-D heteronuclear single-quarter coherence (HSQC) spectra and is shown schematically in an energy diagram. The substitution of Ala for Leu at position 99 of T4 lysozyme (L99A T4L) creates an internal cavity of approximately 150 Å.[4] The cavity can accommodate hydrophobic molecules such as benzene, and detailed X-ray diffraction studies have established that the ground state conformations of wild type and L99A T4L are virtually identical. However, chemical

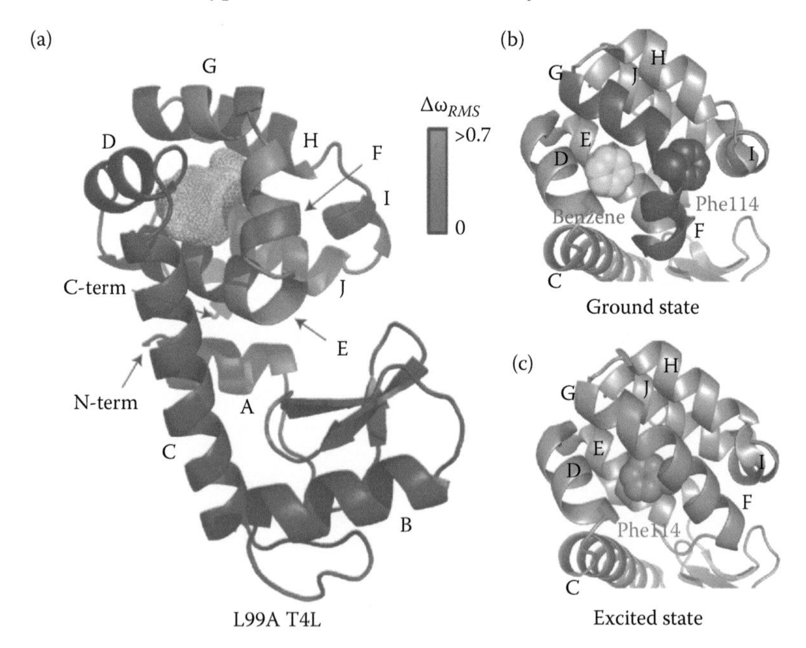


**FIGURE 2.5** **(See color insert.)** An example of unequal populations in mutants of T4 lysozyme (T4L) and the detection of small populations of the minor site. (From Sekhar, A. and L. E. Kay, *Proceedings of the National Academy of Sciences* 110, pp. 12,867–12,874, 2013.) (a) Ground-state X-ray structure of the L99A mutant, color-coded according to the magnitude of chemical shift differences between the ground and excited states. The grey mesh delineates the cavity that results from the L99A mutation. Comparison of the C-terminal domain of L99A T4L in the ground (b) and excited (c) states, highlighting the different orientations of the F and G helices in each of the conformers. The F114 side chain that rotates into the cavity in the excited state and the bound benzene in the ground state are shown using space-filling representations.

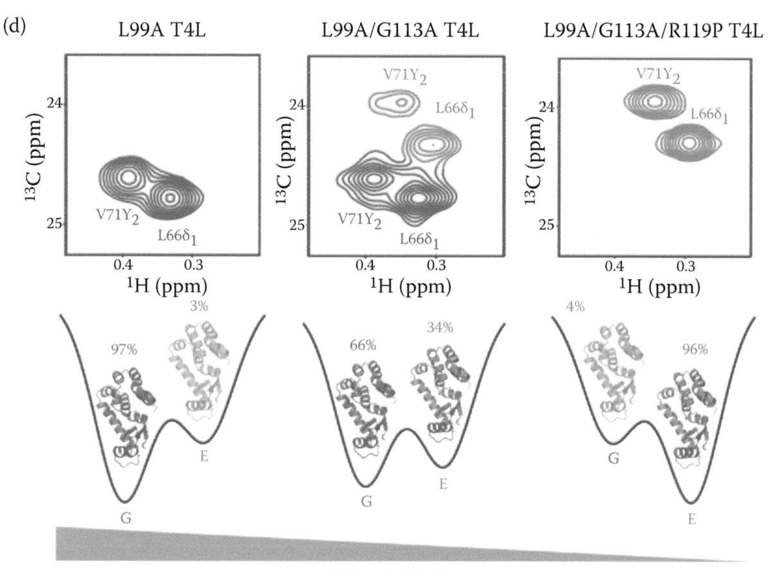


**FIGURE 2.5** **(See color insert.)** (Continued) An example of unequal populations in mutants of T4 lysozyme (T4L) and the detection of small populations of the minor site. (From Sekhar, A. and L. E. Kay, *Proceedings of the National Academy of Sciences* 110, pp. 12,867–12,874, 2013.) (d) The ground and excited state populations can be manipulated by a small number of mutations. Ground (G, left) and excited (E) states become comparable in population in the L99A/G113A construct (center), as seen in 13C–1H correlation spectra. The new set of peaks (red) has chemical shift values that are in excellent agreement with those obtained for state E from fits of the relaxation–dispersion experiments on L99A. (Reprinted by permission from Macmillan Publishers, Ltd.: *Nature (London)*, ref. 20, copyright 2011; From Dr. Ashok Sekhar and Dr. Lewis E. Kay.)

exchange measurements show that L99A T4L is dynamic on the millisecond timescale, unlike the wild-type protein.[20] Benzene derivatives can penetrate into the cavity easily, but the paradox is that solvent cannot. Exchangeable amide protons near the cavity do not exchange with deuterons from a $D_2O$ solvent, even over a period of weeks. This dynamics is the result of exchange between the native state and an alternative conformation that is populated to approximately 3% at room temperature. The small minor site has been referred to as a 'dark site'[21] since it is difficult to observe directly. Even though it is essentially impossible to see this conformation, its structure has been determined from a number of elegant dynamic NMR experiments. These $T_2$ relaxation experiments can give both the magnitude and the sign of the chemical shift difference between the ground and excited states.[22] Even though we cannot observe the excited state directly, this gives us its chemical shifts, and it is now possible to determine protein structures from these data alone. Dynamic NMR allows us to 'see the invisible'.

The equally populated case is somewhat easier to analyse, and the appropriate methods for its study tend to be different from that of the unequal case, particularly when the population of the minor site is only a few percent of the major. When the equilibrium

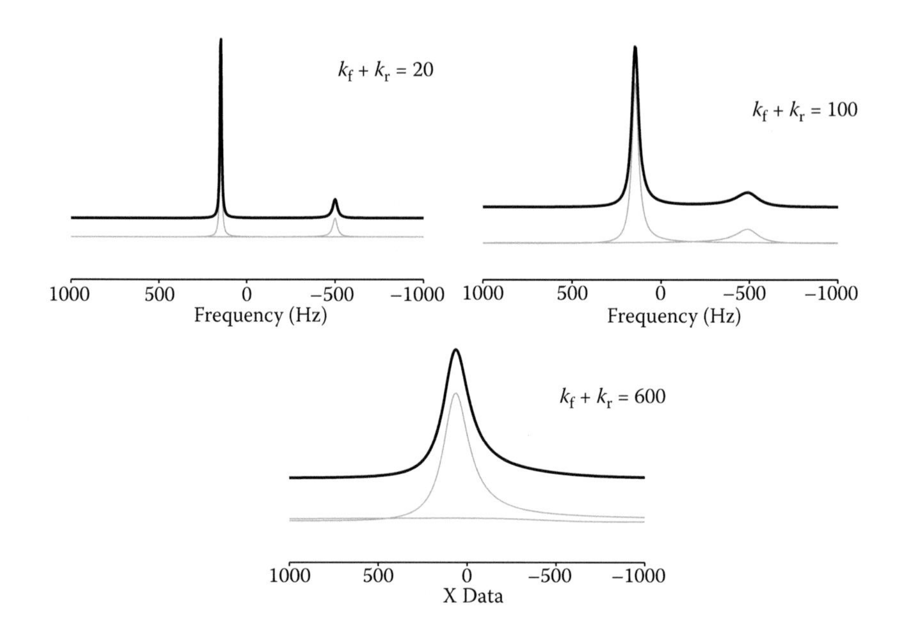


**FIGURE 2.6** Simulations of line shapes for an unequally populated exchange. The population of the minor site is 30% of that of the major site. The rates are specified by the sum of the forward ($k_f$) and the reverse ($k_r$). Note that, to maintain equilibrium, the forward rate (major to minor) is 30% of the reverse rate (minor to major), so the minor site has a broader line. For all the spectra, the composite line shape (heavy line) is made up from two components (lighter lines).

constant is significantly different from 1, the line shape quickly changes from slow exchange to fast exchange (Figure 2.6), and the intermediate case is hard to work with. If the equilibrium constant is $K_{eq}$, which we define as greater than 1, then the minor site has only $1/K_{eq}$ the intensity of the major site. Furthermore, since the system must remain at equilibrium, the rate of nuclei leaving the minor site is $K_{eq}$ times that for the major site, so the minor site is roughly $K_{eq}$ times broader. The combination of these two effects is that the minor site quickly disappears into the baseline as the rate increases.

Many published examples of dynamic NMR are intramolecular reactions – the molecule itself is re-arranging the nuclei in a unimolecular fashion. In this case, the kinetics is true first-order.[23] There are also many intermolecular processes that lead to exchange effects in the NMR spectra. Since this is a bimolecular reaction, the macroscopic reaction follows second-order kinetics. However, regardless of the concentrations, the NMR kinetics remains pseudo–first-order since the spin we are observing is just the Boltzmann population difference, but it exchanges with the bulk. This means that the spectra can be analysed in the same way as in an intramolecular case to give apparent pseudo–first-order rates. Of course, these apparent rates depend on the concentrations, whereas intramolecular cases do not.

Figure 2.7 shows an excellent recent example of intermolecular exchange in which a tellurium jumps between phosphorus-carbon-phosphorus (PCP)-type ligands.[24] Both the $^{31}P$ and $^{1}H$ spectra show classic exchange behaviour (a section of the $^{1}H$ spectrum is shown in Figure 2.8), which can be analysed in the usual fashion, and the

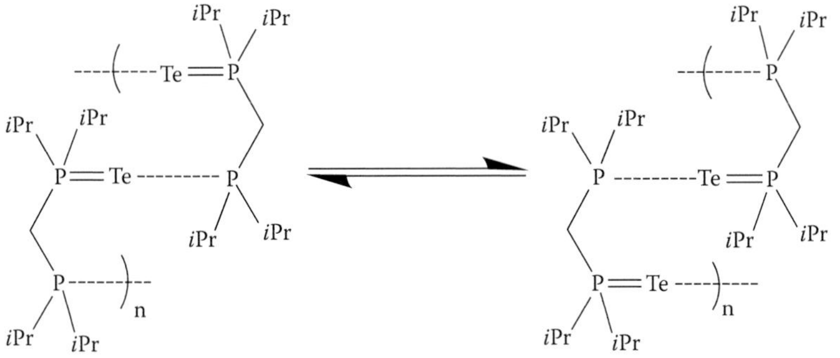

**FIGURE 2.7** Scheme for the intermolecular exchange of tellurium between two PCP ligands in toluene solution. We thank Dr. Philip J. Elder for preparing this diagram. (From Elder, P. J. W. et al., *European Journal of Inorganic Chemistry*, pp. 2867–2876, 2013.)

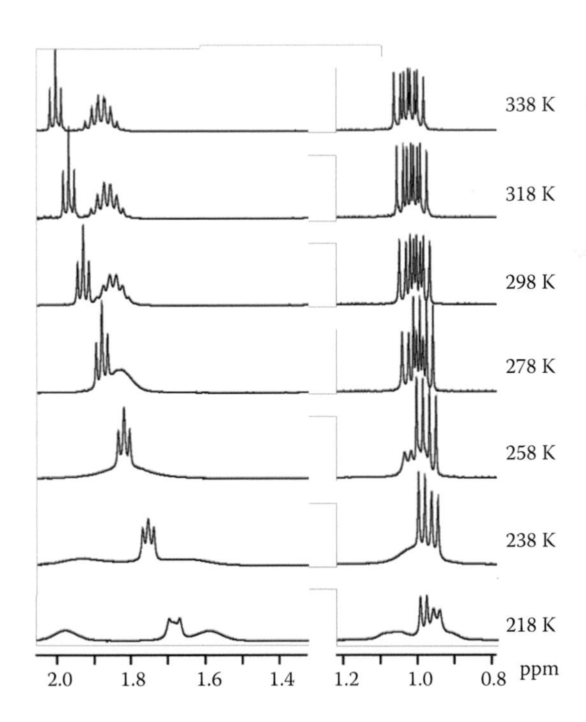


**FIGURE 2.8** Variable-temperature $^1H$ NMR spectra for $TePiPr_2CH_2–PiPr_2$ in $[D_8]$toluene in the temperature range of 218 to 338 K. The portion of the spectra corresponding to $CH_3$ groups is presented at 1/2 vertical scale. (Elder, P. J. W., T. Chivers and R. Thirumoorthi: Experimental and computational investigations of tautomerism and fluxionality in PCP- and PNP-bridged heavy chalcogenides. *European Journal of Inorganic Chemistry*. 2013. pp. 2867–2876. Copyright Wiley-VCH Verlag GmbH & Co. KGaA. Reproduced with permission.)

rates plotted as a function of temperature to give an activation energy (Figure 2.9). However, the apparent rates depend linearly on concentration (Figure 2.10), so the exchange must be intermolecular, and the true kinetics is pseudo–first-order.

The previous examples are cases where the resonance frequency of the spin changes due to differences in the chemical environment, but the physical environment may also play a role. For example, the resonance of the phosphorus nucleus in the head group of a lipid in a phospholipid bilayer depends on the orientation of the bilayer – the so-called 'chemical shift anisotropy'. When a phospholipid diffuses over the surface of a spherical lipid vesicle, its resonance frequency will change, and the $^{31}$P NMR spectrum will show exchange effects.[25]

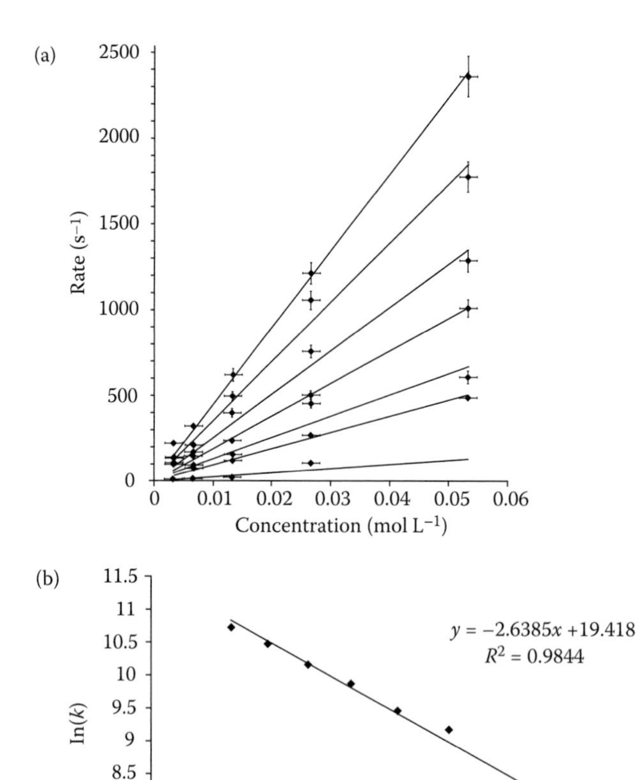


**FIGURE 2.9** (a) Rates for the tellurium transfer process as a function of temperature and concentration. (b) Arrhenius plot of the logarithm of the rate as a function of inverse temperature. (Elder, P. J. W., T. Chivers and R. Thirumoorthi. Experimental and computational investigations of tautomerism and fluxionality in PCP- and PNP-bridged heavy chalcogenides. *European Journal of Inorganic Chemistry*. 2013. pp. 2867–2876. Copyright Wiley-VCH Verlag GmbH & Co. KGaA. Reproduced with permission.)

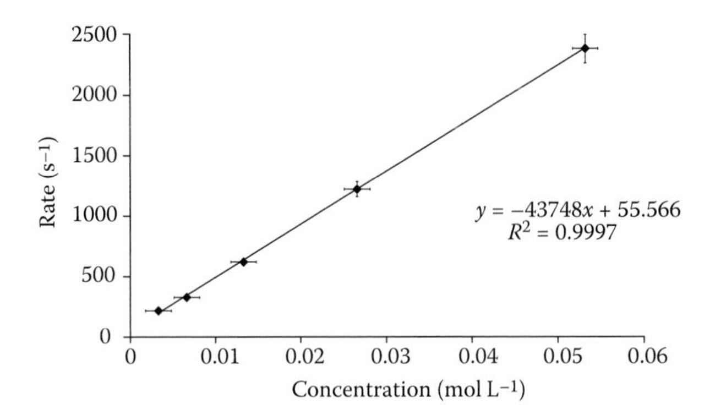


**FIGURE 2.10** Plot of the observed pseudo–first-order rate of the tellurium transfer process (Figure 2.7) as a function of concentration. We thank Dr. Philip J. Elder for preparing this diagram. (From Elder, P. J. W. et al., *European Journal of Inorganic Chemistry*, pp. 2867–2876, 2013.)

### 2.2.2 Slow Exchange

In intermediate exchange, the $T_2$s are short, so complex pulse sequences and multidimensional experiments may not work. However, dynamic effects may not be obvious in a standard 1-D experiment but may show up in a 2-D experiment if the exchange is slow and both sites are distinguishable. In the NOESY experiment (sometimes called 'EXSY' in exchange work, even though the pulse sequence is identical), cross peaks may appear due to exchange. The basis of these experiments is that the relaxation behaviour of a given site may depend on the initial state of another site; the spin–lattice relaxations of the two sites are coupled. This is true in both the case of dipole–dipole relaxation (NOESY) and exchange (EXSY). These experiments are excellent qualitative experiments that may indicate the presence of dynamics and clearly show which lines in the spectra exchange with which. Although values of rates can also be extracted from EXSY spectra, it is the author's personal opinion that there are better ways to extract quantitative information.

Figure 2.2 shows a $^{6}$Li EXSY spectrum of a cathode material for a lithium battery, $Li_3V_2(PO_4)_3$. In this case, the lithium can occupy three distinct sites in the crystal lattice and can jump among these sites, as indicated by the cross peaks. Since the lithium is the charge carrier, this dynamics is crucial to the performance of the material in a battery. This is for an equally populated case, but Figure 2.11 (top trace) shows a common phenomenon when the equilibrium constant is not equal to 1.[26] The figure shows the fluorine spectrum of an organometallic system in which $BF_4^-$ ions are encapsulated in two different pockets of a supramolecular structure; we are, in effect, observing them jumping between the pockets. In this case, the minor site is about 20% of the major (determined by comparing the integrals of the peaks at ~−150.5 and −144.5 ppm). In the EXSY spectrum (Figure 2.11), there are clear cross peaks between the major and minor sites, but the diagonal peak for the minor site is not seen. The appearance of cross peaks is a strong indication of exchange, and we need not worry about the diagonal peak. As we will see in the theory section, it is not

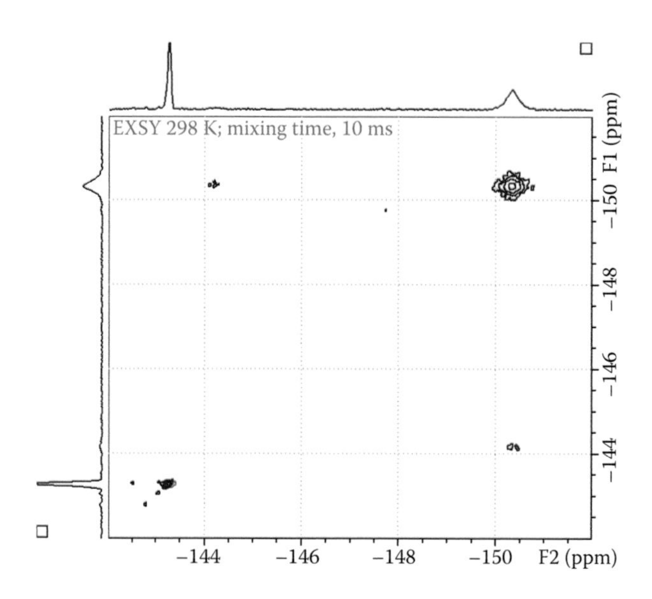


**FIGURE 2.11** $^{19}$F EXSY spectrum at 298 K (0.35 mM in acetone, 470 MHz) of $BF_4^-$ in a $[Pd_4ligand_8]^{8+}$ double-cage structure. The two sites for $BF_4^-$ are unequally populated. The cross peaks between the major and the minor sites are clear, but the diagonal peak for the minor site is not observed (note that the arrows indicate the chemical shift of the signal for the minor site in the 1-D traces). The peak at −143.3 ppm can be ignored. (Data from Dr. S. Freye and Professor G. H. Clever. Freye, S., D. M. Engelhard, M. John and G. H. Clever: Counter-ion dynamics in an interpenetrated coordination cage capable of dissolving AgCl. *Chemistry: A European Journal*. 2013. 19. pp. 2114–2121. Copyright Wiley-VCH Verlag GmbH & Co. KGaA. Reproduced with permission.)

that the cross peaks are unusually strong (they are of the order of $1/K$) but rather that the diagonal peak is unusually weak. There is always a trade-off with EXSY spectra determined by the choice of the mixing time ($t_{mix}$). For simple systems, there are optimal choices,[27] but for complicated systems, long mixing times can be confusing. The estimation of rates is usually based on the assumption that the cross peak builds linearly with mixing time. Similarly, the determination of what exchanges with what assumes that there has been, at most, one exchange event in the mixing time. Short mixing times give the most reliable data, but at the expense of poor signal to noise of the cross peaks.

#### 2.2.2.1 Selective Inversion Relaxation Experiments

The EXSY spectrum is also widely used quantitatively to extract values of rates, but this is not the only way. To get good rate data in the slow exchange region, it is the author's opinion that selective inversion experiments are the most efficient. This experiment is sometimes called the '*zz*' experiment, or the Hoffman–Forsen experiment, after the inventors of a similar experiment on a continuous wave (CW) instrument. In this experiment, a part of the spin system is inverted, and then the spin–lattice relaxation back to equilibrium is observed in a series of 1-D experiments. This is a modified version of the standard inversion–recovery $T_1$ experiment, except that the inversion is selective. Although most sets of initial conditions can be used, the experiment is easiest to understand if one site is inverted and the other is unperturbed. The inverted spin can return to equilibrium by normal spin–lattice relaxation or by exchange with the non-inverted spin. Because of the exchange contribution, the inverted spin relaxes faster than its natural rate, $R_1$. The other site, which was not perturbed, shows a characteristic negative transient. The signal is at equilibrium to start, but then it decreases due to exchange with the inverted spin. Finally, it returns to its equilibrium value. In this experiment, the exchange is balanced against the spin–lattice relaxation rate, and, in principle, both these rates can be extracted from a single experiment. In practice, it is very useful to perform a standard non-selective inversion recovery $T_1$ experiment as well. This experiment is less sensitive to the exchange and gives a reliable estimate of the spin–lattice relaxation rate, which can then be put into the analysis of the selective inversion data. The same non-linear least-squares software[28] can be used to fit both sets of experimental data. In analyzing the non-selective, the exchange rate is held at a reasonable estimate, and the $R_2$ rates are floated. These $R_2$ values can then be fixed, and the exchange rate can be fitted from the selective inversion data. This rate can then be used to improve the $R_2$ analysis, which can then be included in the exchange case. It is the author's opinion that this procedure gives some of the best rate data since there are very few assumptions involved.

Figure 2.12 shows an example of the behaviour referred to above in an organometallic compound. In this case, Cp*(H)Ru=CHR can undergo reversible α-hydride elimination to form Cp*Ru-$CH_2$R, where Cp* is pentamethylcyclopentadiene.[29] The data in Figure 2.12 are for the case where the exchange rate is much faster than the spin–lattice relaxation. Here, the two sites can almost equilibrate before any relaxation has occurred. The inverted site was almost fully inverted, so the non-inverted site comes almost down to zero around the same time as the inverted site comes up to zero. Then, both sites relax at similar slow rates. If the rate is slower, then the bi-exponential character of the inverted site is less clear, but it still relaxes faster than the isolated spin–lattice relaxation rate. For this slower rate, the transient in the non-inverted site stretches out longer and is less deep. Figure 2.13 shows the results of a selective inversion experiment on the lithium battery material used in the EXSY experiment in Figure 2.2. A more general and complicated three-site exchange in a protein system has also been analysed by the selective inversion (or *zz*) experiment.[30] Here, the access to an archaeal form of the proteasome is controlled by a set of N-terminal residues that can be outside the gate or have one of two conformations inside. Again, all the rates and equilibrium populations can be measured. In general, as the exchange rate becomes very slow, the transient becomes almost invisible since the spins have mostly all relaxed before they have a chance to exchange.

One of the earliest slow-exchange experiments was the saturation-transfer experiment (Chapter 3), which could be quite easily done on a CW instrument. In this, a peak was irradiated to saturation with a relatively weak radio-frequency (RF) field. If this site were in exchange with another, then the other peak would have its intensity partly saturated, by exact analogy with the nuclear Overhauser effect. A set of related experiments called 'Chemical Exchange Saturation Transfer' and 'Dark-State Exchange Saturation Transfer'[21,31] follow this sort of procedure but include the effect of RF even when it is off-resonance. A series of experiments is run with the

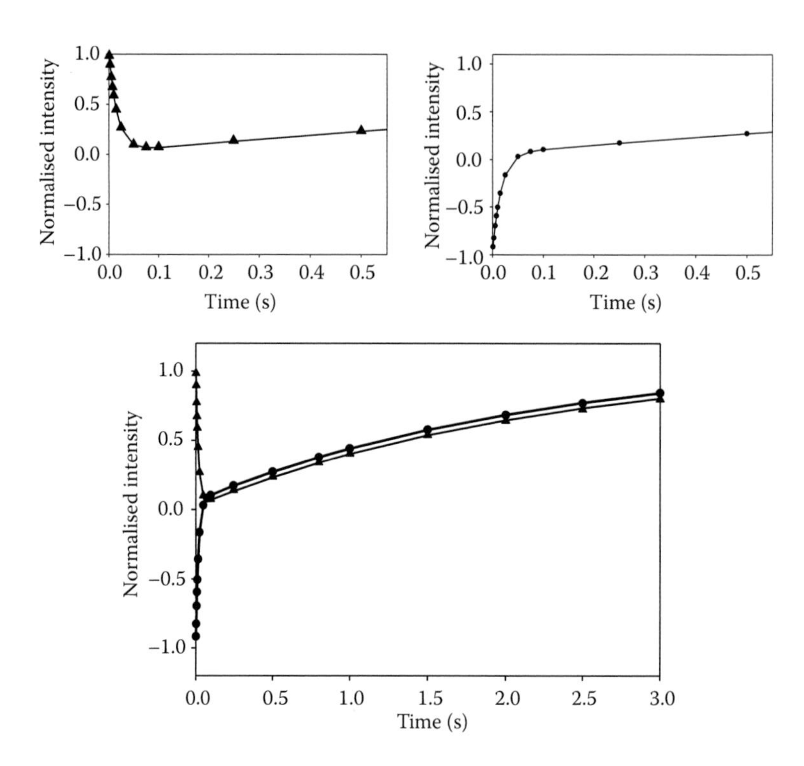


**FIGURE 2.12** Data from a selective inversion experiment on a two-site exchange in an organometallic compound: see reference 29 for details. At the beginning of the experiment, one site is selectively inverted, leaving the other unperturbed. The $z$ magnetisations are then measured at a number of times after the inversion. The top graph shows the behaviour of the non-inverted site, and the bottom, that of the inverted site. The exchange rate in this case is fast compared to the relaxation rate, so the two sites rapidly come to roughly equal values then relax slowly back to their equilibrium values of 1.0. The bottom plot shows all the data, and the two top plots show the inverted and the non-inverted data separately for the first part of the relaxation. (Reprinted with permission from reference 29, pp. 74–83. Copyright 2009 American Chemical Society.)

position of the RF field being moved through the spectrum, and the intensity of the major peak is measured as a function of the RF offset. This can be done in a 1-D fashion (a similar experiment can also be used to measure just the $T_2$ of a single peak[32]), but it can also be embedded in an isotope-filtered multidimensional experiment.

This is particularly useful if the irradiated site has a small population and is 'invisible'[31] or 'dark'.[21] The intensity of the major site will be reduced when the irradiation is near the minor site, and a detailed analysis will give the position of the minor site and the exchange rate. A physical picture, which has a close analogy to the Carr-Purcell-Meiboom-Gill (CPMG) experiment in fast exchange (Section 2.2.3), gives a rationalisation of the effect. If we consider the RF to be on-resonance with the minor site, then spins in that site will precess around the RF field along the $y$ axis. We assume that the RF field is weak enough so that spins in the major site are unaffected, and the spins there remain in their equilibrium state along the $z$ axis. If

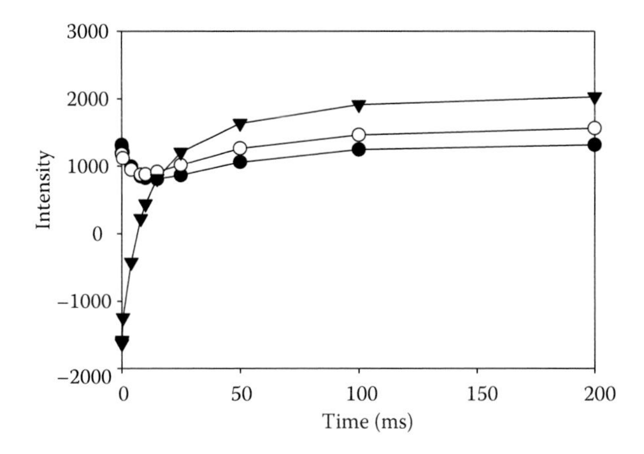


**FIGURE 2.13** Results for $^{6}$Li selective inversion experiment on $Li_3V_2(PO_4)_3$ (cf. Figure 2.2). The low-frequency line in Figure 2.1 was inverted (triangles), and the other two sites (filled and open circles) were monitored as a function of time. We thank Danielle Smiley and Dr. Gillian R. Goward for providing these data.

there is exchange, then a spin will hop from the major to the minor site, where it will precess around *y*. When it hops back after some random interval, its projection along *z* is random, so the *z* magnetisation in the major site will be reduced.

### 2.2.3 Fast Exchange

In fast exchange, we see only a single coalesced line, but there is a very broad and weak second component that we can ignore (Figure 2.14). The width of the observed line (i.e. the spin–spin relaxation rate, $R_2$) depends on the exchange rate. More precisely, it depends on the square of the frequency difference divided by the rate. There are a number of methods for measuring $R_2$ (=$1/T_2$), but the most common is the CPMG experiment. The signal is refocussed by a series of $\pi$ (180°) pulses, and the decay is monitored as a function of time. The important point is that this imposes another timescale on the exchanging system – that determined by the time between the refocussing pulses. If the spins are refocussed quickly, they do not have time to exchange, and the apparent $R_2$ contains little effect from the exchange. If the spacing of the pulses is long, then there is ample time to exchange, and the apparent $R_2$ reflects both the natural $R_2$ (in the absence of exchange) and the exchange effect. As a function of the pulse frequency, the apparent spin–spin relaxation time shows a dispersion (Figures 2.15–2.17). There are a number of ways of doing the experiment, but a common technique is to keep the total time for relaxation constant and to change the number of $\pi$ pulses in that fixed time (Section 6.3.2). The natural $R_2$ contribution to the line width often does not interest us (and it can also complicate the analysis of intermediate exchange spectra). The constant-time experiment means that the contribution from the natural $R_2$ is constant and that the observed dispersion is purely due to chemical exchange (Figure 2.17).

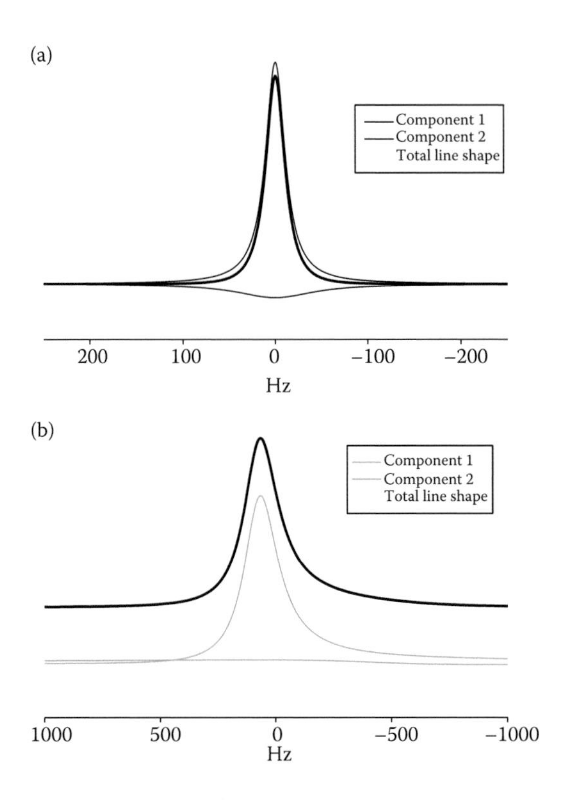


**FIGURE 2.14** Simulations of the components and total line shape for two-site exchange in the fast exchange regime. (a) Equally populated case. (b) Unequal populations where the minor site is 30% of the major site. In practice, the total line shape is effectively a single Lorentzian line closely resembling one of the components, but there is also a broad, out-of-phase, and weak component as well.

A nice rationalisation of the relaxation dispersion has recently appeared.[33] Assume that the major site is on resonance so that, after an excitation pulse along *y*, the major site magnetisation does not precess and remains along the *x* axis. However, a spin in the major site may exchange with the minor site, which is off-resonance. While the spin is in the minor site, it acquires a phase due to its precession so that, when it exchanges back to the major site, it will attenuate the magnetisation there. If a refocussing pulse is applied, the precession in the minor site will be refocussed if no more exchange has occurred. If the spin has jumped back to the major site before the refocussing pulse, then the exchange effects will be observed. Depending on the relative timing of the exchange and the refocussing, more or less attenuation will be seen – the apparent relaxation time will depend on the spacing of the refocusing pulses in the CPMG train.

The molecule 1,6-dipivaloyl-3,4,7,8-tetramethyl-2,5-dithioglycoluril[34] (Figure 2.15) is nominally symmetric, but steric crowding forces it into an asymmetric butterfly-like

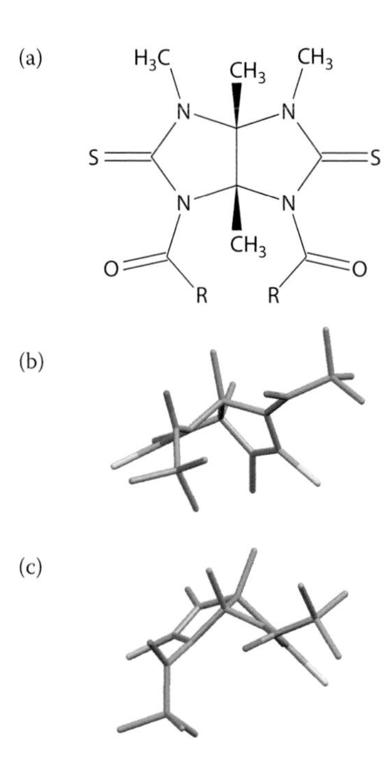


**FIGURE 2.15** Structure (a) and two views (b and c) of the actual geometry of 1,6-dipivaloyl-3,4,7,8-tetramethyl-2,5-dithioglycoluril. The R groups in the structure are tertiary butyl. The two views are more or less along the ring fusion of the two five-membered rings, but with different viewpoints – below and above the average plane of the rings. The molecule can twist into its mirror image at around ambient temperatures. (Reproduced from Bain, A. D. et al., *Canadian Journal of Chemistry* 84, pp. 421–428. ©2006 Canadian Science Publishers or its licensors. With permission.)

conformation. The molecule can flip into the opposite conformation, and in this process, the methyl groups at the ring fusion do not show exchange effects, whereas the N-methyl groups do. Figure 2.16 shows semi-logarithmic plots of the signal decay versus time for three different pulse spacings. The slope of the lines for the ring-fusion methyls is the same, but the N-methyls show significant variations. Figure 2.17 shows a more complete set of data, in which the $R_2$ of the amide proton of residue 4 in the Cro protein from prophage Pfl$_6$[35] is shown as a function of the CPMG frequency (the reciprocal of the time between refocussing pulses).

Another common way of measuring $R_2$ is by measuring the relaxation time in the rotating frame of reference, the $T_{1\rho}$ experiment. In this experiment, the magnetisation is flipped into the $xy$ plane and then spin-locked with an RF field, $B_1$, for some length of time. The signal observed at the end of the spin lock relaxes exponentially at a rate defined by the spin–lattice relaxation time in the rotating frame, $T_{1\rho}$. These experiments have a timescale defined by the RF power in the spin-locking field. During the spin lock, the spins precess around the RF field at a rate given by $\gamma B_1$, where $\gamma$ is the magnetogyric ratio. If this precession rate is comparable to the exchange rate, then

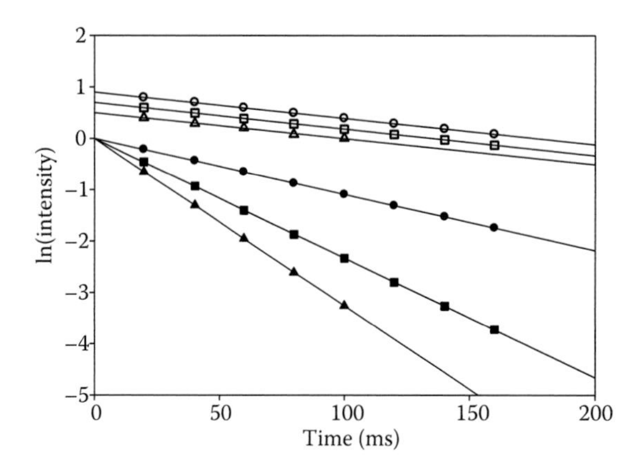


**FIGURE 2.16** Plot of the natural logarithm of the C-methyl signals (open symbols) and N-methyl signals (filled symbols) as a function of refocussing time in a CPMG $T_2$ experiment. The circles correspond to an experiment with 1 ms between refocussing pulses; the squares, 2.5 ms; and the triangles, 10 ms. The lines are linear least-squares fits to the data. The C-methyls are at the ring fusion and are not affected by the exchange. The three sets of data are offset from each other to show that the slope of the curve (proportional to $1/T_2$) is much the same regardless of the pulse spacing. The N-methyls (at the top of the rings in Figure 2.15) exchange with each other. The more time there is between refocussing pulses, the more time there is for exchange and that the $T_2$ is correspondingly shorter (faster relaxation). (Reproduced from Bain, A. D. et al., *Canadian Journal of Chemistry* 84, pp. 421–428. ©2006 Canadian Science Publishers or its licensors. With permission.)

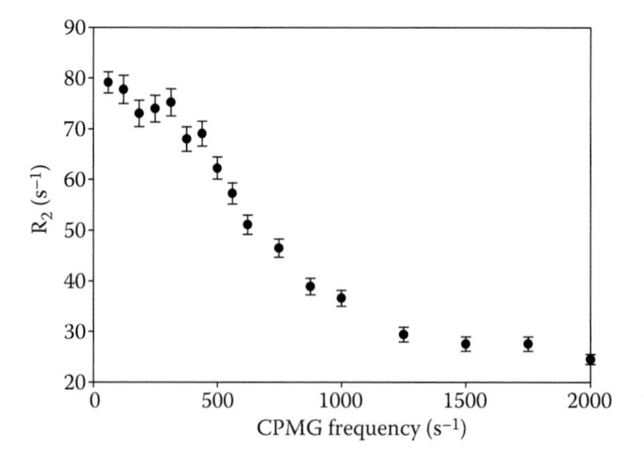


**FIGURE 2.17** $R_2$ of the amide proton of residue 4 in the Cro protein from prophage Pfl$_6$,[35] as a function of the CPMG frequency (the reciprocal of the time between refocussing pulses). The total time between excitation and detection is constant (16 ms), and different CPMG frequencies correspond to different numbers of $\pi$ pulses inserted into this domain. For this case, the population of the minor site was 5%. We thank Dr. Guillaume Bouvignies and Dr. Lewis E. Kay for providing the data for this diagram.

exchange contributes to the total relaxation[36] and shows a dispersion similar to that observed in the CPMG experiment.

Both the CPMG and the $T_{1\rho}$ experiments can provide rate measurement past coalescence, but spectrometer hardware provides quite a limitation.[37] Pulses have a finite length, so short pulse delays lead to substantial duty cycles that can heat the sample and even damage the probe. Similarly, there is a limit to the RF power that one can use in a spin-locking experiment. For typical spectrometers, the upper limits are around a few kilohertz.

## 2.3 KINETICS

If we regard the two chair forms of the cyclohexane derivative as the reagent and the product of a chemical reaction, then we have a reaction as suggested in Figure 2.4. Since the system is at macroscopic equilibrium, the principle of detailed balance must hold: the rate of the forward reaction must equal the rate of the reverse. More specifically, the population of one state (the concentration of the reagent) times the forward rate constant, $k_f$, must equal the population of the other state (product) times the reverse rate constant, $k_r$. The ratio of these rates then gives the equilibrium constant, $K_{eq}$.

$$k_f[\text{reagent}] = k_r[\text{product}]$$
$$K_{eq} = \frac{[\text{reagent}]}{[\text{product}]} = \frac{k_r}{k_f} \tag{2.1}$$

Standard thermodynamics says that the equilibrium constant is determined by the free energy difference between the states and the temperature. In this equation, $\Delta G^0$ is the difference in free energy of formation, $k_B$ is Boltzmann's constant, and $T$ is the absolute temperature.

$$K_{eq} = \exp\left(-\frac{\Delta G^0}{k_B T}\right) \tag{2.2}$$

The rates themselves show a similar dependence on temperature. It was Arrhenius who observed empirically that the rate behaved as

$$k = A \exp\left(-\frac{\Delta E}{k_B T}\right) \tag{2.3}$$

where $A$ is called the 'pre-exponential factor', and $\Delta E$ is the energy of activation. This means that a plot of the natural logarithm of the rate against the reciprocal temperature will give a straight line. Later, Eyring put this observation on a firmer theoretical basis. If we assume that the transition state is in thermodynamic equilibrium with the reagents and products, then the rate can be calculated in terms of fundamental constants and is given by

$$\begin{aligned} k &= \frac{k_B T}{h} \exp\left(-\frac{\Delta G^\dagger}{k_B T}\right) \\ &= \frac{k_B T}{h} \exp\left(-\frac{\Delta H^\dagger}{k_B T}\right) \exp\left(\frac{\Delta S^\dagger}{k_B}\right) \end{aligned} \tag{2.4}$$

where $\Delta H^\dagger$, $\Delta S^\dagger$, and $\Delta G^\dagger$ are the enthalpy, the entropy, and the free energy, respectively, of activation, and $k_B$ is Boltzmann's constant. An Eyring plot is slightly different from an Arrhenius analysis in that the (natural) $\log\left(\frac{k}{T}\right)$ is plotted against the reciprocal temperature. The slope is $-\frac{\Delta H^\dagger}{k_B}$, and the intercept is the (natural) $\log\left(\frac{k_B}{h}\right) + \frac{\Delta S^\dagger}{k_B}$. Normally, we work in molar quantities, so Boltzmann's constant is replaced by the gas constant, $R$.

Note that, typically, rates are measured over quite a narrow range of absolute temperatures, so there is a considerable extrapolation back to $\frac{1}{T} = 0$ in an Eyring plot. This means that, although enthalpies of activation are usually reliable, the entropy of activation may have large biases and errors. If entropies are important (and they often are[16]), then measurements must be taken over a wide range of temperatures, usually with a number of different methods. Figure 2.18 shows the results of some

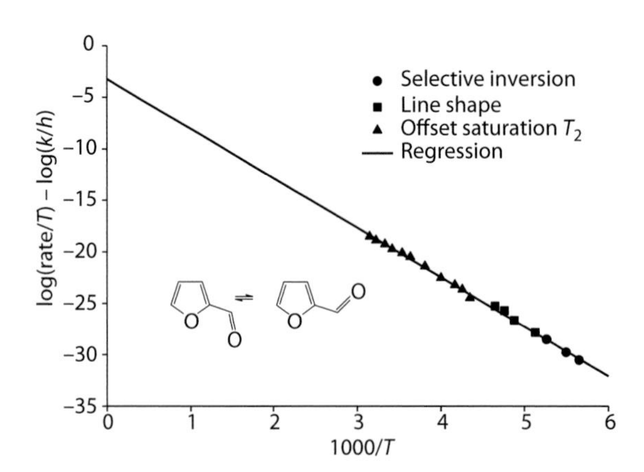


**FIGURE 2.18** Eyring plot of log(rate/temperature) against 1000/temperature for the cis–trans isomerisation of furfural.[38] The exchange rate of the aldehyde proton was measured using $R_2$ offset–saturation methods[32] at high temperature, line shape methods in the intermediate regime, and selective inversion methods at low temperatures. The plot deliberately shows the extrapolation back to $1/T = 0$ (cf. Figure 2.9), which is necessary to estimate the entropy of activation. The enthalpy of activation is given by the slope. (Reprinted with permission from reference 38, pp. 17,338–17,343. Copyright 1995 American Chemical Society.)

measurements on exchange in furfural, which cover a range of 140° in temperature and five orders of magnitude in rate.[38] Even with these data, the figure shows a substantial extrapolation back to the $y$ axis.

## 2.4 THEORY

For a reader who wants to know some of the background behind the effects that we have illustrated, we present some theories requiring some linear algebra. We will try to take a unified approach to the many different manifestations of chemical exchange in NMR. In particular, we will spend a good deal of time with the behaviour of two equally populated sites of a single spin-1/2, solving it in different ways. Once this simple system is understood, more complex problems seem less daunting. This is, by no means, a detailed analysis of all exchange experiments, but it gives some simple examples of how these can be approached. The original discussions of exchange phenomena[39,40] focussed on line shapes in the frequency domain, but with modern pulse spectrometers, a time-domain approach is more appropriate.[41] Comparing the two approaches for the simple case of two equally populated sites is a good introduction to the analysis of more complex experiments.

The classic derivation[39] of the line shapes in Figure 2.1 comes from coupled sets of Bloch equations. A common perception is that the line shape is to be treated as a single entity and is not obviously related to non-exchanging systems. One of the aims here is to show that the exchange line shape is fundamentally no different from a static spectrum. The standard equations also contain a $B_1$ observing field to extract the signal. This is appropriate for systems observed with CW spectrometers. However, the Bloch equations approach is limiting; it does not treat coupled systems, the observing field is not present during acquisition in a modern Fourier transform spectrometer, and the equations give little guidance on the time-domain behaviour essential for Fourier transform spectroscopy. For a general system, a density matrix treatment is necessary,[40,42] but the Bloch equations provide a useful introduction.

The Bloch equations derivation is given, then we show its analogies to a simple time-dependent density matrix calculation. This calculation is a specific example of a general approach applicable to all exchanging systems. The end result is that the line shape for any exchanging system is conceptually very similar to its static spectrum. The line shape is always a sum of individual transitions, but in the exchanging system, the lines may be strongly distorted in position, width, intensity, and phase by the dynamics. These distortions lead to the unusual line shapes. We are familiar with a transition having a frequency and a transition probability. The same is true for a dynamic system, except that the frequency and the transition probability are complex, rather than real, numbers.

### 2.4.1 Bloch Equations in the Frequency Domain

For a single spin-1/2, the Bloch equations, Equation (2.5), predict the behaviour of the magnetisations in the presence of an RF observation field, $B_1$, at frequency $\omega$. The spin has a magnetisation, $M_z$, along the static magnetic field, $B_0$, whose equilibrium value is $M_\infty$. Perpendicular to $B_0$ are two magnetisations, $u$ and $v$, which precess around the static field at the Larmor frequency, $\omega_0 = \gamma B_0$, where $\gamma$ is the magnetogyric ratio. Thus,

$$\frac{du}{dt} = (\omega_o - \omega)v - \frac{u}{T_2}$$

$$\frac{dv}{dt} = -(\omega_o - \omega)u + \gamma B_1 M_z - \frac{v}{T_2} \tag{2.5}$$

$$\frac{dM_z}{dt} = -\gamma B_1 v - \frac{M_z - M_\infty}{T_1}$$

The notation can be simplified by defining a complex magnetisation, $M$:

$$M = u + iv \tag{2.6}$$

For simplicity, we assume that the RF field is weak enough to not change $M_z$ significantly (non-saturating conditions), so it is now a constant, $M_\infty$. In this approximation, the Bloch equations for the $xy$ magnetisations become

$$\frac{dM}{dt} + i\,(\omega_0 - \omega)\,M + \frac{1}{T_2}M = i\gamma B_1 M_\infty \tag{2.7}$$

Now, consider two equally populated sites, A and B, exchanging with each other at a rate, $k$. Assume (for simplicity) that their Larmor frequencies are at +/–δ, so the average frequency is zero. The Bloch equations for the two sites are now

$$\frac{dM_A}{dt} + i(\delta - \omega)M_A + \frac{1}{T_2}M_A = i\gamma B_1 M_{\infty A} - kM_A + kM_B$$

$$\frac{dM_B}{dt} + i(-\delta - \omega)M_B + \frac{1}{T_2}M_B = i\gamma B_1 M_{\infty B} - kM_B + kM_A \tag{2.8}$$

The observable NMR signal is the imaginary part of the sum of the two steady-state magnetisations, $M_A$ and $M_B$, and it should be linear in $B_1$. It is assumed that the $B_1$ field is small, so that the equilibrium $z$ magnetisation is not perturbed. Since the system is equally populated, both the equilibrium $z$ magnetisations are the same, $M_\infty$. The steady state implies that the time derivatives are zero, and a little further calculation (and neglect of $T_2$ terms) gives the NMR spectrum of an exchanging system as in

$$v = \gamma B_1 M_\infty \frac{k(2\delta)^2}{(\delta - \omega)^2(\delta + \omega)^2 + 4\,k^2\omega^2} \tag{2.9}$$

Early work often looked for a coalescence point. At this rate (i.e. temperature), the top of the line shape is flat. In mathematical terms, it is defined by the rate at which both the first and second derivatives of the line shape (with respect to frequency) vanish at average frequency. Some algebra shows that this is when

$$k_{\text{coalescence}} = \frac{\delta}{\sqrt{2}} \tag{2.10}$$

Note that, in this equation, $\delta$ is in radians per second, and the frequency difference between the sites is $2\delta$. Also, there are several different derivations (sometimes differing by factors of 2) of coalescence, so be aware of the parameter definitions and notation. If the Larmor frequencies (in hertz) of the two sites are $\nu_A$ and $\nu_B$, then the rate at coalescence is given by

$$k_{\text{coalescence}} = \frac{2\pi(\nu_A - \nu_B)}{2\sqrt{2}} \tag{2.11}$$

With modern spectrometers and computers, it is quite easy to fit the whole line shape for any particular rate, so the search for the coalescence temperature is now mainly obsolete. In fact, spectra fitted at rates slower than coalescence often give more accurate data since the spectra show more structure. Although this derivation is the standard one in all the older classic textbooks, it is better now to formulate it in the time domain.

### 2.4.2 Bloch Equations in the Time Domain

We now repeat the derivation of the exchange line shape using a time-domain approach. The algebra is more complicated but straightforward and is deliberately given in some detail. The result is the same, of course, but the approach is typical of modern treatments of pulse sequences. It is also formally the same as the methods that deal with complicated systems, such as those with scalar coupling ($J$), which are beyond the Bloch equations. The main tool is a matrix called the 'Liouvillian', which can be derived from the Hamiltonian.[43–45] This, and the algebra involved, may appear daunting, but a straightforward matrix algebra and a computer algebra program, such as Maple (http://www.maplesoft.com) or Mathematica (http://www.wolfram.com), can be very useful here.

In a Fourier transform spectrometer, the $B_1$ field used in the above derivation is superfluous – the system is observed under free precession, after excitation by the pulse. Under these conditions, the coupled Bloch equations, Equation (2.8), can be rewritten in a matrix form, which omits the terms in $B_1$ and $\omega$:

$$\begin{aligned} \frac{d}{dt}\begin{pmatrix} M_A \\ M_B \end{pmatrix} &= \begin{pmatrix} -i\delta - \dfrac{1}{T_2} - k & k \\ k & i\delta - \dfrac{1}{T_2} - k \end{pmatrix} \begin{pmatrix} M_A \\ M_B \end{pmatrix} \\ &= -(i\mathbf{L} + \mathbf{R} + \mathbf{K}) \begin{pmatrix} M_A \\ M_B \end{pmatrix} \end{aligned} \tag{2.12}$$

where the first matrix, **L**, later to be revealed as the Liouvillian, is (in this case)

$$\mathbf{L} = \begin{pmatrix} -\delta & 0 \\ 0 & \delta \end{pmatrix} \tag{2.13}$$

The relaxation matrix, **R** (we have now switched to $R_2 = \frac{1}{T_2}$, $R_1 = \frac{1}{T_1}$), is given by

$$\mathbf{R} = \begin{pmatrix} R_2 & 0 \\ 0 & R_2 \end{pmatrix} \tag{2.14}$$

and the exchange matrix, **K**, is

$$\mathbf{K} = \begin{pmatrix} k & -k \\ -k & k \end{pmatrix} \tag{2.15}$$

Note that **K** has the property that the columns sum to zero, which means that the total number of spins is preserved.

Equation 2.12 represents a set of first-order differential equations, whose formal solution is given by

$$\begin{pmatrix} M_A(t) \\ M_B(t) \end{pmatrix} = \exp\left(-[i\mathbf{L} + \mathbf{R} + \mathbf{K}]t\right) \begin{pmatrix} M_A(0) \\ M_B(0) \end{pmatrix} \tag{2.16}$$

There are many ways to calculate the exponential of a matrix, but the most appropriate in this case is to diagonalise it first to obtain a diagonal matrix, $\mathbf{\Lambda}$, with the eigenvalues, $\lambda_1$ and $\lambda_2$, along the diagonal:

$$\mathbf{U}^{-1}[i\mathbf{L} + \mathbf{R} + \mathbf{K}]\mathbf{U} = \mathbf{\Lambda} \tag{2.17}$$

A standard theorem in linear algebra proves that the matrices that diagonalise a given matrix also diagonalises the exponential of the matrix. Note that the matrix in the exponential is not Hermitian (there are imaginary elements on the diagonal), so some familiar eigenvalue relations no longer hold. The eigenvalues are complex, the eigenvectors are complex, and the inverse of the eigenvector matrix is not simply the complex–conjugate transpose (left and right eigenvectors are different). However, we can still do the calculation by explicit inversion of **U**. Equation 2.16 becomes as follows:

$$\begin{aligned} \begin{pmatrix} M_A(t) \\ M_B(t) \end{pmatrix} &= \mathbf{U}\exp(-\mathbf{\Lambda}t)\mathbf{U}^{-1} \begin{pmatrix} M_A(0) \\ M_B(0) \end{pmatrix} \\ &= \mathbf{U} \begin{pmatrix} e^{-\lambda_1 t} & 0 \\ 0 & e^{-\lambda_2 t} \end{pmatrix} \mathbf{U}^{-1} \begin{pmatrix} M_A(0) \\ M_B(0) \end{pmatrix} \end{aligned} \tag{2.18}$$

We can get the eigenvalues simply by setting up the secular determinant of the matrix [$i$**L** + **R** + **K**] and solving the usual characteristic equation:

$$\begin{vmatrix} -i\delta + R_2 + k - \lambda & -k \\ -k & i\delta + R_2 + k - \lambda \end{vmatrix} = 0 \tag{2.19}$$

The roots of this equation are given by

$$\begin{aligned} \lambda_{1,2} &= (R_2 + k) \pm i\sqrt{\delta^2 - k^2} \\ &= (R_2 + k) \pm \sqrt{k^2 - \delta^2} \end{aligned} \tag{2.20}$$

These eigenvalues are the (complex) frequencies of the lines in the spectrum; the imaginary part gives the oscillation frequency and the real part gives the rate of decay in the time domain. The eigenvalues have been written in two ways to cover two different cases, which will be dealt with in detail later. If $k < \delta$ (slow exchange, the first line of Equation 2.20), then there are two different imaginary frequencies, which become $\pm\delta$ in the limit of small $k$. The width of these lines is given by the real part, $(R_2 + k)$. Note that the exchange not only broadens the lines, but also distorts the phases. In fast exchange, when $k$ exceeds $\delta$, which is half the shift difference, the quantity in the square root in the second line of Equation 2.20 becomes positive, so the roots are pure real. The imaginary part vanishes since both transitions are at average frequency, which we defined to be zero. Note that this critical point is when the rate is a factor of $\sqrt{2}$ faster than coalescence (Equation 2.10). Reeves and Shaw[46] derived this relation by algebraic manipulation in 1970, but it comes out clearly from the time–domain discussion. This calculation shows that, after coalescence, the spectrum is still two lines, but they are both at the average chemical shift (zero, in this case) and have different widths. One line is negative in intensity, and broader, as in Figure 2.14. This corresponds to the eigenvalue $(R_2 + k) + \sqrt{k^2 - \delta^2}$, where the $k$ terms add. As the exchange gets faster, the broad component gets broader and decreases in integrated intensity, so it contributes little to the observed line shape. Relatively soon after coalescence, it apparently vanishes, leaving a single narrow Lorentzian line corresponding to the eigenvalue $(R_2 + k) - \sqrt{k^2 - \delta^2}$ in the limit of fast exchange. When $k$ dominates, then the width of the Lorentzian line is given by

$$(R_2 + k) - k\sqrt{1 - \frac{\delta^2}{k^2}} \approx R_2 + \frac{\delta^2}{2k} \tag{2.21}$$

If the chemical shift difference, $\delta$, is not known, then we cannot extract the rate directly. However, experiments that measure $R_2$ or $T_{1\rho}$ have a timescale associated with them. If the timescale is comparable to the rate, then the apparent $T_2$ will depend on that timescale, so both the rate and the square of the shift difference can be extracted. More sophisticated experiments can even extract the sign as well as the magnitude of the shift difference.

We can proceed with the general derivations, but the algebra is a bit easier if we ignore $R_2$, as we did in Equation 2.9. The intensities of the lines, from Equation 2.18, are determined by the eigenvectors of $[i\mathbf{L} + \mathbf{R} + \mathbf{K}]$. Note that the intensity may be a complex number, whose real and imaginary parts determine both the integrated intensity and the phase. A matrix $\mathbf{U}$ of un-normalised right eigenvectors corresponding to the eigenvalues $k - i\sqrt{\delta^2 - k^2}$ and $k + i\sqrt{\delta^2 - k^2}$ can be calculated (most easily by computer programs for linear algebra) and is given by

$$\mathbf{U} = \begin{pmatrix} k & k \\ i\left(\sqrt{\delta^2 - k^2} - \delta\right) & -i\left(\sqrt{\delta^2 - k^2} + \delta\right) \end{pmatrix} \tag{2.22}$$

Because the original matrix was not Hermitian, the matrix of left eigenvectors (as rows) is the calculated inverse of this matrix, not simply its adjoint (its complex–conjugate transpose).

Some simple linear algebra that shows its inverse is given by

$$\mathbf{U}^{-1} = \begin{pmatrix} \dfrac{\sqrt{\delta^2 - k^2} + \delta}{2k\sqrt{\delta^2 - k^2}} & \dfrac{-i}{2\sqrt{\delta^2 - k^2}} \\ \dfrac{\sqrt{\delta^2 - k^2} - \delta}{2k\sqrt{\delta^2 - k^2}} & \dfrac{i}{2\sqrt{\delta^2 - k^2}} \end{pmatrix} \tag{2.23}$$

We can now substitute these into Equation 2.18. We assume that the observe pulse has taken the $z$ magnetisations of the two sites into the $xy$ plane, and they are equal and just written as 1. Equation 2.18 becomes

$$\begin{pmatrix} M_A(t) \\ M_B(t) \end{pmatrix} = \mathbf{U}\exp(-\boldsymbol{\Lambda}t)\mathbf{U}^{-1}\begin{pmatrix} 1 \\ 1 \end{pmatrix} \tag{2.24}$$

When we open the receiver after the pulse, we observe all of the spins, so our signal is the sum of $M_A(t)$ and $M_B(t)$:

$$M_A(t) + M_B(t) = \frac{\sqrt{\delta^2 - k^2} - ik}{\sqrt{\delta^2 - k^2}} e^{-\lambda_1 t} + \frac{\sqrt{\delta^2 - k^2} + ik}{\sqrt{\delta^2 - k^2}} e^{-\lambda_2 t} \tag{2.25}$$

Note that this is written in a form appropriate for slow rates so that all the terms in the square roots are positive. For each line, there is a complex intensity factor that multiplies the exponential. The denominators of the intensities are the same, the real parts of the numerators are the same, and the imaginary parts have opposite signs, so they have opposite phases.

Fourier transformation produces the frequency–domain spectrum, $S(\omega)$, given by Equation 2.26. This has complex numbers in the denominators, so we rationalise them by bringing them to the numerator to clarify further calculations:

$$\begin{aligned} S(\omega) &= \frac{\sqrt{\delta^2-k^2}-ik}{\sqrt{\delta^2-k^2}} \frac{1}{-k+i\left(\sqrt{\delta^2-k^2}-\omega\right)} + \frac{\sqrt{\delta^2-k^2}+ik}{\sqrt{\delta^2-k^2}} \frac{1}{-k+i\left(-\sqrt{\delta^2-k^2}-\omega\right)} \\ &= \frac{\sqrt{\delta^2-k^2}-ik}{\sqrt{\delta^2-k^2}} \frac{-k-i\left(\sqrt{\delta^2-k^2}-\omega\right)}{k^2+\left(\sqrt{\delta^2-k^2}-\omega\right)^2} + \frac{\sqrt{\delta^2-k^2}+ik}{\sqrt{\delta^2-k^2}} \frac{-k+i\left(-\sqrt{\delta^2-k^2}-\omega\right)}{k^2+\left(\sqrt{\delta^2-k^2}+\omega\right)^2} \end{aligned} \tag{2.26}$$

where the parameters are defined in Equations 2.20 to 2.23. This is the full signal, which has both real and imaginary parts. Normally, we phase the spectrum so that it shows just the real parts; in this case, there is an arbitrary minus sign to make the spectrum positive. Equation 2.27 shows this. We start with the full expression directly from Equation 2.26 then simplify it to give the expression that we have already derived in Equation 2.9.

$$\begin{aligned} \text{Real signal} &= \frac{k}{k^2+\left(\sqrt{\delta^2-k^2}-\omega\right)^2} - \frac{k\left(-\sqrt{\delta^2-k^2}+\omega\right)}{\sqrt{\delta^2-k^2}\left(k^2+\left(\sqrt{\delta^2-k^2}+\omega\right)^2\right)} \\ &\quad + \frac{k}{k^2+\left(\sqrt{\delta^2-k^2}-\omega\right)^2} - \frac{k\left(\sqrt{\delta^2-k^2}+\omega\right)}{\sqrt{\delta^2-k^2}\left(k^2+\left(\sqrt{\delta^2-k^2}+\omega\right)^2\right)} \\ &= \frac{4k\delta^2}{\left(\delta^2-2\omega\sqrt{\delta^2-k^2}+\omega^2\right)\left(\delta^2+2\omega\sqrt{\delta^2-k^2}+\omega^2\right)} \\ &= \frac{k(2\delta)^2}{(\delta-\omega)^2(\delta+\omega)^2+4k^2\omega^2} \end{aligned} \tag{2.27}$$

Substituting the values from Equation 2.22 into Equation 2.26 gives the two-site line shape; the real and imaginary parts give the absorption and dispersion modes. There may well be neater ways of expressing these quantities, but these are the formats from the Maple (http://www.maplesoft.com) output, so they are less susceptible to typographical errors. The first two lines in Equation 2.27 show the two components, which sum to the total line shape that we derived in the steady-state case, Equation 2.9. This is a general approach, written in a way appropriate for the slow-exchange case ($k < \delta$).

When the rate, $k$, is larger than $\delta$ (half the frequency separation of the two sites), the eigenvalues can be written as $k-\sqrt{k^2-\delta^2}$ and $k+\sqrt{k^2-\delta^2}$, where the quantity inside the square root is positive. The matrix of eigenvectors (cf. Equation 2.22) can be written as Equation 2.28.

$$\mathbf{U}=\begin{pmatrix} k & k \\ \sqrt{k^2-\delta^2}-i\delta & -\sqrt{k^2-\delta^2}-i\delta \end{pmatrix} \tag{2.28}$$

The inverse of this matrix is given in Equation 2.29:

$$\mathbf{U}^{-1}=\begin{pmatrix} \dfrac{-\sqrt{k^2-\delta^2}-i\delta}{2k\sqrt{k^2-\delta^2}} & \dfrac{1}{2\sqrt{k^2-\delta^2}} \\ \dfrac{\sqrt{k^2-\delta^2}-i\delta}{2k\sqrt{k^2-\delta^2}} & \dfrac{-1}{2\sqrt{k^2-\delta^2}} \end{pmatrix} \tag{2.29}$$

A similar calculation to the one given above gives the total time–domain signal for fast exchange, Equation 2.30:

$$M_{\mathrm{A}}(t)+M_{\mathrm{B}}(t)=\frac{\sqrt{k^2-\delta^2}+k}{\sqrt{k^2-\delta^2}}e^{-\lambda_1 t}+\frac{\sqrt{k^2-\delta^2}-k}{\sqrt{k^2-\delta^2}}e^{-\lambda_2 t} \tag{2.30}$$

The changing of the real and imaginary parts now shows that both lines have the same position (zero frequency under our definition of $\delta$) but different intensities. In one line, the terms in $k$ add in the limit of large $k$, and in the second, they subtract to give a small (and negative) intensity. After the Fourier transform, we get the expression for the spectrum, as in Equation 2.31. Again, we see the two components in the first two lines and they sum (as they must) to the expression in Equation 2.9.

$$\begin{aligned}\text{Real signal}&=\frac{\delta^2}{\sqrt{k^2-\delta^2}\left(2k^2-\delta^2+\omega^2+2k\sqrt{k^2-\delta^2}\right)}\\&\quad+\frac{\delta^2}{\sqrt{k^2-\delta^2}\left(-2k^2+\delta^2-\omega^2+2k\sqrt{k^2-\delta^2}\right)}\\&=\frac{4k\delta^2}{\left(2k^2-\delta^2+\omega^2+2k\sqrt{\delta^2-k^2}\right)\left(-2k^2+\delta^2-\omega^2+2k\sqrt{\delta^2-k^2}\right)}\\&=\frac{k(2\delta)^2}{(\delta-\omega)^2(\delta+\omega)^2+4k^2\omega^2}\end{aligned} \tag{2.31}$$

### 2.4.3 Bloch Equations with Unequal Populations

Finally, let us consider the case where the two sites are not equally populated. Here, the forward and reverse rates, $k_f$, and $k_r$ are not equal, so the exchange matrix becomes

$$\mathbf{K} = \begin{pmatrix} k_f & -k_r \\ -k_f & k_r \end{pmatrix} \tag{2.32}$$

The eigenvalues of the total Liouvillian are now more complicated:

$$\lambda_1, \lambda_2 = -\frac{1}{2}k_f - \frac{1}{2}k_r \pm \frac{1}{2}\sqrt{k_r^2 + 2k_f k_r + k_f^2 - 4\delta^2 - 4i\delta k_f + 4ik_r\delta} \tag{2.33}$$

Note two things about these eigenvalues: they involve the square root of a complex number, and they can be re-written in terms of the total rate, $k_t$, defined as the sum of the forward and reverse rates:

$$k_t = k_f + k_r \tag{2.34}$$

The exchange matrix can now be re-written in terms of $k_t$ and a factor $p < 1$, which gives the relative intensity of the minor site (the equilibrium constant of the minor site over the major site). It is also convenient to define the frequencies of the two sites so that their weighted average (which is what we observe in fast exchange) is zero:

$$i\mathbf{L} + \mathbf{K} = \begin{pmatrix} i\,p\delta + \dfrac{k_t p}{1+p} & -\dfrac{k_t}{1+p} \\ -\dfrac{k_t p}{1+p} & -i\,\delta + \dfrac{k_t}{1+p} \end{pmatrix} \tag{2.35}$$

From here, it is possible to proceed algebraically,[46] but it is easier to go numerically through the steps that we followed in the equally populated case. There is one point that should be mentioned, however. The initial state (as in Equation 2.24) now includes the unequal populations of the two sites:

$$\begin{pmatrix} M_A(t) \\ M_B(t) \end{pmatrix} = \mathbf{U}\exp(-\Lambda t)\mathbf{U}^{-1}\begin{pmatrix} 1 \\ p \end{pmatrix} \tag{2.36}$$

However, when we observe the total signal (as in Equation 2.25), the contributions from both eigenvalues are equal since the receiver is equally sensitive (per spin) to each site.

The numerical calculations show two out-of-phase components that add to the in-phase total line shape, as before (Figures 2.6 and 2.14), and the minor site is $1/p$ broader than the major site. There is no clear break between the slow and fast regimes as there was for the equally populated case, but the minor site here also loses integrated intensity and becomes mostly negative for fast rates. The combination of a low population for the minor site, a broad line, and this phenomenon of the minor site losing relative integrated intensity makes the minor site vanish for the spectrum very quickly as the rate increases.

### 2.4.4 Density Matrix Treatment

The use of the density matrix may be intimidating, but we hope to show that it is a straightforward and natural way to describe dynamics in NMR. The definition of the density matrix, ρ, for a spin system is given in all the standard references.[42,47–49] The important property is that it contains all the information about the system. For any operator, $\boldsymbol{P}$, the expectation value of that observable is given by

$$\langle \boldsymbol{P} \rangle = trace\left(\boldsymbol{P}^{\dagger}\rho\right) \tag{2.37}$$

In this equation, $\boldsymbol{P}^{\dagger}$ is the adjoint (complex–conjugate transpose) of $\boldsymbol{P}$. We need not go into the derivation or properties of the density matrix here since it is covered in many other works. The necessary concept is that its elements encapsulate all the observables of the spin system, and the result of any particular measurement is given by Equation 2.37.

For a coupled spin system, the complete density matrix description is required, except in some special cases that resemble uncoupled systems. This is because it is the nucleus that changes magnetic environments under the exchange process, not the observed spectral line. In a coupled spin system, there is no longer a one-to-one relation between nuclei and transitions. When the nucleus jumps to the other site, its relation to the transitions will change. The density matrix is needed to describe this properly.

In NMR, we often know the density matrix at the start of the experiment – we usually start at equilibrium. To calculate the density matrix at some later stage, we need its equation of motion. This is given by the Liouville–von Neumann equation:

$$\begin{aligned} \frac{\partial}{\partial t}\rho(t) &= -i[\boldsymbol{H},\rho(t)] \\ &= -i\mathbf{L}\,\rho(t) \end{aligned} \tag{2.38}$$

Note that, in this equation, the first line corresponds to the usual definition of the Liouville–von Neumann equation, in which the time development of the density matrix is given by its commutator with the Hamiltonian, $\boldsymbol{H}$. In the second line, we have moved to Liouville space, the vector space defined by the complete set of linear operators on the spins. In this space, the density matrix becomes a vector, and the action of taking the commutator becomes a matrix in Liouville space, or a super operator, **L**. The details of Liouville space and super-operator calculations[43,44,47,50,51] need not concern us here. The important point is that it is a natural way[40,52] to describe dynamics in NMR.

If we include relaxation, via a Redfield-type matrix, **R**, and exchange, via a Kubo–Sack or related[53–55] matrix, **K**, the equation of the motion of the density matrix becomes Equation 2.39. Note the deliberate resemblance to Equation 2.12:

$$\frac{\partial}{\partial t}\rho(t) = -[i\mathbf{L} + \mathbf{R} + \mathbf{K}]\,\rho(t) \tag{2.39}$$

Now, this is a rigorous equation, valid for any spin system, coupled or uncoupled. It is solved in the same way, via the matrix exponential:

$$\rho(t) = \exp(-[i\mathbf{L} + \mathbf{R} + \mathbf{K}]t)\,\rho(0) \tag{2.40}$$

Another subtlety is that we do not detect the density matrix directly – we normally detect the total magnetisation along the $x$ or $y$ axis. This is represented by the operator, $\boldsymbol{I}_x$, which becomes a vector in Liouville space. Also, the definition of the scalar product in Liouville space is the trace relation, as in Equation 2.37. We use parentheses,[44] rather than angle brackets, to distinguish Liouville space kets, bras, and inner products from the spin–space counterparts. Therefore, the measured signal, $F(t)$ is given by

$$\begin{aligned} F(t) &= (\boldsymbol{I}_x \mid \rho(t)) \\ &= \left(\boldsymbol{I}_x \mid \exp(-[i\mathbf{L} + \mathbf{R} + \mathbf{K}]t)\rho(0)\right) \end{aligned} \tag{2.41}$$

If we have started the experiment with a simple, non-selective $\pi/2$ (or 90°) pulse, the density matrix at time zero is given by $\boldsymbol{I}_x$. We diagonalise the total matrix with a set of eigenvectors, **U**, as before to get

$$\begin{aligned} F(t) &= (\boldsymbol{I}_x \mid \mathbf{U}\exp(-\Lambda t)\,\mathbf{U}^{-1}\,\boldsymbol{I}_x) \\ &= \sum_{\mathrm{i}=1}^{n}\left(\boldsymbol{I}_x^{\dagger}\,\mathbf{U}\right)_{\mathrm{i}}(\mathbf{U}^{-1}\,\boldsymbol{I}_x)_{\mathrm{i}}\exp(-\lambda_{\mathrm{i}}t) \end{aligned} \tag{2.42}$$

In this equation, note that the usual complex conjugate does not appear – the inverse of **U** is $\mathbf{U}^{-1}$, not $\mathbf{U}^{\dagger}$. Equation 2.42 means that the free induction decay (FID) for the exchanging system is always the sum of decaying exponentials since the eigenvalues, $\lambda_{\mathrm{i}}$, have both imaginary and real parts. The intensity is given by the projection of the appropriate eigenvector onto $\boldsymbol{I}_x$.

This leads to a simple, physical picture of the transition probability. Each transition is represented by an eigenvector of the total Liouvillian (including relaxation and exchange). At the start of the experiment, it receives an amount of coherence from the initial $\boldsymbol{I}_x$ given by its projection. It then evolves in time and contributes an amount to the detected signal proportional to its projection along $\boldsymbol{I}_x$. The intensity of the line is given by the product of how much it received at the start of the experiment and how much it contributes to the detected signal. In the absence of dynamics, this calculation simply reproduces the standard transition probability. However, the concept is applicable even in dynamic systems.

We have now solved the two-site, equally populated line shape problem three ways, and that is enough. We have demonstrated the equivalence of the classic derivation to a full density matrix treatment. The density matrix treatment now serves as a prototype for the solution of more complex problems.

#### 2.4.4.1 More Complex Systems

For a two-site case, the kinetic matrix is easy to set up, but larger kinetic schemes are more complex.[19,56] Consider a three-site system. There are now six rates, but not all of them are independent. Thus (note that the columns of the matrix still sum to zero),

$$\mathbf{K} = \begin{pmatrix} k_f^{AB} + k_f^{AC} & -k_r^{AB} & -k_r^{AC} \\ -k_f^{AB} & k_r^{AB} + k_f^{BC} & -k_r^{BC} \\ -k_f^{AC} & -k_f^{BC} & k_r^{AC} + k_r^{BC} \end{pmatrix} \quad (2.43)$$

Since we are dealing with a system at equilibrium, it is easy to measure the relative populations of each site. If we define an equilibrium constant for the AB reaction as

$$K^{AB} = \frac{M_B}{M_A} = \frac{k_f^{AB}}{k_r^{AB}} \quad (2.44)$$

and the others analogously, we can rewrite Equation 2.43 as

$$\mathbf{K} = \begin{pmatrix} K^{AB}k_r^{AB} + K^{AC}k_r^{AC} & -k_r^{AB} & -k_r^{AC} \\ -K^{AB}k_r^{AB} & k_r^{AB} + K^{BC}k_r^{BC} & -k_r^{BC} \\ -K^{AC}k_r^{AC} & -K^{BC}k_r^{BC} & k_r^{AC} + k_r^{BC} \end{pmatrix} \quad (2.45)$$

Furthermore, the equilibrium populations must satisfy the principle of detailed balance, which says that, at equilibrium, the number of spins leaving a site must equal the number coming back:

$$\begin{pmatrix} K^{AB}k_r^{AB} + K^{AC}k_r^{AC} & -k_r^{AB} & -k_r^{AC} \\ -K^{AB}k_r^{AB} & k_r^{AB} + K^{BC}k_r^{BC} & -k_r^{BC} \\ -K^{AC}k_r^{AC} & -K^{BC}k_r^{BC} & k_r^{AC} + k_r^{BC} \end{pmatrix} \begin{pmatrix} M_A \\ M_B \\ M_C \end{pmatrix} = \begin{pmatrix} 0 \\ 0 \\ 0 \end{pmatrix} \quad (2.46)$$

This is a set of linear equations, but not the most convenient form for calculation. One of the equations in this set is redundant since the bottom equation is simply the negative of the sum of the top two. The set of equations can be reduced to

$$\begin{pmatrix} K^{AB}k_r^{AB} + K^{AC}k_r^{AC} & -k_r^{AB} \\ -K^{AB}k_r^{AB} & k_r^{AB} + K^{BC}k_r^{BC} \end{pmatrix} \begin{pmatrix} M_A \\ M_B \end{pmatrix} = \begin{pmatrix} k_r^{AC}M_c \\ k_r^{BC}M_c \end{pmatrix} \tag{2.47}$$

This calculation can be generalised for more complex kinetic schemes so that equilibrium populations and kinetic matrices can be cross-checked.

### 2.4.4.2 Coupled Spin Systems

The formal approach to exchange in a coupled spin system is much the same as described above, but there are some complications. Here, the Bloch equations approach will not work, so the density matrix is essential. One reason is that the transitions in a coupled spin system are made up from the spin wave functions. These wave functions depend on the spectral parameters: shifts, couplings, etc. When a nucleus jumps to a different site, the parameters change, and so do the wave functions. Therefore, coherence associated with a single transition in one site may be distributed among several transitions in the other. In the single-spin case, there is only one transition, so the problem does not arise. For a coupled system, we must calculate the spectrum in each site before we consider the effects of exchange.

The case of exchange in an *AB* spin system, whether mutual or not, serves as a prototype for any exchange in liquids. For any spin system, the spectral analysis (shifts and couplings in the absence of exchange) of each site must be done to set up the Liouvillian. Then, the exchange mechanism must be described to indicate which spin exchanges with which. Once these two matrices in Liouville space are set up, the combined matrix can be diagonalised using standard methods to yield eigenvalues and eigenvectors. The observed FID or spectral line shape can be calculated using Equation 2.42. Figure 2.19 shows the spectra of the aromatic protons in N,N-dimethyl-4-nitrosoaniline, which provides a nice example of exchange in a coupled spin system. For $n$ spin-1/2, the number of possible transitions is given by the binomial coefficient $\begin{pmatrix} 2n \\ n-1 \end{pmatrix}$, which is 56 for 4 spins, as in this case. However, as in the analysis of static spectra, many transitions have very low intensity. As with two uncoupled sites, the transitions are distorted in line width and phase, but each is a single line. Setting up the matrices may be tedious for large spin systems and complex exchange mechanisms, but the formalism is always the same. From this point of view, a coupled spin system is much the same as an uncoupled one, just with larger matrices. Careful adjustment of the parameters can provide a fit of the quality of that shown in Figure 2.20. Spectra such as this, which are well below coalescence and show lots of structure, tend to provide the most accurate rate data. In doing these fits, an important parameter

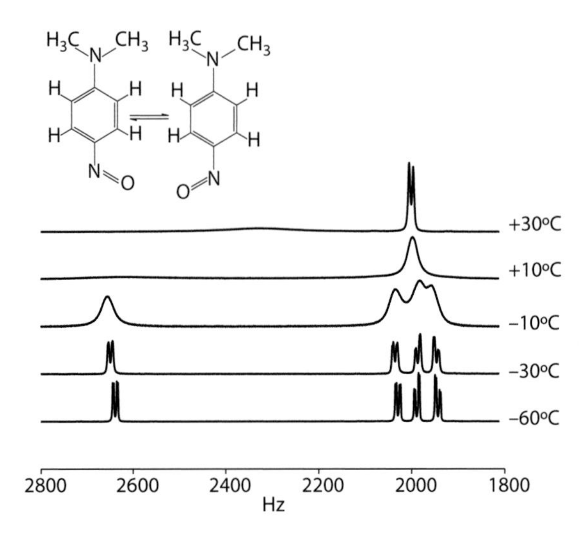


**FIGURE 2.19** Variable temperature proton spectra of the aromatic protons in N,N-dimethyl-4-nitrosoanaline. The proton at high frequency is the one that is close to the nitroso group and has the larger shift difference with its exchange partner. At 30°C, this pair is in intermediate exchange, whereas the pair near the aniline nitrogen has already coalesced into their average spectrum. (Reprinted from *Progress in Nuclear Magnetic Resonance Spectroscopy*, 43, Bain, A. D., Chemical exchange in NMR, pp. 63–103. Copyright 2003, with permission from Elsevier.)

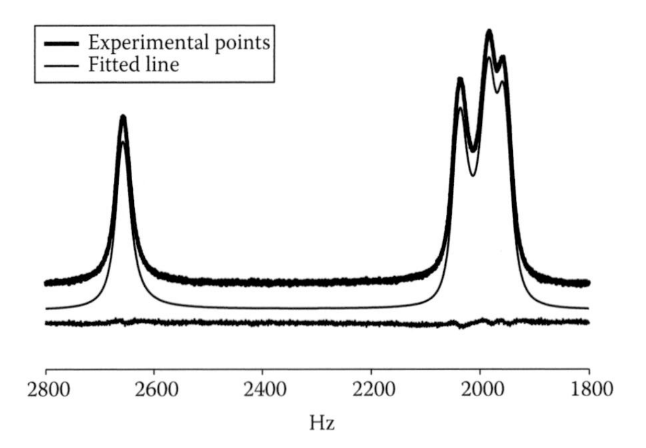


**FIGURE 2.20** Iterative fit of a calculated line shape (middle plot, light line) to the experimental spectrum of N,N-dimethyl-4-nitrosoanaline at −10°C (top plot, offset vertically for clarity). The bottom plot shows the difference between calculated and observed. (Reprinted from *Progress in Nuclear Magnetic Resonance Spectroscopy*, 43, Bain, A. D., Chemical exchange in NMR, pp. 63–103. Copyright 2003, with permission from Elsevier.)

is the chemical shift difference in the absence of exchange. This is an abstract number since peak positions will shift with exchange rates, even though the shifts in the absence of exchange remain the same. A standard technique is to acquire a series of spectra at temperatures well below those where exchange effects are important, as in Figure 2.21. The trend in shift can then be extrapolated to higher temperatures and provide input to a fit such as that in Figure 2.20. For systems of a few spins (<6), such iterative calculated fits are readily available with modern computers and software.

#### 2.4.4.3 Z Magnetisations

So far, we have been dealing mainly with the observable $xy$ magnetisations to look at line shapes in a number of different systems, but the $z$ magnetisations can also be used to probe dynamics. In the Bloch equations derivation, we assumed that the $z$ magnetisations had their equilibrium values. This allowed us to separate their behaviour and concentrate on the observable quantities in the $xy$ plane. The $z$ magnetisations do not precess, but they do relax to their equilibrium values, so it is their deviation from equilibrium that relaxes with the spin–lattice relaxation time, $T_1$. The equations defining the behaviour of the $z$ magnetisations have the form analogous to Equation 2.12:

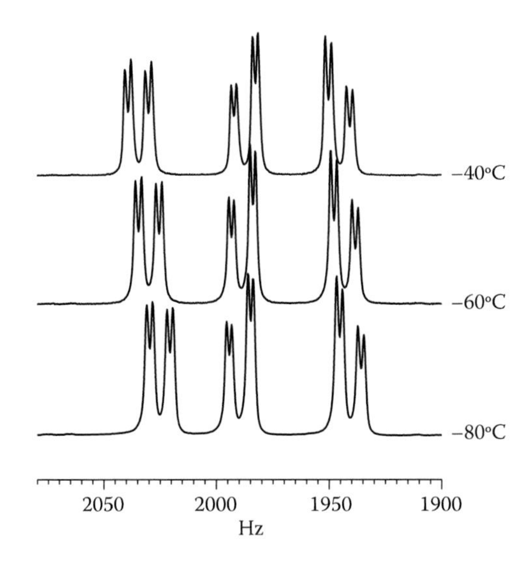


**FIGURE 2.21** Low-temperature proton spectra of the aromatic protons in N,N-dimethyl-4-nitrosoanaline, showing a typical chemical shift drift with temperature. (Reprinted from *Progress in Nuclear Magnetic Resonance Spectroscopy*, 43, Bain, A. D., Chemical exchange in NMR, pp. 63–103. Copyright 2003, with permission from Elsevier.)

$$\frac{d}{dt}\begin{pmatrix} M_{\infty A} - M_{zA} \\ M_{\infty B} - M_{zB} \end{pmatrix} = \frac{d}{dt}\begin{pmatrix} -M_{zA} \\ -M_{zB} \end{pmatrix} = \begin{pmatrix} -\frac{1}{T_1} - k & k \\ k & -\frac{1}{T_1} - k \end{pmatrix}\begin{pmatrix} M_{\infty A} - M_{zA} \\ M_{\infty B} - M_{zB} \end{pmatrix}$$

$$= \frac{d}{dt}\begin{pmatrix} M_{zA} \\ M_{zB} \end{pmatrix} = \begin{pmatrix} -\frac{1}{T_1} - k & k \\ k & -\frac{1}{T_1} - k \end{pmatrix}\begin{pmatrix} M_{zA} - M_{\infty A} \\ M_{zB} - M_{\infty B} \end{pmatrix} \tag{2.48}$$

$$= -(\mathbf{R} + \mathbf{K})\begin{pmatrix} M_{zA} - M_{\infty A} \\ M_{zB} - M_{\infty B} \end{pmatrix}$$

Note that the matrix is now Hermitian, so its eigenvalues and eigenvectors are simpler. The eigenvalues are

$$\lambda_1 = \frac{1}{T_1}$$
$$\lambda_2 = \frac{1}{T_1} + 2k \tag{2.49}$$

and the matrix of eigenvectors (which is its own inverse) is

$$\mathbf{U} = \begin{pmatrix} \frac{1}{\sqrt{2}} & \frac{1}{\sqrt{2}} \\ \frac{1}{\sqrt{2}} & -\frac{1}{\sqrt{2}} \end{pmatrix} \tag{2.50}$$

The solution to the equation is formally the same as before:

$$\begin{pmatrix} M_{zA}(t) - M_{\infty A} \\ M_{zB}(t) - M_{\infty B} \end{pmatrix} = \mathbf{U}\exp(-\Lambda t)\mathbf{U}^{-1}\begin{pmatrix} M_{zA}(0) - M_{\infty A} \\ M_{zB}(0) - M_{\infty B} \end{pmatrix}$$
$$= \mathbf{U}\begin{pmatrix} e^{-\lambda_1 t} & 0 \\ 0 & e^{-\lambda_2 t} \end{pmatrix}\mathbf{U}^{-1}\begin{pmatrix} M_{zA}(0) - M_{\infty A} \\ M_{zB}(0) - M_{\infty B} \end{pmatrix} \tag{2.51}$$

#### 2.4.4.4 Spin–Lattice Relaxation Experiments

We normally probe the relaxation of $z$ magnetisations by perturbing them from equilibrium and then sampling them with a pulse after a number of delays. Again, for the equally populated, two-site case $M_{\infty A} = M_{\infty B} = 1$, we can assume that, following a non-selective inversion pulse, we can write the initial conditions as

$$\begin{pmatrix} M_{zA}(0) - M_{\infty A} \\ M_{zB}(0) - M_{\infty B} \end{pmatrix} = \begin{pmatrix} -2M_{\infty A} \\ -2M_{\infty B} \end{pmatrix} = \begin{pmatrix} -2 \\ -2 \end{pmatrix} \tag{2.52}$$

We follow the steps in Equation 2.51. First, we multiply by $\mathbf{U}^{-1}$:

$$\mathbf{U}^{-1}\begin{pmatrix} M_{zA}(0) - M_{\infty A} \\ M_{zB}(0) - M_{\infty B} \end{pmatrix} = \begin{pmatrix} \frac{1}{\sqrt{2}} & \frac{1}{\sqrt{2}} \\ \frac{1}{\sqrt{2}} & -\frac{1}{\sqrt{2}} \end{pmatrix}\begin{pmatrix} -2 \\ -2 \end{pmatrix} = \begin{pmatrix} -2\sqrt{2} \\ 0 \end{pmatrix} \tag{2.53}$$

This then evolves under the eigenvalues (Equation 2.49) and is multiplied by $\mathbf{U}$:

$$\begin{pmatrix} M_{zA}(t) - M_{\infty A} \\ M_{zB}(t) - M_{\infty B} \end{pmatrix} = \begin{pmatrix} \frac{1}{\sqrt{2}} & \frac{1}{\sqrt{2}} \\ \frac{1}{\sqrt{2}} & -\frac{1}{\sqrt{2}} \end{pmatrix}\begin{pmatrix} \exp\left(-t\frac{1}{T_1}\right) & 0 \\ 0 & \exp\left(-t\left(\frac{1}{T_1}+k\right)\right) \end{pmatrix}\begin{pmatrix} -2\sqrt{2} \\ 0 \end{pmatrix} = \begin{pmatrix} -2\exp\left(-t\frac{1}{T_1}\right) \\ -2\exp\left(-t\frac{1}{T_1}\right) \end{pmatrix} \tag{2.54}$$

Note that there is no appearance of the exchange rate. If we do a non-selective inversion of both sites, then an exchange process simply swaps an inverted spin for another inverted spin, so for an equally populated case, the exchange is invisible.

However, if we do a selective inversion so that one site is inverted but the other is unperturbed, the exchange is revealed. If we invert site A, the initial conditions are

$$\begin{pmatrix} M_{zA}(0) - M_{\infty A} \\ M_{zB}(0) - M_{\infty B} \end{pmatrix} = \begin{pmatrix} -2M_{\infty A} \\ 0 \end{pmatrix} = \begin{pmatrix} -2 \\ 0 \end{pmatrix} \tag{2.55}$$

Multiplication by $\mathbf{U}^{-1}$ gives

$$\mathbf{U}^{-1}\begin{pmatrix} M_{zA}(0)-M_{\infty A} \\ M_{zB}(0)-M_{\infty B} \end{pmatrix} = \begin{pmatrix} \frac{1}{\sqrt{2}} & \frac{1}{\sqrt{2}} \\ \frac{1}{\sqrt{2}} & -\frac{1}{\sqrt{2}} \end{pmatrix}\begin{pmatrix} -2 \\ 0 \end{pmatrix} = \begin{pmatrix} -\sqrt{2} \\ -\sqrt{2} \end{pmatrix} \tag{2.56}$$

and the rest of the evolution is given by

$$\begin{aligned}\begin{pmatrix} M_{zA}(t)-M_{\infty A} \\ M_{zB}(t)-M_{\infty B} \end{pmatrix} &= \begin{pmatrix} \frac{1}{\sqrt{2}} & \frac{1}{\sqrt{2}} \\ \frac{1}{\sqrt{2}} & -\frac{1}{\sqrt{2}} \end{pmatrix}\begin{pmatrix} \exp\left(-t\frac{1}{T_1}\right) & 0 \\ 0 & \exp\left(-t\left(\frac{1}{T_1}+k\right)\right) \end{pmatrix}\begin{pmatrix} -\sqrt{2} \\ -\sqrt{2} \end{pmatrix} \\ &= \begin{pmatrix} \frac{1}{\sqrt{2}} & \frac{1}{\sqrt{2}} \\ \frac{1}{\sqrt{2}} & -\frac{1}{\sqrt{2}} \end{pmatrix}\begin{pmatrix} -\sqrt{2}\exp\left(-t\frac{1}{T_1}\right) \\ -\sqrt{2}\exp\left(-t\left(\frac{1}{T_1}+k\right)\right) \end{pmatrix} \\ &= \begin{pmatrix} -\exp\left(-t\frac{1}{T_1}\right)-\exp\left(-t\left(\frac{1}{T_1}+k\right)\right) \\ -\exp\left(-t\frac{1}{T_1}\right)+\exp\left(-t\left(\frac{1}{T_1}+k\right)\right) \end{pmatrix}\end{aligned} \tag{2.57}$$

Site A relaxes as a sum of two exponentials, so it relaxes faster than the normal spin–lattice relaxation rate. Site B, which was unperturbed, starts out at zero, but then has a negative transient, before finally relaxing back to equilibrium. This transient is very characteristic of the chemical exchange process. The faster relaxation of the inverted site is also characteristic, but separating the sum of two exponential decays is a famously difficult problem, so the transient in the non-inverted site is better data to look for.

We chose a clean selective inversion as the initial condition since that gives the clearest picture. However, this may not be experimentally very feasible, so any differential preparation of one $z$ magnetisation over the other will still show some effects of exchange.

#### 2.4.4.5 NOESY/EXSY 2-D Experiment

The pulse sequence and the physics of the two experiments are identical. The experiment probes the relaxation of the $z$ magnetisations, and cross peaks indicate the relaxation of one site that is affected by the initial state of another. This can happen if there is dipole–dipole relaxation (Overhauser effects) or exchange between the two sites. The difference in name is often used to distinguish whether we are looking for Overhauser effects or exchange phenomena. As before, we set up the Liouville–von Neumann equation for a system with unequal populations:

$$\frac{d}{dt}\begin{pmatrix} M_z^A(\infty)-M_z^A(t) \\ M_z^B(\infty)-M_z^B(t) \end{pmatrix} = -\begin{pmatrix} R_2+k_f & -k_r \\ -k_f & R_2+k_r \end{pmatrix}\begin{pmatrix} M_z^A(\infty)-M_z^A(t) \\ M_z^B(\infty)-M_z^B(t) \end{pmatrix}$$
$$= -(\mathbf{R}+\mathbf{K})\begin{pmatrix} M_z^A(\infty)-M_z^A(t) \\ M_z^B(\infty)-M_z^B(t) \end{pmatrix} \quad (2.58)$$

The eigenvalues are given by Equation 2.59:

$$\lambda_1 = R_2$$
$$\lambda_1 = R_2 + k_f + k_r \quad (2.59)$$

As we did in the line shape calculation, it is convenient to define a total rate, $k_t$:

$$k_t = k_f + k_r \quad (2.34)$$

and a relative population, $p$, of the major and minor sites:

$$p = \frac{k_f}{k_r} = \frac{[\text{minor}]}{[\text{major}]} \quad (2.60)$$

The Liouvillian becomes:

$$\begin{pmatrix} R_2+k_f & -k_r \\ -k_f & R_2+k_r \end{pmatrix} = \begin{pmatrix} R_2+\dfrac{kp}{1+p} & -\dfrac{k}{1+p} \\ -\dfrac{kp}{1+p} & R_2+\dfrac{k}{1+p} \end{pmatrix} \quad (2.61)$$

This is like a saturation-recovery $T_1$ experiment. Before the second pulse (after the $t_1$ evolution time), the $z$ magnetisations are given by

$$\begin{pmatrix} M_z^A(\infty) - M_z^A(t_1) \\ M_z^B(\infty) - M_z^B(t_1) \end{pmatrix} \tag{2.62}$$

As with any relaxation experiment,

$$\begin{pmatrix} M_z^A(\infty) - M_z^A(t_m) \\ M_z^B(\infty) - M_z^B(t_m) \end{pmatrix} = \exp(-(\mathbf{R}+\mathbf{K})t_m) \begin{pmatrix} M_z^A(\infty) - M_z^A(t_1) \\ M_z^B(\infty) - M_z^B(t_1) \end{pmatrix} \tag{2.63}$$

This is solved in the usual way using the diagonalisation of the Liouvillian, with the matrix of eigenvectors:

$$\mathbf{U}^{-1}(\mathbf{R}+\mathbf{K})\mathbf{U} = \mathbf{\Lambda} \tag{2.64}$$

and

$$\exp(-(\boldsymbol{R}+\boldsymbol{K})t_m) = \boldsymbol{U}\exp(-\boldsymbol{\Lambda}t_m)\boldsymbol{U}^{-1} \tag{2.65}$$

The eigenvalues and the associated matrix, $\boldsymbol{U}$, of eigenvectors (as columns) are

$$\begin{matrix} (r) & (r+k) \end{matrix}$$

$$\begin{pmatrix} \frac{1}{\sqrt{1+p^2}} & -\frac{1}{\sqrt{2}} \\ \frac{p}{\sqrt{1+p^2}} & \frac{1}{\sqrt{2}} \end{pmatrix} \tag{2.66}$$

Note that the original matrix was not symmetric and that the eigenvectors are not orthogonal. We have to calculate the inverse of $\boldsymbol{U}$ explicitly, but this is straightforward:

$$\boldsymbol{U}^{-1} = \begin{pmatrix} \frac{\sqrt{1+p^2}}{1+p} & \frac{\sqrt{1+p^2}}{1+p} \\ -\frac{\sqrt{2}p}{1+p} & \frac{\sqrt{2}}{1+p} \end{pmatrix} \tag{2.67}$$

Putting all of these together, we find that the expressions for the diagonal and cross peaks are

$$\begin{aligned}
\text{diagonal (major)} &= \frac{\exp(-r\,t_\text{m}) + p\exp(-(r+k)\,t_\text{m})}{1+p} M^\text{A} \\
\text{diagonal (minor)} &= \frac{p\exp(-r\,t_\text{m}) + \exp(-(r+k)\,t_\text{m})}{1+p} M^\text{B} \\
\text{cross} &= \frac{\exp(-r\,t_\text{m}) + \exp(-(r+k)\,t_\text{m})}{1+p} M^\text{B} \\
&= p\frac{\exp(-r\,t_\text{m}) + \exp(-(r+k)\,t_\text{m})}{1+p} M^\text{A}
\end{aligned} \tag{2.68}$$

To first order in the mixing time, $t_\text{m}$, the ratio of the cross peak to the diagonal peak of the major site is

$$\frac{\text{cross}}{\text{major}} \approx \frac{k}{1+p} t_\text{m} \tag{2.69}$$

If we have equal populations, then $p = 1$, and since $k$ is defined as the sum of the forward and reverse rates, the numerator is twice the rate. In this case, the familiar result is shown that the slope of the graph of the ratio against the mixing time is the rate.

## 2.5 SUMMARY

We have assumed that there is dynamics in most systems, but the question becomes whether this is important, interesting, or useful. Are the typical suspects present in the chemical system? Amides, large rings, and a vast number of organometallic compounds have intra- or intermolecular processes that go on at typical spectrometer temperatures. Perhaps the first step is to go looking for evidence of dynamics. Are some lines in the spectrum unusually broad? If this is temperature dependent, then this is strong evidence for dynamics. Because chemical exchange usually has a higher activation energy than most processes that cause relaxation, exchange effects are much more temperature dependent. What are the timescales available to you? Over how wide a temperature range can the molecule be studied? Do you want a qualitative picture of what the reaction is, or do you need quantitative rates as a function of temperature? Is it an intermolecular process, in which the apparent rate will depend on concentration? Can you observe a minor site directly? These are just a few of the questions that arise.

Perhaps the place to start is to run spectra at a number of temperatures, then perhaps a quick EXSY experiment. Both these are easy to do and can give strong indications of what is going on. The choice of experiment then depends on the system – whether it is in fast, intermediate, or slow exchange, and whether it is equally or unequally populated. There are many experiments available on modern spectrometers, and we have given some examples of their use.

## REFERENCES


1. Gutowsky, H. S. and C. H. Holm. 1956. Rate processes and nuclear magnetic resonance spectra: Part II. Hindered internal rotation of amides. *Journal of Chemical Physics* 25, p. 1228–1234.
2. Fauconnier, T., A. D. Bain, P. Hazendonk, R. A. Bell and C. J. L. Lock. 1998. Structure and dynamics of azapropazone derivatives studied by crystallography and nuclear magnetic resonance. *Canadian Journal of Chemistry* 76, p. 426–430.
3. Hansen, M. R., R. Graf and H. W. Spiess. 2013. Solid-state NMR in macromolecular systems: Insights on how molecular entities move. *Accounts of Chemical Research* 46, p. 1996–2007.
4. White, M. A., R. E. Wasylishen, P. E. Eaton, Y. Xiong, K. Pramod and N. Nodari. 1992. Orientational disorder in solid cubane: A thermodynamic and C-13 NMR study. *Journal of Physical Chemistry* 96, p. 421–425.
5. Cahill, L. S., R. P. Chapman, J. F. Britten and G. R. Goward. 2006. Li-7 NMR and two-dimensional exchange study of lithium dynamics in monoclinic $Li_3V_2(PO_4)_{(3)}$. *Journal of Physical Chemistry* B 110, p. 7171–7177.
6. Moreneo, P. O., C. Suarez, N. S. True and C. B. LeMaster. 1992. Gas-phase NMR study of the degenerate cope rearrangement of bullvalene. *Journal of Physical Chemistry* 96, p. 10,206–10,212.
7. Jackman, L. M. and F. A. Cotton. 1975. *Dynamic Nuclear Magnetic Resonance Spectroscopy*. New York: Academic Press.
8. Palmer, A. G., M. J. Grey and C. Wang. 2005. Solution NMR spin relaxation methods for characterising chemical exchange in high–molecular-weight systems. *Methods in Enzymology* 394, p. 430–465.
9. Sekhar, A. and L. E. Kay. 2013. NMR paves the way for atomic level descriptions of sparsely populated, transiently formed biomolecular conformers. *Proceedings of the National Academy of Sciences* 110, p. 12,867–12,874.
10. Skorupska, E. A., R. B. Nazarski, M. Ciechanska, A. Jozwiak and A. Klys. 2013. Dynamic $^1$H NMR spectroscopic study of hindered internal rotation in selected N,N-dialkyl isonicotinamides: An experimental and DFT analysis. *Tetrahedron* 69, p. 8147–8154.
11. Amini, S. K., M. Tafazzoli, H. A. Jenkins, G. R. Goward and A. D. Bain. 2009. Measurement and density functional calculations of C-13 and N-15 NMR chemical shift tensors of a push–pull ethylene. *Canadian Journal of Chemistry* 87, p. 563–570.
12. Sandström, J. 1982. *Dynamic NMR Spectroscopy*. London, United Kingdom: Academic Press.
13. Sandström, J. 1983. Static and dynamic stereochemistry of push–pull and strained ethylenes. *Topics in Stereochemistry* 14, p. 83–181.
14. Kleinpeter, E. and A. Frank. 2009. Distinction of push–pull effect and steric hindrance in disubstituted alkynes. *Tetrahedron* 65, p. 4418–4421.
15. Kleinpeter, E., U. Bolke and J. Kreicberga. 2010. Quantification of the push–pull character of azo dyes and a basis for their evaluation as potential non-linear optical materials. *Tetrahedron* 66, p. 4503–4509.
16. Ababneh-Khasawneh, M., B. E. Fortier-McGill, M. E. Occhionorelli and A. D. Bain. 2011. Solvent effects on chemical exchange in a push–pull ethylene as studied by NMR: The importance of entropy. *Journal of Physical Chemistry* A 115, p. 7531–7537.

17. Bain, A. D., M. Baron, S. K. Burger, V. J. Kowalewski and M. B. Rodriguez. 2011. Interconversion study in 1,4 substituted six-membered cyclohexane-type rings: Structure and dynamics of trans-1,4-dibromo-1,4-dicyanocyclohexane. *Journal of Physical Chemistry* A 115, p. 9207–9216.
18. Cremer, D. and J. A. Pople. 1975. A general definition of ring puckering coordinates. *Journal of the American Chemical Society* 97, p. 3754–3758.
19. Bain, A. D., R. A. Bell, D. A. Fletcher, P. Hazendonk, R. B. Maharajh, S. Rigby and J. F. Valliant. 1999. NMR studies of chemical exchange among five conformers of a 10-membered ring compound containing two amide bonds and a disuphide. *Journal of the Chemical Society, Perkin Transactions* 2, p. 1447–1454.
20. Bouvignies, G., P. Vallurpalli, D. F. Hansen, B. E. Correia, O. Lange, A. Bah, R. M. Vernon, F. W. Dahlquist, D. Baker and L. E. Kay. 2011. Solution structure of a minor and transiently formed state of a T4 lysozyme mutant. *Nature (London)* 477, p. 111–114.
21. Fawzi, N. L., J. Ying, R. Ghirlando, D. Torchia and G. M. Clore. 2011. Atomic-resolution dynamics on the surface of amyloid-beta protofibrils probed by solution NMR. *Nature (London)* 480, p. 268–274.
22. Korzhnev, D. M., T. L. Religa, W. Banachewicz, A. R. Fersht and L. E. Kay. 2010. A transient and low-populated protein-folding intermediate at atomic resolution. *Science* 329, p. 1312–1316.
23. Laidler, K. J. 1965. *Chemical Kinetics*. New York: McGraw Hill.
24. Elder, P. J. W., T. Chivers and R. Thirumoorthi. 2013. Experimental and computational investigations of tautomerism and fluxionality in PCP- and PNP-bridged heavy chalcogenides. *European Journal of Inorganic Chemistry* p. 2867–2876.
25. Macdonald, P. M., Q. Saleem, A. Lai and H. H. Morales. 2013. NMR methods for measuring lateral diffusion in membranes. *Chemistry and Physics of Liquids* 166, p. 31–44.
26. Freye, S., D. M. Engelhard, M. John and G. H. Clever. 2013. Counter-ion dynamics in an interpenetrated coordination cage capable of dissolving AgCl. *Chemistry: A European Journal* 19, p. 2114–2121.
27. Perrin, C. L. 1989. Optimum mixing time for chemical kinetics by 2-D NMR. *Journal of Magnetic Resonance* 82, p. 619–621.
28. Bain, A. D. CFIT can be downloaded from the website http://www.chemistry.mcmaster.ca/bain/.
29. Rankin, M. A., D. F. MacLean, R. McDonald, M. J. Ferguson, M. D. Lumsden and M. Stradiotto. 2009. Probing the dynamics and reactivity of a stereochemically non-rigid Cp*Ru(H)(κ(2)-*P*,carbene) complex. *Organometallics* 28, p. 74–83.
30. Religa, T. L., R. Sprangers and L. E. Kay. 2010. Dynamic regulation of archaeal proteasome gate opening as studied by TROSY NMR. *Science* 328, p. 98–102.
31. Bouvignies, G. and L. E. Kay. 2012. Measurement of proton chemical shifts in invisible states of slowly exchanging protein systems by chemical exchange saturation transfer. *Journal of Physical Chemistry* B 116, p. 14,311–14,317.
32. Bain, A. D., W. P. Y. Ho and J. S. Martin. 1981. A new way of measuring NMR spin–spin relaxation times (T2). *Journal of Magnetic Resonance* 43, p. 328–330.
33. Vallurpalli, P., G. Bouvignies and L. E. Kay. 2012. Studying invisible excited protein states in slow exchange with a major state conformation. *Journal of the American Chemical Society* 134, p. 8148–8161.
34. Bain, A. D., H. Chen and P. H. M. Harrison. 2006. Studies of structure and dynamics in a nominally symmetric twisted amide by NMR and electronic structure calculations. *Canadian Journal of Chemistry* 84, p. 421–428.
35. Bouvignies, G., P. Vallurpalli, M. H. J. Cordes, D. F. Hansen and L. E. Kay. 2011. Measuring H-1(N) temperature coefficients in invisible protein states by relaxation dispersion NMR spectroscopy. *Journal of Biomolecular NMR* 50, p. 13–18.

36. Deverell, C., R. E. Morgan and J. H. Strange. 1970. Studies of chemical exchange by nuclear magnetic relaxation in the rotating frame. *Molecular Physics* 18, p. 553–559.
37. Vallurpalli, P., G. Bouvignies and L. E. Kay. 2011. Increasing the exchange timescale that can be probed by CPMG relaxation dispersion NMR. *Journal of Physical Chemistry* B 115, p. 14,891–14,900.
38. Bain, A. D., G. J. Duns, F. Rathgeb and J. Vanderkloet. 1995. A study of chemical exchange in unequally populated systems by novel NMR methodologies: Application to the cis–trans isomerisation in furfural. *Journal of Physical Chemistry* 99, p. 17,338–17,343.
39. Carrington, A. and A. D. McLachlan. 1967. *Introduction to Magnetic Resonance*. New York: Harper & Row.
40. Binsch, G. 1969. A unified theory of exchange effects on nuclear magnetic resonance line shapes. *Journal of the American Chemical Society* 91, p. 1304–1309.
41. Bain, A. D. 2003. Chemical exchange in NMR. *Progress in Nuclear Magnetic Resonance Spectroscopy* 43, p. 63–103.
42. Lynden-Bell, R. M. 1967. The calculation of line shapes by density matrix methods. *Progress in Nuclear Magnetic Resonance Spectroscopy* 2, p. 163–204.
43. Fano, U. 1964. Liouville representation of quantum mechanics with application to relaxation processes. In E. R. Caianiello, ed. *Lectures on the Many Body Problem* 2, p. 217–239.
44. Banwell, C. N. and H. Primas. 1963. On the analysis of high-resolution nuclear magnetic resonance spectra: I. Methods of calculating NMR spectra. *Molecular Physics* 6, p. 225–256.
45. Bain, A. D. and B. Berno. 2011. Liouvillians in NMR: The direct method revisited. *Progress in Magnetic Nuclear Resonance Spectroscopy* 59, p. 223–244.
46. Reeves, L. W. and K. N. Shaw. 1970. Nuclear magnetic resonance studies of multi-site chemical exchange: I. Matrix formulation of the Bloch equations. *Canadian Journal Chemistry* 48, p. 3641–3653.
47. Ernst, R. R., G. Bodenhausen and W. Wokaun. 1987. *Principles of Nuclear Magnetic Resonance in One and Two Dimensions*. Oxford, United Kingdom: Clarendon Press.
48. Slichter, C. P. 1989. *Principles of Magnetic Resonance*. Berlin, Germany: Springer-Verlag.
49. Fano, U. 1957. Description of states in quantum mechanics by density matrix and operator techniques. *Reviews of Modern Physics* 29, p. 74–93.
50. Jeener, J. 1982. Super-operators in magnetic resonance. *Advances in Magnetic and Optical Resonance* 10, p. 1–51.
51. Bain, A. D. 1988. The superspin formalism for pulse NMR. *Progress in Nuclear Magnetic Resonance Spectroscopy* 20, p. 295–315.
52. Bain, A. D. and G. J. Duns. 1996. A unified approach to dynamic NMR based on a physical interpretation of the transition probability. *Canadian Journal of Chemistry* 74, p. 819–824.
53. Kubo, R. 1954. Note on the stochastic theory of resonance absorption. *Journal of the Physical Society of Japan* 9, p. 935–944.
54. Sack, R. A. 1958. A contribution to the theory of the exchange narrowing of spectral lines. *Molecular Physics* 1, p. 163–167.
55. Kaplan, J. I. and G. Fraenkel. 1972. Effect of molecular reorganisation on nuclear magnetic resonance line shapes: Permutation of indices method. *Journal of the American Chemical Society* 94, p. 2907–2912.
56. Anderson, J. E. and J. E. T. Corrie. 1992. The rotation-dominated ring inversion/nitrogen inversion/rotation process in N-acyloxy-2,2,6,6,tetramethylpiperidines: A dynamic NMR study. *Journal of the Chemical Society, Perkin Transactions* 2, p. 1027–1031.

## FURTHER READINGS

Bain, A. D. 2003. Chemical exchange in NMR. *Progress in Nuclear Magnetic Resonance Spectroscopy* 43, p. 63–103.

Binsch, G. 1969. A unified theory of exchange effects on nuclear magnetic resonance line shapes. *Journal of the American Chemical Society* 91, p. 1304–1309.

Ernst, R. R., G. Bodenhausen and A. Wokaun. 1987. *Principles of Nuclear Magnetic Resonance in One and Two Dimensions*. Oxford, United Kingdom: Clarendon Press, 610 pp.

Jackman, L. M. and F. A. Cotton. 1975. *Dynamic Nuclear Magnetic Resonance Spectroscopy*. New York: Academic Press, 660 pp.

Johnson, C. S. 1965. Chemical rate processes and magnetic resonance. *Advances in Magnetic and Optical Resonance* 1, p. 33–102.

Kaplan, J. I. and G. Fraenkel. 1980. *NMR of Chemically Exchanging Systems*. New York: Academic Press, 165 pp.

Kowalewski, J. and L. Maler. 2006. *Nuclear Spin Relaxation in Liquids: Theory, Experiments, and Applications*. Boca Raton: CRC Press, Taylor and Francis Group, 426 pp.

Laidler, K. J. 1965. *Chemical Kinetics*. New York: McGraw Hill, 566 pp.

Orrell, K. G., V. Sik and D. Stephenson. 1990. Quantitative investigations of molecular stereodynamics by 1-D and 2-D NMR methods. *Progress in Nuclear Magnetic Resonance Spectroscopy* 22, p. 141–208.

Sandström, J. 1982. *Dynamic NMR Spectroscopy*. London, United Kingdom: Academic Press, 226 pp.

Tycko, R., ed. 1994. *Nuclear Magnetic Resonance Probes of Molecular Dynamics*. Boston: Kluwer Academic Publishers, 550 pp.

# 3 NMR in Ligand Binding Studies

*Michelle L. Rowe, Jane L. Wagstaff and Mark J. Howard*

## CONTENTS

3.1 Introduction .......... 64
3.2 Saturation in NMR .......... 64
3.2.1 What Is Saturation? .......... 64
3.2.2 Saturation in Solvent Suppression .......... 65
3.2.3 NOE Difference Experiment .......... 65
3.3 STD Nuclear Magnetic Resonance .......... 68
3.3.1 STD and Protein–Ligand Binding .......... 68
3.3.2 $k_{off}$ in STD NMR .......... 71
3.3.3 STD NMR Experimental Approach .......... 73
3.3.4 Basic Analysis of STD NMR Data .......... 76
3.3.5 Selective Excitation Pulses for Protein Saturation in STD NMR .......... 80
3.3.6 Saturation Times in STD NMR Experiments .......... 83
3.3.7 STD NMR and Ligand Binding Competition .......... 87
3.3.8 2-D STD NMR .......... 88
3.3.9 STD NMR and Peptide-Based Ligands .......... 91
3.3.10 STD NMR and Very Large Proteins .......... 96
3.3.11 Transfer Methods Involving Water: WaterLOGSY .......... 97
3.3.12 Not STD NMR: Using $^{19}F$ in Ligand-Observe Screening .......... 100
3.4 tr-NOE and Transferred RDC .......... 101
3.4.1 Exchange tr-NOE .......... 102
3.4.1.1 Illustrations of et-NOE Use .......... 104
3.4.2 Transferred RDC .......... 105
3.4.2.1 Basis of RDCs .......... 105
3.4.2.2 Hints and Tips for Obtaining RDCs .......... 108
3.4.2.3 Illustrations of RDC Use .......... 109
3.5 Protein-Based Observation of Ligand Binding .......... 110
3.5.1 Chemical Shift Perturbations in Drug Discovery .......... 110
3.5.2 Quantifying Chemical Shift Perturbations .......... 114
3.5.3 Binding Affinities from Chemical Shift Perturbations .......... 115
3.6 Summary .......... 118
References .......... 118

## 3.1 INTRODUCTION

NMR spectroscopy has been used for many years to study the effect of one molecule interacting with another, but it was with the inception of the research conducted by Fesik et al. at Abbot Laboratories[1] that NMR fragment-based drug discovery (FBDD) was recognised as a widely applicable screening method.

The basis of any NMR approach to drug discovery is to identify (small-molecule) ligands that bind to a protein (or target, in general) of interest; these ligands may then act as lead-type systems for future drug design. In this chapter, we will provide an overview of some of the principle NMR-based ligand binding analysis methods, first categorising these as either (1) ligand-observe or (2) protein-observe methods. As the names suggest, ligand-observe methods monitor ligand NMR resonances to ascertain whether binding to a target takes place, and protein-observe monitors protein resonances to the same effect. The majority of this chapter (Sections 3.3 and 3.4) is concerned with ligand-observe methods, not only because this involves small molecules that produce simple NMR spectra, but also because ligand-observe is the primary method for drug screening in industry. Protein-observe is also discussed but is for the most part limited to Section 3.5, at the end of the chapter.

In conducting ligand binding studies, the popular NMR approach, which continues to prove most versatile, is saturation transfer difference nuclear magnetic resonance (STD NMR); consequently, both the theory and practical considerations surrounding this technique are considered in detail in this chapter. It is not only relatively simple to conduct as an NMR experiment, but also capable of providing powerful results of a quantitative nature. While explaining the approach, our aim is to also provide some practical insights to enable researchers to implement these techniques on any instrumentation available to them.

The description of the STD NMR approach is followed by consideration of other magnetisation transfer methods used in ligand binding studies, namely the transferred nuclear Overhauser effect (tr-NOE) and residual dipolar coupling (RDC) measurements. Finally, the focus moves to outlining the protein-observe approach using chemical shift mapping.

As a final note, it is not appropriate to cover every aspect of NMR-based fragment screening in this chapter. Rather, we refer you to a recent review covering the construction of a fragment library, developing the screen, and what to do following a screen,[2] and some articles on a variety of biophysical approaches to drug discovery.[3–5]

## 3.2 SATURATION IN NMR

In this section, we set the scene for the STD NMR approach by briefly introducing saturation as used in solvent suppression and the steady-state NOE experiment, using the latter technique to introduce 'difference' NMR.

### 3.2.1 What Is Saturation?

In simple terms, saturation in NMR is the effect of equalising the spin-state populations of nuclei under the influence of a static magnetic field ($B_o$). If there is no difference in population between spin states prior to the application of a read pulse, no signal is observed. For a spin-1/2 nucleus, two degenerate spin states are created when it is placed in a magnetic field, and the summation of the unequal population of these states defines the bulk magnetisation ($M_z$), which is perturbed by radio frequency pulses to produce NMR signals. We can visualise changes in this magnetisation over four 90° (or $\pi/2$) pulse (p1)-acquisition (aq here and $t_2$ in Chapter 1) steps as a function of recovery time (d0 here, RD in Chapter 1) permitted between the pulses (Figure 3.1) to illustrate the manifestation of saturation of spin states. Clearly, the sum of the acquisition time (aq) and d0 needs to be sufficient to permit the return of the magnetisation vector to the $z$ axis (the direction of the applied magnetic field); in practice, this needs to be of the order of approximately $5T_1$.[6] If this is not the case, saturation will be apparent, and signal intensity will be diminished.

While other pulse angles may be used to induce saturation, only a 90° nutation of magnetisation will provide complete and efficient saturation, and therefore, careful calibration of 90° pulses is necessary for the saturation methods that follow in this chapter.

### 3.2.2 Saturation in Solvent Suppression

Using saturation to provide solvent suppression was once the most common method for removing unwanted solvent peaks. The actual technique, referred to as 'pre-saturation' because the approach saturates the solvent resonance prior to detection of any required resonances, works by the application of an radio frequency (RF) pulse of low power over a period of 1 to 2 s. A low-power pulse, by definition, has a narrow excitation bandwidth, and the long application period is an attempt to force the solvent nuclear spin-state populations to be equal. Pre-saturation does a remarkable job when you consider that it works very effectively when observing $^1H$ NMR signals from a 1 mm sample in a water-based solvent where the effective $^1H_2O$ concentration approaches 55 M and the subsequent $^1H$ concentration approaches 110 M. A demonstration of pre-saturation is shown for L-tryptophan (L-Trp) in an aqueous buffer in Figure 3.2. Of course, while suppressing the solvent signal, the same effect will also result for any protons that resonate at or near the solvent resonance frequency. To minimise this, power levels are kept as low as possible, increasing the duration of its application. This is all well and good unless the protons of interest are involved in chemical exchange with the solvent (Chapter 2). Such labile protons suffer from the unwanted effect of reduced signals, and it can be quite severe in some cases (Figure 3.2b); in fact, this is saturation transfer. Fortunately, there are other solvent suppression schemes such as WATERGATE[7] (Section 6.3.1) and excitation sculpting,[8,9] which can prevent excessive loss of signal for labile protons (Figure 3.2c). Finally, it is worth noting that saturation of labile proton signals via pre-saturation can be useful – it is an elegant way of removing unwanted broad alcohol OH resonances in aqueous co-solvents such as trifluoroethanol (TFE).

### 3.2.3 NOE Difference Experiment

For many years, this experiment was fundamental to NMR-based structural elucidation in organic chemistry, and the NOE difference experiment is introduced

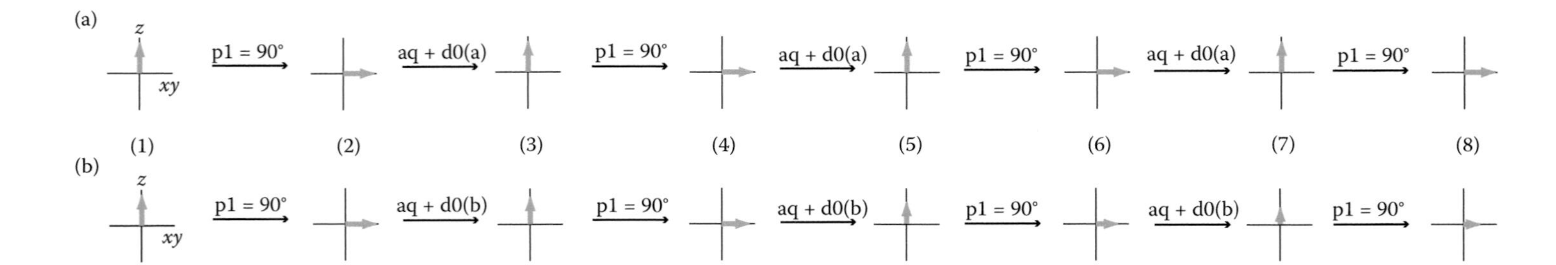


**FIGURE 3.1** Schematic showing the effect on the magnetisation vector, shown as the grey arrow, for a single pulse-acquire delay experiment (a) and when successive scans prevent full recovery between continuous pulse acquisition (b). The vertical axis is the *z* axis, and the horizontal line represents the transverse *x*/*y* axes. Diagrams (a1) and (b1) are equivalent and indicate the start of the experiment, and both experiments have the same 90° pulse (p1) and acquisition time (aq) but d0(a) >> d0(b). The NMR signal acquired would be proportional to the horizontal grey arrow following each pulse at points 2, 4, 6 and 8.

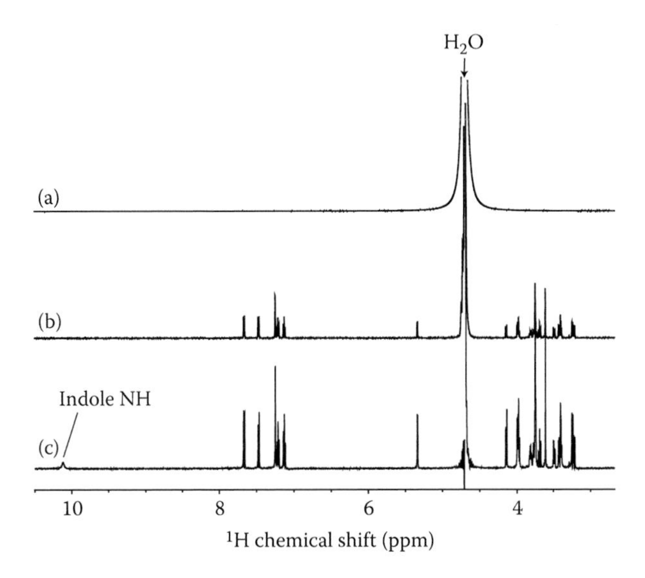


**FIGURE 3.2** The $^1H$ single pulse acquire spectrum (a), pre-saturation for 2 s on the water resonance (b), and excitation-sculptured $^1H$ 1-D (c). Conditions: 298 K of 1-mM L-Trp in PBS at pH 7.2. The sample contains 5% $D_2O$ for spectrometer locking.

here to highlight the use of viewing NMR data as a 'difference' result rather than a pure addition of free induction decays (FIDs) following acquisition. The NOE difference experiment works by pre-saturation (often referred to as 'irradiation') of a target resonance and monitoring the intensity change of other resonances in the molecule due to a steady-state NOE interaction between nuclei that are close in space. The intensity changes in the resonances of neighboring nuclei are of the order of a few percent of the total resonance intensity, making them difficult to identify and measure from a single 1-D NMR experiment. This issue is solved by using 'difference spectroscopy' where a spectrum without specific saturation (the 'off-resonance' experiment) is subtracted from the NOE-enhanced ('on-resonance') spectrum. This process is also sometimes called 'double-resonance NMR', which describes the use of a specific second irradiation frequency. It is important to note that, in the 'off-resonance' experiment, a 'saturation' pulse is applied but outside of the normal spectral window. Most modern spectrometers acquire both data sets simultaneously in an interleaved fashion to ensure equivalent experimental conditions for both the on/off data acquisitions, primarily to alleviate issues surrounding the small heating effect associated with the application of RF saturation pulses. This interleaving of 'on-resonance' and 'off-resonance' data acquisitions to create a difference spectrum is also utilised in STD NMR, but the nature of the saturation pulse is very different. NOE difference experiments use a square-wave pulse of low power to saturate a particular resonance in the spectrum; as will be discussed in the following, more elegantly shaped pulses are utilised in STD NMR.

## 3.3 STD NUCLEAR MAGNETIC RESONANCE

STD NMR was first reported in 1999[10] and has been heavily utilised as a cornerstone method for NMR-based FBDD. There are several excellent STD NMR reviews available on this topic,[11,12] including a review of applications within drug discovery[13] and comparing STD with other NMR methodologies with similar applications.[14,15]

### 3.3.1 STD AND PROTEIN–LIGAND BINDING

STD NMR is a ligand-based NMR detection method for registering and measuring the interactions of small molecules with protein targets; as such, it has a distinct advantage of being an NMR method with no upper molecular weight limit on the protein target to be studied. In addition, STD NMR is popular because of its ease of implementation, and the need for relatively low concentrations of protein, typically, micromolar amounts of protein compared to millimolar quantities of ligand, ensuring the ligand is in large excess for reasons to be explained in the following.

The approach of STD NMR data acquisition is similar to that seen for the NOE difference experiment and involves the collection of two data sets: one with saturation on-resonance ($STD_{on}$) and one with saturation off-resonance ($STD_{off}$), and a difference spectrum is created by subtraction of $STD_{off}$ from $STD_{on}$. What differentiates STD from NOE difference is both the nature and the intended target of the on-saturation pulse. STD NMR experiments use a selective shaped pulse that only saturates specific $^1$H resonances in the protein without exciting any $^1$H resonances of the ligand. This is possible because the envelope of a protein spectrum covers a wider chemical shift range than that of a typical small molecule (ligand), as illustrated in Figure 3.3. The primary target for an STD selective pulse are upfield-shifted methyl protons. We will go on to show that the L-Trp/bovine serum albumin (BSA) system used in Figure 3.3 is ideal for demonstrating the fundamental features of STD NMR.

Selective saturation of methyl $^1$H resonances in the protein is only the beginning of the $STD_{on}$ experiment. This selective excitation of protein methyl protons is applied for approximately 5 s (or longer) and acts as a reservoir from which this saturation-based excitation is transferred throughout the proton network in the protein, which, of course, includes every potential binding pocket or site. The mechanism by which the saturation is transferred so extensively is called 'spin diffusion' and is common in large macromolecules. Furthermore, the efficiency of spin diffusion depends on the molecular correlation time $\tau_c$, which, in turn, depends on the viscosity of the solvent medium, the temperature of the sample, and most significantly, the molecular weight of the molecule being studied. As the molecular weight of a protein increases, its $\tau_c$ will become longer, and spin diffusion will become more efficient. This spin-diffused saturation can then be transferred to a bound ligand, thus perturbing the detected ligand signals. Hence, STD NMR is one of the few solution-state protein NMR methods where a large molecular weight is an advantage and not a hindrance.

If a ligand molecule binds while the protein is selectively saturated, and some saturation is transferred to it prior to ligand dissociation (timescale dependent), the ligand will retain this saturation-perturbation during the detection period. It is crucial that the ligand dissociates from the protein to make the saturation transfer

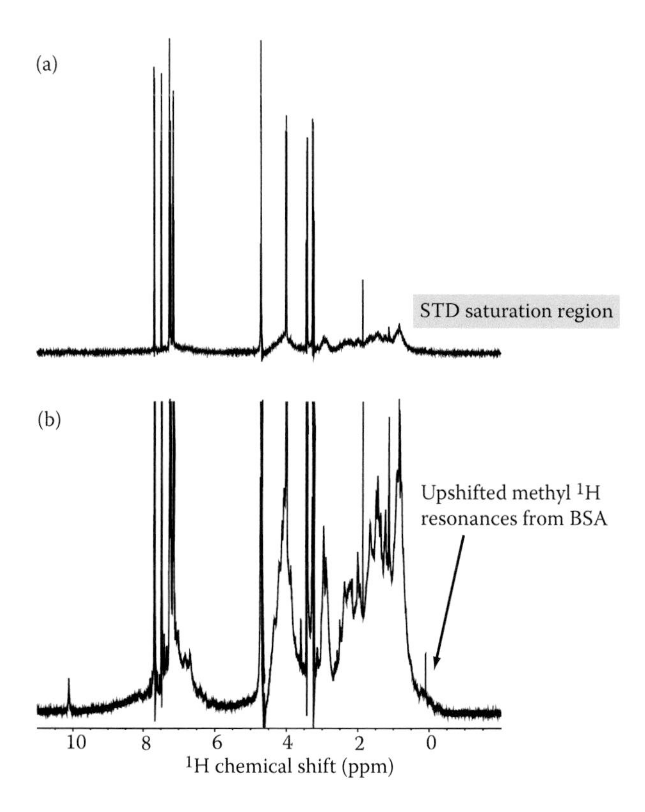


**FIGURE 3.3** $^1$H NMR spectrum of a typical STD NMR sample containing 1-mM L-Trp and 20-μM BSA; L-Trp is in 50 times excess, PBS buffer. The L-Trp resonances are clearly visible in (a), and a vertical axis expansion of this spectrum in (b) highlights the wide protein $^1$H envelope of chemical shifts. The upshifted methyl groups that are targeted in STD saturation are highlighted.

visible because the STD detects the perturbation in the free (hence, sharp) ligand resonances as a result of binding during protein saturation. The $STD_{on}$ experiment is shown schematically in Figure 3.4a, where a model protein molecule is saturated and transfer occurs to a ligand that is in an exchange equilibrium with the protein, with a dissociation constant, $K_D$, defined by the on and off kinetic rates, such that $K_D = k_{off}/k_{on}$ (Section 3.3.2). We soon see that the off-rate, $k_{off}$, is important for the difference detection of saturation transfer. For now, it is sufficient to know that most STD-detected binding events involve a protein/ligand equilibrium, with $K_D$ values in the millimolar-to-nanomolar range and $k_{off} > 1\ s^{-1}$.

To obtain a difference spectrum, a second NMR spectrum needs to be recorded with the saturation 'off-resonance' ($STD_{off}$), with selective excitation moved to a point far from the protein $^1$H signals (Figure 3.4b). This is usually achieved by moving the saturation pulse to –30 ppm or even greater distances from the main spectral envelope. As with the NOE difference 'off-resonance' spectrum, the $STD_{off}$ spectrum should be the same as a standard 1-D $^1$H NMR spectrum of the ligand, hence why it is also often referred to as the control ($STD_{ctrl}$). Once again, to produce a reliable difference spectrum ($STD_{diff}$), the $STD_{on}$ and $STD_{off}$ spectra are normally acquired in

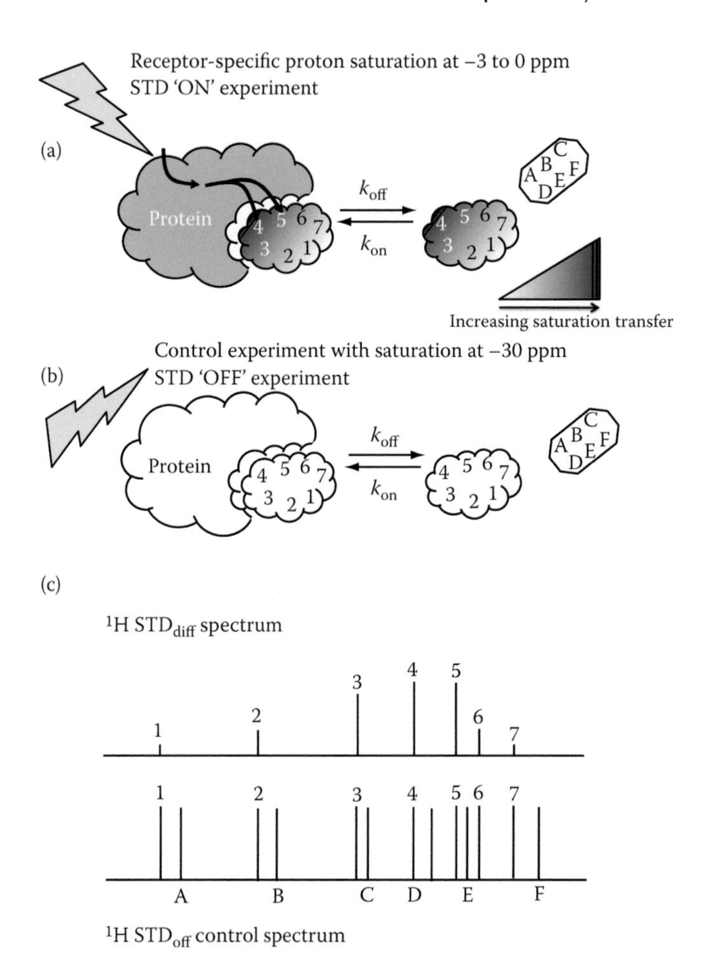


**FIGURE 3.4** Schematic representation of $^1$H STD NMR using a binding 'cloud' ligand and a non-binding 'lozenge' ligand to a protein receptor, showing the $STD_{on}$ experiment (a) and the $STD_{off}$ experiment (b). The 'cloud' ligand is in bimolecular association exchange. The $STD_{diff}$ result (c) identifies positive binding and the ligand binding orientation for the binding 'cloud' ligand and protons 1 to 7 and no resonances depicting the absence of binding for the 'lozenge' ligand with protons A to F. $STD_{off}$ spectrum (c) should be the same as the standard $^1$H NMR spectrum.

an interleaved fashion. The hypothetical results shown in Figure 3.4 demonstrate that STD NMR data serve two purposes: (1) only resonances for protons of the molecule ('cloud' ligand) that comes close to the protein are perturbed, and (2) the extent of perturbation of the bound ligand signals depends on the proximity of the different regions of the ligand to the protein. Thus, $STD_{diff}$ intensities increase in the order 1/7 → 2/6 → 3 → 4/5 regions, with 4/5 being more intimately associated with the protein target. This is the first illustration of the STD NMR approach being used as a 'screen'.

Although the STD NMR experiment has similarities to the NOE equivalent, it also has several idiosyncrasies that require careful consideration. Thus, we will now take a detailed view of the role of the equilibrium off-rate ($k_{off}$) on STD and an overview of the STD NMR experimental approach, including the basic quantitative evaluation of such data. Once these areas are established, we will walk through the role and optimisation of parameters such as the selective excitation pulse and saturation time. Consideration of the STD approach will finish with an exploration of the use of STD to study competitive binding, an introduction to more advanced STD approaches, and the use of water-mediated saturation transfer for ligand binding studies.

### 3.3.2 $k_{off}$ in STD NMR

We need to remember that resonance lines in the difference spectrum ($STD_{diff}$) are only observed when the $STD_{on}$ and $STD_{off}$ spectra are different, and this only happens when saturation is transferred from the protein to the ligand during the $STD_{on}$ experiment. Consequently, the $STD_{on}$ and $STD_{off}$ NMR experiments must be optimised to measure $^1H$ resonances from the free ligand, not those either from the protein background or, more importantly, from ligands that remain bound to the protein. This means that an $STD_{on}$ experiment detects ligands that have not bound to the protein at all or those that have bound and dissociated from the protein; the latter class of ligands will experience perturbed magnetisation from saturation transfer. The $STD_{off}$ experiment will also measure non-binding ligands or those that have bound and dissociated, but as there is no saturation in this case, both classes of ligands provide the data without perturbation and are essentially the same. Now, it should be straightforward to see what happens when the $STD_{diff}$ is created – bound ligands become partially saturated themselves, and once dissociated, this is detected. The dissociation event is an equilibrium that can be expressed as follows (Equation 3.1):

$$P + L \underset{k_{off}}{\overset{k_{on}}{\rightleftharpoons}} PL \tag{3.1}$$

where $P$ is the protein, $L$ is the ligand, and $PL$ is the protein–ligand complex.

The dissociation constant, $K_D$, for this process is defined as follows:

$$K_D = \frac{[P].[L]}{[PL]} = \frac{k_{off}}{k_{on}} \tag{3.2}$$

The seminal review by Meyer and Peters[11] used Equations 3.1 and 3.2 to predict $k_{off}$ over a range of dissociation constants by assuming a diffusion-controlled association rate (i.e. on-rate, $k_{on}$) of approximately $1 \times 10^7$ s$^{-1}$ M$^{-1}$. Their result for $K_D$ values of 1 mM, 1 μM, and 1 nM predict $k_{off}$ values of 10,000 s$^{-1}$, 10 s$^{-1}$, and 0.01 s$^{-1}$, respectively. When the $K_D$ of a binding event is below the nanomolar range, STD NMR becomes ineffective for ligand binding observation because, although the ligand becomes saturated, it fails to dissociate in sufficient numbers for detection as the saturation-perturbed free state. Hence, the quoted operational limits for studying protein–ligand interactions by STD NMR is the millimolar-to-nanomolar dissociation constant ($K_D$) range. $k_{on}$ will also affect the ability to detect an STD NMR effect. If you consider a $K_D$ of 1 nM with $k_{on}$ that varies by three orders of magnitude between $1 \times 10^7$ and $1 \times 10^4$ $s^{-1}$ $M^{-1}$, the $k_{off}$ will equally vary between $1 \times 10^{-2}$ and $1 \times 10^{-5}$ $s^{-1}$. There is clearly a balancing act between $k_{on}$ and $k_{off}$, and it is worth remembering that smaller $k_{on}$ values can explain why an STD approach will fail for $K_D$ values higher than 1 nM. Small $k_{on}$ values are most likely due to the use of larger ligands that diffuse slowly or if an interaction involves a large protein conformational change upon ligand binding. These two events are worth remembering if you attempt STD NMR but fail to observe a difference spectrum while operating with a binding event known to be in the micromolar-to-nanomolar range.

What we have effectively described in the previous paragraph is that a binding event will produce a positive $STD_{diff}$ result only when $K_D > 1$ nM, with caveats regarding $k_{on}$ and $k_{off}$. Thus, when there is a binding event outside these conditions when $K_D << 1$ nM, a negative $STD_{diff}$ spectrum will arise, which could be confused with supporting what is actually a non-binding event (such as that demonstrated in Figure 3.4 for the 'lozenge' ligand). Any concerns relating to this may be alleviated by recording Carr-Purcell-Meiboom-Gill (CPMG) data for the system (Section 6.3.2 and in the following) or by running a small number of 1-D $^1H$ experiments on samples containing a range of ligand concentrations but with a constant protein concentration. For example, you can vary the ligand concentration between 0.05 and 1 mM and maintain 10 μM of protein in each sample. Use identical sample buffer and NMR acquisition conditions for each 1-D $^1H$ data set and create a plot of intensity (for a well-resolved ligand peak) with concentration of ligand. The resulting plot should be linear as the intensity is directly proportional to the ligand concentration, but if the plot deviates from linearity, the ligand is most likely binding to the protein. The deviation from linearity will become more apparent as ligand concentration falls because the sub-nanomolar $K_D$ binding event would drive a larger proportion of the ligand present to bind to the protein. This method also has the potential to create non-linear intensity/ligand concentration plots in the presence of chemical exchange line broadening (Chapter 2), which can manifest in $K_D$ processes greater than 1 nM, with small $k_{on}$ rates that can also impair STD data acquisition. In such cases, observing the ligand alone as a function of concentration would prove useful in confirming subtle but repeatable differences. It is possible to invert this approach and use constant ligand and variable protein concentration where a plot of intensity (or chemical shift) with ligand concentration provides an occupancy curve (Section 3.5). However, protein can be in short supply and may also not be soluble at millimolar concentrations, making the variable ligand method more cost effective and practical.

When the dissociation constant and off-rate are favourable, numerous ligands will cycle on and off the protein many times during the saturation period, with the effect that this increases the total number of saturation-perturbed free ligands in solution. This highlights the value of large off-rates, and we have noted that the same protein target, when interrogated with ligands across a $K_D$ range, provides larger $STD_{diff}$ signal intensities for millimolar-to-micromolar binding events. It is very difficult to describe a 'sweet spot' of $K_D$, $k_{off}$ and $k_{on}$ for optimal STD NMR because even for a single protein with multiple ligands, each of these properties is fluid and unique to each ligand–protein pair.

In the context of ligand binding studies, the CPMG experiment is employed as it enables a change in ligand resonance line widths as a result of protein binding to be observed. Typical spin echo times used for such studies are between 400 and 600 ms. To ascertain ligand binding using the CPMG approach requires two samples containing identical concentrations of ligand but with and without protein. When a ligand binds to a protein, it will experience a line-broadening effect, even if it binds and dissociates, and this will be seen as a reduction of the ligand resonance intensity. This happens because the protein $T_2$ is much shorter than the ligand $T_2$, and this difference influences a change in the free-ligand $T_2$ when the protein is present. As $T_2$ is related to resonance line width (Chapter 2 and Section 3.3.3), it manifests as a reduction in resonance height when $T_2$ becomes smaller in a CPMG experiment of constant spin echo time. The CPMG does have its own caveats, including the need to create two near-identical samples with respect to ligand concentration and physical properties, such that any changes observed can be attributed to the presence or absence of protein. However, CPMG does provide a useful supplement to both STD NMR (Section 3.3.3) and WaterLOGSY approaches (Section 3.3.11) as they all operate under very similar protein–ligand conditions. As a result, CPMG is mentioned periodically throughout the subsequent sections of this chapter.

### 3.3.3 STD NMR Experimental Approach

The STD NMR experimental approach is summarised schematically in Figure 3.5, and there are striking similarities to the pulse scheme for the NOE difference approach. However, the STD sequence as shown has two additional operations, and these are both effectively cosmetic but crucial components to the experiment. Water suppression is required because STD NMR data sets are acquired from samples

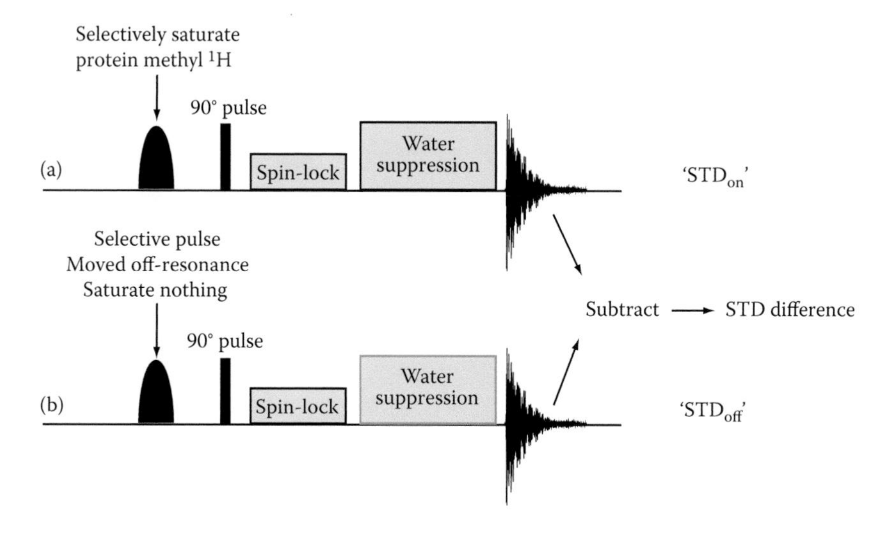


**FIGURE 3.5** Generalised pulse sequence schemes for a 1-D $^1$H STD NMR experiment showing STD$_{on}$ (a) and STD$_{off}$ (b) experiments and approach to the subtraction of these data sets to provide STD$_{diff}$.

in aqueous media, and the water signal needs to be reduced. The spin lock is used to dramatically reduce the broad background protein signal to provide a relatively flat baseline for ligand resonance detection (cf. Figure 3.3 where protein signals are apparent). The difference in $^1$H $T_2$ utilised here can be orders of magnitude with the free-ligand $T_2$ values approaching 1 s or more, whereas the protein can have $T_2$ values of tens of milliseconds or less ($R_2$ values of <1 s$^{-1}$ and >100 s$^{-1}$ for free ligand and protein, respectively). Regarding the spin lock, note the earlier statement that protein and bound ligand $^1$H nuclei will have relatively small $T_2$ values. When the ligand binds to the protein, it is fused and adopts the slow tumbling characteristics of the protein. We have mentioned that these characteristics drive spin diffusion and promote the spread of saturation across the protein and onto the ligand, but this slow motion also reduces $T_2$ to the point that these changes manifest as resonances broadened beyond detection for high–molecular weight species (Chapter 2). This is the crux of the molecular weight issue that faces solution protein NMR – as a protein increases in size, its tumbling slows, $T_2$ gets smaller, and the NMR resonance line width increases. There are many other caveats and facets to this conundrum, but it is clear that what aids spin diffusion and saturation transfer is not complementary to the direct NMR detection of proteins. On a practical note, as a rule of thumb, 30 ms for the STD experimental spin lock works well in most cases, but if your protein is over 100 kDa, you can drop the spin lock to 10 to 20 ms. Conversely, if your protein is small, you will have to increase the spin lock toward 50 ms or the limit set by your spectrometer manufacturer.

Having outlined the basic approach in Figure 3.6, we show the actual pulse sequence used to produce data to be presented later. All practical information

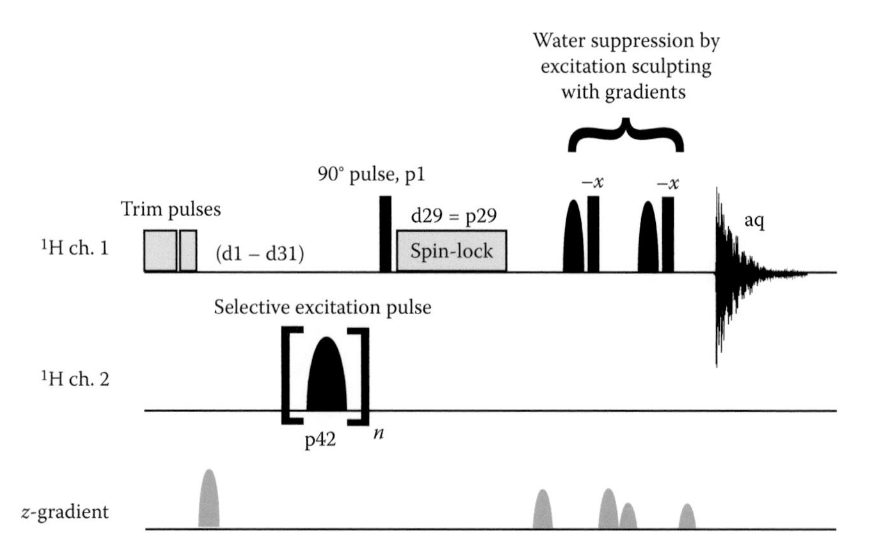


**FIGURE 3.6** Pulse sequence diagram for a 1-D $^1$H STD NMR experiment based on the pulse sequence stddiffesgp.3 as used on our Bruker Avance III 14.1-Tesla (600-MHz $^1$H) spectrometer. The equivalent elements to those shown in Figure 3.5 are labelled, and specific parameters are discussed in the text.

referred to in the STD NMR section is specific to our spectrometer (a Bruker 600-MHz spectrometer). Therefore, we recommend you to contact your manufacturer and discuss routing and settings with an NMR applications specialist if in doubt.

With the disclaimer out of the way, Figure 3.6 may look vastly more complicated than Figure 3.5, but we will see that it is extremely similar despite the plethora of pulses that now exist. The application of the selective shaped excitation pulse via channel 2, away from the remainder of $^{1}H$ pulses, and the detection of the FID in channel 1 tend to minimise artefacts resulting from the shaped pulse. In the past, we have successfully applied this selective shaped excitation pulse on the same channel as other pulses and acquisition on both Bruker and Varian/Agilent spectrometers without issue, but if your spectrometer is capable of driving two $^{1}H$ channels, the method in Figure 3.6 should provide 'cleaner' results. The action of pulsed field gradients on the $z$ axis ($G_z$) is to purge unwanted signal and drive water suppression using the excitation sculpting method. The selective shaped excitation pulse for saturation, defined as p42 via channel 2, is shown bracketed because it is applied $n$ times over the entire saturation period of several seconds. The saturation period is defined using a delay time, d20, and the loop $n$ is calculated within the pulse sequence as an integer where $n$ = d20/p42. The power applied to the p42 selective shaped excitation pulse is very low and should not cause any damage on modern spectrometers; do check your duty cycle and power level limits if you are unsure. The value calculated for $n \times$ p42 is actually the pre-saturation time, d31; however, as the shaped pulse p42 << d20, then d31 ≈ d20. Note that the labelling of these delays follows the Bruker convention. At the start of each scan, two 'trim pulses' exist to de-focus any residual magnetisation that has dispersive phase.

As directed at the outset, the sum of the relaxation delay (d1, in this case; d0 in Figure 3.1) and acquisition time (aq) should be somewhere between 1 and 5 times the longest $^{1}H$ longitudinal relaxation time ($T_1$) value. Figure 3.6 identifies the delay after the trim pulses and before selective saturation as [d1 – d31], which is [relaxation delay – actual saturation time] and brings with it two important points. First, for the sequence to work, clearly, d1 needs to be longer than d31. Second, if you plan on running different saturation delays, it is best to ensure either that d1 >> d31 or that the difference in these times is constant across all your experiments. Having d1 >> d31 covers the argument that you should maintain a long experimental recycling delay to ensure complete magnetisation recovery, but when saturation itself can be 5 s or more, this creates over 10 s per scan for a $T_1$ of only 1 s. The entire experiment would suddenly become lengthy and less justifiable. It is worth remembering that the saturation pulse does not perturb free ligand – this only happens at the 90° hard pulse. Therefore, the most practical and time-efficient option is to keep (d1 – d31) constant such that the total time between scans following acquisition (aq) and before the excitation/detection pulse (p1) is also constant. We will see later that using constant d1 with d31 variable over a range where $d31_{max}$ < d1 produces exceptionally good STD build-up curves over reasonable experimental time periods. It should be noted that, as acquisition times (to ensure adequate digital resolution) approach 2 s, this is adequate for efficient relaxation of protein protons. To ensure sufficient magnetisation equilibrium across the board prior to acquiring data from successive scans, we would also recommend running with 8 to 16 dummy scans.

This leaves two parameters to discuss: the spin-lock time and the actual definition of the selective shaped saturation pulse. It has already been noted that the typical spin-lock time is 30 ms. The selective saturation pulse can come in various guises and will be discussed in more detail in Section 3.3.5 where we demonstrate the use of two types of shaped pulses: Gaussian 90° and Eburp2 90°. The user can select the pulse to be used (e.g. Gaus1.1000 or Eburp2.1000), provide the software with an intended pulse width (e.g. 50 ms, to be inputted at 50,000 μs), and obtain the suggested power level for the pulse. The software provides the power level as a $\gamma B_1$ value that needs to be converted to watts or decibels of attenuation for input in the spectrometer, and this is demonstrated in Section 3.3.5.

Finally, the selective pulse is applied on-resonance and off-resonance via the use of an irradiation frequency list called up by the pulse program.

### 3.3.4 Basic Analysis of STD NMR Data

To describe the process of STD data analysis, we need to interrogate a data set such as the one shown in Figure 3.7. These data were collected with 512 scans (256 interleaved acquisitions each for $STD_{off}$ and $STD_{on}$, respectively) over approximately 44 min using a 50 ms Eburp2 selective pulse applied at 0 ppm ($STD_{on}$) and –30 ppm ($STD_{off}$) for 2.5 s in each scan, with a d1 delay of 4.0 s. The data were treated with a 1 Hz exponential multiplier prior to zero filling to increase digital resolution. We begin our analysis of Figure 3.7 by qualitatively viewing the $STD_{off}$ and $STD_{diff}$ spectra with respect to the aromatic $^1$H resonances between 7.0 and 8.0 ppm and the aliphatic resonances between 3.0 and 4.0 ppm. Please refer to Table 3.1 for resonance assignments. All $^1$H resonances from the ligand across the $STD_{off}$ control spectrum (Figure 3.7a) are approximately of equal area and intensity. The resonance at 7.26 ppm from the H2 proton is a singlet; hence, its intensity is higher. In contrast, the expanded $STD_{diff}$ spectrum in Figure 3.7c emphasises the L-Trp resonances as having unequal intensity. This is due to differing proximity of the parts of the L-Trp ligand to the BSA protein when it binds; the L-Trp aromatic region in $STD_{diff}$ is over twice the intensity as the aliphatic region for Hα and Hβ protons. The relative intensities across the ring resonances also appear to have changed, suggesting a preferential orientation of the indole ring when L-Trp interacts with BSA. This qualitative analysis is helpful, but a truly quantitative analysis provides the ultimate result for comparing the saturation transfer effectiveness to each proton. This is achieved by calculating the STD amplification factor ($STD_{amp}$) for each resonance – the ratio of its intensities from $STD_{diff}$ and $STD_{off}$ ($STD_{ctrl}$) multiplied by the ligand excess, as defined in the following equation:

$$STD_{amp} \text{ of resonance } x = \left( \frac{STD_{diff} \text{ intensity of resonance } x}{STD_{off} \text{ intensity of resonance } x} \right) \times \text{ligand excess} \tag{3.3}$$

This is usually simplified to

$$STD_{amp} = \left( \frac{I - I_o}{I_o} \right) \times \text{ligand excess} \tag{3.4}$$

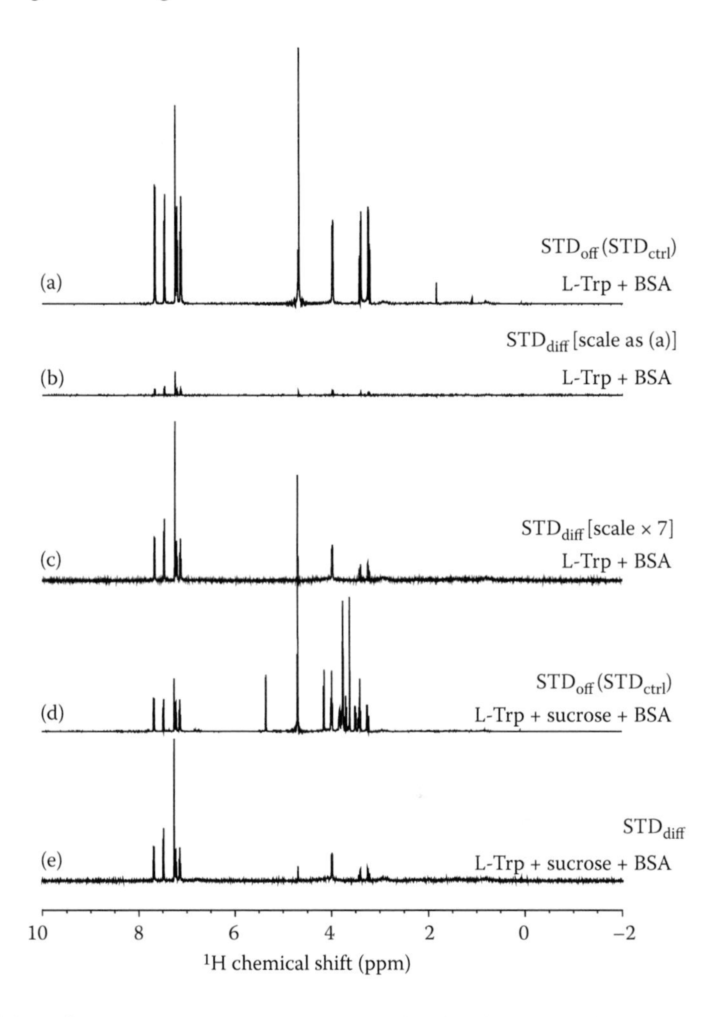


**FIGURE 3.7** $^1$H STD NMR experimental data showing STD$_{off}$ (a), STD$_{diff}$ shown on the same scale as STD$_{off}$ (b), and STD$_{diff}$ with an expanded vertical scale (c) for a 1-mM L-Trp ligand with 20 μM BSA in $D_2O$-based PBS. The STD NMR spectra utilising a 0.5-mM sucrose negative control ligand with L-Trp/BSA is shown as STD$_{off}$ (d) and STD$_{diff}$ (e). The residual HDO resonance is observed at approximately 4.7 ppm. The sucrose anomeric proton is clearly visible at 5.36 ppm in (d) but is absent in (e), which confirms non-binding.

where the intensities are defined according to STD$_{on}$ = $I$, STD$_{off}$ = $I_o$. Therefore, STD$_{diff}$ = $(I - I_o)$. The L-Trp/BSA system in Figure 3.7 has been characterised as a model single-site system;[16] therefore, the ligand excess here would be [L-Trp/BSA = 1 mM/0.02 mM = 50]. Table 3.1 shows the calculation of STD$_{amp}$ for all L-Trp resonances and also collates this information as a percentage of the maximum amplification factor.

Table 3.1 confirms that the largest saturation transfer exists to the H2 indole proton followed by the H7 indole proton. These two protons are on one side of the long axis of the indole group, and it is likely that this side of the group is the primary

**TABLE 3.1**
**STD Amplification Factors ($STD_{amp}$) and Relative Percentage of the Maximum $STD_{amp}$ for All L-Trp Ligand $^1$H Resonances from Experimental Data Shown in Figure 3.7**

| L-Trp Resonance | Chemical Shift (ppm) | $I_{STD}$ (arbitrary units) | $I_o$ (arbitrary units) | $STD_{amp}$ | STD Percent of Maximum |
|---|---|---|---|---|---|
| H4 | 7.68 | 3,654,428 | 64,880,390 | 2.82 | 45.8 |
| H7 | 7.48 | 4,579,597 | 51,645,427 | 4.43 | 72.1 |
| H2 | 7.26 | 6,117,525 | 49,714,288 | 6.15 | 100.0 |
| H6 | 7.21 | 3,871,815 | 63,627,899 | 3.04 | 49.5 |
| H5 | 7.14 | 3,729,445 | 62,830,652 | 2.97 | 48.2 |
| Hα | 3.99 | 4,672,776 | 68,968,227 | 3.39 | 55.1 |
| Hβ | 3.43/3.24 | 4,289,906 | 159,946,035 | 1.34 | 21.8 |

binding face. The aliphatic Hα proton provides similar data to H4, H5, and H6, but the Hβ protons experience the lowest transfer efficiency. For simplicity, the intensities of each of the prochiral Hβ proton were summed for the calculations. This was deliberate to illustrate that, as long as intensities $I$ and $I_o$ are measured for the same regions, proton number is accounted for in data handling. We notice here that having two β protons has not distorted the outcome of the experiment and placed unnecessary bias on the result, and that if the intensities were halved and shared between both β protons, the result would be the same. Equally, if one prochiral β proton exhibited a dramatic change from $STD_{off}$ to $STD_{diff}$ when compared to the other, we would gain information regarding the stereoselectivity of ligand binding to this chemical group. This is not the case in L-Trp/BSA, and we see that both β protons provide very similar intensities with respect to each other in Figure 3.7a and c. This ability of STD NMR to rank ligand protons based on proximity to the protein via the efficiency of saturation transfer is the basis of group epitope mapping (GEM), which provides definitive information regarding ligand binding orientation, a concept known within drug discovery as revealing the binding mode of the ligand. The GEM application of STD NMR was first reported in 2001 for the 120-kDa protein lectin *Ricinus communis* agglutinin I (RCA120) with saccharide ligands (Section 7.2.3), including D-galactoside.[17] Despite this innovation, STD NMR is still widely used in a more qualitative context as a 'yes'/'no' hit confirmation to find chemical fragments that bind to new protein drug targets. This is demonstrated in Figure 3.7d and e, which show the $STD_{off}$ and $STD_{diff}$ results for the same L-Trp/BSA system shown in Figure 3.7a to c with addition of 0.5-mM sucrose. Comparing $STD_{off}$ spectra, it is easy to identify sucrose $^1$H resonances, and it is clear that an overlap of these resonances with those from L-Trp Hα and Hβ exists. The ideal sucrose resonance to monitor is 5.36 ppm and is from the anomeric proton (Section 7.2.2). The $STD_{diff}$ spectrum (Figure 3.7e) shows how effective STD NMR is at removing ligand resonances that do not bind to the protein. Saturation transfer will not influence any non-interacting ligands, and their resulting $STD_{on}$ and $STD_{off}$ resonances should subtract completely and cancel out. This experiment involving L-Trp/sucrose/

BSA is strongly recommended when you start working with STD NMR because it is extremely important to prove that your STD experiments can identify non-binding as much as it can identify positive binding (Section 3.3.9).

Before turning to STD NMR optimisation, it would be useful to divulge some further practical information with respect to our basic experiment and the sample used to create the Figure 3.7/Table 3.1 data. Figure 3.8 is composed of four spectra, with Figure 3.8a and c being identical to Figure 3.7a and c. Figure 3.8b is the $STD_{diff}$ spectrum processed with identical phase to the $STD_{off}$ spectrum (Figure 3.8a), and it is inverted because $STD_{diff}$ is produced from ($STD_{on}$ – $STD_{off}$), and the revelation is that saturation transfer creates an intensity reduction in the $STD_{on}$ data set when compared to $STD_{off}$. This is to be expected, but this negative result is introduced here

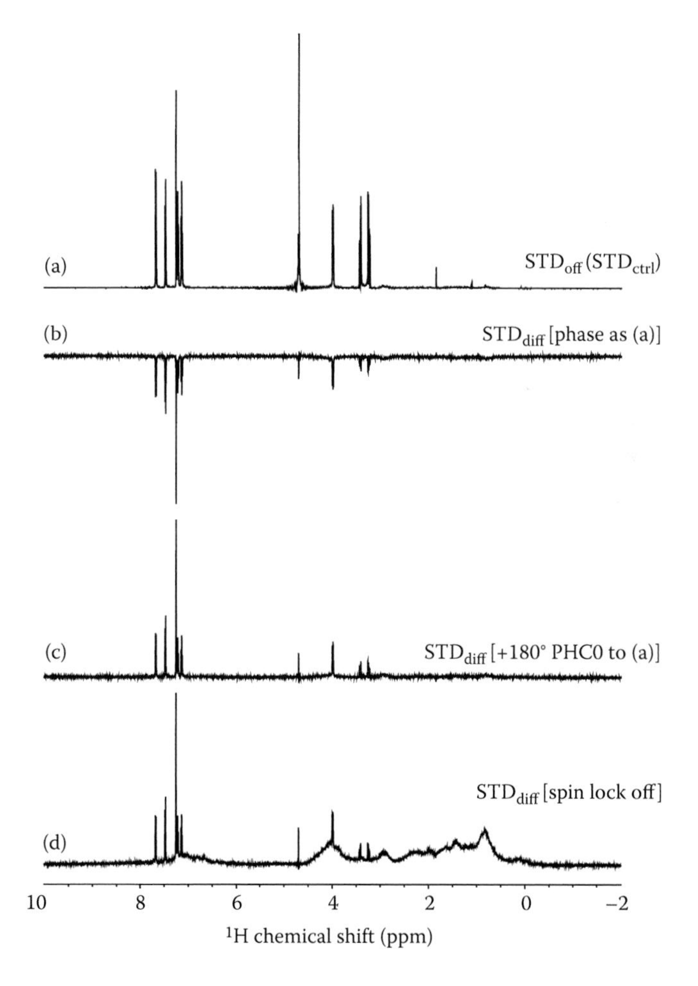


**FIGURE 3.8** $^1$H STD NMR experimental data for the standard 1-mM L-Trp/20-μM BSA sample in $D_2O$ from Figure 3.7 showing $STD_{ctrl}$ (a), $STD_{diff}$ with identical spectrum phase as $STD_{ctrl}$ (b), $STD_{diff}$ with 180° zero-order phase correction (c), and $STD_{diff}$ acquired identically as (a) to (c), with the spin lock inactivated in the pulse sequence (d).

to suggest that this as an excellent initial test of any STD data set that you acquire: if your $STD_{off}$ and $STD_{diff}$ spectra do not exhibit a 180° phase difference, then there is something amiss with your acquisition. Figure 3.8b and c show the effect of a 180° zero-order phase correction on the $STD_{diff}$ data to produce a positive spectrum. Figure 3.8d is an $STD_{diff}$ spectrum created under identical conditions to Figure 3.8b and c, but the spin lock has been turned off to illustrate the role of spin locking in reducing protein resonances. Figure 3.8d reiterates that integrating data created without a spin lock would propagate large errors through any $STD_{amp}$ calculations subsequently made and may invalidate any result, especially if you want to use the GEM approach.

### 3.3.5 Selective Excitation Pulses for Protein Saturation in STD NMR

The selective excitation pulse is one of the fundamental elements of the STD NMR experiment and has to be capable of specifically saturating the protein and initiating the process by which saturation is transferred across the protein and to ligand molecules that bind to it. In addition, the selective saturation should not perturb any ligand resonances. As mentioned in Section 3.3.1, the targets of this selective saturation pulse are upfield-shifted protein methyl $^1H$ resonances, and the original STD approach used a 90° Gaussian pulse applied for 50 ms to achieve this. The selectivity of such pulses is defined as the width of excitation or, more analytically, its excitation profile. All NMR pulses have excitation profiles, and pulses come in two broad categories: hard and soft. Hard pulses, such as 90° and 180° square-wave pulses found in many sequences, have a very broad bandwidth because they are expected to excite many nuclei over a wide chemical shift range. Soft pulses can also have defined nutation angles such as 90° or 180°, but 'soft' implies that these pulses have a finite excitation width that is typically smaller than the entire chemical shift range of the nucleus being irradiated. Therefore, by definition, the selective pulse for STD is 'soft'. As well as being 'hard' and 'soft', pulses can also come in different shapes, for example, square, Gaussian, burp, and snob, to name a few. The majority of non-square wave pulses are designed for 'soft' pulse applications to limit unwanted excitation. This is because a soft square pulse creates irradiation sidebands in addition to the central excitation band.[18] The central excitation band selectively excites over a narrow region, but the unwanted irradiation sidebands excite at other regions of the spectrum at intervals around the central excitation point. With respect to STD NMR, a soft square pulse has the potential to create unwanted sidebands that excite ligand resonances and produce false results. Shaped pulses are designed to minimise these sidebands and are preferentially used. The concept and theory of shaped NMR pulses can be quite complex, but there are some excellent resources to guide the user, including approaches to simulate and compare pulse design and operation.[18–20]

In addition to there being different shaped pulses, the duration of the pulse can also be varied. This should not be confused with the STD saturation time. We will explore shaped pulses that are individually repeatedly applied for between 10 and 50 ms. The effects of running 10, 20, 30, 40 and 50 ms Eburp2 and Gaussian pulses are shown for several $STD_{diff}$ spectra in Figure 3.9 for the L-Trp/BSA system.

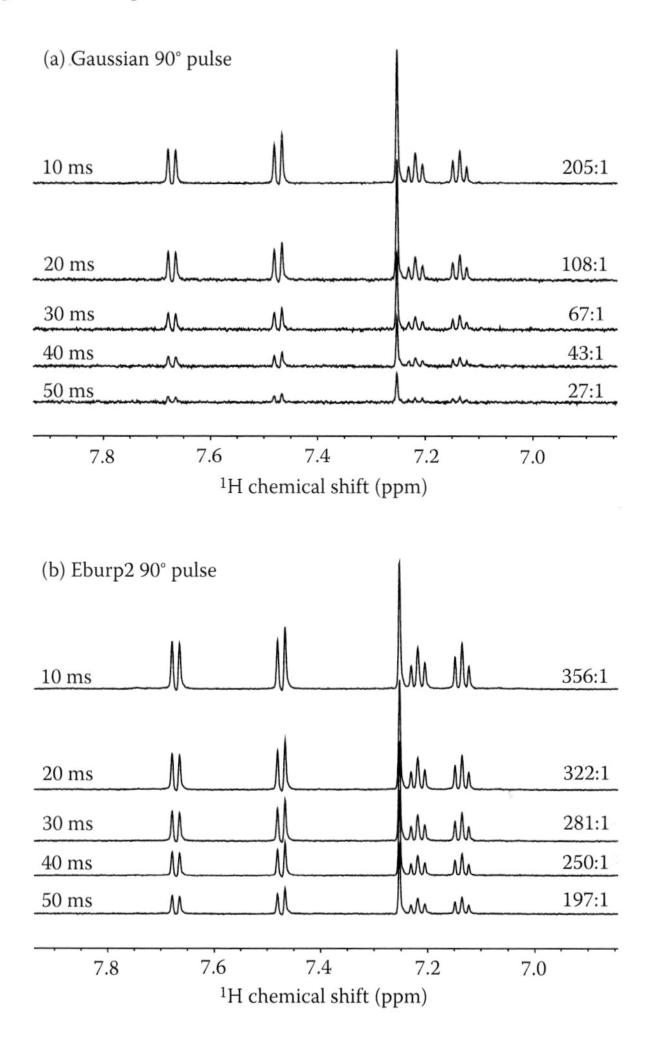


**FIGURE 3.9** $^{1}$H STD NMR experimental data showing the aromatic region of $STD_{diff}$ spectra for the standard 1 mM L-Trp/20 μM BSA sample in $D_2O$, with Gaussian selective pulses (a) and Eburp2 selective pulses (b) of varying lengths: 10, 20, 30, 40 and 50 ms. All experiments were acquired with identical acquisition parameters, including a total saturation time of 2 s, $STD_{off}$ set to –30 ppm, and $STD_{on}$ set to 0 ppm over a period of 51 min to acquire each $STD_{on}/STD_{off}$ data set of 512 scans each. The s/n ratio was measured for the tallest resonance H2 at 7.26 ppm.

Figure 3.9 demonstrates the crucial nature of selecting the correct shaped pulse for STD NMR data acquisition. Historically, STD NMR data have typically been obtained using 50 ms Gaussian pulses that provide a minimal risk to ligand excitation and a suitably narrow band for the saturation of protein methyl protons.[10,11] This approach does work (Figure 3.9a), but it is not particularly sensitive as it provides a signal-to-noise (s/n) ratio of 27:1. This sensitivity issue, which is especially problematic with less sensitive room temperature probes, can be circumvented to a degree by adopting shorter Gaussian pulse lengths as shown across the $STD_{diff}$ spectra in Figure 3.9a. Reducing the Gaussian pulse length from 50 to 10 ms, but maintaining the $STD_{off}/STD_{on}$ position, provides a dramatic 7.7 times improvement in the s/n ratio. This seems to produce the desired result for very little cost, but this change has to be applied with caution so as to not accidentally excite ligand resonances, resulting in a 'false-positive' STD. Shortening the Gaussian pulse makes its excitation band wider, and a 10 ms Gaussian pulse will excite over approximately 600 Hz, whereas the 50 ms Gaussian pulse excites over approximately 120 Hz (equating to ~1 and ~0.2 ppm, respectively, on a 14.1-Tesla/600-MHz spectrometer). This effect is illustrated in Figure 3.10. Experimental results are shown in Figure 3.9 – setting $STD_{on}$ to 0 ppm places the nearest L-Trp ligand resonances at 3.24 ppm downfield from the shaped pulse, which should be safe using a 10 ms Gaussian pulse.

Despite its positive attributes, the Gaussian pulse can become problematic in STD NMR at shorter pulse widths. Therefore, a pulse that provides the same benefits as the Gaussian pulse but with less of the associated risks is required; this is where the Eburp2 pulse provides additional utility. As a direct comparison, Figure 3.9b shows the same exercise of variable pulse length for the Eburp2 pulse, with the result that a 50 ms Eburp2 pulse provides virtually the same s/n ratio as a 10 ms Gaussian pulse. There is also the opportunity to shorten the Eburp2 pulse to provide even more s/n, but the improvement between 50 and 10 ms is 1.8 times compared to 7.7 times for the Gaussian. Therefore, we recommend using a 50 ms Eburp2 for STD NMR, and

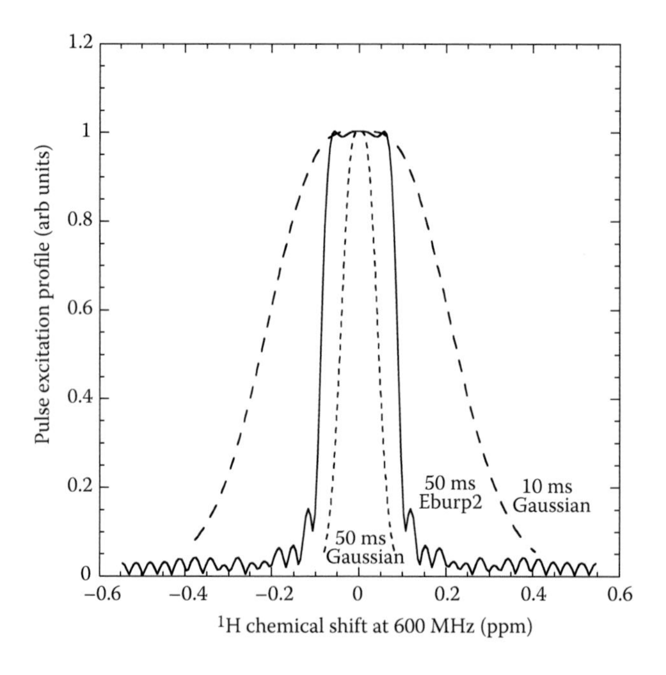


**FIGURE 3.10** Excitation profiles of a 50 ms Eburp2 pulse (————), a 50 ms Gaussian pulse (- - - -), and a 10 ms Gaussian pulse (– – – –) using a parts-per-million scale to show width as applied on a 14.1-T/600-MHz spectrometer.

this conclusion is in agreement with a recent study looking at optimising excitation schemes for STD NMR.[21] The reason why the Eburp2 pulse provides improved $STD_{diff}$ data lies in its different excitation profile (Figure 3.10). In our STD NMR experiments, 0 ppm on Figure 3.10 represents the point at which the pulse is applied on-resonance or off-resonance as dictated in the frequency list, which is set up before the experiment commences. The 50 ms Gaussian pulse has a narrow tip of optimal excitation focussed around ±0.02 ppm of center; this would create a very narrow and specific saturation in an STD NMR experiment. In contrast, the 50 ms Eburp2 pulse has a flat optimal excitation maximum over approximately ±0.07 ppm, which would enable saturation of protein methyl groups over a wider range of parts-per-million (ppm) values than the Gaussian 50 ms pulse. Ultimately, the Eburp2 pulse is more efficient at saturating many protein methyl resonances than the 50 ms Gaussian pulse. This is further confirmed when comparing $STD_{diff}$ data in Figure 3.9 for the 10 ms Gaussian pulse with the 50 ms Eburp2 pulse; the s/n ratios from these two pulses are very similar due to comparable efficiencies of saturation. Take a look at the optimal excitation regions in Figure 3.10 where the excitation profiles are close to 1.0 for both the 10 ms Gaussian pulse and the 50 ms Eburp2 pulse. The problem with the 10 ms Gaussian pulse is the wider excitation limits compared to the 50 ms Eburp2. However, the Gaussian pulse does have its uses. For example, the NOE difference experiment (Section 3.2.3) ideally implements a pulse that can saturate a single resonance in a spectrum of multiple resonances, and the narrow excitation tip of a 50 ms Gaussian pulse is ideal for this purpose.

Another parameter worth investigating with respect to selective shaped pulses for STD NMR is the on-saturation position. Figure 3.11 illustrates the effect of saturation position when a 50 ms second Eburp2 pulse is applied over a 2 s saturation period. The Eburp2 off-resonance position is left constant at –30 ppm, but the on-resonance position is moved from between −4 and 0 ppm together with the result of misplacing the pulse immediately adjacent to a ligand resonance at +3 ppm. These data highlight the necessity of exciting protons from protein methyl groups to facilitate transfer. The +3-ppm on-resonance saturation spectrum was included to demonstrate the effect of the pulse interacting with the ligand. It may initially appear that this spectrum is not significantly different to that with saturation at 0 ppm, but look more closely at the Hβ resonances (3.43/3.24 ppm); these are now more intense than the Hα resonances (3.99 ppm). A sweep of saturation points is, therefore, recommended to detect the safest on-resonance setting that does not distort relative intensities but provides the most efficient saturation transfer and, hence, the greatest s/n (Figure 3.11).

### 3.3.6 Saturation Times in STD NMR Experiments

The effect of variable saturation time for the L-Trp/BSA system is shown in Figure 3.12 and demonstrates the role of saturation time across all ligand $^1H$ resonances. This figure was created using STD amplification factors ($STD_{amp}$) for each resonance calculated via the method described in Section 3.3.4 using Equations 3.3 and 3.4, and the resulting graph of $STD_{amp}$ versus saturation time is known as an STD build-up curve. Here, $STD_{amp}$ is used as a measure of the intensity observed in the $STD_{diff}$ spectrum for a variety of saturation times from 0.5 to 4 s. It may

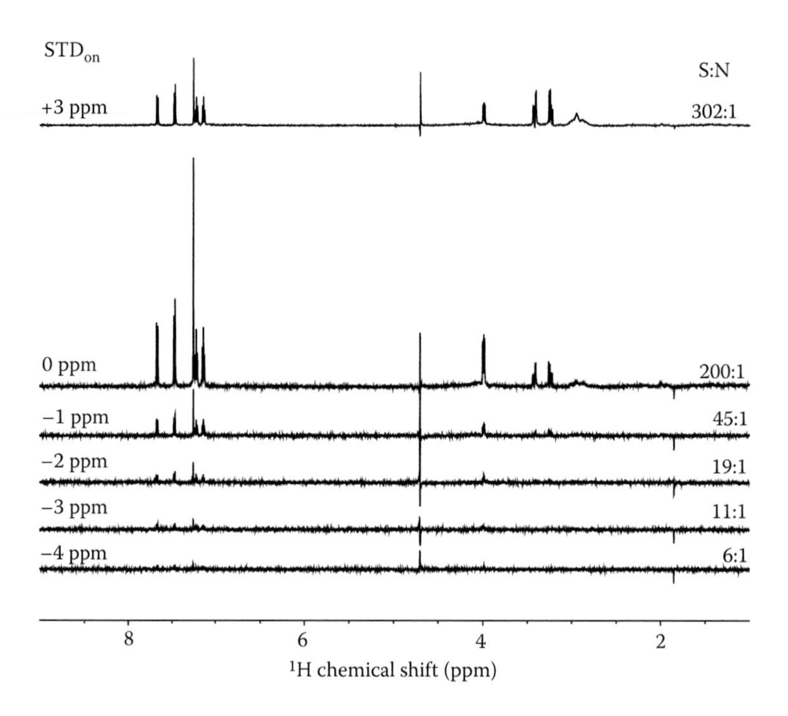


**FIGURE 3.11** $^{1}$H STD NMR experimental data showing the aromatic region of STD$_{diff}$ spectra for the standard 1 mM L-Trp/20-μM BSA sample in $D_2O$ using 50 ms Eburp2 selective pulses with a variety of on-resonance positions. All experiments were acquired with the identical acquisition parameters, including a total saturation time of 2 s, STD$_{off}$ set to –30 ppm, and STD$_{on}$ set to 0 ppm over a period of 51 min to acquire each STD$_{on}$/STD$_{off}$ data set of 512 scans each. The s/n ratio was measured for the tallest resonance H2 at 7.26 ppm.

be possible to saturate beyond 4 s as long as this is within the recommended limits set by your spectrometer manufacturer. Clearly, increasing the saturation time increases the signal observed, but as expected, all intensities tend toward a plateau. Furthermore, protons H2 and H7 from L-Trp are very much within the build-up phase even at saturation times of 4 s. As commented earlier, it is common to present STD$_{amp}$ data as a percentage of maximum, but Figure 3.12 indicates that the relative proportion of STD$_{amp}$ for Hβ will diminish for successive saturation times greater than 3 s. In Section 3.3.4, we introduced relative STD$_{amp}$ values and noted their use in GEM to inform upon ligand binding orientation. However, as the STD$_{amp}$ is saturation time dependent, it is advised to obtain data over a range of times and create the STD build-up curves for each new study.

Furthermore, if you are in possession of a suite of STD build-up curves for your ligand, you can apply a very simple but powerful quantitative analysis that addresses the issue of STD build-up variability, as demonstrated with known inhibitors to the protein human thymidylate synthase.[22]

STD$_{amp}$ data plotted against saturation time can be fitted for each curve to Equation 3.5 using any suitable software package (e.g. Kaleidagraph v.4.1 was used to fit the data in Figure 3.12):

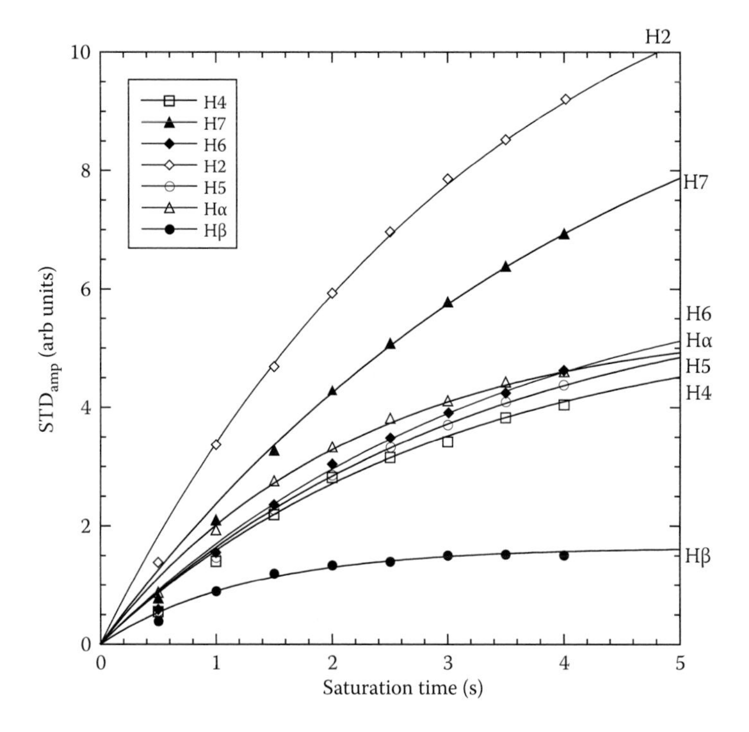


**FIGURE 3.12** The variation of STD amplification factor with saturation time for all L-Trp resonances in $STD_{diff}$ spectra for the standard 1 mM L-Trp/20 μM BSA sample in $D_2O$ using 50 ms Eburp2 selective pulses placed at 0 and –30 ppm for on-saturation and off-saturation, respectively.

$$STD_{amp} = STD_{amp\,max}\left(1 - e^{-k_{STD}t}\right) \tag{3.5}$$

where $STD_{max}$ and $k_{STD}$ define the plateau of $STD_{amp}$ and the STD rate constant, respectively. These coefficients can then be used in Equation 3.6 to obtain the slope for each curve when saturation time is zero ($STD_{FIT}$):

$$STD_{FIT} = k_{STD}.STD_{max} \tag{3.6}$$

The relative $STD_{FIT}$ is, therefore, a realistic measure of the proximity of each ligand proton to the protein and an improved measure of STD data for GEM analysis (Table 3.2).

Comparing relative contributions of L-Trp protons using percent maximum for $STD_{FIT}$ (Table 3.2) and single-point $STD_{amp}$ determination (Table 3.1) highlights that the most significant interaction with BSA is through the H2 proton, regardless of the analysis method used. However, it is clear from the $STD_{FIT}$ approach that protons H4 and Hβ are significantly underestimated using the single-point $STD_{amp}$ method. Apart from the early plateau of $STD_{amp}$ for Hβ, it is not immediately apparent that a discrepancy may exist between the analysis methods and suggests that this

**TABLE 3.2**
**$STD_{max}$, $k_{STD}$, $STD_{FIT}$, and Percentage $STD_{FIT}$ for All L-Trp Ligand $^1$H Resonances from Experimental Data Shown in Figure 3.12**

| L-Trp Resonance | Chemical Shift (ppm) | $STD_{max}$ (arbitrary units) | $k_{STD}$ ($s^{-1}$) | $STD_{FIT}$ | $STD_{FIT}$ Percent of Maximum |
|---|---|---|---|---|---|
| H4 | 7.68 | 5.484 | 0.4580 | 2.512 | 64.0 |
| H7 | 7.48 | 11.526 | 0.2296 | 2.646 | 67.5 |
| H2 | 7.26 | 13.095 | 0.2995 | 3.922 | 100 |
| H6 | 7.21 | 6.1631 | 0.3081 | 1.899 | 48.4 |
| H5 | 7.14 | 5.5635 | 0.3315 | 1.844 | 47.0 |
| Hα | 3.99 | 6.6479 | 0.2937 | 1.953 | 49.8 |
| Hβ | 3.43/3.24 | 1.6366 | 0.7893 | 1.292 | 32.9 |

quantitative STD approach should be employed wherever possible if relative distance information between protein and ligand is sought.

It is not always practical to run multiple saturation time experiments to create a build-up curve for $STD_{FIT}$ analysis; fortunately, there is an alternative approach. GEM considering relaxation of the ligand (GEM-CRL) uses the longitudinal (or spin–lattice) relaxation time ($T_1$) of each ligand proton to modify its associated $STD_{amp}$ value to provide a semi-quantitative result (QSTD), which is obtained using Equation 3.7:[23]

$$QSTD = \frac{STD_{amp}}{T_1} \tag{3.7}$$

The theory behind this approach and comparisons with the standard STD approach are provided in reference 23, in a complementary manner to the description given in Section 3.3.1. The GEM-CRL approach is very elegant due to its simplicity and requires only a single $STD_{amp}$ value for each ligand proton, that is, it is only necessary to measure a single STD NMR experiment together with a measure of the $^1$H $T_1$ value of all ligand protons (most readily using an inversion–recovery experiment[24,25] [Chapter 1], a driven-equilibrium single pulse,[26] or any other acceptable method). Once QSTD values have been determined, they can be compared and ranked using percent maximum as shown for $STD_{amp}$ and $STD_{FIT}$. The computational modelling used in the GEM-CRL example referred to[23] makes reference to the known structure of the protein being studied. Similar computational analyses are applicable to STD data, provided that a structure file is available for the protein in the presence and absence of ligand. The analysis uses an approach called 'complete relaxation and conformational exchange matrix analysis' (CORCEMA) and has a saturation transfer version called 'CORCEMA-ST'.[27]

GEM-CRL can be very useful when it is impractical to acquire many STD data sets due to time and/or sample constraints. We have not provided data for the GEM-CRL approach applied to the L-Trp/BSA system for three reasons: (1) the method is ideally suited to STD experiments with at least 100-fold ligand excesses; (2) the L-Trp/BSA has been previously described as both a specific binding[28] and a non-specific binding[29] system (non-specific binding leads to poor correlations between GEM-CRL and $STD_{FIT}$ methods); and finally, (3) the GEM-CRL approach appears to work best when the STD recycling delay (d1–d31 in Figure 3.6) is at least three times the longest $T_1$ relaxation time of the ligand (with $T_1$ values determined between 2 to 6 s in $D_2O$, this is prohibitive for the L-Trp/BSA system; if working with $H_2O$-based buffers, then GEM-CRL is ideal). As a final important note, our own laboratory experience of comparing GEM-CRL and $STD_{FIT}$ methods using ligand–protein systems that exhibit a known specific binding mode is that the approaches produce complementary quantitative STD results. Perhaps using GEM-CRL and $STD_{FIT}$ together is an appropriate approach for differentiating non-specific positive binding events in STD NMR.

### 3.3.7 STD NMR and Ligand Binding Competition

Any competitive inhibition event can be characterised using the Cheng and Prusoff relationship[30] shown in the following for a protein, *P*, ligand, *L*, and inhibitor, *I*. The protein can interact with either the ligand or the inhibitor to create two equilibrium relationships shown as Equations 3.1 and 3.8:

$$P + I \rightleftharpoons PI \tag{3.8}$$

We have seen that Equation 3.1 can be defined in terms of a dissociation constant $K_D$ (Equation 3.2), and the protein–inhibitor interaction can also be defined as an equilibrium in Equation 3.9:

$$K_i = \frac{[P].[I]}{[PI]} \tag{3.9}$$

and, consequently,

$$K_D = \frac{[L].K_i}{IC_{50} - K_i} \tag{3.10}$$

where $IC_{50}$ is the concentration at which the protein is 50% inhibited and can be found using STD NMR (or another method) by titrating an inhibitor with fixed concentrations of both protein and ligand. Although this appears straightforward, more complicated situations arise, which can dramatically affect the result of binding assays conducted by STD NMR (or any complementary method). You need to be aware that, if the ligand–inhibitor combination suffers from (1) independent binding sites, that is, non-competitive inhibition; (2) secondary-site allosteric inhibition, where binding at one binding site causes a structural rearrangement that alters the other binding site; or (3) secondary-site allosteric enhancement, then it is necessary to carefully consider your experimental approach.[11]

Despite the issues mentioned above, monitoring competitive binding of two ligands to a single target has been utilised by many different spectroscopic methods, including NMR. Here, we describe how it is possible to use STD NMR to monitor

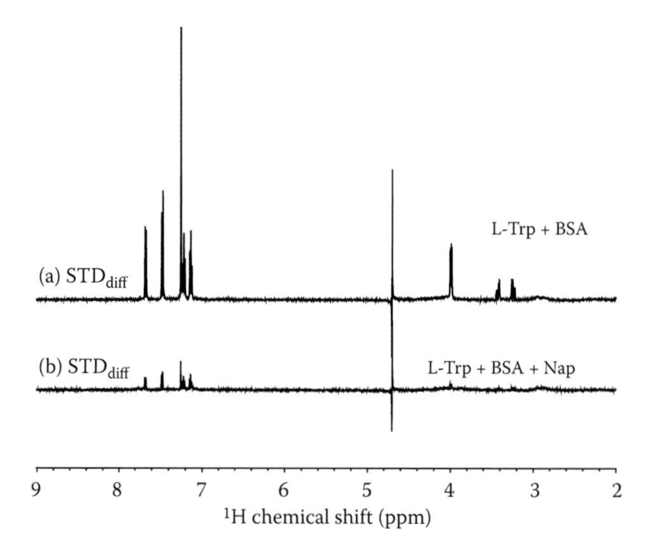


**FIGURE 3.13** $^1$H STD$_{diff}$ spectra for the 1 mM L-Trp/20 μM BSA system alone (a) and with 0.1-mM naproxen inhibitor (b). Both data sets were acquired identically with a total of 256 scans (interleaved 128 STD$_{on}$ + 128 STD$_{off}$), a 4 s relaxation delay and a 2 s saturation period obtained by cycling a 50 ms Eburp2 pulse with on/off saturation set to 0/–30 ppm.

binding affinity. To demonstrate this using our L-Trp/BSA system, we introduce a competitor molecule, naproxen (Nap). Naproxen is a non-steroidal, anti-inflammatory molecule that is known to bind to BSA with 1000-fold higher affinity than L-Trp.[16] The competitive binding of L-Trp/naproxen to BSA or human serum albumin (HSA) has been studied by STD NMR and WaterLOGSY (Section 3.3.11).[16,29]

The general effect on the STD$_{diff}$ spectrum of having a competitive ligand present or absent is shown in Figure 3.13 for our sample system: 1-mM L-Trp and 20-μM BSA system but with ±0.1-mM naproxen. The drop in STD$_{diff}$ intensity of the L-Trp resonances due to competition by naproxen is quite remarkable. Naproxen has been reported as having a dissociation constant ($K_D^{Nap}$) for BSA in the sub-micromolar region compared to L-Trp that has a $K_D^{L\text{-}Trp}$ between 30 and 320 μM. The lower concentration and high affinity of naproxen makes its NMR signals essentially absent from the STD$_{diff}$ spectrum shown in Figure 3.13.

Further information regarding quantitative applications of STD NMR, including the use of build-up curves, *in silico* simulations, molecular recognition, revealing kinetics of binding, and determination of $K_D$, can be found in the excellent review by Angulo and Nieto.[31]

### 3.3.8 2-D STD NMR

The STD approach can readily be imported to 2-D NMR experiments, with the saturation transfer step becoming integral to the preparation period and the difference element incorporated in the process of acquisition. The evolution period ($t_1$) is not interfered with, so the correlation sought is retained. Therefore, STD-NOESY,

STD-TOCSY, and STD-heteronuclear single-quantum coherence (HSQC) experiments will also measure NOEs, spin-system correlations, or proton-heteronuclear correlations between ligand protons. However, the correlated results observed are of a direct consequence of the nature of the ligand when it is bound to the protein. This is a route into determining the bound conformation of the ligand. 2-D STD experiments are not only useful for observing ligand conformation, but being 2-D NMR data sets, they also reduce overlap issues and enable the deconvolution of resonances that crowd regions of a 1-D $^{1}H$ spectrum.[10,11,32–35] There is, however, one issue with these experiments that has to be aired. Individual 1-D $^{1}H$ STD NMR experiments are typically acquired over a 0.5 to 2 h period and, sometimes, longer, depending on the sample; thus, 2-D STD data acquisition can be quite a lengthy process. In reality, 2-D STD works by pushing NMR sensitivity boundaries to their limit and utilises the advantage of cryogenically cooled probes. The result of STD-NOESY, STD-TOCSY, and STD-$^{1}H$–$^{13}C$ HSQC experiments for our 1 mM L-Trp/20 μM BSA sample is shown in Figure 3.14. L-Trp does not undergo any radical conformational change upon binding BSA; therefore, the 2-D STD results look remarkably similar to the standard 2-D NMR data for L-Trp with BSA being absent. However, despite the illustration that the L-Trp/BSA system can be used to set up 2-D STD experiment, the data highlights H4–H6 STD-NOESY contacts and changes in the relative intensities of STD-TOCSY resonance for H4–H5 and H4–H6, which could suggest coupling constant changes between tryptophan ring protons, which could be due to the different conformation of the bound form. In addition, the STD-HSQC data set further supports the aromatic region and Hα as the main binding regions, with Hβ being absent in Figure 3.14e. Standard NOESY, TOCSY, and HSQC data sets were acquired over 3.0, 2.5, and 4.5 h, but their 2-D STD counterparts took 9, 8.5, and 24 h, respectively, and were also acquired with lower F1 resolution. In addition, the STD-HSQC data were processed with 32 points of linear prediction to circumvent the poor 64-point F1 resolution. Figure 3.14 also illustrates that 2-D STD data analysis requires the standard correlation spectrum for comparison because any 2-D STD experiment is a difference-only acquisition; the $STD_{on}$ and $STD_{off}$ for the 2-D is subtracted within the pulse sequence via phase cycling, and the Fourier-transformed result is the $STD_{diff}$. These 2-D STD experiments are usually available in standard pulse sequence libraries and usually have delay and pulse parameters identical to the 1-D $^{1}H$ counterpart. Always make sure that you have observed a standard 1-D $^{1}H$ $STD_{diff}$ spectrum before moving to 2-D STD and set up the STD components (saturation time, shaped pulse, etc.) identically to the 1-D STD experiment. The 2-D aspects of the sequence should mirror a typical 2-D set-up, but it is usually necessary to run the 2-D STD with more scans and fewer F1 points. For example, Figure 3.14 STD-NOESY and STD-TOCSY were acquired with 2048 × 256 points and 32 scans per slice, whereas the STD-$^{1}H$–$^{13}C$-HSQC was acquired with 2048 × 64 points and 512 scans per slice. The STD-$^{1}H$–$^{13}C$-HSQC is particularly tough because you are trying to observe saturation transfer to $^{1}H$ nuclei that are attached to the 1.1% natural abundance $^{13}C$ nuclei in the ligand. Clearly, this can be dramatically improved if you have $^{13}C$-enriched ligand. It is worth noting that the STD-NOESY and standard NOESY mixing times used in Figure 3.14 were 250 and 500 ms, respectively, and chosen to coincide with the different NOE build-ups experienced by the ligand in each experiment. The bound ligand in the STD-NOESY will experience efficient NOE

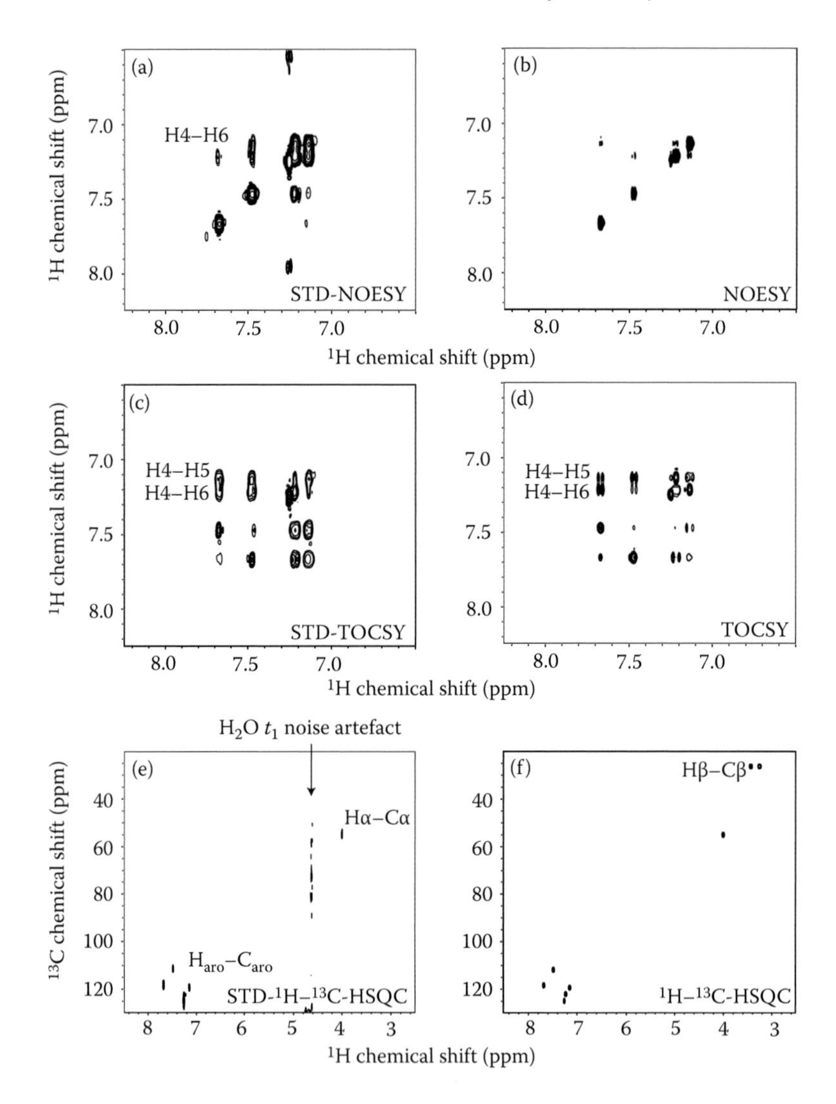


**FIGURE 3.14** 1 mM L-Trp/BSA ($D_2O$, standard buffer) STD-NOESY with a 250 ms mixing time (a), STD-TOCSY with a 60 ms spin lock (c), and STD-$^1$H–$^{13}$C-HSQC (e), together with 1 mM L-Trp NOESY with a 500 ms mixing time (b), TOCSY with a 60 ms spin-lock time (d), and $^1$H–$^{13}$C HSQC (f). The aromatic proton region is shown in (a) to (d), and the entire spectrum is shown in (e) and (f). STD data sets were all acquired using a 2 s saturation period obtained by cycling a 50 ms Eburp2 pulse with on/off saturation set to 0/–30 ppm.

build-up associated with the large molecular weight of the protein–ligand complex, and a shorter mixing time should be used. The free-ligand standard NOESY requires a longer mixing time, and despite this longer period, you can see (Figure 3.14a and b) that the STD-NOESY contains more through-space contact data despite having a shorter mixing time. This is to be expected, especially as the bound ligand should normally adopt a preferred conformation in the binding pocket of the protein.

### 3.3.9 STD NMR and Peptide-Based Ligands

Peptide-based ligands provide an excellent natural target to many biological macromolecules that are part of the general protein, receptor, and enzyme families. Therefore, with the proviso regarding acceptable $K_D$ values, STD NMR should provide an ideal method to detect an interaction. Furthermore, STD NMR can provide binding information regarding the peptide–protein interaction and information on the molecular bases of specificity that drives such a process. The example used in this section involves the interaction of small peptides with the receptor protein integrin αvβ6.[35–39] Integrin αvβ6 is a member of the αv class of integrins that bind a specific amino acid motif comprised by arginine-glycine-aspartic acid known as an 'RGD motif'. Our initial structural study of three such peptides highlighted that they were all helical yet displayed different binding affinities; the length of the helix was, however, linked to affinity. The issue with this hypothesis was the need to use a structural stabiliser, TFE, for helix formation; thus, there was a need to establish whether the helices survived when bound to the integrin. STD NMR was used for GEM mapping, and thus, we confirmed a helical bound conformation for all peptides.[36]

The main issue confronting our peptide–protein STD NMR study was the degree of resonance degeneracy. This is demonstrated in Figure 3.15 for the peptide A20-FMDV2 with the amino acid sequence NAVPNLRGDLQVLAQKVART. Figure 3.15 shows how crowded both $^{1}H$ $STD_{off}$ and $STD_{diff}$ spectra are. Clearly, this problem could be circumvented using 2-D STD, but these data were collected in our pre-cryogenically cooled probe days and, our 1-D STD acquisition times were between 8 and 12 h using Gaussian pulses. Instead, we adopted a method of 'binning' resonances, a common data reduction technique. This did mean that a particular intensity for a multiplet was given to more than one resonance in more than one amino acid; the labels shown in Figure 3.15 for $^{1}H$ groups as well as individual resonances reflect this. The same binning procedure was used for $STD_{off}$ and $STD_{diff}$ to enable the calculation of $STD_{amp}$ for each resonance bin. Finally, the sum of all $STD_{amp}$ values for each individual amino acid was obtained, and this was normalised for the number of protons that each amino acid possessed. This was to ensure that amino acids such as leucine, which contains 12 protons, do not bias the summed $STD_{amp}$ data with respect to amino acids such as glycine, which contain only 3 protons. The result still provided the greatest $STD_{amp}$ factors for leucine residues that confirmed the primary binding motif as a RGDLXXL motif, with the LXXL forming the N-terminal section of a helix upon binding to the integrin. Despite this apparent success at confirming the helix formation upon binding, we were concerned that the leucine Hδ were being excited by the 9.4 ms Gaussian on-saturation pulse that was set to −2.5 ppm. This was tested by moving the on-resonance pulse from −2.5 to +10.5 ppm to excite protein amide protons and initiate saturation transfer across protein from these protons instead of those in methyl groups. This enabled confirmation that the leucine was a key contact to the integrin and that the Gaussian pulse at −2.5 ppm was optimal. If you are conducting STD NMR studies using $H_2O$ buffer, switching the on-resonance saturation to +10 to 12 ppm is an excellent method to test for accidental excitation. Look for a result where the $STD_{diff}$ intensities are all lower

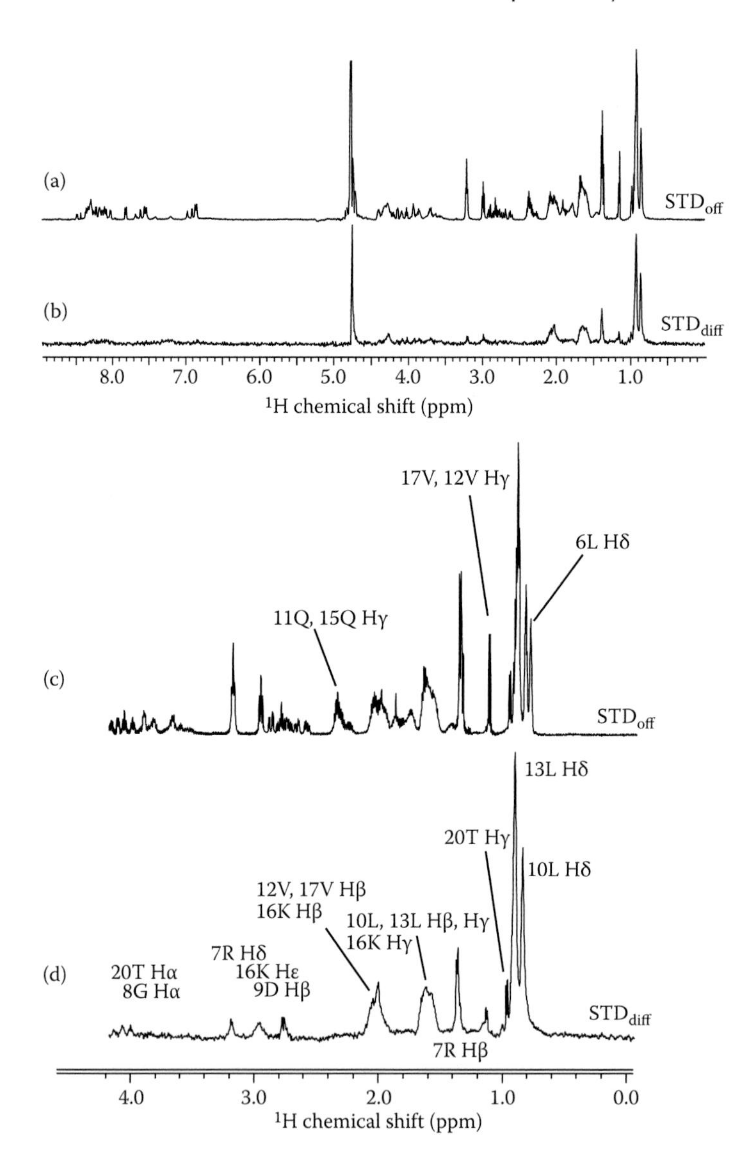


**FIGURE 3.15** $^{1}H$ $STD_{off}$ spectra (a and c) and $STD_{diff}$ spectra (b and d) for the peptide ligand A20-FMDV2 (2 mM), with the protein receptor integrin αvβ6 (28 μM) at 283 K. Data were obtained using 9.8 ms Gaussian pulses applied at –2.5 and –70 ppm for on-resonance and off-resonance, respectively. The sample buffer used was 25 mM phosphate buffered saline with 100 mM sodium chloride at pH 6.5 and included 1 mM $Mg^{2+}$ and 0.5 mM $Ca^{2+}$ to ensure that the integrin was activated. (c) and (d) show expansion of the aliphatic regions of (a) and (b) with key amino acid residue assignments.

but the relative proportions within the same data set are the same. HN-mediated saturation transfer is never going to be as efficient as using methyl protons, hence, the overall drop in intensity for $STD_{diff}$. As an additional point, protein–peptide interactions studied by STD NMR should also be accompanied by control experiments. Our peptide–integrin example used three additional control options: (1) the addition of ethylenediaminetetraacetic acid (EDTA), which removed calcium and magnesium, deactivated the integrin, and resulted in a 100-fold drop in $STD_{diff}$ intensity; (2) the control $STD_{diff}$ spectrum using the A20-FMDV2 peptide and a different protein; and (3) the reverse of the second control using integrin αvβ6 and a peptide from a completely different project.[40] Both control experiments provided data showing minimal to no binding of the A20-FMDV2 and integrin αvβ6 to their control agents.

The peptide A20-FMDV2 is so called because it originates from the type O1 bovine fetal serum (BFS) capsid protein VP1 of foot-and-mouth disease virus (FMDV), and bovine αvβ6 is a known receptor for the disease in cattle;[41] this spurred another project but, this time, utilising 2-D STD techniques despite issues relating to sensitivity. The HSQC approach was adopted because of the large degree of overlap still present in homonuclear 2-D $^{1}H$ experiments such as NOESY and TOCSY. This approach has been demonstrated as useful for screening glycopyranosides with agglutinin[32] (Section 7.2), although the peptides being used provided much more complex spectra. Recombinantly producing peptides in *E. coli*, rather than having them synthetically produced, provided a means of counteracting the NMR sensitivity issue involving natural abundance $^{13}C$.[38] The recombinant approach allows the *E. coli* to be grown on $^{13}C$-labelled glucose to provide a peptide product with more than 90% incorporation of the $^{13}C$ isotope once purified; this labelling method is common for producing $^{13}C$- and $^{15}N$-enriched protein samples for NMR-based structural biology but is less common for peptides. Details of the protocol for making $^{13}C$- and $^{15}N$-labelled peptides can be found in the associated reference.[38] The STD-$^{1}H$–$^{13}C$-HSQC with $^{13}C$-enriched peptides is very successful. The resolution obtained is shown in Figure 3.16 for uniformally $^{13}C$-enriched A20-FMDV2 peptide with integrin αvβ6. The figure also demonstrates the utility of a cryogenically cooled probe where the STD-HSQC experiment time was reduced to 3 h, compared with 12 h in the original study.[35] The data sets shown in Figure 3.16 can be used to provide $STD_{amp}$ values for each proton attached to a carbon that registers a resonance in the STD-$^{1}H$–$^{13}C$-HSQC. It can also be used to provide $STD_{off}$ intensities ($I_o$) (Figure 3.16a) if it was acquired using the same sample that generates the STD-HSQC in Figure 3.16b. $STD_{amp}$ calculations should be completed as shown earlier, but the intensities from $STD_{off}$ should be multiplied to reflect the scan number difference between the HSQC and STD-HSQC experiments. Referring to the legend for Figure 3.16, $STD_{off}$ intensities should be multiplied by 32/4 = 8 times to be correctly scaled. As with 1-D STD data, the results are usually shown as a percentage of the largest $STD_{amp}$, and this is shown for the A20FMDV2/integrin system in Figure 3.17a. Figure 3.17a is a histogram of $STD_{amp}$ factors for all A20FMDV2 resonances, which can be resolved in the STD-$^{1}H$–$^{13}C$-HSQC as percentages of the maximum $STD_{amp}$, that is, from the Hβ proton from valine-12. Amino acids that display $STD_{amp}$ more than 50% of the maximum value are highlighted on the structure of A20FMDV2 in Figure 3.17b to demonstrate this approach providing GEM information. In this case,

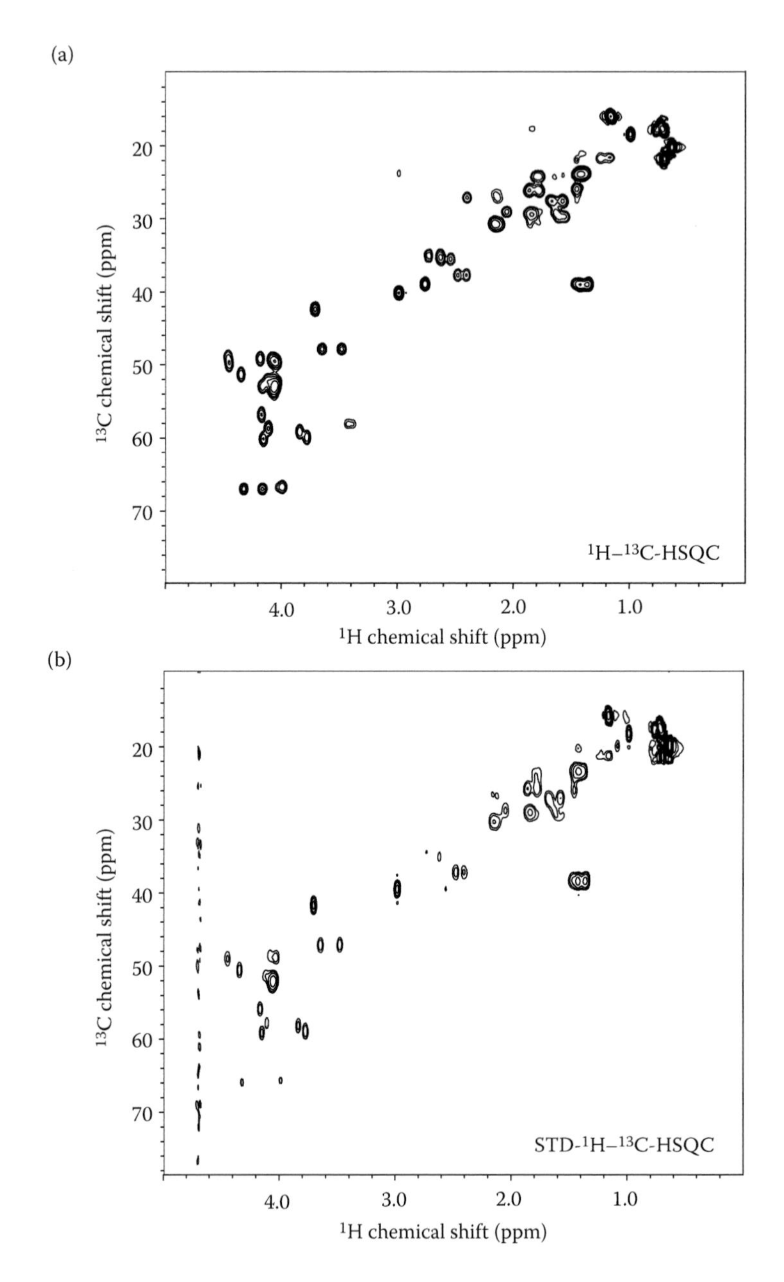


**FIGURE 3.16** Regions of the $^1$H–$^{13}$C HSQC (a) and STD-$^1$H–$^{13}$C-HSQC for 2 mM U-$^{13}$C–A20-FMDV2 + 5 μM integrin αvβ6 using a 10 ms Gaussian pulse applied at –3 and –30 ppm for on-resonance and off-resonance, respectively (b). The standard HSQC spectrum was acquired in 30 min (2048 × 256, ns = 4), and the 2-D STD spectrum was acquired in 3 h (2048 × 128, ns = 32) using a cryogenically cooled probe, with the sample at 283 K using an identical buffer system to that described in Figure 3.15.

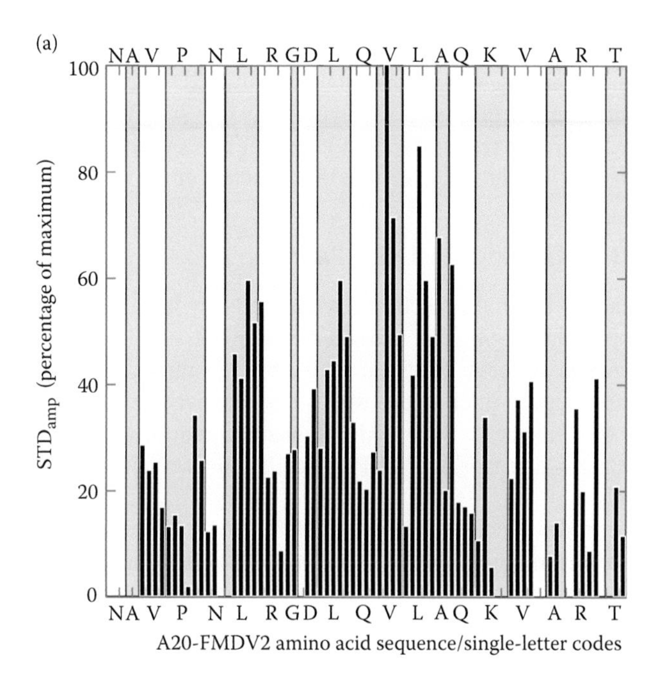


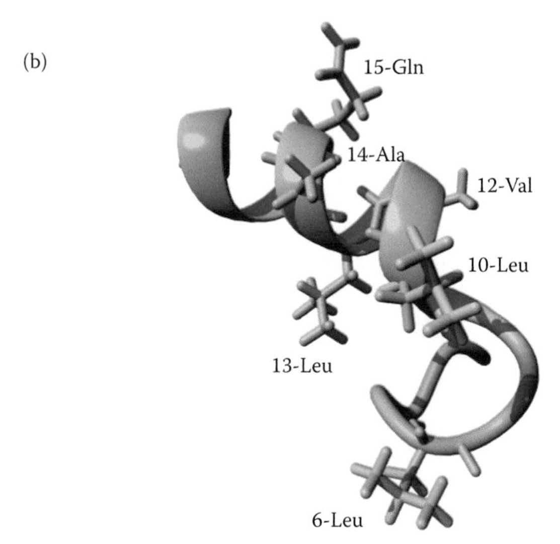


**FIGURE 3.17** STD amplification factors ($STD_{amp}$) taken for 2 mM peptide A20-FMDV2 in the presence of 5 μM integrin αvβ6 (a) and amino acid residues shown as sticks with $STD_{amp}$ more than 50% of the maximum value labelled on the TFE structure of A20-FMDV2 (b) to support helix formation of the peptide when it binds to the integrin.

it was surprising that the primary RGD amino acid motif was not primarily involved in the integrin interaction, which suggests that non-RGD interactions are important for the αvβ6 integrin. Furthermore, the hypothesis was supported that peptides such as A20FMDV2 bound integrin αvβ6 in a helical conformation despite having no structure when free in solution, except in the presence of the co-solvent TFE.[42]

### 3.3.10 STD NMR and Very Large Proteins

An obvious place to begin this section is to define 'very large protein systems'. STD NMR works as a ligand-observe method for viewing small-molecule interactions with soluble protein systems, but there are several examples where STD NMR has been used to study ligand interactions with virus-like particles (VLPs) and receptors on the surface of whole cells. STD NMR acquisition requires modification from the standard approach when working VLPs or cells, and we now describe the recommended alterations.

Saturation transfer double-difference (STDD) NMR was first reported by Meyer et al. in 2005 regarding the direct observation of a cyclic-peptide ligand binding to a protein-receptor integrin αIIbβ3 present in liposomes and directly on the surface of live cells.[43] STDD differs from STD by obtaining a double-difference spectrum where a cell-only STD spectrum is subtracted from a 'cell + ligand' STD spectrum to identify only the bound ligand resonances. The power to observe small-molecule interactions with live cells has potential for drug screening involving physiological systems. In addition, the cellular αIIbβ3 study demonstrated that the ligand displayed higher affinity for the protein receptor on cells than when within the liposome model system. STDD has found many applications beyond the integrin system, including a recent screen of a marine natural product library against human cannabinoid G-protein–coupled receptors CB1 and CB2.[44] Despite the obvious power of cell-based STDD, there are concerns that it can be limited to specific cell or tissue types and may suffer from sedimentation effects in an NMR tube. However, there have been reports that STDD is possible with membrane proteins in living cells using high-resolution magic angle spinning.[45] If you are taken by the concept of STDD, then this approach has been further advanced with its own specific double-difference microcoil NMR probe that allows the acquisition of signals from two different samples at once.[46] The set-up allowed for separation of binding and non-binding components of the spectrum through phase cycling and was demonstrated by separating $^{1}H$ octanoic acid resonances to confirm interaction with human serum albumin as opposed to those from the non-binding ligand, glucose.

STD NMR of VLPs has recently emerged as a valuable method for studying the interactions of different saccharide molecules involved in viral recognition (Section 7.2). The first example involved the study of histo-blood group antigens (HBGA) as ligands toward rabbit haemorrhagic disease virus (RHDV).[47] This study also highlighted the need for a modified STD NMR set-up for studying VLPs because of the extremely broad proton resonances of the VLP that extend well beyond the normal proton window, thus, where the off-resonance pulse should be placed. The HBGA/RHDV study demonstrated an elegant method where the off-resonance frequency was varied between ±300 ppm in an attempt to avoid accidental saturation. This may seem excessive, but significant unwanted saturation was observed with the VLPs when the off-resonance point was set to ± 50 ppm, and all data analysed were collected using −4 and +300 ppm on-resonance and off-resonance saturation, respectively. This indirect saturation by the off-resonance pulse is clearly an issue for any large species having a long rotational correlation time, and care should be taken to minimise the effect.[47,48] There have been many examples of VLP-ligand analyses by STD NMR, but it is worth mentioning the recent studies of VLPs from influenza and the attempts to understand preferential binding of glycans to the protein hemagglutinin and their role in sub-type identification[49] and potential mechanisms for cross-species infection.[50]

### 3.3.11 Transfer Methods Involving Water: WaterLOGSY

WaterLOGSY is based on the enhanced protein hydration observed through gradient spectroscopy (ePHOGSY)-NOE experiment, which transfers magnetisation via NOE and spin diffusion from bulk water molecules via the protein to the ligand. WaterLOGSY is a ligand-observe transfer method that is also used by pharmaceutical and drug screening companies to search for potential new chemical fragments for drug design and discovery. WaterLOGSY, or LOGSY as it is known in screening circles, is typically used in a large-scale ligand library screen of thousands of chemical fragments alongside both STD and CPMG experiments to create three results for each cocktail of ligands under interrogation. Ligands are usually screened in cocktail mixtures of between 4 and 10 ligands to maximise spectrometer time and output. The WaterLOGSY experiment, like CPMG, has the potential of highlighting false positive hits due to non-specific binding or aggregation; therefore, it serves to complement STD NMR results. If you engage in discussions with NMR screening specialists, you will always find that each have their favourite experiment. Some groups prefer using STD with WaterLOGSY/CPMG to confirm, whereas others use WaterLOGSY with STD/CPMG to confirm, and a detailed discussion of the pros, cons, and complementation of each method would create an interesting chapter elsewhere.

WaterLOGSY for screening purposes uses two identical experimental data acquisitions: one for the ligand (or cocktail of ligands) in an aqueous buffer alone and the second with identical proportions of ligand in the presence of the protein target. Therefore, the WaterLOGSY approach requires two samples to be prepared instead of the one sample required for STD NMR. Considering CPMG in the mix for a moment, this is also akin to WaterLOGSY such that CPMG data also need to be acquired using the same two-sample approach: one with ligands and the other with ligands and protein. Luckily, CPMG and WaterLOGSY utilise a significant ligand excess just as in STD NMR; therefore, you only ever need two samples to conduct all three experiments: ligand alone for CPMG/WaterLOGSY control and ligand plus protein for CPMG/WaterLOGSY/STD. The point to remember is to make sure that your two samples are identical in terms of ligand concentration and physical conditions (buffer, pH, etc.) and to prepare your samples in water and $D_2O$. The reason for discussing sample preparation highlights the advantage of running WaterLOGSY/STD/CPMG concurrently using only two samples. The price to pay is in the need for exemplary water suppression for all experiments.

A schematic representation of the WaterLOGSY experiment is shown in Figure 3.18a, and it is worth reading the original research papers and associated review of WaterLOGSY and complementary methods.[15,51,52] The representation shown complements the STD schematic in Figure 3.4 and highlights the similarities and differences between the approaches. As a ligand-observe method, WaterLOGSY is like STD NMR and requires a ligand excess to create a significant perturbed ligand population for positive result. WaterLOGSY perturbs the bulk water magnetisation and enables this to be transferred through the protein–ligand complex and to the free ligand in a selective manner. The experiment measures the change in the sign of NOE cross-relaxation, with its associated result being that non-binding ligand resonances appear with opposite sign than those from binding ligands. This result is shown in Figure 3.18 for the L-Trp/sucrose system in the absence (b) and presence (c)

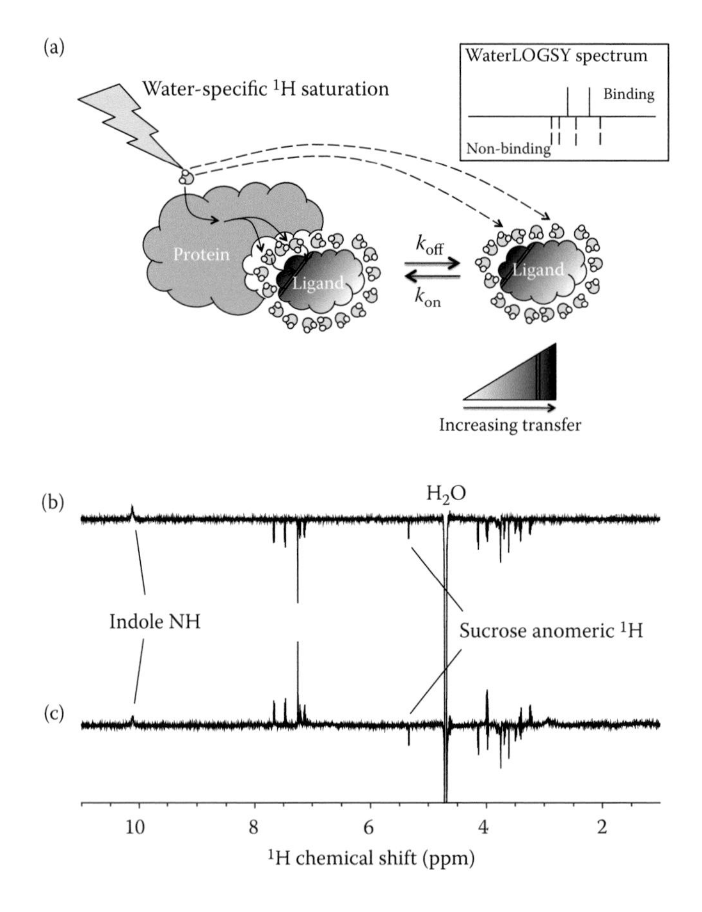


**FIGURE 3.18** Schematic representation of the WaterLOGSY experiment (a) and WaterLOGSY NMR spectra of 1 mM L-Trp and 0.5 mM sucrose in 95% $H_2O$/5% $D_2O$ PBS buffer in the absence (b) and presence (c) of BSA. A change in sign of each LOGSY resonance confirms L-Trp/BSA interaction. The separation of binding/non-binding resonances for L-Trp/sucrose can be clearly seen in (c), with sucrose being the non-binding ligand. Compare this data to the equivalent STD results in Figure 3.7.

of BSA to demonstrate both a positive and non-binding interaction. Stockman and Dalvitt's review[15] provides an excellent qualitative overview of the theoretical basis behind WaterLOGSY, but as summarised here, the method utilises water molecules in unique molecular associations to drive the differences observed through NOE cross-relaxation. These unique associations are shown schematically in Figure 3.18a, and a sample with protein and ligands will contain 'free' water and water associated with the protein surface, binding site, and ligand. The residence time of water molecules in protein cavities varies over the nanosecond-to-microsecond timescale, and therefore, water can be 'in residence' over a period of time much longer than the molecular tumbling correlation time of the protein. The cross-relaxation rate that gives rise to NOE is sensitive to residence time, and at residence times greater than approximately 300 ns at 600 MHz, NOE changes sign. What effectively happens is that the protein-associated water protons take on the NOE characteristics of the protein, and protein NOEs are of opposite sign to small-molecule NOEs (i.e. NOE effects for small molecules are positive, whereas those for large molecules are negative). We have come across this phenomenon before within the STD experiment where the ligand-bound/free size differential creates the $STD_{diff}$ effect where $STD_{on}$ is less than $STD_{off}$ by being of opposite sign and creates a negative difference result. In WaterLOGSY, the change in sign is transferred through the protein and back to the water associated with bound ligand before it dissociates and is detected. If a ligand does not bind, it does not experience any change in sign of NOE. In addition, the WaterLOGSY experiment boosts the signal intensity for bound ligands through chemical exchange (Chapter 2) of selectively excited water protons with a multitude of labile protons on the protein. Transfer of this magnetisation through the protein to the water in the binding site will utilise spin diffusion in a manner similar to STD NMR. Ultimately, WaterLOGSY is a NOE-based experiment where selective excitation of the water signal is followed by 0.5 to 2 s of NOE mixing to facilitate transfer of bulk water magnetisation to the protein. As WaterLOGSY requires ligand dissociation, it has a drawback as in STD NMR, and we refer you back to that section that discusses the need to appreciate $k_{on}$, $k_{off}$ and $K_D$. As a final point, due to WaterLOGSY being NOE sign dependent, there is another non-binding/non-interaction condition that can exist in WaterLOGSY where no signal exists in the experiment in the presence or absence of protein. This is when the correlation characteristics of the ligand are at the $\omega\tau_c = 1.12$ ($\omega\tau_c \approx 1$) condition, where $\omega$ is the spectrometer frequency and $\tau_c$ is the correlation time. This is a very special condition on an NMR spectrometer and is the exact point where NOE passes from positive to negative. If your WaterLOGSY experiment happens to involve a non-binding ligand that invokes $\omega\tau_c \approx 1$, you will see no resonances from it because NOE $\approx 0$. If you suspect this condition being active in your sample, try lowering the temperature to move the correlation time and repeat your experiment.

Figure 3.18 highlights the need for two samples and two data sets to differentiate binding from non-binding. Here, the WaterLOGSY NMR experiment was acquired identically over 21.5 min for each sample with 32 k points and 256 scans. The Fourier-transformed processed result was phased identically for Figure 3.18b and c, with positive signals supporting an interaction. Using sucrose as a negative control ligand is exceptionally useful when setting up WaterLOGSY with L-Trp/BSA because the majority of the disaccharide ring protons are in the chemical shift range of 3.4 to 4.2 ppm and overlap with the L-Trp Hα and Hβ. A correctly set WaterLOGSY should allow differentiation of the L-Trp Hα/Hβ from the sucrose ring protons as they should phase with opposite sign as in Figure 3.18c. The sucrose anomeric proton at 5.3 ppm provides further confirmation that the disaccharide shows no positive interaction as this resonance remains negative in both experiments. L-Trp protons invert for the LOGSY sequence in the presence of BSA protein, but the indole NH is positive in both experiments because this proton is in chemical exchange with protons from water. The observation of the NH 'double-positive' resonance highlights the need for two samples because a positive WaterLOGSY result can be due to protein–ligand interaction or chemical exchange of protons in the ligand with those in water. It is only comparison of the WaterLOGSY result in the presence and absence of protein that can confirm interaction or chemical exchange.

We have been deliberately vague regarding the experimental set-up of the WaterLOGSY experiment because the NMR sequence is not usually provided with current NMR spectrometers, and we recommend that you contact your spectrometer manufacturer's application support for details of available WaterLOGSY sequences. The pulse sequence used in Figure 3.18 is an 'application sequence' with our own modifications, including a 30-ms spin lock to remove the protein background as in the STD sequence. Other parameters of interest are a 20 ms Gaussian 180° pulse to selectively invert the water signal, a 1.2-s NOE mixing period, and excitation sculpting for water suppression. Our personal experiences have found excitation sculpting far superior to WATERGATE 3-9-19 suppression. As mentioned earlier, water suppression is crucial in this experiment to ensure a flat baseline for correct phasing and identification of interaction/non-interaction resonances. The 20 ms Gaussian 180° pulse works very well, and Figure 3.18b and c illustrates that the use of this pulse at a water resonance of 4.7 ppm has little distortion effect on the L-Trp Hα or sucrose resonances in close proximity.

WaterLOGSY can be used for quantitative evaluation, and we have already referred to its use in serum albumin binding for the study of BSA binding sites with different ligands.[16] A modification of the basic approach called 'Aroma WaterLOGSY' was reported in 2010 where water magnetisation is not perturbed following the initial inversion pulses to provide enhanced relaxation and optimisation that is reported to provide double the sensitivity of the standard sequence.[53] Also, it is worth noting the solvent accessibility, ligand binding, and mapping of ligand orientation by NMR spectroscopy (SALMON) method for future reference. SALMON is a modified WaterLOGSY approach that can be used to probe for bulk water accessibility to the ligand and derive bound orientation of the ligand.[54]

### 3.3.12 Not STD NMR: Using $^{19}F$ in Ligand-Observe Screening

This short section is to introduce the possibility of using a nucleus other than $^{1}H$ to conduct ligand-observe screening; in this case, it is $^{19}F$. The references and examples within this section do not involve saturation transfer or methods akin to STD or WaterLOGSY but offer the opportunity for you to think 'outside the box' and for some interesting alternatives to standard saturation transfer methods.

Fluorine is being incorporated into many new-generation pharmaceuticals and is proving to be an important tactic for maximising pharmacokinetic properties, resulting in it being used in $^{19}$F NMR screening[55–57] (Section 7.2.3.1) as well as being utilised in the study of biological events.[58] $^{19}$F is 100% abundant, with only $^{1}$H and $^{3}$H exhibiting higher sensitivity on a nucleus-by-nucleus basis. It also has a large chemical shift range that is very sensitive to local microenvironments; these properties lend themselves well to running cocktails of several ligands. There are two approaches that are currently employed for $^{19}$F screening: direct observation of fluorinated ligands and competition experiments. The direct observation approach involves acquiring ligand spectra in the presence and absence of protein; the free ligand exhibits a long transverse (or spin–spin) relaxation (long $T_2$), but any binding to the protein will shorten the ligand $^{19}$F $T_2$[59] such that the $^{19}$F ligand signal is reduced upon binding. Competition experiments utilise a cocktail of ligands that include 'unknown' binders, a negative control, and a 'spy' molecule that competes in a similar fashion to the naproxen/L-Trp STD example seen in Section 3.3.7. This approach is quite powerful but only works when the 'unknown' ligands out-compete the 'spy' for the protein being investigated. Another method is to utilise a CPMG spin echo and $^{1}$H decoupling during acquisition in an experiment called fluorine chemical shift anisotropy (CSA) and exchange screening. This approach utilises the large $^{19}$F CSA relaxation mechanism to provide a method that is sensitive to a much lower binding affinity threshold.[60]

The practical note regarding these $^{19}$F approaches is to ensure that you have the correct hardware to conduct these experiments. In the first instance, you require an NMR probe with $^{19}$F detection and, ideally, a probe than can also decouple $^{1}$H at the same time as acquiring $^{19}$F. In addition, few routine NMR spectrometers are sold with the hardware configuration to drive such a probe, and thus, most tend to share a single set of linear amplifier components for $^{19}$F/$^{1}$H. This means that you can pulse-acquire $^{19}$F or $^{1}$H but not use both simultaneously. This is not an insurmountable problem, but it would mean a purchase of some extra hardware.

## 3.4 tr-NOE AND TRANSFERRED RDC

Saturation transfer experiments generally involve small ligands that are not expected to alter their structure upon binding. However, larger potential ligands such as carbohydrates or peptides have free structures that can significantly differ compared to the bound state; this can make the interpretation of standard STD data more complicated. As we have seen in Section 3.3.9, more elaborate and complex 2-D STD methods have to be employed to understand structurally significant binding effects. If you are planning on studying one of these more complex ligand binding events by NMR, it may be worth considering one of the following little-exploited, but powerful, methods of analysis: the tr-NOE and transferred RDCs (Section 7.2.2.2). This section offers an insight into the breadth of information that can be gleaned via using these methods. In each section, the reader will be directed to materials that will provide a more in-depth treatment of both sample preparation and experimental set-up.

Regarding NOE, it is worth mentioning an interesting theoretical and practical point. Although we have seen that the NOE process drives saturation transfer in

STD NMR and WaterLOGSY experiments to produce distance-based information between the target and the ligand, you must be aware that two 'flavors' of NOE exist. The NOE difference experiment (Section 3.2.3), STD NMR (Section 3.3), and WaterLOGSY (Section 3.3.11) all utilise the steady-state NOE build-up approach where a nucleus is irradiated to saturation and maintained in that state for many seconds. However, the commonly known 1-D and 2-D NOESY (and ROESY) NMR experiments that provide equivalent through-space information have no such saturation process and utilise the alternate NOE phenomenon, the kinetic NOE (or transient NOE).

### 3.4.1 Exchange tr-NOE

Regardless of the mechanism for generating through-space data (steady-state or kinetic), it is reasonable to assume that such 'through-space' interactions between a ligand and a target protein can be studied in a similar way, and indeed, we have seen this in Section 3.3 for STD NMR. In this section, we are only concerned with the kinetic NOE where distance information can also be obtained via the build-up of NOE signals (cross-peak volumes in the 2-D case). The kinetic NOE is proportional to the inversed sixth power of the distance between the two cross-relaxing nuclei (typically, two protons), which is mediated by dipole–dipole cross-relaxation. For more general theory and application of NOE within structural biology, the following review is a good place to start,[61] although we can also recommend Neuhaus and Williamson's comprehensive text on NOE.[62]

The exchange tr-NOE (et-NOE), to save confusion with the standard transient NOE (tr-NOE), was first used more than 40 years ago in the observation of peptides and hormones binding to the protein bovine neurophysin,[63,64] allowing structural characterisation of the bound state of the ligand. The information about the bound conformation of ligand gathered by et-NOE is observed on the molecule after it has dissociated from its target. In an analogous fashion to STD NMR experiments, the et-NOE requires a protein–ligand system exhibiting an off-rate within a specific range. NOESY cross-peaks are formed when the ligand is in a bound state, which results in their ability to infer structural characteristics of the ligand that may differ from the unbound state. In this respect, the et-NOE experiment is not dissimilar to the STD-NOESY described in Section 3.3.8. However, et-NOE uses the kinetic NOE alone, but STD-NOESY creates saturation transfer to the ligand via the steady-state NOE and then provides structural distance information using the kinetic NOE. The overall result is very much the same, but you may find your particular system more receptive to one method over the other. Some background information regarding the practicalities in using the et-NOE is now summarised. The first requirement is for fast exchange (Chapter 2) between bound and free states. This is typified by one set of averaged resonances for the ligand when in the presence of the target protein, where exchange between the two states is faster than the chemical shift difference in Hz. Also, ligands with large numbers of protons are ideal; however, isotopically labelled binding partners can be used. In addition, the disassociation rate ($k_{off}$) must also be greater than the length of time that magnetisation remains in the free state, that is, $k_{off} >> R_1^{free}$, where $R_1^{free} = 1/T_1^{free}$ and $T_1^{free}$ is the longitudinal relaxation time for each nucleus in the free ligand. As with

STD NMR, a large ligand excess is required for an et-NOE experiment. Typically, the ligand should be 10 to 50 times the concentration of the protein target. Also, as with STD NMR, only a small amount of protein, or a large macromolecular complex such as a membrane or nucleic acid, is required (micromolar amounts). Another parallel with STD NMR is that larger proteins provide better et-NOE results, and the reason behind this is identical to STD NMR saturation transfer efficiency. To confirm ligand et-NOE peaks that are a result of its structure when bound to the protein, a control experiment in the absence of protein must be completed. Finally, weak, non-specific binding of the ligand to target molecule can result in NOEs being observed. Competition experiments involving the addition of a second tight-binding ligand to compete with the ligand of interest can reduce such NOE peaks.

The et-NOE method uses a standard NOESY experiment and relies on the difference in size of the target protein or complex and the ligands being investigated. In a standard NOESY experiment, the rate of NOE build-up is fast as the molecular correlation time $\tau_c$ increases (Figure 3.19a); large target proteins provide fast

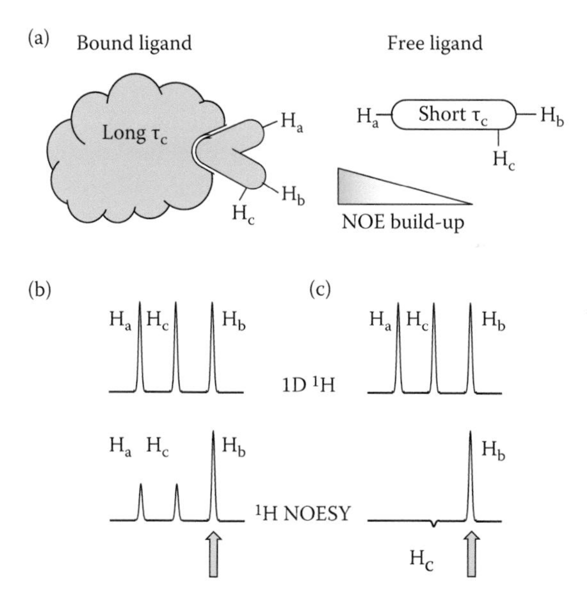


**FIGURE 3.19** (a) The exchange tr-NOE long correlation times associated with large molecules provide efficient NOE build-up (shown in grey), which extends from the protein to any ligand bound. Small molecules have shorter correlation times where smaller NOEs also take longer to build-up. Ligand binding can also alter its secondary structure, moving protons to closer proximity than would be possible in the free ligand state, as illustrated by the altered distance between protons '$H_a$' and '$H_b$'. NOE difference-based selective excitation of the ligand proton '$H_b$', as shown by the arrow, produces two different spectra in the presence (b) or absence (c) of target protein. In the presence of protein (b), excitation of proton b gives two positive NOE peaks, one to proton $H_c$, which is a near neighbour of $H_b$, and one to proton $H_a$, which, on binding, is brought closer to b. In the absence of the target protein, excitation of proton $H_b$ (c) results in only one negative peak corresponding to its near neighbour proton $H_c$, where the small, negative peak is a result of the short correlation time of the free ligand.

NOE build-up, but small ligands provide slow build-up. As already noted, there is a relationship between the size of a molecule and its correlation time, so in general terms, the larger the molecule, the larger the correlation time, and so the faster the build-up of NOE. Recall an issue with WaterLOGSY (Section 3.2.11) where data can be absent if $\omega\tau_c = 1.12$ ($\omega\tau_c \approx 1$) exists. This also applies to et-NOE, and some small molecules can fall into the transition between positive and negative NOE, which results in this null point. et-NOE occurs when the ligand binds to the target then dissociates in a manner such that the off-rate is greater than the longitudinal relaxation rate: $k_{off} > R_1$. In such cases, the NOE signals created when the ligand was bound to the target will remain with the free molecule. Fast exchange between bound and free ligands when combined with the large ligand excess provides a signal boost by providing many molecules in the system with et-NOE. NOE data collected in the presence of the protein target must be analysed together with the control experimental data. Although the occurrence of NOE peaks for free-ligand protons is limited under standard experimental conditions, the existence of bound orientation ligand et-NOE peaks can be confirmed by repeating the experiment in the absence of protein. If the et-NOE experiment is combined with selective saturation of a particular resonance, as in the NOE difference experiment (Section 3.2.3), the binding result can easily be observed through the ligand (Figure 3.19b).

Regarding off-rates, the same caveats as with the STD NMR experimental approach exist (Section 3.3.2) where a relatively tight-binding ligand will exhibit a $k_{off}$ outside the range of the et-NOE experiment. This will prevent the detection of ligand-bound NOEs of interest. Many experimentalists using the et-NOE method discuss the problems of spin diffusion in the collection and interpretation of et-NOE; this can create issues regarding the quantitative nature of et-NOE and how interpretation equates to specific distances between nuclei. These structural differences when a ligand binds are crucial for drug discovery. As mentioned in Section 3.3, spin diffusion is the transfer of magnetisation across neighboring atoms and is problematic as it alters NOE peak intensity and, thus, compromises distance estimations based on these. This is not just an issue for large proteins as spin diffusion through the ligand can also occur. There are a number of steps that can be taken to reduce the effect of spin diffusion, including modified pulse sequences with shorter pre-saturation and NOE mixing times.[65] It is best practice with any NOE-based experiment to collect data over a variety of saturation times for steady-state NOE–driven experiments or a variety of NOE mixing times for kinetic/transient NOE experiments to gauge whether build-up occurs in line with relative cross-relaxation rates for each nucleus under investigation.

#### 3.4.1.1 Illustrations of et-NOE Use

et-NOE has been used to study a number of different ligand-target interactions using both 1-D and 2-D NOESY experiments. Binding of small molecules to HSA has been investigated, showing not only NOE cross-peaks created upon the binding of ligand to HSA, but also peaks created between two ligands – caprylate and dansylglycine – which bind HSA in neighbouring sites. These ligands do not interact in the absence of HSA.[66] The interaction of a cyclic peptide with the sequence motif arginine–glycine–aspartic acid (RGD) to the cell surface receptor integrin α5β1 has been reported using $^{15}N$ isotopic enrichment of the ligand and a 2-D HSQC–tr-NOESY NMR (note that this is still exchange transferred) experiment.[67] Isotopic enrichment of the ligand allows for an edited pulse sequence that improves spin diffusion suppression by selectively irradiating enriched protons within the ligand. Additional NOE peaks are seen in the spectrum in the presence of integrin that, during structure calculation of the peptide, reveal an altered secondary structure to the free peptide. This bound structure will, therefore, aid in the design of improved ligands. Another application of et-NOE has been in the investigation of aggregation inhibitors in the formation of β-amyloid peptide fibrillation.[68] The approach exploited the inversion of signal that occurs between ligand protons in the presence or absence of the target macromolecule. Three possible aggregation inhibitors – melatonin, thioflavin T, and oleuropein – were subjected to specific proton excitation in 1-D NOESY experiments. Adjacent protons received NOE transfer; however, in the presence of amyloid aggregates under identical experimental conditions, the NOE peak inverted. This could only occur if the build-up of NOE had occurred while the ligand was bound to the aggregate and subjected to exchange tr-NOE magnetisation. A control experiment with sucrose, known to not bind amyloid fibrils, in the presence and absence of aggregates showed no such change in signal intensity. By completing these experiments using selective irradiation in 1-D, significant time is saved over the more complete 2-D NOESY experiments, an important point if the screening of large numbers of ligands is required.

### 3.4.2 Transferred RDC

RDCs are now used extensively in the determination of protein and nucleic acid structures. Once again, we will provide an overview of the types of systems that can be investigated by transferred RDC and explain some of the theory and sample preparation that is required to collect them. Further reading of a recent extensive review on the use of transferred-RDC experiments is highly recommended.[69]

#### 3.4.2.1 Basis of RDCs

A dipolar coupling occurs as the result of the interaction between the local magnetic fields of two adjacent nuclei (Figure 3.20a), the value of which is dependent on the internuclear vector and the external magnetic field strength. Under the usual solution-state conditions, these dipolar couplings are cancelled out by averaging due to the isotropic tumbling of the molecule. A small portion or RDC can, therefore, be reinstated by weak alignment of the molecule in solution. The alignment is termed 'weak' as only a small population, 1 in a 1000 molecules, needs to be aligned. The RDC value will then report on the angle between the vector and the external magnetic field. If multiple RDC values are collected for a molecule, long-range distance restraint information can be determined, which can be used alongside more traditional NOE distance restraints in structure calculation.

In the study of protein–ligand interactions, RDC measurements can be very useful. If the target protein is successfully aligned, then any interacting ligand will adopt the same overall alignment relative to the external field when it binds (Figure 3.20b). Any non-binding ligand will be free in solution with negligible RDCs formed. RDCs can

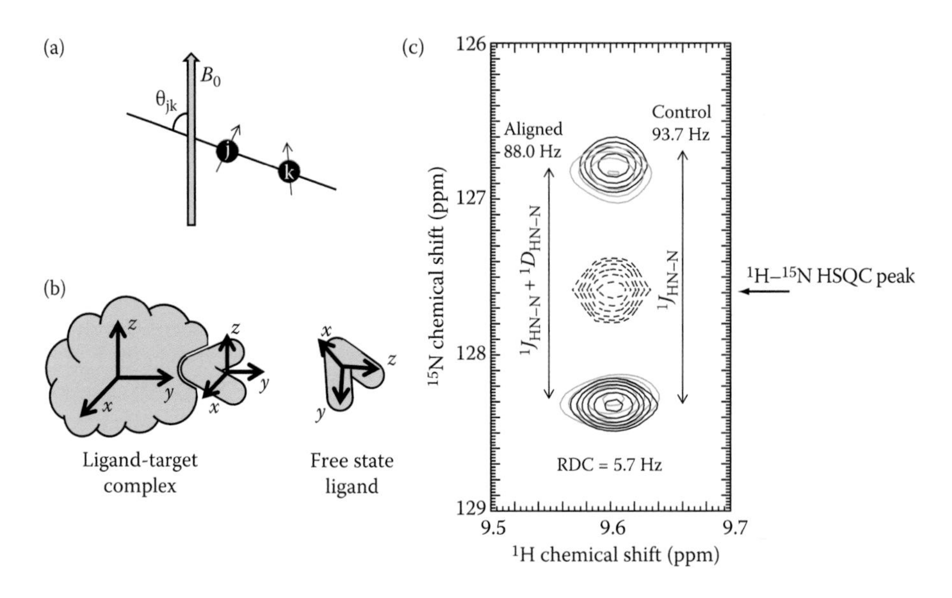


**FIGURE 3.20** (a) The transferred RDC. The RDC occurs in weakly aligned media between two adjacent nuclei, j and k, due to the influence of their respective magnetic moments in the presence of an external field $B_0$. (b) The internuclear vector with the field is shown as $\theta_{jk}$. The transferred RDC occurs when a ligand binds to a target molecule that is already weakly aligned; therefore, a different ligand alignment tensor is produced when complex or free in solution. (c) An example of $^1H_N$–$^{15}$N RDC measurement by subtracting the aligned scalar coupling from the grey resonances ($^1J_{HN-N} + {}^1D_{HN-N}$) from the control black resonances ($^1J_{HN-N}$). Each RDC measurement is centered on the standard HSQC resonance, and RDCs are obtained using a modified HSQC experiment to view couplings in the $^{15}$N (F1) dimension in the presence and absence of a weak-alignment medium.

be obtained for a number of different nuclei pairs, each with their specific internuclear vectors that provide unique RDC information that is structurally useful. For example, in a double-labelled protein sample (with $^{13}$C and $^{15}$N enrichment), HN–N, HN–CO, CA–CB, HA–CA, CA–CO, and CO–N can be measured. This also highlights the potential use of RDC as both a protein-observe and ligand-observe approach to ligand binding. The approach for measuring RDCs typically involves deconvolution of the small RDC ($^1D$) from a larger spin coupling ($^1J$) for the nucleus pair of interest such that two spectra are obtained in the presence and absence of alignment media reporting ($^1J + {}^1D$) and $^1J$, with the difference between these two data sets giving the $^1D$ RDC.

An example of a $^1$HN–$^{15}$N RDC data set is shown in Figure 3.20c. The protein $^1$H–$^{15}$N HSQC experiment mentioned is described further in Section 3.5.1 for those unfamiliar with its operation. The NMR experiment is a standard $^1$H–$^{15}$N HSQC, with the $^1$H decoupling removed during the $^{15}$N 2-D evolution period. This splits the single HSQC peak into two (black peaks), where the HSQC peak would be at their center, and the distance between the split peaks is the scalar coupling ($^1J_{HN-N}$). The same experiment collected in the presence of the alignment media gives a slightly different coupling (grey peaks) that is essentially [$^1J_{HN-N} + {}^1D_{HN-N}$]. The RDC values calculated will be both negative and positive; in Figure 3.20c, the RDC would be negative as $[^1J_{HN-N} + {}^1D_{HN-N}] < [^1J_{HN-N}]$. When your protein creates $^1H-^{15}N$ HSQC data that are dispersed, it is relatively easy to measure both $^1J_{HN-N}$ and $[^1J_{HN-N} + {}^1D_{HN-N}]$, but for crowded systems, you can use modified experiments that provide either the top or bottom resonance of the coupled pair. This approach, called 'in-phase, anti-phase' or 'IPAP'[70] is extremely useful, and most new spectrometers come with IPAP sequences as standard. An alignment or order tensor can be generated using this value because a vector will be aligned with or against the external magnetic field at the maximum positive value, and at 90° at the maximum negative value. An RDC value of zero occurs when the vector is aligned at 57.4° to the field, that is, the magic angle. Collecting more than one type of RDC is recommended because, in molecules where the studied couplings have a known geometric relationship, confidence and accuracy in structure determination can be gained.[71] There are a number of different alignment media that have been successfully used to collect RDCs, including polyacrylamide gels, filamentous phage, bicelles, lamelle, and lanthanide ions, and the media used should be chosen to best fit the protein–ligand system that is being investigated (Section 7.2).

#### *3.4.2.1.1 Alignment Media*

A variety of chemical and biochemical entities are used for the purpose of inducing partial alignment of molecules to enable RDC measurement. A number of these are described in the following.

*3.4.2.1.1.1 Polyacrylamide Gels* As an inert, charge-neutral media, polyacrylamide gels produce alignment based on steric hindrance, although, if necessary, charged gels can be made. A dried gel plug (5%–7% acrylamide) is hydrated with the protein and ligand in a buffer solution, and alignment is achieved by either squashing a gel of a diameter smaller than that of the NMR tube (with a standard Shigemi tube) or stretching a gel with a larger diameter into a specially adapted NMR tube (kits available at New Era Enterprises Inc., U.S.A.). The squash or stretch of the gel alters the shape of the cross-linked acrylamide structures within the gel, allowing a uniform, partial alignment to occur. The degree of alignment can be modified with higher- or lower-percentage gels. Larger proteins may need a smaller amount of acrylamide to prevent line broadening, but these gels are less robust than the higher-percentage gels. Advantageously, the protein and the ligand can be easily recovered from the gel by dialysis.

*3.4.2.1.1.2 Filamentous Pf1 Phage* A popular alignment media, phage, can be bought ready to use or produced in your laboratory by infection of a *Pseudomonais aeruginosa* culture with a small amount of phage. Typically, phage are used at a concentration of 5 to 10 mg/mL, and the strength of alignment is increased with higher concentrations. The phage align within the magnetic field because the uniformly packed α-helical coat proteins of these massive protein assemblies (up to 38 MDa) give them large anisotropic magnetic susceptibility and so naturally align with the external magnetic field. Alignment of the protein–ligand complex is both steric and electrostatic, which, again, can be modulated by altering the salt content of the buffer used. The phage are negatively charged, and as such, they provide a useful alignment media for negatively charged proteins, and RNA and DNA molecules. As with the polyacrylamide gels, protein–ligand sample is recoverable; however, the viscous nature of the phage samples make these much more difficult to prepare and handle. It is worth investing in a suitable positive displacement pipette if you plan working the Pf1 phage as an RDC alignment medium.

*3.4.2.1.1.3 Bicelles* The alignment of molecules using bicelles is done so by combining different ratios of the phospholipids DMPC (1,2-dimyristoyl-sn-glycero-3-phosphocholine) and DHPC (1,2-dihexanyl-sn-glycero-3-phosphocholine). At low temperatures, the lipid mixture is liquid and transparent, meaning that sample preparation must take place at a low temperature, preferably on ice. When the sample is placed in the magnet at a higher temperature, a phase change occurs within the lipid bicelle (phase change begins at ~25°C, and experiments are collected at 38°C–45°C), and the solution becomes viscous and slightly opaque and then becomes transparent at higher temperatures. This phase change creates the alignment of the bicelles, which occurs perpendicular to the external magnetic field. The alignment of the target protein–ligand complex is steric because of the bicelles' overall neutral charge. This was one of the first methods used to partially align molecules for RDC measurements, and bicelles formed from different ratios of phospholipids can be used to modulate the orientation of the target molecule.

*3.4.2.1.1.4 Lammelle* A relatively inexpensive alignment source, lammelle are actually polyethylene glycol ether (PEG) bilayers, formed by mixing the PEG with alcohol. They are prepared by dropwise addition of alcohol followed by vortexing, which will change a biphasic solution to a milky liquid, to a viscous, semi-translucent solution at the correct ratio. A commonly used combination is 3% to 5% $C_{12}E_5$ (N-dodecyl pentaethylene glycol ether) with hexanol, which has a neutral charge and is, therefore, suitable for most biomolecules. The lammelle can also be made with positive or negative charges by doping the sheets with charged lipids. Different lengths of PEG work best at different experimental temperatures, so it is worth experimenting to see which works best for your system.

*3.4.2.1.1.5 Lanthanides* Lanthanide ions, including terbium and thulium, have been used to create weak partial alignments for RDC measurements because they have anisotropic magnetic susceptibility. They are more complicated to use than more traditional media because of the need to bind them to the protein of interest either with a peptide tag added to the N- or C-terminus of your target protein or via a cysteine residue (native or mutated into the protein sequence). If you use a tag to orientate the target protein, it must be engineered to be rigid within the context of the rest of the protein; otherwise, alignment of the whole protein–ligand complex cannot be achieved. This, clearly, can create issues with potential N- or C-terminal tags as these regions of many proteins are normally unstructured and far from rigid.

### 3.4.2.2 Hints and Tips for Obtaining RDCs

To confirm that your sample has aligned in your medium of choice, you can check the effect of the medium on the NMR water signal within the sample. Water molecules will become aligned in the water shells surrounding the alignment structure created in the medium. As deuterium oxide is invariably present as a lock solvent, obtain a $^{2}H$ 1-D spectrum of your sample, and you should observe a double $^{2}H$ peak due to $D_2O$ alignment in the sample. When checking the result, it is worth noting that the greater the degree of alignment, the larger the splitting in the $^{2}H$ peak, and so the larger the RDC. An excellent example of this can be seen in the work of Prestegard et al.,[71] but it is easily recreated in your own spectrometer. However, as a word of caution, should you wish to use RDCs in the protein-observe mode, larger molecules in steric alignment media need a lower alignment; otherwise, their peaks become too broad to measure.

Some media, such as polyacrylamide gels, also provide their own resonance peaks in NMR experiments. This issue that, for example, manifests in both $^{1}H$ and $^{15}N$ NMR spectra of gel-based media is tantamount to interference on top of the data that you want to collect. The only circumvention of this issue is to use high concentrations (0.5–1 mM) of your protein.

Furthermore, each alignment medium has its own idiosyncrasies and will require optimisation for each system studied. As such, we recommend that you allow plenty of time to test data acquisition and perfect sample preparation before collecting data with precious protein–ligand samples.

Once you have collected and calculated your RDCs, there are a number of different software packages that can be used in the analysis of data. These include REDCAT[72] and MODULE,[73] which allow back-calculated RDC values from starting protein data bank (PDB) files to be calculated, compared, and reconciled with the experimental data and enable the relative orientation and structure of the target–ligand complex to be determined.

#### 3.4.2.3 Illustrations of RDC Use

There are many examples of RDC measurements being used to aid protein and macromolecule structure calculation, and the collection of target–ligand RDC measurements follows the same methodology. One example is in the analysis of the carbohydrate trimannoside binding to mannose-binding protein (MBP) (Section 7.2). The 53-kDa homotrimer protein MBP was enriched with $^{15}N$ and HN–N couplings measured in the presence and absence of bicelles to create the partial alignment. Natural abundance HC–C measurements were acquired for the three-ringed ligand in the presence and absence of aligned MBP.[74] The order tensors of both the protein and ligand were established, and a structure of the bound complex was derived.

N-acetyl-glucosaminyltransferase V (GnTV) is a possible drug target in the treatment of cancer because of its role in the production of branched oligosaccharides (Section 7.2). The control of branching affects cell invasion properties and the potential of metastasis, and GnTV ligand binding has been studied by et-NOE and STD.[75] Using phage as the alignment medium, the binding of two carbohydrate ligands, UDP-GlucNAc and βGlucNAc (1,2)αMan(1,6)βMan(OR), in adjoining sites to the protein GnTV was investigated using transferred RDCs.[76] Again, RDCs for the ligands were measured using natural abundance $^{13}C$. As with the design of new ligands for HSA facilitated by et-NOE data, this is an example of structure activity relationship determination (SAR) by NMR (Section 3.5.1).

The use of transferred RDCs has perhaps been less well exploited than exchange tr-NOEs because, unlike NOE, the RDC values for the ligand are more dominated by the free than the bound. With this in mind, the more traditional methods of RDC sample preparation have also been adapted to increase the bound state nature of ligands in RDC alignment. The adaptation requires a greater association of the target protein with the alignment media, which, in turn, allows the bound state of the ligand to increase in the average ligand RDC values. Two methods of enhancing the bound state have been proposed, the first by the association with bicelle medium being increased by the modification of the target protein with hydrophobic alkyl tail to the N- or C-terminus through modification of a cysteine residue.[77] Alternatively, alignment bicelles can be altered to contain nickel that can then bind His-tagged proteins. This circumvents the need to chemically modify the target protein as many proteins are expressed and purified with this kind of affinity tag.[78] Adapting the experiment with increased association must be handled with care as too much association will render the target molecule peaks too broad for accurate analysis.

## 3.5 PROTEIN-BASED OBSERVATION OF LIGAND BINDING

Once binding interactions have been identified as part of the drug discovery process, further investigations are usually carried out to determine whether ligands are binding competitively or cooperatively. This process applies to both natural and drug discovery ligands regarding the determination of the binding affinity of each 'hit'. Together with functional and pharmacokinetic studies, this will help determine if the new target has the desired inhibitory function required and if the required dose can be delivered in vivo to make it into a potential drug candidate.

NMR can play an important role in this stage of drug discovery via experiments that utilise chemical shifts. Monitoring and inferring from changes in chemical shift can be of great benefit. In the last section of this chapter, we introduce the detail of protein-observe methods as an alternative to the ligand-observe methods explained previously in Section 3.3. Advances in spectrometer hardware, such as higher field systems and the development of cryogenically cooled probes, provide increased resolution and sensitivity as well as opportunity to use tailored pulse sequences for specific experiments. This all coincides to create the possibility of obtaining high-quality NMR spectra of proteins at concentration levels in the micromolar range, with data acquisition taking only a few minutes.[79] As with screening using STD NMR and WaterLOGSY methods, cocktails of 10 or more ligands can be used in each protein-observe experiment followed by deconvolution of any of the cocktails that contain a binding ligand to find the 'hit'. As with ligand-observe experiments, this approach increases throughput dramatically and so places protein-observe NMR-based screening within the desired throughput range of true high-throughput screening methods.

### 3.5.1 Chemical Shift Perturbations in Drug Discovery

Protein-observe methods are frequently used to characterise ligand binding events due to the ability of such experiments to provide information on both the structural location of the binding site on the protein and binding affinities. The caveat to the former point is the need to have the NMR spectrum assigned and the structure of the protein solved; this is further discussed later. However, even when no chemical shift assignments are available to determine the location of the binding site, such experiments can provide binding affinity data and enable easy identification of competitive binding. As we all know, the NMR chemical shift reports on and is sensitive to the local electronic environment of each individual nucleus. If there is a change in the local environment, for example, during ligand binding, the chemical shift will report a change that is described and known as a secondary chemical shift ($\Delta\delta$); this shift is different from the chemical shift in the original state. An appreciation of the concept of nuclear shielding by electrons and factors that contribute to each chemical shift will help the reader to understand why the secondary shift occurs; such information is readily available in a host of standard NMR textbooks. Such factors include the effect of nearby paramagnetic nuclei, the anisotropy of neighbouring functional groups (ring current associated with aromatic groups being a special example of this), and the electrostatic effects of charged groups. All of these factors consequentially increase or decrease the local electron density, and so the level of shielding provided to the nucleus in question is altered, and this, in turn, has an effect on the chemical shift.

During ligand binding there is a high probability that a nucleus from the protein will be subjected to at least one of the above factors and so will experience a perturbation of its chemical shift. For example, hydrophobic binding sites often bind ligands containing aromatic rings that give rise to relatively large perturbations. Useful reviews on this subject that include summaries of several case studies are available.[11,80–82]

For those unfamiliar with protein NMR spectroscopy, the reader is introduced to perhaps the most common experiment used in NMR studies of proteins, 2-D $^{1}H$–$^{15}N$ HSQC, in which a correlation is observed between any $^{1}H$ directly bonded to a $^{15}N$ nucleus. The directly bonded correlation is provided by 'tuning' the HSQC experiment through the use of the $^{1}J_{HN-N}$ coupling constant. $^{1}H$–$^{15}N$ HSQCs lend themselves well to the determination of ligand binding sites in proteins for a variety of reasons. First, they usually provide well-dispersed spectra, that is, exploiting both $^{1}H$ and $^{15}N$ frequency dimensions. This does depend on the size/molecular weight of the protein in question and its folding; folded/globular proteins result in spectra with better chemical shift dispersion. Second, as the building block for proteins is the singly nitrogen (for the most part) containing amino acid, $^{1}H$–$^{15}N$ HSQC shows a single correlation (i.e. one resonance peak) for each residue. The amino acid proline is cyclic and does not possess a backbone NH group, and the amino acid side chains of tryptophan, glutamine and asparagine contain NH and $NH_2$ functionalities, which may also be detected. Under favourable conditions, arginine and histidine $NH_x$ groups can be seen, but they are normally in fast exchange (Chapter 2) with water protons and so broaden beyond detection. These side chain functional groups can be very useful as they are often involved in ligand binding. If they grace their presence in your $^{1}H$–$^{15}N$ HSQC, make the most of it, and use all side chains as well as backbone resonances to monitor ligand binding. Chemical shift changes upon ligand binding are easily observed in a $^{1}H$–$^{15}N$ HSQC by comparison with similar data recorded in the absence of ligand (Figure 3.21).[83] It is worth remembering that the natural isotope of nitrogen is $^{14}N$, and although $^{14}N$ is NMR active, it has a nuclear spin $I = 1$, and the relatively large quadrupole moment that is 10 times that of $^{2}H$ prevents this isotope from being useful for high-resolution NMR (Chapter 5). As a result, isotopic enrichment with the spin = 1/2 $^{15}N$ isotope is necessary because it is only 0.364% naturally abundant. Isotopic enrichment can be achieved using a variety of expression systems that recombinantly express the protein from DNA. Such organisms used include the popular *E. coli* and *P. pastoris* systems or even eukaryotic cells and cell free systems. Reviews that include a description of the basic approaches to isotopic enrichment for protein NMR spectroscopy are available.[84,85]

$^{1}H$–$^{15}N$ HSQC has another advantage as chemical shifts within are sensitive to hydrogen bonding and will report directly should the NH group in question be directly involved in hydrogen bonding. Furthermore, the NH groups will also be perturbed if the adjacent peptide bond carbonyl group from the preceding amino acid residue forms a new hydrogen bond with the ligand. It is important with any experiment to ensure that the pH is kept constant, and this is especially important with $^{1}H$–$^{15}N$ HSQCs. Protonation of any amide group involved in hydrogen bonds can cause a significant change to the chemical shift value[86] and can result in incorrect indicators of binding. $^{1}H$–$^{15}N$ HSQC is so sensitive that it can also detect ionisation changes of amino acid side chain groups with pKa values close to the operational pH range.

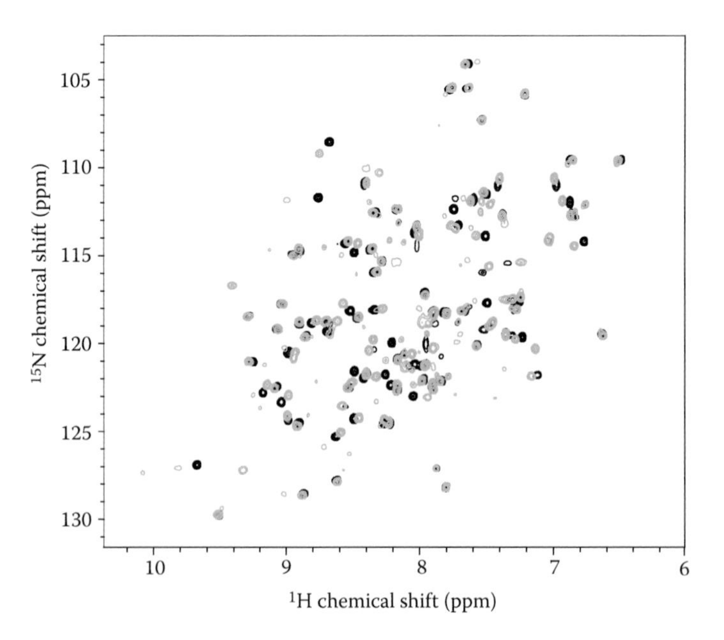


**FIGURE 3.21** An example of chemical shift perturbations $^{1}H$–$^{15}N$ HSQC of 0.4 mM unbound protein CobE substrate-carrier protein involved in cobalamin biosynthesis (black), overlaid with a second $^{1}H$–$^{15}N$ HSQC of CobE in the presence of 0.8 mM ligand hydrogenobyrinic acid (grey). Some resonances are seen to exhibit secondary shifts upon binding to the ligand; however, many remain unchanged, suggesting little or no overall change in structure.

Primary drug screening is possible using protein-observe methods. Chemical shift perturbations are elegantly employed in the methodology of 'SAR by NMR'[1] in which $^{1}H$–$^{15}N$ HSQC experiments are carried out to identify the primary target. These initial results are followed by experiments in which structurally similar ligands are investigated to optimise binding and help determine the binding orientation.[87] A second screen is then carried out to find ligands that bind in a second binding site. A second round of optimisation is carried out, and the binding site and orientation is determined, followed by linkage of the two individual ligands to form a large ligand with higher binding affinity. The development of this SAR-by-NMR approach by Fesik et al. was a landmark event when it was reported toward the end of the last century.[1]

Although $^{1}H$–$^{15}N$ HSQC is extremely helpful in characterising ligand binding events and is, by far, the most common approach used, it does not always lend itself well to reporting hydrophobic interactions, which have a smaller effect on the backbone NH chemical shifts. In addition, it is not able to detect interactions involving proline residues. In these situations, it may be useful to look at an $^{1}H$–$^{13}C$ HSQC. This approach is not employed as routinely as the $^{1}H$–$^{15}N$ HSQC as amino acids contain multiple correlations via CH, $CH_2$ and $CH_3$ groups that result in the protein $^{1}H$–$^{13}C$ HSQC being more complex than the $^{1}H$–$^{15}N$ counterpart. This complexity leads to spectral overlap and difficulty in interpretations. Also, while the chemical shift range is large, $^{13}CH$ chemical shift differences are usually smaller than those seen for $^{15}NH$ and generally more difficult to observe.[88] Also, obtaining full $^{13}C$ assignments for a protein is a much more arduous process and is rarely undertaken unless NMR-based structure determination is planned. For a typical 15 kDa protein, $^{1}H$–$^{15}N$ assignment can take as little as 1 to 2 weeks once data are collected, but complete $^{1}H$–$^{13}C$ assignment of a protein can run into months of analysis time. For those interested in the assignment process, we refer you to reference 89.

Over and above the inherent assignment problems, the necessity for isotopic enrichment is an additional issue. Modern spectrometer hardware and the high level of sensitivity that can be achieved using cryogenically cooled probe systems mean that, in favourable circumstances, it is possible to collect experiments without isotopic enrichment.[90] Band-selective optimised flip-angle short-transient (SOFAST)-HMQC is a method of acquiring 2-D data very quickly over a short period of total acquisition time, usually using recycle delays of around 100 ms without the necessity for isotope enrichment. They work because the SOFAST approach uses small flip angles to prevent saturation and loss of signal (cf. Section 3.2.1 and Figure 3.1).

An alternative $^{13}C$-based experiment that can give useful binding information is the HNCO; this provides a single correlation for the backbone of each amino acid and so does not suffer from spectral overlap. The correlations observed in the HNCO are between the amide $^{1}HN$ proton, $^{15}N$, and $^{13}CO$ of the peptide bond. This suggests a 3-D experiment; however, it can be reduced into 2-D correlating the $^{1}HN$ and $^{13}CO$, but beware, you still need the $^{15}N$ enrichment in addition to $^{13}C$ because the experiment requires transfer via $^{15}N$ to work. The carbonyl $^{13}CO$ chemical shift is, like the $^{15}NH$, sensitive to hydrogen bonding effects. Care must be taken in interpretation, however, as the CO group that the experiment reports on is for the (i-1) amino acid residue to the HN to which it is correlated.

The problem of spectral overlap can be somewhat reduced by the labelling of only methyl groups in residues such as valine, leucine, and isoleucine, employing clever use of synthetic metabolic precursors.[91–94] An equivalent experiment that works to reduce the number of resonances seen in a $^{1}$H–$^{15}$N HSQC to those of solvent-exposed amides only is the solvent exposed amides with TROSY (SEA-TROSY).[95] It is worth mentioning here, however, that as proteins get larger with increasing numbers of amino acid resonances, there is also a corresponding increase in transverse relaxation due to the increased number of protons that act as 'magnetisation sinks'. It, therefore, becomes increasingly necessary to produce proteins where many of the C$^{1}$H moieties are replaced by C$^{2}$H, thus decreasing the number of 'sinks'. This approach of deuteration of the protein adds more pressure on isotopic enrichment for protein detection methods, and although cumbersome and costly, it is sometimes the only way to achieve protein-observe with species with molecular weights around or in excess of 30 kDa.

Despite these challenges, the many different experiments available to observe proteins do provide different unique angles to view the binding interaction of a ligand. Ultimately, the most applicable route will depend on the protein under investigation, and so this section is merely a brief taste of the most common approaches.

### 3.5.2 Quantifying Chemical Shift Perturbations

For any of the 2-D experiments listed previously, such as the $^{1}$H–$^{15}$N HSQC, the chemical shift differences in each dimension, that is, the $^{1}$H and $^{15}$N, are usually combined by simple use of Pythagoras' theorem to calculate the 'diagonal'. This forms the basis of the process known as 'chemical shift mapping'. If assignments are available for only one of the forms, usually for protein without ligand, a nearest neighbour approach is used to represent a minimum threshold of the secondary shift.[96] In cases where there is a known structure of the protein under investigation, residues with a high level of chemical shift difference upon addition of ligand can be highlighted (or mapped) on the structure to indicate the location of the ligand binding site. What is defined as a 'significant' chemical shift difference for binding site mapping is open to interpretation, but this is usually achieved by statistical analysis of the data where significance is identified from mean and standard deviation, although there are several ways of identifying an appropriate cut-off value.[80] In performing the Pythagorian calculations, the question of how to weight the shifts from different nuclei arises; for example, a typical chemical shift range for amide protons is between 6 and 11 ppm – a spread of 5 ppm. However, a typical range for amide $^{15}$N is between 100 and 130 ppm, providing a spread of 30 ppm. This would suggest that $^{15}$N chemical shifts are adjusted by a factor of 1/6 (or 0.167) to make them comparable to the $^{1}$H shift range. There is no 'gold standard' way of treating these values, and so approaches often simply use an adjustment factor based on the ratio of the range of shifts for the protein being investigated as illustrated earlier. There is often a different adjustment used for glycine residues that have a different $^{15}$N chemical shift range from all other amino acids. An example of a combined approach would be to scale $^{15}$N chemical shifts to 0.14 for all amino acids except glycine, which is scaled to 0.2.[96] The general scaling factor approach leads to an equation of the form

$$\Delta\delta = \sqrt{(\Delta^1\text{H})^2 + \left(\frac{\Delta^{15}\text{N}}{\alpha}\right)^2} \tag{3.11}$$

where $\Delta\delta$ is the combined change in chemical shift, $\Delta^1H$ is the change in proton chemical shift in parts per million, $\Delta^{15}N$ is the change in $^{15}N$ chemical shift in parts per million, and $\alpha$ is the adjustment factor. Equation 3.11 can easily be adapted for use in $^1H$–$^{13}C$ HSQC experiments using an appropriate adjustment factor, for example, with $\alpha = 0.3$.[80] In addition, Equation 3.11 can be adapted to combine the results from several experiments.[97] Ultimately, the change in chemical shift in hertz can be used 'as is', which is equivalent to using a 0.1-ppm weighting factor for $^{15}N$ and a 0.25-ppm weighting factor for $^{13}C$. The key is to remain consistent in your approach when analysing related data sets.

### 3.5.3 Binding Affinities from Chemical Shift Perturbations

When chemical shift perturbations are observed across a series of experiments where ligand concentration is varied while protein concentration is kept constant, the result provides quantitative information on the binding affinity of the protein–ligand pair. Referring back to Section 3.3.2 and Equations 3.1 and 3.2 reminds us that the binding affinity is assessed by the dissociation constant, $K_D$. In the case of a simple 1:1 binding event, at equilibrium, $K_D$ is equal to the ratio of the off and on rates of ligand binding. As also discussed in Section 3.3.2, the on-rate is highly influenced by the probability of molecular collisions and so is related to the diffusion of both protein and ligand molecules in solution and can vary from around $10^4$ to $10^{11}$ $s^{-1}$ $M^{-1}$ for purely diffusion-controlled events.[11] For interactions that involve a partially buried binding site or require structural changes in the protein, the on-rate can be significantly different.

The magnitude of the off-rate is important in the quantification of binding events. Conditions of 'fast-exchange' (Chapter 2), where $k_{off}$ is much greater than the difference in the chemical shifts of free and bound proteins in hertz ($k_{off} >> \Delta Hz$), give rise to a gradual tracking of the resonance from the free/unbound position to that of the bound position upon addition of increasing amounts of ligand (Figure 3.22a and c). Under 'slow-exchange' (Chapter 2) conditions, where $k_{off} < \Delta Hz$, addition of ligand causes the resonance of the free protein to gradually decrease in intensity, with a corresponding increase in intensity of the resonance of the bound form (Figure 3.22b and d). The following paragraphs relate to conditions of fast exchange.

During a series of ligand titration experiments, known concentrations of protein, $P$, and ligand, $L$, are necessary. Under equilibrium, a proportion of each will form either bound or unbound/free species, as shown in Equations 3.12 and 3.13:

$$[L] = [L]_f + [PL] \tag{3.12}$$

and

$$[P] = [P]_f + [PL] \tag{3.13}$$

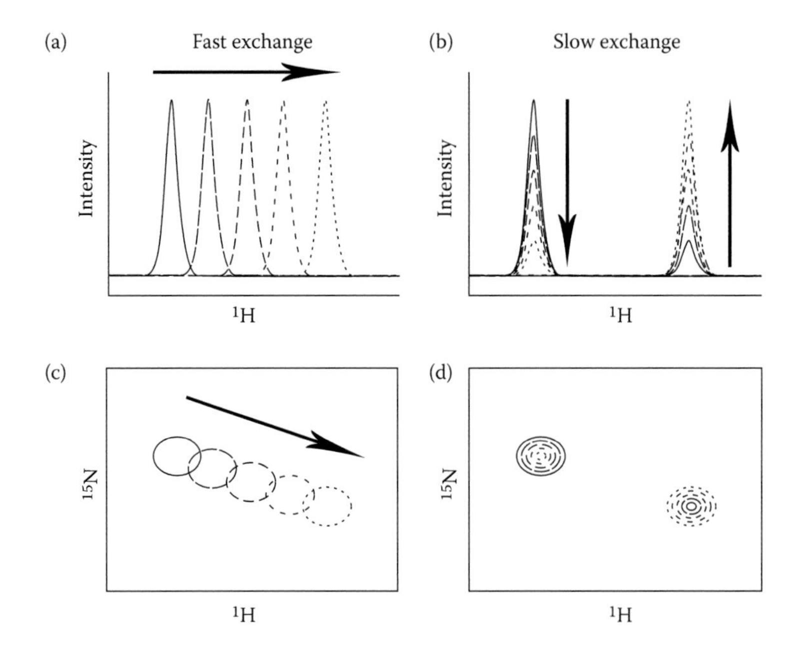


**FIGURE 3.22** The effect of increasing ligand concentration for fast (a and c) and slow (b and d) exchange rate binding events in both 1-D and 2-D experiments. Under fast exchange, peaks move gradually from the unbound position (solid line) to the bound position (fine-dotted line). In some cases where the exchange rate is nearing the frequency of the difference in shift, there may be some broadening of intermediate states. Under slow exchange conditions, the peak intensity of the unbound state decreases, whereas the peak intensity of the bound state increases upon addition to ligand. The arrows represent the direction of change.

Under conditions of fast exchange, the chemical shift position is a fraction-weighted average of the positions of the free and bound chemical shifts, so $\Delta\delta_{obs}$ is dependent on *PL*, which is itself determined by *L*, *P*, and $K_D$ such that

$$\Delta\delta_{obs} = \Delta\delta_{max} \frac{(K_D + [L] + [P]) - \sqrt{((K_D + [L] + [P])^2 - 4[P][L])}}{2[P]} \tag{3.14}$$

A plot of chemical shift difference against the ratio of ligand to protein will allow fitting of the results to obtain the $K_D$ value. Figure 3.23a shows theoretical plots created using Equation 3.14 where the same concentration of protein is used for titration experiments with ligands of varying $K_D$ values. When high-affinity ligands are investigated, the chemical shift seen nears the saturation point where all binding sites are occupied, and so the maximum chemical shift perturbation is easily determined, as in the case of the 5 μM curve. However, when lower-affinity ligands are under investigation, saturation point is hard to achieve, and the error in the fitted $K_D$ value is likely to be relatively large, as in the 500 μM curve.

In cases where a secondary binding event takes place at the same or different binding site with a second binding affinity, the trajectory (resonance tracking) of

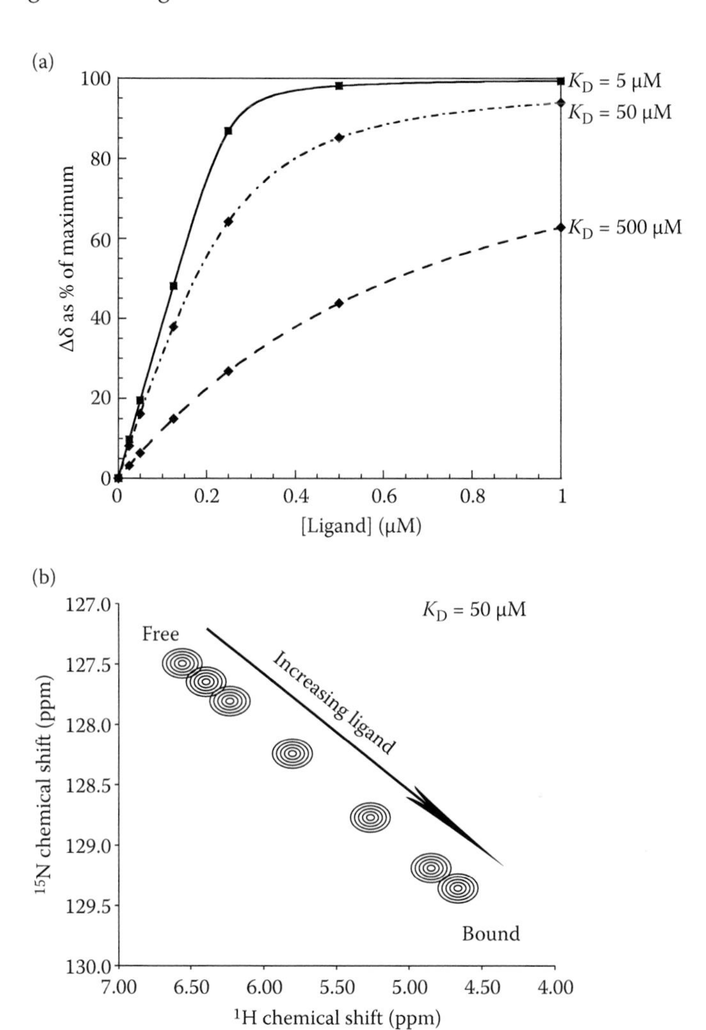


**FIGURE 3.23** (a) Theoretical curves for dissociation constant determination for titration of 0.25 mM protein with ligands of different $K_D$ against the chemical shift difference expressed as a percentage of the maximum. (b) Representative $^{1}H$–$^{15}N$ HSQC overlay of the $K_D = 50$ μM case, with a representative $\Delta\delta_{max}$ of 2 ppm in both $^{15}N$ and $^{1}H$ dimensions. Sampling positions are as represented by diamonds in (a). It can be seen that the chemical shift difference approaches its maximum value with the ligand:protein ratio of 4:1 with 1 mM ligand.

a resonance in the $^{1}H$–$^{15}N$ HSQC will show deviation from a straight line and can appear bent or even curve back on itself. Wright et al. have shown that it is possible to use singular value decomposition to determine the location of two distinct binding sites and quantify the $K_D$ of each site independently.[45]

Of course, the calculation of $K_D$ values from 1-D NMR spectra is theoretically possible where spectral resolution is sufficient to provide information on individual nuclei. This is often not the case in $^{1}H$ NMR spectra unless small ligand molecules are being investigated. Also, in many cases, the relaxation effect upon binding to large and relatively slowly tumbling protein targets can cause significant line broadening and complicate the interpretation of the data. In such cases, it may be possible to use the dependence of peak intensity on the added ligand concentration in a similar way to chemical shift perturbations to extract binding affinity information.

## 3.6 SUMMARY

The aim of this chapter was to introduce the concepts of using NMR spectroscopy to study ligand interactions with protein targets with a view to these methods being associated with drug screening and drug discovery. The majority of the chapter deals with the theory and practicalities of STD, including enhancements and related technologies such as multi-dimensional STD and WaterLOGSY. These are fundamental methods used in a wide variety of early-stage, fragment-based drug screening, but they can also be applied to specific protein target studies in academic and industrial settings. We hope that the overview of practical STD NMR, including working through the L-Trp/BSA system, provides a useful repository for you to try these experiments in your laboratory. The latter part of the chapter was designed to make you aware of other methods from both ligand-observe and protein-observe angles that you can utilise to enrich your knowledge and understanding of target–ligand interactions. Practically, and from experience, we have found that studies of protein–ligand complexes by NMR benefit from a variety of approaches, and it pays to try different methods that complement and support your hypotheses. We wish you the best of luck with your experiments but finish with the advice that practice makes perfect and be prepared for the odd disappointment and sleepless nights because, as we have mentioned, these experiments can be not only technically challenging from an NMR viewpoint, but also critical to the physical and interaction properties of the system that you are trying to study. If you can gain additional information regarding off-rates or $K_D$ of your binding event from the literature or from alternative methods such as isothermal calorimetry or surface plasmon resonance, it does give you confidence to find the molecular details using NMR spectroscopy.

## REFERENCES

1. Shuker, S. B. et al. 1996. Discovering high-affinity ligands for proteins: SAR by NMR. *Science* 274, p. 1531–1534.
2. Lepre, C. A. 2011. Practical aspects of NMR-based fragment screening. *Methods in Enzymology* 493, p. 219–239.
3. Ciulli, A. et al. 2006. Probing hotspots at protein–ligand binding sites: A fragment-based approach using biophysical methods. *Journal of Medicinal Chemistry* 49, p. 4992–5000.
4. Silvestre, H. L. et al. 2013. Integrated biophysical approach to fragment screening and validation for fragment-based lead discovery. *Proceedings of the National Academy of Science U.S.A.* 110, p. 12,984–12,989.
5. Ciulli, A. 2013. Biophysical screening for the discovery of small-molecule ligands. *Methods of Molecular Biology* 1008, p. 357–388.
6. Claridge, T. 2009. *High-Resolution NMR Techniques in Organic Chemistry*, 2nd ed. Oxford, United Kingdom: Elsevier, p. 21–22.

7. Piotto, M., V. Saudek and V. Sklenar. 1992. Gradient-tailored excitation for single-quantum NMR spectroscopy of aqueous solutions. *Journal of Biomolecular NMR* 2, p. 661–665.
8. Hwang, T. L. and A. J. Shaka. 1995. Water suppression that works: Excitation sculpting using arbitrary waveforms and pulsed field gradients. *Journal of Magnetic Resonance Series A* 112, p. 275–279.
9. Stott, K. et al. 1995. Excitation sculpting in high-resolution nuclear magnetic resonance spectroscopy: Application to selective NOE experiments. *Journal of the American Chemical Society* 117, p. 4199–4200.
10. Mayer, M. and B. Meyer. 1999. Characterisation of ligand binding by saturation transfer difference NMR spectroscopy. *Angewandte Chemie International Edition* 38, p. 1784–1788.
11. Meyer, B. and T. Peters. 2003. NMR spectroscopy techniques for screening and identifying ligand binding to protein receptors. *Angewandte Chemie International Edition* 42, p. 864–90.
12. Viegas, A. et al. 2011. Saturation transfer difference (STD) NMR: A simple and fast method for ligand screening and characterisation of protein binding. *Journal of Chemical Education* 88, p. 990–994.
13. Bhunia, A., S. Bhattacharjya and S. Chatterjee. 2012. Applications of saturation transfer difference NMR in biological systems. *Drug Discovery Today* 17, p. 505–513.
14. Peng, J. W., J. Moore and N. Abdul-Manan. 2004. NMR experiments for lead generation in drug discovery. *Progress in Nuclear Magnetic Resonance Spectroscopy* 44, p. 225–256.
15. Stockman, B. J. and C. Dalvit. 2002. NMR screening techniques in drug discovery and drug design. *Progress in Nuclear Magnetic Resonance Spectroscopy* 41, p. 187–231.
16. Fielding, L., S. Rutherford and D. Fletcher. 2005. Determination of protein–ligand binding affinity by NMR: Observations from serum albumin model systems. *Magnetic Resonance in Chemistry* 43, p. 463–470.
17. Mayer, M. and B. Meyer. 2001. Group epitope mapping by saturation transfer difference NMR to identify segments of a ligand in direct contact with a protein receptor. *Journal of the American Chemical Society* 123, p. 6108–6117.
18. Freeman, R. 1998. Shaped radio-frequency pulses in high-resolution NMR. *Progress in Nuclear Magnetic Resonance Spectroscopy* 32, p. 59–106.
19. Freeman, R. 1988. *A Handbook of Nuclear Magnetic Resonance*. Harlow, United Kingdom: Longman, p. 207–215.
20. Freeman, R. 1998. *Spin Choreography: Basic Steps in High-Resolution NMR*. New York: Oxford University Press, p. 143–186.
21. Homer, J. and J. K. Roberts. 1990. The use of a single non-acquisition pulse sequence to achieve equilibrium prior to the despot determination of $T_1$. *Journal of Magnetic Resonance* 87, p. 141–143.
22. Begley, D. W., S. X. Zheng and G. Varani. 2010. Fragment-based discovery of novel thymidylate synthase leads by NMR screening and group epitope mapping. *Chemical Biology and Drug Design* 76, p. 218–233.
23. Kemper, S. et al. 2010. Group epitope mapping considering relaxation of the ligand (GEM-CRL): Including longitudinal relaxation rates in the analysis of saturation transfer difference (STD) experiments. *Journal of Magnetic Resonance* 203, p. 1–10.
24. Freeman, R. and H. D. W. Hill. 1969. High-resolution studies of nuclear spin–lattice relaxation times. *Journal Chemical Physics* 51, p. 3140.
25. Vold, R. L. et al. 1968. Measurement of spin relaxation in complex systems. *Journal of Chemical Physics* 48, p. 3831.

26. Homer, J. and M. S. Beevers. 1985. Driven-equilibrium single-pulse observation of $T_1$ relaxation: A re-evaluation of a rapid new method for determining NMR spin–lattice relaxation times. *Journal of Magnetic Resonance* 63, p. 287–297.
27. Rama Krishna, N. and V. Jayalakshmi. 2006. Complete relaxation and conformational exchange matrix analysis of STD NMR spectra of ligand–receptor complexes. *Progress in Nuclear Magnetic Resonance Spectroscopy* 49, p. 1–25.
28. Lucas, L. H. et al. 2003. Transferred nuclear Overhauser effect in nuclear magnetic resonance diffusion measurements of ligand–protein binding. *Analytical Chemistry* 75, p. 627–634.
29. Ji, Z. S., Z. X. Yao and M. L. Liu. 2009. Saturation transfer difference nuclear magnetic resonance study on the specific binding of ligand to protein. *Analytical Biochemistry* 385, p. 380–382.
30. Cheng, Y. and W. H. Prusoff. 1973. Relationship between inhibition constant (K1) and concentration of inhibitor which causes 50 percent inhibition (I50) of an enzymatic reaction. *Biochemical Pharmacology* 22, p. 3099–3108.
31. Angulo, J. and P. M. Nieto. 2011. STD NMR: Application to transient interactions between biomolecules: A quantitative approach. *European Biophysics Journal* 40, p. 1357–1369.
32. Vogtherr, M. and T. Peters. 2000. Application of NMR-based binding assays to identify key hydroxy groups for intermolecular recognition. *Journal of the American Chemical Society* 122, p. 6093–6099.
33. Maaheimo, H. et al. 2000. Mapping the binding of synthetic disaccharides representing epitopes of chlamydial lipopolysaccharide to antibodies with NMR. *Biochemistry* 39, p. 12,778–12,788.
34. Haselhorst, T. et al. 2007. STD NMR spectroscopy and molecular modelling investigation of the binding of N-acetylneuraminic acid derivatives to rhesus rotavirus VP8* core. *Glycobiology* 17, p. 1030–1030.
35. Wagstaff, J. L. et al. 2010. Two-dimensional heteronuclear saturation transfer difference NMR reveals detailed integrin αvβ6 protein–peptide interactions. *Chemical Communications* 46, p. 7533–7535.
36. Dicara, D. et al. 2007. Structure–function analysis of Arg–Gly–Asp helix motifs in αvβ6 integrin ligands. *Journal of Biological Chemistry* 282, p. 9657–9665.
37. Dicara, D. et al. 2008. Foot-and-mouth disease virus forms a highly stable, EDTA-resistant complex with its principal receptor, integrin αvβ6: Implications for infectiousness. *Journal of Virology* 82, p. 1537–1546.
38. Wagstaff, J. L., M. J. Howard and R. A. Williamson. 2010. Production of recombinant isotopically labelled peptide by fusion to an insoluble partner protein: Generation of integrin αvβ6 binding peptides for NMR. *Molecular BioSystems* 6, p. 2380–2385.
39. Wagstaff, J. L. et al. 2012. NMR relaxation and structural elucidation of peptides in the presence and absence of trifluoroethanol illuminates the critical molecular nature of integrin αvβ6 ligand specificity. *RSC Advances* 2, p. 11,019–11,028.
40. Marchante, R. et al. 2013. Structural definition is important for the propagation of the yeast [PSI+] prion. *Molecular Cell* 50, p. 675–85.
41. Jackson, T. et al. 2000. The epithelial integrin αvβ6 is a receptor for foot-and-mouth disease virus. *Journal of Virology* 74, p. 4949–4956.
42. Povey, J. F. et al. 2007. Comparison of the effects of 2,2,2-trifluoroethanol on peptide and protein structure and function. *Journal of Structural Biology* 157, p. 329–338.
43. Claasen, B. et al. 2005. Direct observation of ligand binding to membrane proteins in living cells by a saturation transfer double-difference (STDD) NMR spectroscopy method shows a significantly higher affinity of integrin α(IIb)β(3) in native platelets than in liposomes. *Journal of the American Chemical Society* 127, p. 916–919.

44. Pereira, A. et al. 2009. Functional cell-based screening and saturation transfer double-difference NMR have identified haplosamate A as a cannabinoid receptor agonist. *ACS Chemical Biology* 4, p. 139–144.
45. Airoldi, C. et al. 2011. Saturation transfer difference NMR experiments of membrane proteins in living cells under HR-MAS conditions: The interaction of the SGLT1 co-transporter with its ligands. *Chemistry* 17, p. 13,395–13,399.
46. Bergeron, S. J. et al. 2008. Saturation transfer double-difference NMR spectroscopy using a dual solenoid microcoil difference probe. *Magnetic Resonance Chemistry* 46, p. 925–929.
47. Rademacher, C. et al. 2008. NMR experiments reveal the molecular basis of receptor recognition by a calicivirus. *Journal of the American Chemical Society* 130, p. 3669–3675.
48. Rademacher, C. and T. Peters. 2008. Molecular recognition of ligands by native viruses and virus-like particles as studied by NMR experiments. *Bioactive Lipids Conference* 273, p. 183–202.
49. McCullough, C. et al. 2012. Characterisation of influenza hemagglutinin interactions with receptor by NMR. *PLoS One* 7(7): e33958. doi:10.1371/journal.pone.0033958.
50. Haselhorst, T. et al. 2008. Avian influenza H5-containing virus-like particles (VLPs): Host-cell receptor specificity by STD NMR spectroscopy. *Angewandte Chemie International Edition* 47, p. 1910–1912.
51. Dalvit, C. et al. 2001. WaterLOGSY as a method for primary NMR screening: Practical aspects and range of applicability. *Journal of Biomolecular NMR* 21, p. 349–359.
52. Dalvit, C. et al. 2000. Identification of compounds with binding affinity to proteins via magnetisation transfer from bulk water. *Journal of Biomolecular NMR* 18, p. 65–68.
53. Hu, J., P. O. Eriksson and G. Kern. 2010. Aroma WaterLOGSY: A fast and sensitive screening tool for drug discovery. *Magnetic Resonance in Chemistry* 48, p. 909–911.
54. Ludwig, C. et al. 2008. SALMON: Solvent accessibility, ligand binding, and mapping of ligand orientation by NMR spectroscopy. *Journal of Medical Chemistry* 51, p. 1–3.
55. Dalvit, C. et al. 2001. Rapid acquisition of $^{1}$H and $^{19}$F NMR experiments for direct and competition ligand-based screening. *Magnetic Resonance in Chemistry* 49, p. 199–202.
56. Dalvit, C. and A. Vulpetti. 2011. Fluorine–protein interactions and F-19 NMR isotropic chemical shifts: An empirical correlation with implications for drug design. *ChemMedChem* 6, p. 104–114.
57. Lepre, C. A., J. M. Moore and J. W. Peng. 2004. Theory and applications of NMR-based screening in pharmaceutical research. *Chemical Reviews* 104, p. 3641–3676.
58. Chen, H. et al. 2013. F-19 NMR: A valuable tool for studying biological events. *Chemical Society Reviews* 42, p. 7971–7982.
59. Berggren, G. et al. 2010. Synthesis and characterisation of low-valent Mn complexes as models for Mn catalases. *Dalton Transactions* 39, p. 11,035–11,044.
60. Dalvit, C. et al. 2003. Fluorine NMR experiments for high-throughput screening: Theoretical aspects, practical considerations, and range of applicability. *Journal of the American Chemical Society* 125, p. 7696–7703.
61. Williamson, M. P. 2009. Applications of the NOE in molecular biology. *Annual Reports on NMR Spectroscopy* 65, p. 77–109.
62. Neuhaus, D. and M. P. Williamson. 2000. *The Nuclear Overhauser Effect in Structural and Conformational Analysis*, 2nd ed. Weinheim, Germany: Wiley-VCH, p. 178–185.
63. Balaram, P., A. Bothner-By and E. Breslow. 1973. Nuclear magnetic resonance studies of the interaction of peptides and hormones with bovine neurophysin. *Biochemistry* 12, p. 4695–4704.
64. Balaram, P., A. Bothner-By and J. Dadok. 1972. Negative nuclear Overhuaser effects as probes of macromolecular structure. *Journal of the American Chemical Society* 94, p. 4015–4017.

65. Post, C. B. 2003. Exchange-transferred NOE spectroscopy and bound ligand structure determination. *Current Opinion in Structural Biology* 13, p. 581–588.
66. Lucas, L. H., K. E. Price and C. K. Larive. 2004. Epitope mapping and competitive binding of HSA drug site II ligands by NMR diffusion measurements. *Journal of the American Chemical Society* 126, p. 14,258–14,266.
67. Zhang, L. et al. 2002. Receptor-bound conformation of an α(5)β(1) integrin antagonist by (15)N-edited 2-D transferred nuclear Overhauser effects. *Journal of the American Chemical Society* 124, p. 2862–2863.
68. Benaki, D. et al. 2009. Detection of interactions of the beta-amyloid peptide with small molecules employing transferred NOEs. *Journal of Peptide Science* 15, p. 435–441.
69. Jain, N. U. 2009. Use of residual dipolar couplings in structural analysis of protein–ligand complexes by solution NMR spectroscopy. *Micro and Nano Tech Bionalysis: Methods and Protocols* 544, p. 231–252.
70. Ottiger, M., F. Delaglio and A. Bax. 1998. Measurement of J and dipolar couplings from simplified two-dimensional NMR spectra. *Journal of Magnetic Resonance* 131, p. 373–378.
71. Prestegard, J. H., C. M. Bougault and A. I. Kishore. 2004. Residual dipolar couplings in structure determination of biomolecules. *Chemical Reviews* 104, p. 3519–3540.
72. Valafar, H. and J. H. Prestegard. 2004. REDCAT: A residual dipolar coupling analysis tool. *Journal of Magnetic Resonance* 167, p. 228–241.
73. Dosset, P. et al. 2001. A novel interactive tool for rigid-body modelling of multi-domain macromolecules using residual dipolar couplings. *Journal of Biomolecular NMR* 20, p. 223–231.
74. Jain, N. U., S. Noble and J. H. Prestegard. 2003. Structural characterisation of a mannose-binding protein–trimannoside complex using residual dipolar couplings. *Journal of Molecular Biology*, 328, p. 451–462.
75. Macnaughtan, M. A. et al. 2007. NMR structural characterisation of substrates bound to N-acetylglucosaminyltransferase V. *Journal of Molecular Biology* 366, p. 1266–1281.
76. Pierce, M. et al. 1987. Activity of UDP-GlcNAc:α-mannoside β(1,6) N-acetylglucosaminyltransferase (GnT V) in cultured cells using a synthetic trisaccharide acceptor. *Biochemical and Biophysical Research Communications* 146, p. 679–684.
77. Zhuang, T., H. Leffler and J. H. Prestegard. 2006. Enhancement of bound-state residual dipolar couplings: Conformational analysis of lactose bound to galectin-3. *Protein Science* 15, p. 1780–1790.
78. Seidel III, R. D., T. Zhuang and J. H. Prestegard. 2007. Bound-state residual dipolar couplings for rapidly exchanging ligands of His-tagged proteins. *Journal of the American Chemical Society* 129, p. 4834–4839.
79. Hajduk, P. J. et al. 1999. High-throughput nuclear magnetic resonance–based screening. *Journal of Medical Chemistry* 42, p. 2315–2317.
80. Williamson, M. P. 2013. Using chemical shift perturbation to characterise ligand binding. *Progress in Nuclear Magnetic Resonance Spectrometry* 73, p. 1–16.
81. Stockman, B. J. and C. Dalvit. 2002. NMR screening techniques in drug discovery and drug design. *Progress in Nuclear Magnetic Resonance Spectrometry* 41, p. 187–231.
82. Takeuchi, K. and G. Wagner. 2006. NMR studies of protein interactions. *Current Opinion in Structural Biology* 16, p. 109–17.
83. Deery, E. et al. 2012. An enzyme-trap approach allows isolation of intermediates in cobalamin biosynthesis. *Nature Chemical Biology* 8, p. 933–940.
84. Gardner, K. H. and L. E. Kay. 1998. The use of $^{2}H$, $^{13}C$, $^{15}N$ multi-dimensional NMR to study the structure and dynamics of proteins. *Annual Review of Biophysics and Biomolecular Structure* 27, p. 357–406.
85. Acton, T. B. et al. 2011. Preparation of protein samples for NMR structure, function, and small-molecule screening studies. In Lawrence, C. K., ed. *Methods in Enzymology*. London, United Kingdom: Academic Press, p. 21–60.

86. Tomlinson, J. H. et al. 2010. Structural origins of pH-dependent chemical shifts in the B1 domain of protein G. *Proteins* 78, p. 3000–3016.
87. Medek, A. et al. 2000. The use of differential chemical shifts for determining the binding site location and orientation of protein-bound ligands. *Journal of the American Chemical Society* 122, p. 1241–1242.
88. Carbajo, R. J. et al. 2005. Structure of the F1 binding domain of the stator of bovine F1Fo-ATPase and how it binds an alpha sub-unit. *Journal of Molecular Biology* 351, p. 824–838.
89. Cavanagh, J. et al. 2007. *Protein NMR Spectroscopy: Principles and Practice*, 2nd ed. London, United Kingdom: Academic Press, p. 781–817.
90. Quinternet, M. et al. 2012. Unravelling complex small-molecule binding mechanisms by using simple NMR spectroscopy. *Chemistry* 18, p. 3969–3974.
91. Goto, N. K. et al. 1999. A robust and cost-effective method for the production of Val, Leu, Ile (delta 1) methyl-protonated $^{15}$N-, $^{13}$C-, $^{2}$H-labelled proteins. *Journal of Biomolecular NMR* 13, p. 369–374.
92. Gardner, K. H. and L. E. Kay. 1997. Production and incorporation of $^{15}$N, $^{13}$C, $^{2}$H ($^{1}$H-δ1 methyl) isoleucine into proteins for multi-dimensional NMR studies. *Journal of the American Chemical Society* 119, p. 7599–7600.
93. Rosen, M. K. et al. 1996. Selective methyl group protonation of perdeuterated proteins. *Journal of Molecular Biology* 263, p. 627–636.
94. Hajduk, P. J. et al. 2000. NMR-based screening of proteins containing $^{13}$C-labelled methyl groups. *Journal of the American Chemical Society* 122, p. 7898–7904.
95. Pellecchia, M. et al. 2001. SEA-TROSY (solvent-exposed amides with TROSY): A method to resolve the problem of spectral overlap in very large proteins. *Journal of the American Chemical Society* 123, p. 4633–4634.
96. Williamson, R. A. et al. 1997. Mapping the binding site for matrix metalloproteinase on the N-terminal domain of the tissue inhibitor of metalloproteinases-2 by NMR chemical shift perturbation. *Biochemistry* 36, p. 13,882–13,889.
97. Schumann, F. H. et al. 2007. Combined chemical shift changes and amino acid-specific chemical shift mapping of protein–protein interactions. *Journal of Biomolecular NMR* 39, p. 275–289.

# 4 Diffusion
## *Definition, Description and Measurement*

*Scott A. Willis, Timothy Stait-Gardner, Amninder S. Virk, Reika Masuda, Mikhail Zubkov, Gang Zheng and William S. Price*

**CONTENTS**

4.1 Introduction ... 125
4.2 Measuring Diffusion Using NMR ... 129
4.2.1 Relaxation, Gradients and Echoes ... 129
4.2.2 Note on the Effect of Gradients ... 136
4.3 Diffusive Averaging Effects ... 136
4.3.1 Association and Polydispersity ... 136
4.3.2 Obstruction ... 139
4.3.3 Binding Equilibria ... 142
4.4 Diffusion and Ion/Molecule Radius in Solution ... 144
4.4.1 Stokes–Einstein–Sutherland Equation ... 144
4.4.2 Hydration and Solvation ... 145
4.4.2.1 Hydration of Non-Electrolytes ... 146
4.4.2.2 Hydration of Ions ... 146
4.4.2.3 Hydrated Radius by Experiment ... 154
4.5 NMR Measurement Interference ... 155
4.5.1 Eddy Currents ... 156
4.5.2 Field Inhomogeneities ... 158
4.5.3 Solvent Suppression ... 160
4.5.4 Convection ... 162
4.6 Concluding Remarks ... 163
References ... 163

## 4.1 INTRODUCTION

Studying the translational (diffusive) motion of molecules in solution provides immense amounts of information about the environment that the motion is occurring in as well as information about binding equilibria and any restrictions or boundaries present. Nuclear magnetic resonance (NMR) is a powerful tool for studying these motions, and this chapter will cover some of the basic aspects of NMR diffusion measurements and their application to studying free diffusion, obstructed systems, and systems where aggregation and binding events are present, and will also introduce some of the experimental complications and their resolutions. We start by describing what is meant by 'diffusion'.

Diffusion is the random translational motion of particles and, as such, is an incoherent motion. Hence, diffusion is differentiated from the coherent motion of molecules and bulk fluid flow (Figure 4.1). However, both can be successfully measured by NMR.[1] There are two common forms of diffusion: free diffusion (or self-diffusion) and mutual diffusion (Figure 4.1).[2]

Free diffusion (or self-diffusion) refers to the stochastic molecular motions at thermal equilibrium in pure/uniform solution and is characterised by a self-diffusion coefficient $D$ ($m^2s^{-1}$), which is related to mean-squared displacement (MSD), $\langle R^2 \rangle$, by

$$\langle R^2 \rangle = 2nDt \tag{4.1}$$

where $t$ is the time over which diffusion occurs and $n$ is the dimensionality of the system (i.e. 1, 2 or 3; for a single dimension, say, along the $z$ axis, the MSD might be written as $\langle Z^2 \rangle$).

In contrast to self-diffusion, mutual diffusion, which is also known as 'inter-diffusion', 'concentration diffusion' or 'transport diffusion', is where the molecular motion is driven by chemical potential gradients and concentration differences (Figure 4.1). NMR techniques, such as diffusion-weighted magnetic resonance imaging (MRI) and pulsed gradient spin echo (PGSE) NMR (see later), can be used to measure self-diffusion and MRI merely by observing the time dependence of concentration to measure mutual diffusion. Note that mutual and self-diffusion coefficients are only equivalent at infinite dilution (e.g. see Figure 1 in reference 3). Some typical values for

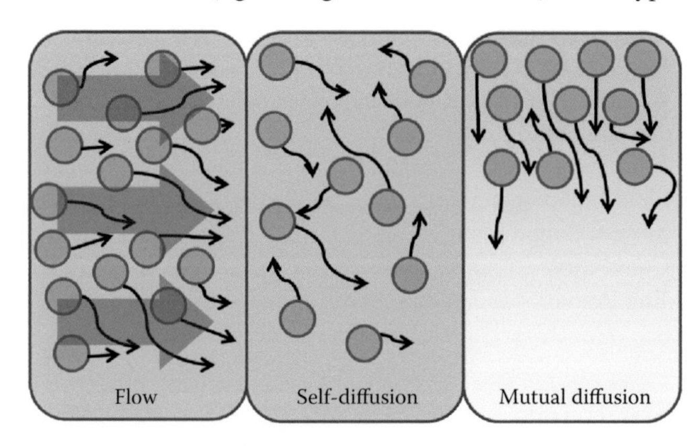


**FIGURE 4.1** Pictorial representation of the difference between flow, self-diffusion, and mutual diffusion. The large arrows overlaid on the depiction of flow show the direction of bulk fluid movement (i.e. flow). The uniform background color in the depiction of self-diffusion indicates a uniform solution, while the gradient color in the background of the depiction of mutual diffusion indicates a concentration gradient with the highest concentration at the top of the figure (i.e. darker in color).

self-diffusion coefficients are 2.5 to $0.5 \times 10^{-10}$ m$^2$ s$^{-1}$ (5000–115,000-g mol$^{-1}$ linear polyvinyl acetate in $CDCl_3$ at 25°C; extrapolated infinite dilution values),[4] $5.2 \times 10^{-10}$ m$^2$ s$^{-1}$ (sucrose in water at 25°C and infinite dilution),[5,6] $1.05 \times 10^{-10}$ m$^2$ s$^{-1}$ (lysozyme in 90%:10% $H_2O$:$D_2O$ at 25°C and 1.5 mM),[7] and $1.33 \times 10^{-9}$ m$^2$ s$^{-1}$ ($Na^+$ in water at 25°C and infinite dilution),[8–10] but these are dependent on temperature, solvent, and concentration, and would be different in restricted/obstructed systems, where binding equilibria or polydispersity result in averaging of the diffusion coefficients.

Self-diffusion can be divided into further categories: restricted, unrestricted, isotropic, and anisotropic. Restricted diffusion is diffusion that takes place in confining geometries, such as porous rock or biological cells, with the timescale of measurement long enough that molecular interactions with the boundaries are significant (Figure 4.2). As a brief note, the effects of restricting geometries depend on the

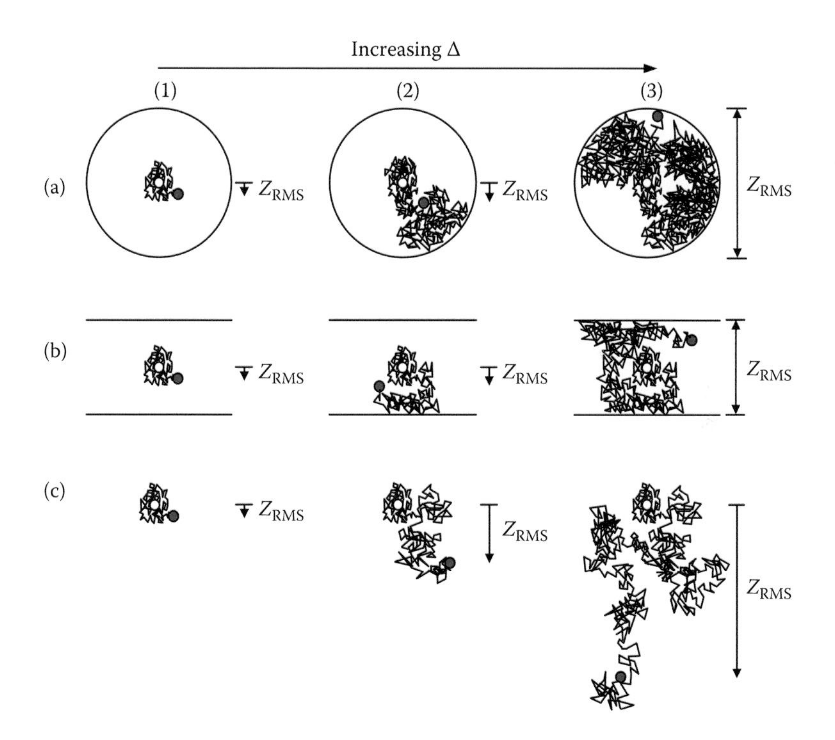


**FIGURE 4.2** A generalised view of restricted diffusion with increasing diffusion time and the resulting RMSD. (a) The restricting geometry is a spherical pore with reflecting boundaries (i.e. the molecule is not transported through the boundary or relaxed by the boundary). (b) The diffusion is restricted in the vertical direction by two infinitely long planes with reflecting boundary conditions. (c) The case for free isotropic diffusion. The effect on the RMSD along the $z$ axis (i.e. $Z_{RMS}$ or $\sqrt{\langle Z^2 \rangle}$) is shown for increasing diffusion time, $\Delta$, for the short time limit (1), intermediate time limit (2), and long time limit (3). The displacement is shown for the particle starting at the light circle and ending at the dark circle; note that, in the long time limit, the displacement shown is the maximum possible for the restricting geometry. The random Brownian motion is indicated with the erratic line, with the start and end positions of the molecule/particle shown as ○ and ●, respectively.

measurement timescale, which consequently affect the measured diffusion coefficient (and displacement). Examples of diffusion in a sphere (with impermeable and reflecting boundaries), diffusion between two infinitely long impermeable reflecting planes, and free diffusion are given in Figure 4.2. For free diffusion, the MSD varies linearly with time (Equation 4.1) and reflects the diffusion coefficient for all time.[11] However, for restricted diffusion, the displacement is not a linear function of time, and the diffusion coefficient is time dependent or dependent on the size and shape of the restricting geometry; therefore, the measured diffusion coefficient depends on the diffusion time ($\Delta$) and is referred to as the 'apparent diffusion coefficient'.[11–14]

By considering Figure 4.2, the effects of the restriction can be described for three timescales. The different timescales can be described by comparing the diffusion length (i.e. the root MSD, [RMSD] during $\Delta$, i.e. the square root of the MSD) and a length scale, which describes the structure,[11–14] for example, the radius of the sphere in Figure 4.2a or the separation of the planes in Figure 4.2b. The first timescale (Figure 4.2a (1), b (1), c (1)) is known as the short time limit,[12] or the unrestricted case where $\Delta$ is very short.[13] In this case, $\Delta$ is so short that the diffusing particles do not reach the boundaries, and so the diffusion coefficient is unrestricted and is the same as that measured for the free diffusion, which, at infinite dilution, is denoted $D_0$ (i.e. the MSD scales linearly with time; Equation 4.1).[11,12] The second timescale (Figure 4.2a (2), b (2), c (2)) is known as intermediate times,[12] or short times, with the diffusion length approaching the length scale of the restricting geometry.[11] In this case, the number of particles feeling the restriction increases with time, and the MSD does not scale linearly with time; as such, the measured diffusion coefficient is the apparent (time dependent/effective) diffusion coefficient, which is less than that for free unrestricted diffusion (i.e. $D_{apparent}(t) < D_0$).[11–13] Situations where the MSD does not scale linearly with time are also referred to as 'anomalous diffusion'.[12,14] $D_{apparent}(t)$ for intermediate times can be related to the restricting geometry properties, for example, surface-to-volume ratio of the pore[11–13] and surface relaxivity (i.e. enhanced decay of the signal in the NMR spectrum).[11,13] The final timescale (Figure 4.2a (3), b (3), c (3)) is the long time limit and can be considered in two ways: the pores are completely closed, and so no particles can escape, or the pores are open and connected to other pores.[11,13,15] For the long time limit with completely restricting geometry (e.g. the sphere with reflecting walls), the diffusion length is independent of the diffusion time and the diffusion coefficient in the unrestricted system (i.e. $D_{apparent}(t) \rightarrow 0$ or is $\ll D_0$ as $t \rightarrow \infty$; the MSD displacement for the unrestricted system will be much greater than the length scale of the restricting geometry), and the MSD reflects the length scale of the restricting geometry.[11–13] For the long time limit with connected pores (i.e. not complete restriction), there may be cases where the diffusion length is greater than the length scale of the restricting geometry.[12] In this case, $D_{apparent}(t)$ plateaus out, and the reduction from $D_0$ is related to the connectivity and tortuosity (a description of the path length around obstacles).[12,13] $D_{apparent}(t)$ can also be used to find the permeability of the boundaries.[11,13]

In isotropic systems, the diffusion behaviour is rotationally invariant with no preference for diffusion along one direction over another. Diffusion in such systems can be characterised by a scalar $D$. The diffusion in anisotropic systems, on

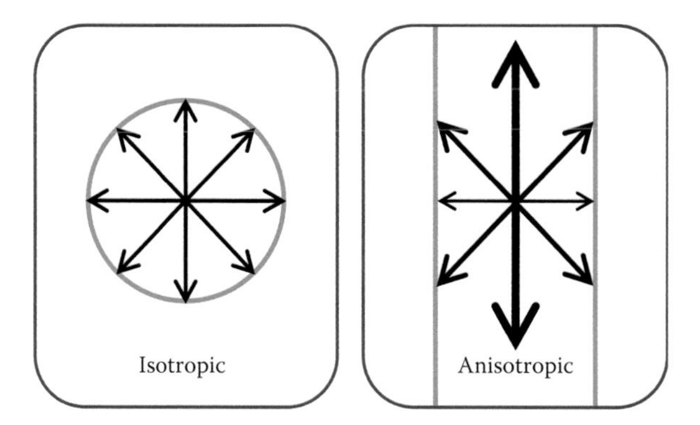


**FIGURE 4.3** Depiction of the difference between isotropic diffusion and anisotropic diffusion (simplified 2-D depiction of 3-D diffusion) with a boundary, such as a membrane, and the black arrows show the direction of diffusion, with the arrow thickness representing the size of the apparent diffusion coefficient along that direction (a thicker arrow indicates a higher apparent diffusivity).

the other hand, cannot be characterised by a single value, and diffusion shows a directional preference (Figure 4.3). It has to be characterised by a 3 × 3 symmetric matrix called the 'diffusion tensor'. Both isotropic diffusion and anisotropic diffusion can be found in systems exhibiting either free diffusion or restricted diffusion. For example, certain liquid crystals align with the magnetic field, and free diffusion within such a system shows a directional preference.

## 4.2 MEASURING DIFFUSION USING NMR

There are numerous methods available for measuring diffusion (i.e. other than NMR methods), and these are summarised in Table 4.1. To measure the translational self-diffusion coefficient of a chemical species, it is necessary to non-invasively label the molecules of interest so that their motion may be monitored. NMR provides a very powerful technique for doing this by enabling fine manipulations of the nuclear spin states of NMR-sensitive nuclei. Since a nuclear spin state has a truly negligible effect on the motion of its parent molecule, the labelling can be said to be non-invasive and does not perturb delicate thermodynamic interactions.

### 4.2.1 Relaxation, Gradients and Echoes

The two existing methods for measuring self-diffusion coefficients in the solution state using NMR are via $T_1$ (Chapter 1) relaxation measurements or via PGSE NMR (which is sometimes referred to as 'NMR diffusometry' or, especially when the results are presented in a 2-D mode, with the diffusion coefficient on one axis and chemical shift on the other, 'diffusion-ordered spectroscopy') methods. Relaxation measurements are favoured where diffusion occurs in the picosecond to nanosecond timescale, while PGSE NMR (Section 6.3.2) is favoured in longer timescales (millisecond to second) due to its robustness and better accuracy. Relaxation measurements

**TABLE 4.1**
**Summary of a Range of Methods Used for Measuring Diffusion and the Type of Diffusion Measured**

| Method | Type of Diffusion | Typical Concentration Range | References |
|---|---|---|---|
| Infinite couple | Mutual[a] | – | [16–18] |
| Capacity intermittent titration techniques | Mutual[a] | – | [19] |
| Secondary ion mass spectroscopy | Self[a] | – | [20–22] |
| Electron microprobe analysis | Mutual[a] | – | [22,23] |
| Auger electron spectroscopy | Mutual[a] | – | [24,25] |
| Diaphragm cell | Self and mutual[b] | Varies with time | [26,27] |
| Ultracentrifugation | Self and mutual | 1 mg/mL–5 mg/mL (varies with time) | [28,29] |
| Taylor dispersion | Mutual[b] | Varies with time | [16,30] |
| Spinning disc | Mutual[b] | Varies with time | [31] |
| Fluorescence spectroscopy | Mutual | 1 fM–1 μM | [32] |
| Attenuated total reflection infrared spectroscopy | Mutual | 0.1 mM–0.1 M | [33] |
| Guoy interferometer | Mutual | 0.1 mM–5 M | [34,35] |
| Rayleigh or Mach–Zhender interferometer | Mutual | 0.2 mM–5 M | [36–38] |
| Dynamic light scattering | Mutual[c] | Depends on the particle size (0.3 nm–100 μm) | [39,40] |
| Capillary methods | Self | 5 mM–0.2 M | [41,42] |
| Small-angle neutron scattering | Self[d] | 10 mg/mL–30 mg/mL | [43,44] |
| NMR/MRI diffusion | Self and mutual[e] | 1 mg/L upward and multi-component systems | [11,45–47] |

[a] Measures diffusion in solids.
[b] Accurate concentration difference between two species needs to be known.
[c] Depends on concentration fluctuations.
[d] Limited to a scale of 0.01 to 10 nM and single scattering events.
[e] Can measure diffusion coefficients in the range of $10^{-6}$ to $10^{-15}$ $m^2$ $s^{-1}$ in a mixture of species of various molecular weights (i.e. can measure the diffusion coefficients of multiple species in a mixture at once); only limitation is being able to detect a sufficiently strong NMR signal.

provide the reorientational correlation time, $\tau_c$, of a probe molecule, which depends on the molecular size via

$$\tau_c = \frac{4\pi\eta R_H^3}{3k_B T} \tag{4.2}$$

where $k_B$ is the Boltzmann constant (J $K^{-1}$), $T$ is the temperature (K), $\eta$ is the viscosity of the solvent (Pa s), and $R_H$ is the hydrodynamic (or Stokes) radius (m) of the diffusing particle and can be related to the self-diffusion coefficient of the molecule by using the Stokes–Einstein–Sutherland equation (i.e. Equation 4.22 or 4.23).

PGSE NMR methods for measuring diffusion are explained in some detail in the following in terms of NMR pulse sequences. Discussion of phase cycling[48] is eschewed as it does not add to a conceptual understanding of the sequences. However, it is cogent to phase-cycle the sequences as this eliminates artefacts originating from radio frequency (RF) inhomogeneity, imperfectly calibrated RF pulses, imperfections in the quadrature detector, and so on. To sense translational motion (i.e. to measure diffusion), spatial discrimination is needed, and for that reason, the NMR pulse sequences employ magnetic field gradients to distinctly label spins in different locations within the sample.

Various PGSE NMR pulse sequences exist for measuring diffusion; of these, the simplest and the most common pulse sequences are the PGSE (the Stejskal and Tanner sequence, a modified Hahn spin echo; Figure 4.4)[46] and the pulsed gradient stimulated echo (PGSTE)[49] sequences. The mathematical treatment of the spin echo attenuation data from this sequence was developed by Stejskal and Tanner in 1965.[46]

The PGSE pulse sequence shown in Figure 4.4 is composed of five important elements: (1) an initial $\pi/2$ RF pulse, (2) a gradient of duration $\delta$ and magnitude $g$ (T $m^{-1}$), (3) a waiting period $\Delta - \delta$ (in the middle of which is a $\pi$ pulse), (4) another gradient with the same duration and magnitude as the first, and (5) acquisition that begins at the echo maximum (note that the first half of the echo is not shown in Figure 4.4). Figure 4.5 gives a simplistic depiction of the events occurring during the PGSE sequence for three nuclear spin magnetic moments with and without diffusion. (An important note: Here, a $\pi/2$ pulse along the $x$ axis rotates $z$ magnetisation into the $-y$ direction; this convention is used in key texts that include a more mathematical description of NMR. The convention adopted in the remainder of this book, and indeed, in the majority of basic NMR texts, is that a $\pi/2$ pulse along the $x$ axis

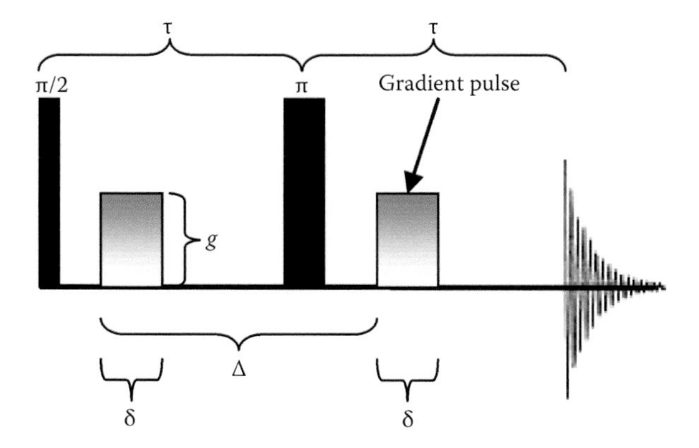


**FIGURE 4.4** The Stejskal and Tanner sequence (a modified Hahn spin echo).[46] The black rectangles represent the RF pulses in the NMR experiment. $\Delta$ is the diffusion time (s), $g$ is the strength of the gradient magnetic field (which is typically varied during the diffusion measurement) (T $m^{-1}$), $\delta$ is the duration of the gradient pulse (s), and $2\tau$ is the echo time (s), where the minimum echo time accessible is $(\Delta + \delta)/2$.

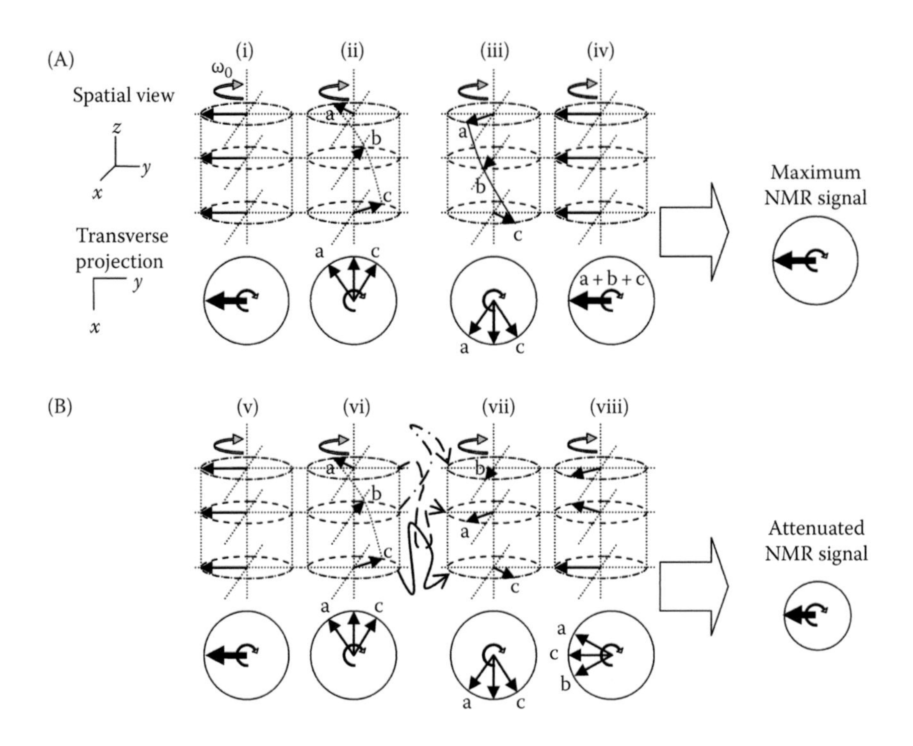


**FIGURE 4.5** A simplistic depiction of the events occurring during the Stejskal and Tanner sequence for three nuclear spin magnetic moments adapted from Price[14] and Johnson[50]. $B_0$ is in the positive $z$ direction, and all the pulses are in accordance with the sequence. Also, note that the diagram was done for the case of a negative gradient; however, if it was done for a positive gradient (as in the references), the same result is achieved. After the $\pi/2$ RF pulse, the net magnetisation precesses in the transverse plane (i and v); after the first gradient pulse, the phase angles are different for each position in the direction of the gradient (ii and vi); after $\Delta$ and the $\pi$ RF pulse, the phase angles are reversed (iii and vii). (A) Without diffusion during $\Delta$, the positions of each of the vectors (a, b, c) are the same after the $\pi$ RF pulse, and so, after the second gradient (identical to the first) is applied, all phase angles are refocussed, and the signal is maximum (iv). (B) With diffusion during $\Delta$, the positions of each of the vectors (a, b, c) is different after the $\pi$ RF pulse, and so, after the second gradient is applied, the phase angles are not completely refocussed, and the signal is attenuated (viii). In i, iv, v and viii, all spins have the same precession frequency (i.e. $\omega_0$), but for ii, iii, vi and vii, the precession frequency is spatially dependent (i.e. $\omega_0(z)$), and for a constant gradient, $\omega_0(z)$ is linearly dependent on $z$.

rotates the magnetisation into the $+y$ direction, is used elsewhere). In turn (Figures 4.4 and 4.5), (i) the RF pulse flips the magnetisation across the whole sample onto the $xy$ plane, (ii) the magnetisation is wound up into a helix by the gradient pulse, (iii) the molecules carrying the magnetisation diffuse during the period $\Delta - \delta$; also, the chirality of the helix is flipped by application of the $\pi$ pulse, (iv) the magnetisation helix will have attenuated as a result of diffusion, but to see the degree of attenuation, the helix must be unwound so that the signal can be acquired; this is the purpose of this gradient pulse (note that the effect of this gradient pulse is opposite to the first courtesy of the $\pi$ pulse), and (v) the signal is acquired. In the absence of diffusion, the effect of the second gradient pulse is to unwind the helix, ultimately giving a signal with attenuation resulting only from relaxation.

Steps i and v (in Figure 4.5) are straightforward, but an understanding of steps ii to v requires further explanation. During step ii, application of the gradient pulse (assume that it is along $z$) causes the transverse magnetisation to precess more slowly for lower $z$ and more quickly for higher $z$, the result being that the magnetisation is wound into a helix, with a pitch proportional to the 'area' of the gradient pulse (i.e. $\delta g$ for a rectangular pulse). During step iii, the molecules carrying the magnetisation diffuse for $\Delta - \delta$ and cause the helix to attenuate with the degree of attenuation dependent on the pitch of the helix – finely pitched helices will attenuate much faster due to the molecules having to diffuse only a small distance to reach areas where the magnetisation is substantially different, whereas during the same diffusion time, loosely pitched helices attenuate less.

Also, during step iii, there is a $\pi$ pulse that has the purpose of flipping the magnetization, turning a left-handed helix into a right-handed helix, and vice versa. This means that, during step iv, rather than winding the helix up further, the next gradient pulse will unwind the helix. The magnetisation is now in a coherent state and will induct a current in the receiver coil. Notice too that acquisition begins when all chemical shifts have been refocussed, at time $2\tau$. The sequence is repeated multiple times with different diffusion encoding gradient strengths, $g$. All other parameters are left unchanged, which means that the attenuation of the signal from transverse relaxation is identical for all iterations. The attenuation from diffusion, however, is different for the different gradient values for the reasons explained above, with higher gradients giving more finely pitched helices with resultant increase in diffusive attenuation. Even though the attenuation due to diffusion and the attenuation due to relaxation can be separated, the maximum timescale of the diffusion experiment is still dependent on relaxation.

The equations governing the evolution of nuclear magnetisation and diffusion in a magnetic field are the Torrey–Bloch equations,[51] solving them for the PGSE sequence (with rectangular gradient pulses), and calculating the resulting signal magnitude (i.e. $S(g, \Delta)$), normalised with respect to the signal for $g = 0$ (i.e. $S(0, \Delta)$) for a single, freely diffusing species allows the normalised signal attenuation, $E(g, \Delta)$, to be defined as

$$\begin{aligned} E(g,\Delta) = \frac{S(g,\Delta)}{S(0,\Delta)} &= \exp\left(-\gamma^2 g^2 D \delta^2 \left(\Delta - \frac{\delta}{3}\right)\right) \\ &= \exp\left(-4\pi^2 q^2 D \left(\Delta - \frac{\delta}{3}\right)\right) \\ &= \exp(-bD) \end{aligned} \tag{4.3}$$

where the $b$ term is known as the 'gradient or diffusion weighting factor' (i.e. $b = \gamma^2\delta^2 g^2(\Delta - \delta/3)$) or, simply, as the $b$ value. $q$ is used to characterise the gradient pulses and is the inverse of the pitch of the magnetisation helix, which is the area of the gradient pulse weighted by $\gamma/2\pi$ (i.e. $q = \gamma g \delta/2\pi$, with units of $m^{-1}$). Note that $S(g, \Delta)$

is the total signal at time $2\tau$ (i.e. $S(g, 2\tau)$) and that the term $S(0, \Delta)$ or $S(0, 2\tau)$ is the signal when the gradient strength is 0 T m$^{-1}$ and is proportional to the $T_2$ relaxation (Chapters 1, 3, 5 and 6) during the experiment for the PGSE sequence but will be different for other pulse sequences. The diffusion coefficient is the diffusion measured along the direction of the gradient vector, **g**, and, for isotropic samples, is the same when measured in any direction.

Generally, an NMR diffusion experiment consists of collecting spectra for different values of $g$, $\delta$, or $\Delta$, followed by non-linear regression (to avoid unequal weighting of noise) to the corresponding attenuation equation. As an example, to measure diffusion using PGSE, simply run the sequence a number of times, say, 16, for different values of $g$. Plot $E(g, \Delta)$ versus $b$, which gives a single exponential decay (for a single, freely diffusing molecule). Then, using a Levenberg–Marquardt non-linear least squares algorithm, fit $A\exp(-bD)$ to the data, with $A$ and $D$ being the fit coefficients; after fitting, $A$ should come out close to unity, and $D$ is the measured diffusion coefficient. The integral of the NMR peak is the preferred value to be fitted with the corresponding attenuation equation (Equation 4.3). Many other processing and experimental considerations can be found in the literature[11,50] as it is important to be aware of all potential sources of error so that artefacts or anomalies can be identified and rectified (or at least accounted for).

The values of the parameters $g$, $\delta$, $\Delta$ and $\tau$ to use in the experiment are relatively straightforward to choose. Typically, the duration $\delta$ is on the order of a few milliseconds. The highest gradient value should cause about 90% attenuation, which means that $\exp(-bD)$ should be approximately 0.1 for maximum $b$, and hence, $b_{max} = \gamma^2\delta^2 g_{max}{}^2(\Delta - \delta/3) \approx 2.3/D$. Thus, having a rough estimate of $D$ is useful before running the experiment. $\Delta$ should not be so long that the signal will have disappeared from $T_2$ relaxation; 100 ms is a good choice for many experiments; choosing anything on the order of a few milliseconds would be an appropriate choice for $\delta$, and this, thus, determines $g_{max}$. Equally spaced $g$ values between zero and $g_{max}$ are often chosen (while considered *comme il faut*, this is not *de rigueur*). Actually, for the $g = 0$ T m$^{-1}$ value, the sequence is not as robust, and there is more potential for inaccuracies from imperfections in the execution of the pulse sequence. It is better to start with a $g$ value that is close to but not quite zero, that is, 1% of $g_{max}$ is certainly sufficient.

While it is common to use rectangular gradient pulses in the NMR diffusion sequence, this is an idealisation. Perfect rectangular pulses are physically impossible due to gradient coil inductance, and thus, other gradient pulse shapes may be beneficial to the experiments. If non-rectangular gradient pulse shapes are utilised, then a correction to the attenuation equation must be included as the solution for the Stejskal and Tanner sequence also depends on gradient shape.[11,52,53] However, pulse shapes close enough to rectangular are achievable, and Equation 4.3 is perfectly acceptable.

Another closely related sequence is the PGSTE pulse sequence[49] (Figure 4.6). This differs in that the $\pi$ pulse is split into two $\pi/2$ pulses, which are moved as far apart from each other as possible. The sequence works much the same way except that the magnetisation is stored longitudinally for the bulk of the execution

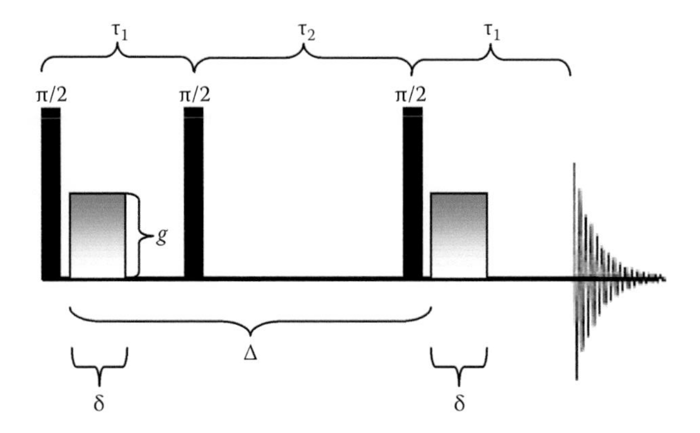


**FIGURE 4.6** The PGSTE pulse sequence.[49] $2\tau_1 + \tau_2$ is the echo time for this sequence, and $\tau_1$ is only limited by $\delta$, enabling the magnetisation to be stored longitudinally during $\tau_2$ (i.e. for most of $\Delta$).

of the sequence (i.e. during $\tau_2$) and is, therefore, only affected by longitudinal relaxation. The PGSTE pulse sequence[49] (Figure 4.6) is often recommended for macromolecules, or situations where $T_1 >> T_2$ and where $T_2$ is very short, so that the spin–spin relaxation becomes less of a problem for attenuation. All the parameters have the same definition as for Figure 4.4, and indeed, the normalised attenuation equation is identical to the right-hand side of Equation 4.3 at the time of acquisition.

For the case of diffusion in an anisotropic system of uniform alignment (e.g. uniformly aligned fibers, ellipsoidal compartments, cylinders), the PGSE NMR signal attenuation is dependent on the sample orientation or the direction of **g**.[11,54–56] Now, the diffusion coefficient is given by the diffusion tensor (e.g. see reference 57), and the gradient is described by the gradient vector (which need not be in the same direction as $B_o$), and so the attenuation equations (Equation 4.3) become:[11,58–62]

$$\begin{aligned} E(\mathbf{b}) &= \exp\left(-\gamma^2 \mathbf{g}^{\mathrm{T}} \cdot \mathbf{D} \cdot \mathbf{g}\delta^2 \left(\Delta - \frac{\delta}{3}\right)\right) \\ &= \exp(-\mathbf{b}:\mathbf{D}) \\ &= \exp\left(-\begin{pmatrix} b_{xx} & b_{xy} & b_{xz} \\ b_{yx} & b_{yy} & b_{yz} \\ b_{zx} & b_{zy} & b_{zz} \end{pmatrix} : \begin{pmatrix} D_{xx} & D_{xy} & D_{xz} \\ D_{yx} & D_{yy} & D_{yz} \\ D_{zx} & D_{zy} & D_{zz} \end{pmatrix}\right) \end{aligned} \tag{4.4}$$

where ':' is the generalised dot product; **b** is a symmetric matrix (like the diffusion tensor, **D**) and, for the case of rectangular gradient pulses, has the elements $b_{ij} = \gamma^2\delta^2 g_i g_j(\Delta - \delta/3)$, where $i$ and $j$ represent the $x$, $y$ or $z$ directions for Cartesian coordinates; and $\mathbf{g} = [g_x, g_y, g_z]^{\mathrm{T}}$. Other details of diffusion tensor analysis can be

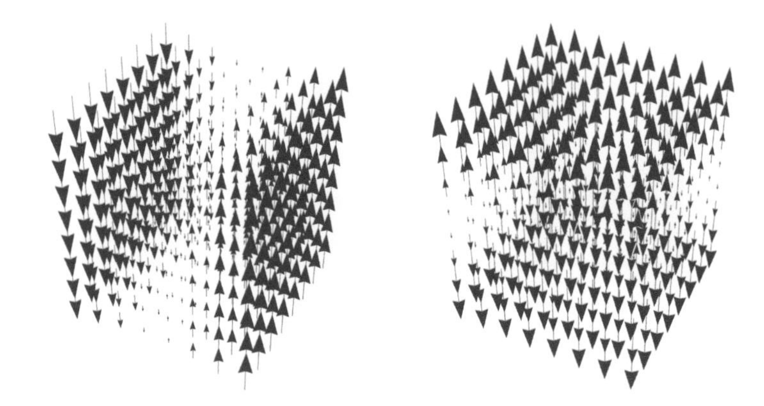

**FIGURE 4.7** A depiction of a magnetic gradient. The gradient is along the $x$ axis on the left and the $z$ axis on the right. Notice that the magnetisation always points along the $z$ axis. Only the $z$ component of the magnetisation varies when a gradient is applied, and the effect of this is to cause a linear variation in Larmor frequency along the gradient direction. The magnetic field represented by these arrows is superimposed on the static magnetic field (the arrow for which would be many orders of magnitude bigger but is still also pointing upward).

found elsewhere.[57] It is usually the case that anisotropic systems are not uniformly aligned, and there exists a large number of different orientation domains/anisotropic structures in the sample (i.e. a powder distribution).[14,63] In such cases, it becomes necessary to perform a powder average over all orientations (or assume that, for a uniformly oriented domain, the gradient direction is randomly oriented over the domain).[14,63]

### 4.2.2 Note on the Effect of Gradients

The magnetic gradient coil is located inside the NMR probe and, as its name suggests, it is used to create a constant gradient in the $z$ component of the magnetic field (recall that magnetism is a vector quantity); this gradient in the $z$ component is usually along the $z$ direction but can be along any other direction if triple axis gradients are used. Note that the gradient is in the $z$ component of the field but can be along any direction; Figure 4.7 makes this concept clear.

## 4.3 DIFFUSIVE AVERAGING EFFECTS

### 4.3.1 Association and Polydispersity

NMR diffusion measurements are a unique tool for studying the binding of biomolecules, ranging from self-associating systems to small ligand–protein associations at low concentrations (e.g. references 7, 64–68). Undoubtedly, while the technical abilities for measuring diffusion with modern NMR spectrometers have improved, a major change is in the realisation of the experimental limitations and the limitations of the currently applied models for analysing diffusion data from associating systems.[11,65,69]

Associating systems such as proteins and polymers tend to either occur as, or form with increasing concentration, polydisperse systems and this size polydispersity presents a particular problem to the analysis of diffusion measurements.[70–72] Since the diffusion coefficient of a molecule is related to its molecular weight through its effective hydrodynamic radius via the Stokes–Einstein–Sutherland equation,[73–76] diffusion measurements open up the possibility of working with mixtures of different species and polydisperse systems, although diffusion in polydisperse systems with the same chemical species are more difficult due to other averaging effects (e.g. references 70, 71). However, note that the Stokes–Einstein–Sutherland equation[73–76] is also subject to several assumptions, and the best choice of radius for simulations or predictions is not always that obtained from the usual forms of the Stokes–Einstein–Sutherland equation (see Section 4.4).[73–76]

For simple mixtures (with non-overlapping resonances in the NMR spectrum), the diffusion coefficient of each component in the mixture can be used to differentiate between different resonances, which can be used to resolve the spectra (i.e. the spectrum for each component is, in effect, deconvoluted).[77] In fact, every NMR diffusion method (e.g. PGSE) can be used to weigh the NMR spectra so that the resulting NMR spectrum shows selected components in a mixture (e.g. applying a diffusion weighting to a spectrum of large and small molecules can be used to remove the peaks associated with the smaller faster diffusing species. This method is particularly useful for studying macromolecules and other samples with multiple overlapping peaks.[78,79]) (Chapter 6). An example of an attenuation plot for overlapped resonances for two species is shown in Figure 4.8.

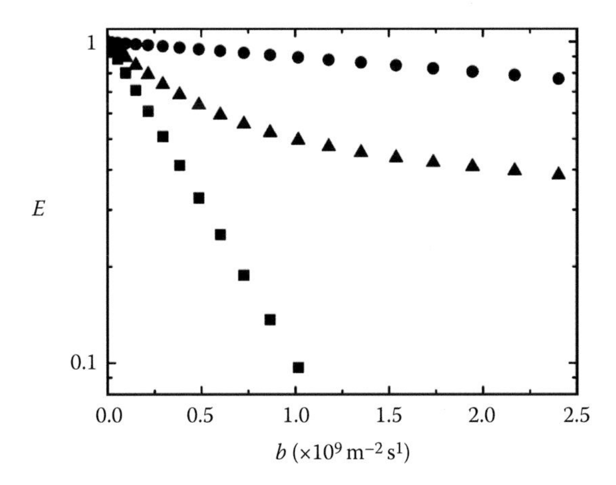


**FIGURE 4.8** Simulated $^{1}$H echo attenuation plots for a fast-diffusing species (■) such as water, with a diffusion coefficient of $2.3 \times 10^{-9}$ m$^2$ s$^{-1}$, and a slow-diffusing species (●) such as lysozyme, with diffusion coefficient of $1.1 \times 10^{-10}$ m$^2$ s$^{-1}$, and the resulting attenuation curve if the two individual attenuations are summed (▲), for example, if the resonances of fast- and slow-diffusing species are overlapped and in equal concentration. The simulation was performed using Equation 4.3, with $\gamma = 2.6752 \times 10^8$ rad s$^{-1}$ T$^{-1}$, $\Delta = 200$ ms, $\delta = 2$ ms and $g$ ranging from 0 to 0.2052 T m$^{-1}$.

The analysis of polydisperse system data is simple when different species appear as isolated resonances in the spectrum. However, in most real systems (e.g. aggregating proteins and other polymers), the resonances of different-sized species will overlap. The analysis of overlapped resonances in polydisperse systems is complicated.[11,70,71] In the case of a polydisperse sample (i.e. with species of different molecular weights), magnetisation of *i*th species is directly proportional to $MW_i n_i$, where $MW_i$ is the molar mass of the *i*th aggregate species and $n_i$ is the number of such aggregates present. The signal attenuation of $N$ freely diffusing species with individual diffusion coefficient $D_i$, and spin–spin relaxation time, $T_{2i}$, is given by

$$E=(g,\Delta)=\frac{S(g,\Delta)}{S(0,\Delta)}=\frac{\sum_i^N MW_i n_i \exp\left(\frac{-2\tau}{T_{2i}}\right)\exp(-bD_i)}{\sum_i^N MW_i n_i \exp\left(\frac{-2\tau}{T_{2i}}\right)} \tag{4.5}$$

Polydisperse systems such as proteins are further complicated by obstruction effects since the average spacing between molecules and aggregates is much less than the RMSD of the molecules over the timescale of the NMR diffusion measurements (i.e. $\Delta$), which means that the molecules will collide with one another. Lysozyme is the most studied protein, and its properties have been widely published (Figure 4.9).[7,40,44,64,67,80–87] When the average space between the molecules is smaller than the RMSD, there is a high probability for the molecules to collide numerous

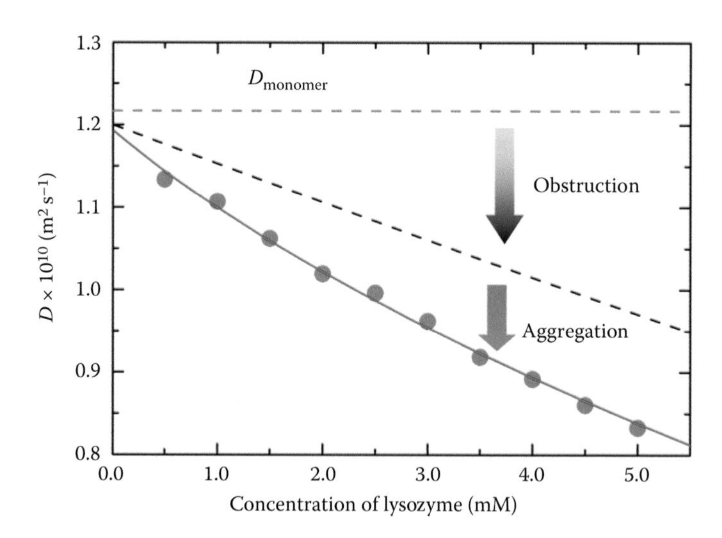


**FIGURE 4.9** An example of lysozyme diffusion decreasing with increasing concentration due to aggregation and obstruction effects. Shown is the experimental data (●), the diffusion coefficient for a monomer without obstruction or aggregation (- -), the predicted diffusion coefficient as per Tokuyama's obstruction model for monomeric lysozyme (- -) and a model including combined obstruction and aggregation effects (—).[7]

times during the measured diffusion time. These collisions will cause the measured diffusion coefficient of any oligomeric species to be reduced further (i.e. in addition to the change caused by aggregation) due to the obstruction.[7,64] The effects of obstruction are present even in the absence of aggregation and result in reducing the diffusion coefficients of the different species on the microscopic scale.[7,64] Including obstruction effects into Equation 4.5 but neglecting the relaxation terms (i.e. assuming that any difference in the relaxation properties of different-sized aggregates is negligible) gives the combined diffusion averaging effect for the different oligomers and is referred to as 'the weight-averaged diffusion coefficient', $\langle D \rangle_w^c$:

$$\langle D \rangle_w^c = \frac{\sum_i^N MW_i n_i D_i^c}{\sum_i^N MW_i n_i} \tag{4.6}$$

where the superscript $c$ denotes the effects of obstruction (i.e. obstruction from crowding).

In addition to microscopic averaging effects, the measured diffusion coefficient may be more strongly weighted to the smaller-sized aggregates since large aggregates or high–molecular weight oligomers become NMR invisible due to relaxation weighting, leaving only the small aggregates NMR visible.[7,64] Such NMR diffusion data can provide a means to study the change in aggregation over time.[7,64] However, for a truly polydisperse system, relaxation terms must be included. Therefore, when measuring diffusion in solutions with increasing concentration, it is necessary to be able to differentiate between the effects of association and obstruction to accurately model and account for changes in the diffusive behaviour.

### 4.3.2 Obstruction

Diffusion is affected by the presence of aggregates and other obstructions; for that reason, it is common to determine the ratio of the diffusion in the presence of obstructions, $D_{\text{Obstructed}}$, to the diffusion in the absence of obstructions, $D_{\text{Unobstructed}}$. This ratio is known as the obstruction factor, $O_D$. The nature of obstruction will determine what menaings $O_D$, $D_{\text{Obstructed}}$, and $D_{\text{Unobstructed}}$ have, and this is something that needs to be considered when describing obstruction factors (e.g. the structural obstruction factor would be the ratio of the diffusion along two axes in a system of periodic aggregates (e.g. aligned fibers), while the total obstruction factor would be the diffusion along one axis relative to the diffusion at infinite dilution (i.e. $D_o$), or there might be obstruction not only from a system of large aggregates (e.g. fibers), but also from the matrix surrounding them. Examples of systems where obstruction factors have been studied are crowded macromolecular solutions and gels,[7,11,64,88–98] aggregate systems/suspensions,[12,99–117] systems with periodic obstacles/gaps[118] and membrane systems.[119–125] Similar representations may also be found when describing diffusion in other porous systems.[11,13,103,126–136] Further to this, such ratios also appear when dealing with thermal (e.g. references 137–139) and electrical conductivity (e.g. references 139–141) since all three phenomena are described by the same physical laws.

A specific example of obstruction models for solutions of aggregating proteins (i.e. crowded solutions) is given here. The Stokes–Einstein–Sutherland equation only holds at infinite dilution, with large particle interactions being ignored and the solute considered to be much larger than the solvent molecules, which results in a one-body problem.[88] Similarly, in a macromolecular solution (e.g. cell cytoplasm, polymer solution, protein solution, emulsions) where the concentration of large particles is significant, interactions between the particles give a complicated many-body problem.[11] In this situation, smaller molecules (e.g. water, glucose) have to skirt around the larger and generally irregularly shaped, 'obstructing' molecules, which lengthens the diffusion path of the smaller molecule.[142,143] Obstruction from different molecules in the solution causes the diffusion coefficients to be reduced by obstruction factors (Note: different size molecules may have different obstruction factors), which, for crowded solutions, depend on the shape, volume fraction, and spatial and orientational distribution of the obstructing molecules.[88,101,144,145]

The mathematical problem of determining the obstruction factor for solutions of different-sized molecules/particles involves solving the diffusion equation under appropriate boundary conditions. Unfortunately, the diffusion path and coordinates of obstructing particles are very complicated. Despite their mathematical complexity, existing models used for accounting for obstruction are, nevertheless, simplistic compared to the physical reality. Approximate solutions are known for very simple cases, and most of them are true for only very low–volume fractions.[88,101,102,144–148] A number of different approximations using different models for obstruction in crowded solutions (i.e. solutions of different-sized species due to the aggregation of proteins) are shown in Figure 4.10.

Jönsson et al.[101] derived $O_D$ for a system containing evenly spaced, monodisperse particles in a cell including effects from obstruction and binding as

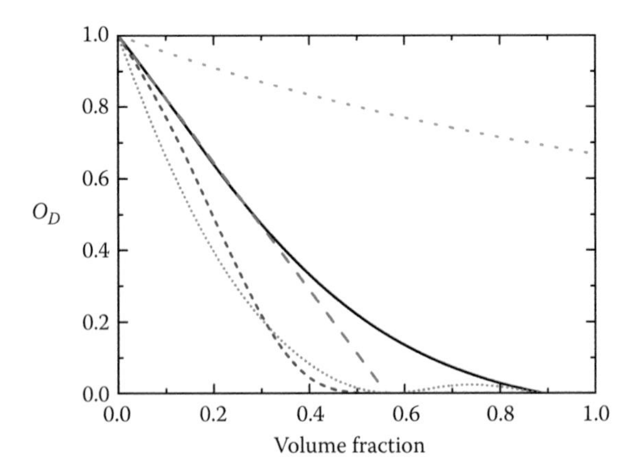


**FIGURE 4.10** Simulations of $O_D$ as a function of volume fraction of the diffusing molecule using the models of Jönsson et al.[101] (- - - - -), Han and Herzfeld[88] (- - - - -), Lekkerkerker and Dhont[148] (— —), Tokuyama and Oppenheim[144,147] short-range (——) and long-range (••••••) interactions.

$$O_D = \frac{1}{1-\left(1-\frac{C_1}{C_2}\right)v_p}\left(\frac{1-\beta v_p}{1+\frac{\beta v_p}{2}}\right) \tag{4.7}$$

where

$$\beta = \frac{D_2C_2 - D_1C_1}{D_2C_2 + 0.5D_1C_1} \tag{4.8}$$

$v_p$ is the volume fraction of monodisperse particles; and $C_i$ and $D_i$ (where $i$ is 1 or 2) are the concentration and self-diffusion coefficients, respectively, of spherical particles ($i$ = 1) and water/solvent surrounding the particles ($i$ = 2). If binding effects are neglected and $C_1 \to 0$, then Equation 4.7 reduces to

$$O_D = \frac{D(t)}{D_0} = \frac{1}{1+v_p/2} \tag{4.9}$$

Thus, the obstruction factor is 2/3 at $v_p = 1$ (i.e. when the system is completely filled with obstructing particles).

Clark et al.[146] modified the equations of Wang[145] and calculated the reduction in the solvent diffusion coefficient in the presence of macromolecules, but note that the following equations are corrected for the typographical error in Clark's original paper;[146] comparing Equations 4 and 6 in Clark's paper[146] and Equation 36 in Wang's paper[145] shows that the $f$ factor of these models should simply be as defined in Equation 4.11 and not $f = (1 - f)$, as implied by the typographical error in reference 146. Thus,

$$O_D = \frac{(1-\alpha' v_h)(1-f)}{1-v_h} \tag{4.10}$$

where the shape factor $\alpha'$ and the volume fraction of solvated macromolecule $v_h$ were defined by Wang,[145] and

$$f = \frac{w}{1-w}\delta_h \tag{4.11}$$

where $\delta_h$ is the grams of bound water per gram of anhydrous macromolecule, and $w$ is the weight fraction of anhydrous macromolecules in solution.[145]

Han and Herzfeld[88] derived a relationship for a hard sphere diffusing among hard spherocylinders:

$$O_D = \exp\left(-\frac{\Delta r}{R}\left(3\frac{v_p}{1-v_p} + \frac{9}{2}\frac{v_p^2}{(1-v_p)^2} + \frac{9}{4}\frac{v_p^3}{(1-v_p)^3}\right)\right) \tag{4.12}$$

where $\Delta r$ is the step size and $R$ is a radius of spherocylinder.

Tokuyama and Oppenheim[144,147] used the Navier–Stokes equations to develop a model for the dynamics of the concentrated hard-sphere suspensions of interacting Brownian particles, with both hydrodynamic and direct interactions. The obstruction factor for both short- and long-range interactions were, respectively, derived as

$$O_{D,\text{Short}} = 1/[1 + H(v_p)],3 \tag{4.13}$$

and

$$O_{D,\text{Long}} = \frac{\left(1 - \frac{9v_p}{32}\right)}{1 + H(v_p) + \frac{v_p/v_p^0}{\left(1 - v_p/v_p^0\right)^2}} \tag{4.14}$$

The exact form of the function $H(v_p)$ is given in references 144, 147.

Most of these models have considered diffusing particles as hard spheres or as spherocylinders, but it is important to note that none of these models have accounted for the presence of aggregation processes. In reality, all diffusing particles are neither spherical nor ellipsoidal. Molecular shapes of diffusing particles are far more complex than spheres or ellipsoids, and often, the effects from hydration, charge, and the rugosity of the surface need to be considered. Furthermore, it is very difficult to do a 'control' experiment to separate the effects of obstruction from aggregation/association. Therefore, better models of obstruction must be established to allow for the reduction in the diffusion coefficient resulting from association to be disentangled from that resulting from obstruction; otherwise, the full information content of the experiment is not accounted for. The simplest and most general approach is the numerical solution of the obstruction factor; however, this requires an appropriate choice for the shape/radius of the diffusing species to be made (see Section 4.4).

### 4.3.3 Binding Equilibria

Diffusion is also affected by any binding equilibria that may be present. For a two-site binding/exchange model (Figure 4.11) with either large diffusion paths (for long-range diffusion behaviour in systems with two different regions) or fast exchange (between free and bound sites; Chapters 2 and 3), the diffusion coefficient is averaged between the two diffusion coefficients (but note that it may be averaged among more than two populations). For binding events, these are the bound and free diffusion coefficients (i.e. in the absence of any other obstructions/restrictions/reductions) and is given by[11,136,149–159]

$$D_{\text{Pop}} = (1 - P_B)D_F + P_B D_B \tag{4.15}$$

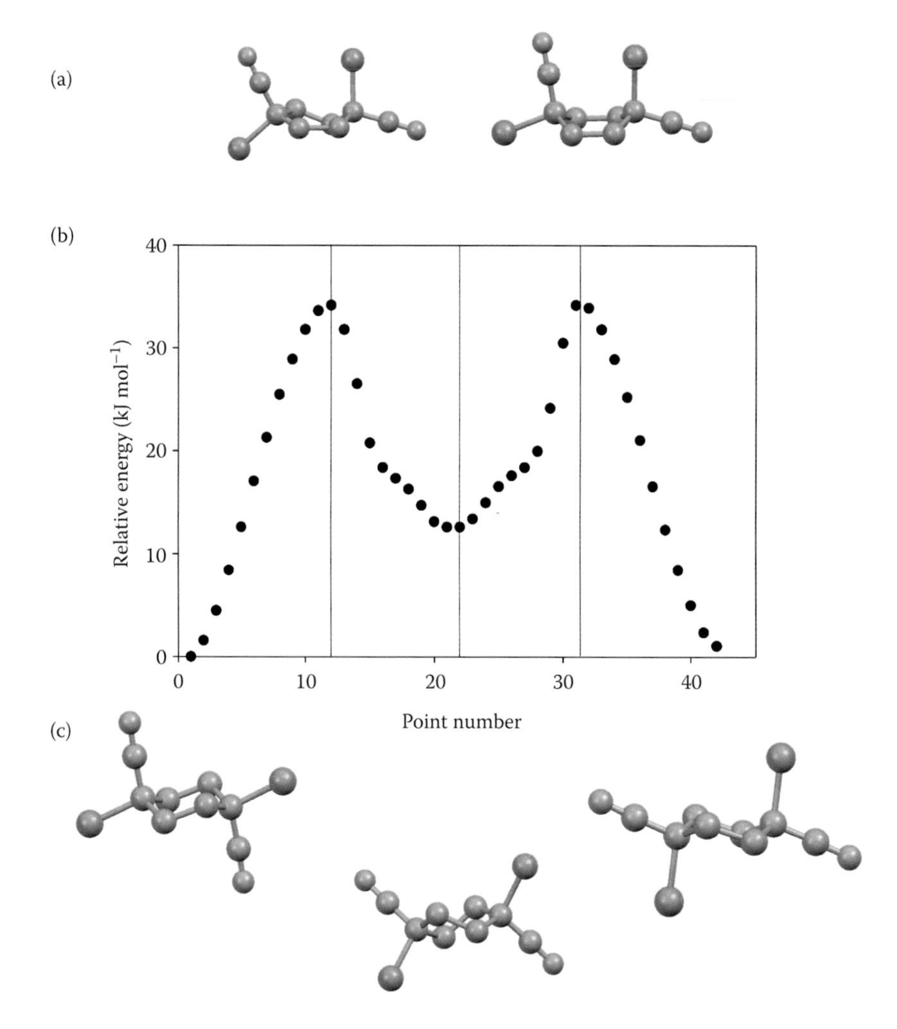


**FIGURE 2.4** Schematic of the reaction coordinate for the exchange in trans-1,4-dibromo-dicyanocyclohexane. Each point represents a geometry along the reaction coordinate, and the vertical line in the middle of the plot shows the locations of the two transition states and the metastable intermediate. (a) Shows the geometries of the two transition states, (b) plot of energy along the reaction coordinate (minimum energy pathway) and (c) shows the geometry of the initial states, the intermediate and final state. (Reprinted with permission from Bain, A. D., M. Baron, S. K. Burger, V. J. Kowalewski and M. B. Rodriguez. Interconversion study in 1,4 substituted six-membered cyclohexane-type rings: Structure and dynamics of trans-1,4-dibromo-1,4-dicyanocyclohexane. *Journal of Physical Chemistry* A 115, pp. 9207–9216. Copyright 2011 American Chemical Society.)

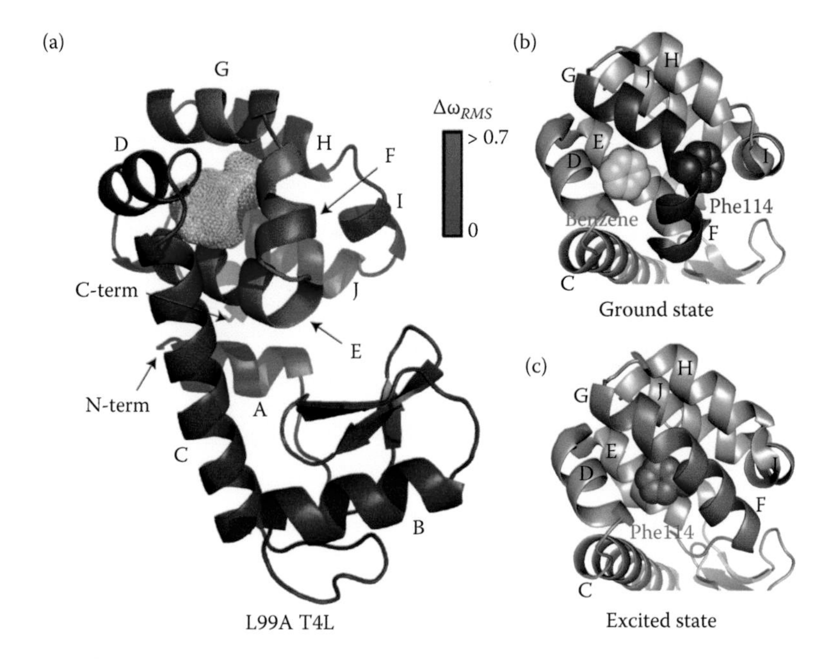


**FIGURE 2.5** An example of unequal populations in mutants of T4 lysozyme (T4L) and the detection of small populations of the minor site. (From Sekhar, A. and L. E. Kay, *Proceedings of the National Academy of Sciences* 110, pp. 12,867–12,874, 2013.) (a) Ground-state X-ray structure of the L99A mutant, color-coded according to the magnitude of chemical shift differences between the ground and excited states. The grey mesh delineates the cavity that results from the L99A mutation. Comparison of the C-terminal domain of L99A T4L in the ground (b) and excited (c) states, highlighting the different orientations of the F and G helices in each of the conformers. The F114 side chain that rotates into the cavity in the excited state and the bound benzene in the ground state are shown using space-filling representations.

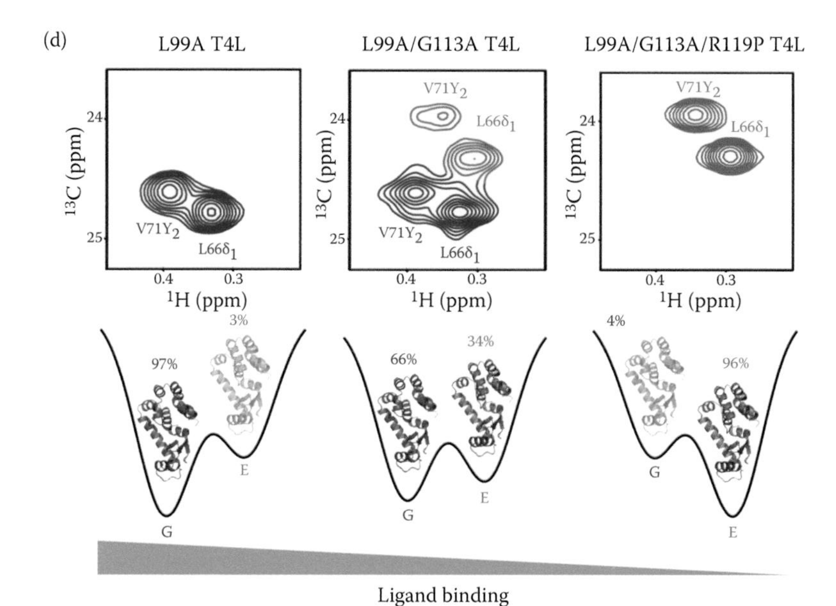


**FIGURE 2.5** (Continued) An example of unequal populations in mutants of T4 lysozyme (T4L) and the detection of small populations of the minor site. (From Sekhar, A. and L. E. Kay, *Proceedings of the National Academy of Sciences* 110, pp. 12,867–12,874, 2013.) (d) The ground and excited state populations can be manipulated by a small number of mutations. Ground (G, left) and excited (E) states become comparable in population in the L99A/G113A construct (center), as seen in 13C–1H correlation spectra. The new set of peaks (red) has chemical shift values that are in excellent agreement with those obtained for state E from fits of the relaxation–dispersion experiments on L99A. (Reprinted by permission from Macmillan Publishers, Ltd.: *Nature (London)*, ref. 20, copyright 2011; From Dr. Ashok Sekhar and Dr. Lewis E. Kay.)

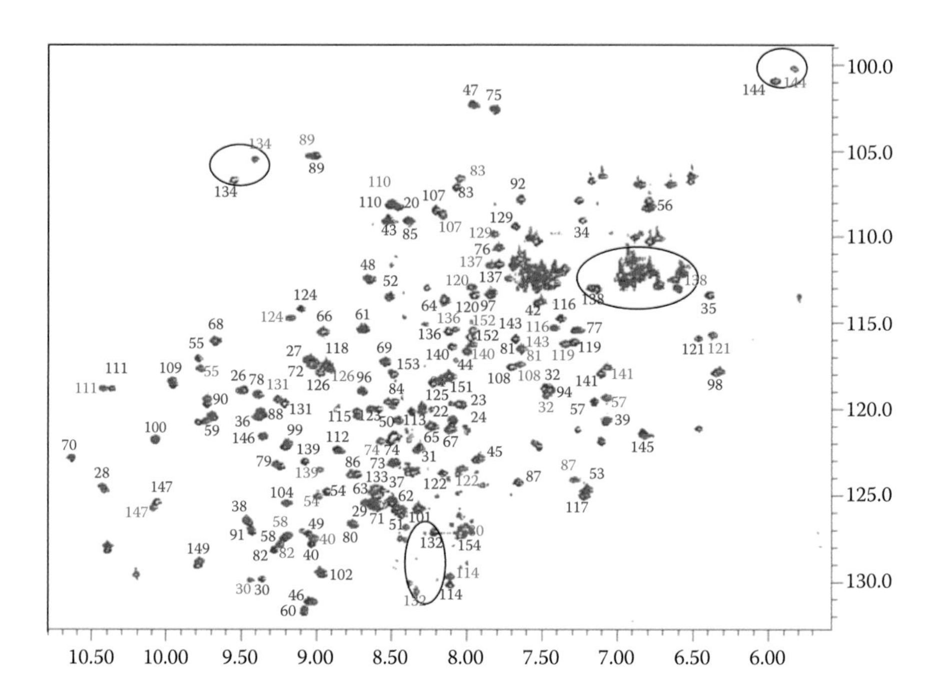


**FIGURE 7.24** The $^{15}N$–$^{1}H$ HSQC spectra of free (blue) and bound (red) FGF-1 are overlaid. Areas with the largest perturbation, indicating greatest contact with the hexasaccharide, are circled. (Reproduced from Canales-Mayordomo, A. et al., *Journal of Biomolecular NMR* 35, pp. 225–239, 2006. With permission.)

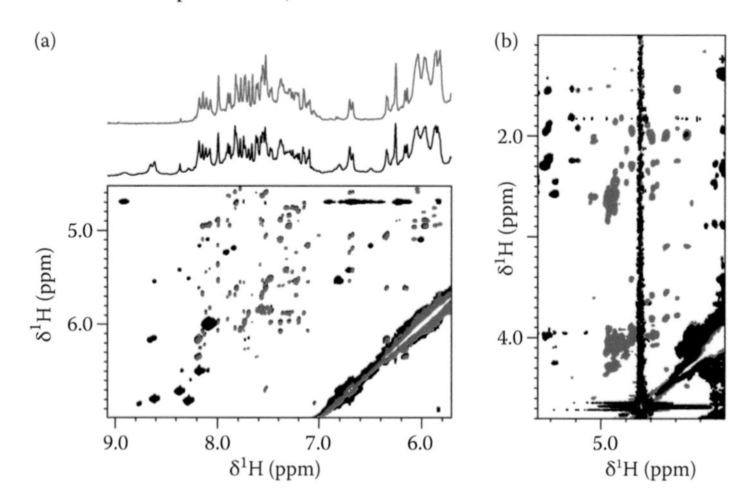


**FIGURE 7.33** Overlay of 500-MHz 2-D NOESY NMR data for a ligand complex of 15-mer duplex DNA d(CGACTAGTCTAGACG) • d(CGTCTAGACTAGTCG) at 25°C in 99% $D_2O$ (red) and 90% $H_2O$/10% $D_2O$ (black). (a) Reduction in the number of NOEs arising from the absence of amino-proton resonances in the 99% $D_2O$ solution. (b) Bleaching out of cross-peaks in the vicinity of the solvent resonance (90% $H_2O$/10% $D_2O$, black) compared with the presence of cross-peaks in the same region for the same DNA complex sample dissolved in 99% $D_2O$.

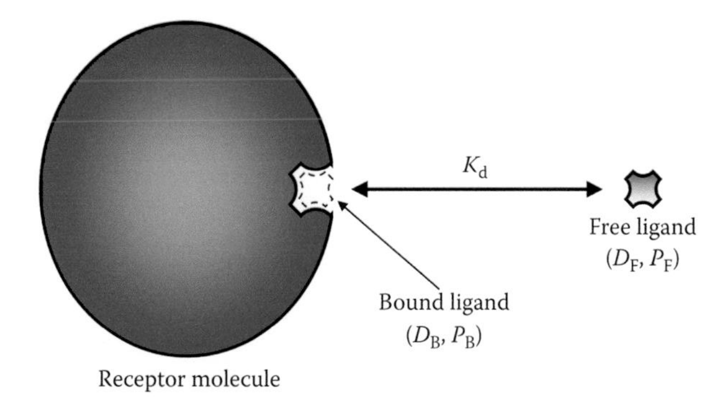


**FIGURE 4.11** Schematic of two-site binding between a ligand and a receptor molecule. The receptor molecule could be a large protein or macromolecule or something much smaller such as a crown ether. The binding is characterised by the equilibrium dissociation constant, $K_d$. The bound (B) and free (F) ligands have population $P$ and diffusion coefficient $D$. Note that, while only one binding site on the receptor molecule is shown, it is possible that there are several binding sites that may or may not be equivalent (i.e. have the same or different $K_d$), which affect the equilibrium diffusion coefficient.

where $D_{Pop}$ is the experimentally observed ('apparent') diffusion coefficient and the population weighted average, $D_F$ is the free (i.e. unbound) diffusion coefficient, $D_B$ is the bound diffusion coefficient, and $P_B$ is the bound population/mole fraction.

For binding events, the value of $P_B$ can be calculated from the number of equivalent binding sites ($n_{Bind}$), the total concentration of receptor molecules ($C_R$), the concentration of ligand molecules ($C_L$), and the dissociation equilibrium constant for the complex ($K_d$) by[11,156,157]

$$P_B = \alpha - \sqrt{\alpha^2 - \beta} \tag{4.16}$$

where

$$\alpha = \frac{C_L + n_{Bind}C_R + K_d}{2C_L} \tag{4.17}$$

and

$$\beta = \frac{n_{Bind}C_R}{C_L} \tag{4.18}$$

If the dissociation equilibrium constant is calculated, then the association (or binding) equilibrium constant, $K_a$, can be easily obtained (as $K_a = 1/K_d$; note that $K_a$ is the ratio of the forward and reverse reaction rates). From this, the standard

Gibbs free energy change associated with the binding equilibrium may be calculated via[6,160,161]

$$\Delta G^0 = -RT \ln (K_a) \quad (4.19)$$

where $R$ is the gas constant (J $K^{-1}$ $mol^{-1}$). A large positive value for $\Delta G^0$ indicates that the reaction is disfavoured.

If both obstruction and binding occur in a system, then both effects contribute to the averaging of the diffusion coefficient, and so[110,111]

$$D_{\text{apparent}} = O_D[(1 - P_B)D_F + P_B D_B] \quad (4.20)$$

and if the bound diffusion coefficient is neglected (e.g. reference 117), then Equation 4.20 becomes

$$D_{\text{apparent}} = O_D[(1 - P_B)D_F] \quad (4.21)$$

This can also be written to include different obstruction factors for the bound and free states,[113] other influences,[117] or the contributions from each obstruction present (i.e. more than one $O_D$ if the total obstruction factor is not used).

## 4.4 DIFFUSION AND ION/MOLECULE RADIUS IN SOLUTION

In this section, an extensive review is provided for the various models, which may be used in analysing diffusion data as it relates to the radius of simple ions and larger molecules. An understanding of these factors will help the reader to be more critical of the results of their diffusion experiments (and are particularly useful when using the radius for simulations). However, for those more interested in the practicalities, you are directed to the Section 4.5, in which some relevant pulse sequences are described.

### 4.4.1 Stokes–Einstein–Sutherland Equation

The Stokes–Einstein–Sutherland equation relates the diffusion of a particle, specifically the diffusion at infinite dilution, to its hydrodynamic size:[73–76]

$$D_0 = \frac{k_B T}{f_s} \quad (4.22)$$

where $f_s$ is the frictional coefficient (J s $m^{-2}$), which, for a solution, depends on the viscosity (assuming that the viscous drag is proportional to the velocity, which is not the case when there is a temperature gradient[162]). The frictional coefficient is also a function of the shape of the diffusing species, and the simplest and crudest approximation of the shape is a sphere,[11,66] but formula for other shapes can be found in the literature.[11,16,163–168] Examples of applying other shape factors to linear and star polymers, and porphyrine systems can be found in references 4, 169, but for a spherical particle, the Stokes–Einstein–Sutherland equation becomes[73–76]

$$D_0 = \frac{k_B T}{\chi \pi \eta R_H} \tag{4.23}$$

where the coefficient, $\chi$, reflects the nature of the solute–solvent interaction (e.g. stick [$\chi = 6$] or slip [$\chi = 4$] conditions). The stick condition can be likened to static friction in that there is a large restraint on the motion, while the slip condition can be likened to kinetic friction in that the objects slide over one another.[170] The common (or 'stick') Stokes–Einstein–Sutherland equation for translational diffusion (with spherical particle shape) may only be applicable for large spherical solutes.[171] The equation has been applied to the self-diffusion of small molecules; however the stick condition needs to be swapped for the slip condition.[171]

### 4.4.2 Hydration and Solvation

For complete prediction of the hydrodynamic radius, the solvation needs to be considered. In solution, the solvent is bound to the diffusing molecule, and as such, moves with the molecule; hence, the hydrodynamic size includes a hydration layer.[11,172] For a spherical molecule, $R_H$ can be related to solvation by[11,66]

$$R_H = \left( \frac{3M}{4\pi N_A} (\bar{v}_{solute} + g_{solvent} \bar{v}_{solvent}) \right)^{1/3} \tag{4.24}$$

where $M$ is the molecular weight of the solute (g mol$^{-1}$), $\bar{v}_{solute}$ is the partial specific volume of the solute, $\bar{v}_{solvent}$ is the partial specific volume of the solvent, and $g_{solvent}$ is the weight fraction of solvent per solute (0.3–0.4 is commonly used for proteins[11,66,172]).

However, as the typical form of the Stokes–Einstein–Sutherland equation (i.e. Equation 4.23) relies on the assumptions that the diffusing molecule is spherical and is much larger than the solvent (i.e. the solvent is a continuum) and that the viscous drag is proportional to the velocity, the choice of radius for use with other calculations or simulations needs careful consideration.

For very small molecules in solution (i.e. having similar or smaller volume compared to the solvent), the slip condition might be more appropriate.[171,173,174] Consideration of the results given by Ravi and Ben-Amotz[171] shows that, when the volume of the solute is smaller or equal to that of the solvent molecules (for their work, the solvent was hexane with a molecular volume of ~110 Å$^3$), the appropriate condition for describing the friction coefficient is the slip condition, but as the volume of the solute gets larger, the slip condition does not adequately describe the friction coefficient, and the stick condition becomes more suitable. It has also been noted that the density of the solvent is also important.[174] Evans, Tominaga, and Davis[174] provide a test value to see if the Stokes–Einstein–Sutherland equation is suitable, and it is given by[174]

$$R_{\text{Solvent-Solute}} = n(r_{\text{Solvent}} + r_{\text{Solute}})^3 \tag{4.25}$$

where $n$ is the solvent's number density (for $H_2O$ at 25°C, this is ~$3.34 \times 10^{28}$ m$^{-3}$ using Avogadro's number times the density divided by the molecular weight), $r_{\text{Solvent}}$ and $r_{\text{Solute}}$ are the radii of the solvent and solute molecules, respectively, calculated from the excluded volume, assuming a spherical molecule. The suitability of the Stokes–Einstein–Sutherland equation improves with increasing values of $R_{\text{Solvent-Solute}}$ (below 1.5, it may not be suitable, and if it is less than 1, the deviation from the Stokes law increases sharply; note that there is a change from the slip to stick conditions on increasing $R_{\text{Solvent-Solute}}$).[174] Hence, for molecules diffusing in $H_2O$ (i.e. $r_{\text{Solvent}}$ = 1.4 Å; NB: 1.4 Å is the commonly used radius for a water molecule in the bulk[175–177]), $r_{\text{Solute}}$ needs to be at least 2.15 Å for the Stokes–Einstein–Sutherland equation to be suitable. Another way to consider this is that, when the size of the spherical molecule is comparable to that of the solvent and the distance between solvent molecules, the correct radius deviates from the Stokes radius.[178] Proposed corrections to the stick form of the Stokes–Einstein–Sutherland equation or other models can be found in the literature.[171,174,178,179]

#### 4.4.2.1 Hydration of Non-Electrolytes

A comparison of the volume calculated using the effective hydrodynamic radius (i.e. determined experimentally by viscosity studies, for example) to the anhydrous volume from molecular models is useful for estimating the hydration for small non-electrolyte molecules.[178] The effective hydrodynamic radius from viscosity studies is the radius of a neutral sphere with the same properties as the solvated molecule (i.e. the molecule including hydration) and is dependent on temperature.[178] The effect of temperature on the hydration of molecules/ions can be related to the Hofmeister effect.[180,181] The Hofmeister series classifies the molecules and ions as either kosmotropes (either ionic/polar or non-ionic/non-polar and are water-structure makers; small ions with high surface charge density or highly polar molecules bind water strongly compared to the water–water interaction, whereas neutral non-polar molecules are hydrated but bind weakly compared to the water–water interaction; the strongly binding kosmotropes are called 'polar kosmotropes', and the binding of non-ionic molecules depends on the total number of hydrogen-bonding interactions) or chaotropes (water-structure breakers, large ions with low surface charge density, which bind water less strongly compared to the water–water interaction).[180,181]

#### 4.4.2.2 Hydration of Ions

For small ions in water, the more accurate size is the hydrated/solvated radius ($r_{\text{Hyd}}$).[179,182–186] This can be seen when considering that the diffusion of an ion increases with increasing ionic crystal size,[187] in contrast to what is expected from the Stokes–Einstein–Sutherland equation (see Table 4.2 for a comparison of the diffusion coefficient for the small ions to the crystal ionic size, $r_c$, and compare this to the trend for the neutral molecules where diffusion decreases with increasing $r_c$). The hydration and solution properties of ions are complicated phenomena (i.e. dependent on the ion type/charge, unhydrated size, viscosity effects, temperature, presence of other ions, concentration etc.), and as such, the following discussion should not be considered exhaustive, and more information can be found in the given references (excellent reviews and collection of data and theories regarding ion properties are given by Marcus[186] and Bockris and Reddy[188]). The brief notes given here, specifically those for dilute aqueous solutions, are intended as a guide for describing the radius of diffusing molecules/ions.

It is both the size of the ions compared to the size of the solvent and their charge that cause the discrepancy between the hydrodynamic radius, the crystal ionic, and the hydrated radius.[179,181,183,184,186,187,189] The size of ions in solution depends on the extent of hydration,[179,182–185,187] which is related to the charge density and the Hofmeister series,[180,181] with strong hydration of multivalent ions.[181] It is generally considered that an ion in solution moves with the hydration/solvation, and so the effective radius may be larger than the bare ion radius,[183,184,186–188] and that the friction experienced by a moving ion is different to that given by the Stokes–Einstein–Sutherland equation and related to the effect of the ion on the bonding/mobility (i.e. structure making/breaking) of local water molecules (i.e. resulting in different local viscosity or density).[184–194] Ion-pair formation also occurs for ions in solution, depending on the sizes and types of the ions involved.[181,184,185]

#### *4.4.2.2.1 Hydration Number and Hydration Shells*

There is a number of different kinds of measurement techniques to describe hydration, and there are differences between static hydration/structure/number, thermodynamic hydration number, and dynamic hydration;[182,185,188] moreover, hydration numbers are typically specific for the method used to determine them.[186] The terms 'hydration number' and 'coordination number' are used in the literature to describe the hydration of ions, but sometimes, they are given as static values (i.e. based on coordination and fixed arrangement in space; useful for discussing reaction mechanisms[185]), thermodynamic values (i.e. based on entropy and energy considerations), or dynamic values (i.e. those that are bound long enough and move with the ion during the diffusion to be included in the size of the ion).[185,187,188] Hence, for moving ions, it is more appropriate to use the dynamic hydration number (i.e. persistent coordination number, primary hydration number), which includes only those water molecules bound to the ion long enough to be part of the diffusion of the ion.[183,184,187,188] The radius of the moving ion may be between the outer radius of the static hydrated ion and the radius of the unhydrated ion.[187] A second hydration number for the dynamic case could also be described, which is related to the number of water molecules around the ion that are affected by it.[188] While the static case is related to the number of water molecules in the coordination structure/hydration sphere, the dynamic case gives an averaged hydration due to residence times (i.e. a mean value for the number of water molecules),[187] and so the hydration number is not necessarily an integer value. Furthermore, depending on the strength of the ion–water interaction, the hydration number measured is not always the same or the hydration shell may not be

**TABLE 4.2**
**Comparison of Different Radii for Small Solutes in $H_2O$ and, Unless Otherwise Noted, These Values Are for 25°C**

| Molecule/Ion | $D_0$ (× $10^{-9}$ $m^2$ $s^{-1}$) | $r_{H, Stick}$[a] (Å) | $r_{H, Slip}$[a] (Å) | $r_c$[b] (Å) | $r_v$[c] (Å) | $r_{Hyd}$[d] (Å) | $r_{Mol}$[e] (Å) | $N_{Hyd, Stick-Mol}$ ($N_{Hyd, Slip-Mol}$) [$N_{Hyd, Stick-c}$] {$N_{Hyd, Slip-c}$} ⟨$N_{Hyd, Hyd-c}$⟩ $N_{Hyd, Lit.}$[f] |
|---|---|---|---|---|---|---|---|---|
| Glucose | 0.69[g] | 3.55[g,h] | 5.33[h] | 3.6[k] | 3.8[i] | – | 3.24 | 3–5 |
| (e.g. pyranose |  | 3.6[i,j] |  | 3.58[l] | 4.2[j] |  |  | (42–43) |
| form) |  |  |  |  | 3.7[l] |  |  | [−1–1] |
|  |  |  |  |  |  |  |  | {38–39} |
| Sucrose | 0.52[g] | 4.7[g,h] | 7.07[h] | 4.4[l,m] | 5.0[i] | – | Min, 3.71 | 7–22 |
|  |  | 4.4[i] |  |  | 5.2[j] |  | Max, 4.02 | (105–117) |
|  |  | 4.6[j] |  |  | 4.67[l] |  |  | [0–10] |
|  |  |  |  |  |  |  |  | {97–104} |
| $Na^+$ | 1.27[n] | 1.93[n] | 2.90[n] | 0.97[u] | – | 3.58[s] | – | [1–3][x] |
|  | 1.33[o,p] | 1.84[r,s,t] | 2.77[r,t] | 0.95[s] |  | 1.79[v] |  | {7–9}[x] |
|  | 1.86[q] | 1.32[q] | 1.98[q] | 1.02[p,u,v,w] |  | 2.18[w] |  | ⟨3–4⟩[y] |
|  |  |  |  |  |  |  |  | ⟨16–17⟩[z] |
|  |  |  |  |  |  |  |  | 2–9[aa] |
| $Mg^{2+}$ | 0.71[o,p] | 3.45[r,t] | 5.18[r,t] | 0.70[u] | – | 4.28[s] | – | [14–16] |
|  |  | 3.47[s] |  | 0.65[s] |  | 3.00[v] |  | {50–51} |
|  |  |  |  | 0.72[p,u,v,w] |  | 2.99[w] |  | ⟨9–10⟩[y] |
|  |  |  |  |  |  |  |  | ⟨28–29⟩[z] |
|  |  |  |  |  |  |  |  | 5–16[aa] |
| $Cu^+$ | 1.2[p] | 2.04[t] | 3.06[t] | 0.77[w] | – | 2.33[w] | – | [2–3] |
|  |  |  |  | 0.96[p] |  |  |  | {10–11} |
|  |  |  |  |  |  |  |  | ⟨4–5⟩ |
|  |  |  |  |  |  |  |  | 3–5[aa] |

| | | | | | | | | |
|---|---|---|---|---|---|---|---|---|
| $Cu^{2+}$ | 0.73[o] | 3.36[r] | 5.04[r] | 0.72[s,u] | – | 4.10[s] | – | [12–15] |
| | 0.71[p] | 3.25[s] | 5.18[t] | 0.73[p,u,w] | | 2.97[w] | | {46–51} |
| | | 3.45[t] | | | | | | ⟨9–10⟩[y] |
| | | | | | | | | ⟨24–25⟩[z] |
| | | | | | | | | 9–13[aa] |
| $Fe^{2+}$ | 0.72[p] | 3.44[s] | 5.11[t] | 0.72[u] | – | 4.28[s] | – | [14–15] |
| | | 3.41[t] | | 0.75[s] | | 2.91[w] | | {48–49} |
| | | | | 0.78[p,u,w] | | | | ⟨8–9⟩[x] |
| | | | | | | | | ⟨28–29⟩[z] |
| | | | | | | | | 7–13[aa] |
| $Fe^{3+}$ | 0.60[p] | 4.06[s] | 6.13[t] | 0.64[u] | – | 4.57[s] | – | [24–25] |
| | | 4.09[t] | | 0.60[s] | | 3.53[w] | | {83–84} |
| | | | | 0.65[p,u,w] | | | | ⟨15–16⟩[y] |
| | | | | | | | | ⟨34–35⟩[z] |
| | | | | | | | | 15–17[aa] |

[a] Stokes (hydrodynamic) radius for stick or slip conditions assuming a spherical particle.

[b] Crystallographic radius or bare ion radius.

[c] Viscometric radius.

[d] Hydrated radius for the ions only.

[e] Radius for an ‘anhydrous’ molecule calculated from the Connolly solvent–excluded volume using ChemBioDraw Ultra, ChemBio3D Ultra and ChemPropStd (ChemBioOffice, 2010, v.12.0.2.1076, Cambridge Soft) with the probe radius of that of water (1.4 Å) and assuming a spherical molecule. 1.4 Å is the commonly used value for the radius of a water molecule in the bulk,[175–177] but it could be smaller near polar groups on surfaces and larger near non-polar groups on surfaces as discussed in reference 175. However, it was noted that the results would sometimes vary for the same molecule (depending on the energy minimization), and so these values should only be considered a rough guide. Models were not done for the smaller ions.

(*continued*)

**TABLE 4.2 (Continued)**
**Comparison of Different Radii for Small Solutes in $H_2O$ and, Unless Otherwise Noted, These Values Are for 25°C**

[f] Estimated hydration number, $N_{Hyd}$. This uses $N_{Hyd,i-j} = (r_i^3 - r_j^3)/r_W^3$, where $r_W$ is the radius of a water molecule and is 1.4 Å, and $r_i$ and $r_j$ correspond to $r_{H,Stick}$, $r_{H,Slip}$, $r_{Mol}$, $r_c$ or $r_{Hyd}$, depending on the $i$ and $j$ in the column label ($r_j$ is the 'anhydrous' value). The radii used in the calculation are indicated by the type of brackets. The values are given as an approximate range of the whole number of water molecules (but note as described in text the value does not need to be an integer because of dynamic averaging; in calculating the ranges using all radii, numbers were rounded up for the maximum and down for the minimum values and if only one value resulted the range covers the fraction value). The values are only calculated for the results for 25°C. $N_{Hyd,Lit.}$ are values from the literature (i.e. see aa). Because of the method of calculation and rounding, the value was sometimes negative, which is obviously unrealistic (but those where the calculated value was negative, the value was small and so could be considered to mean no hydration [i.e. $N_{Hyd} = 0$]).

[g] Reference 5, but there is a typographical error for glucose (i.e. it is given in the reference as 3.65 Å, but recalculation of the Stokes radii using the $D_0$ values and the viscosity of $H_2O$[212] at 25°C gives 3.55 Å). See also reference 6 for more.

[h] Calculated using $D_0$ from g and the viscosity of $H_2O$[195] at 25°C.

[i] The spherical particle radii from reference 196. Note that the crystallographic results of this reference were not used as they give the dimensions of the three axes of the molecule which are hard to convert to an equivalent spherical radius for comparison without making assumptions on the shape to find the volume.

[j] Reference 178; note that the model values of this reference represent an average dimension of the molecule after measuring the largest and smallest dimensions from a physical model and are not used here (and only the $r_v$ values are) since the equivalent spherical molecule radius from other means, such as Stokes radii, are better compared to the equivalent spherical molecule radius based on the Connolly solvent–excluded volumes (i.e. e).

[k] Using the data in references 197–199 for D-glucose (both α and β forms are similar). The result for the equivalent spherical particle radius can be calculated using the crystallographic unit cell volume and number of molecules, 766.45 Å$^3$ (average of all results given for α-D-glucose) and 787.37 Å$^3$ (β-D-glucose) for the unit cell of four molecules; dividing the volume by four; and finding the radius of a sphere for the resulting volume, or by the average density value of those given (i.e. 1.567 g cm$^{-3}$ for α-D-glucose and 1.5275 g cm$^{-3}$ for β-D-glucose) as follows: the molecular weight is 180.16 g mol$^{-1}$, giving 190.88 Å$^3$ per molecule for α-D-glucose and 195.86 Å$^3$ per molecule for β-D-glucose, and then finding the radius required for a sphere of this volume.

[l] Reference 200 at 25°C.

[m] Using the data in references 201, 202 for sucrose. The result for the equivalent spherical particle radius can be calculated using the crystallographic unit cell volume and number of molecules, 715.04 Å$^3$ for the unit cell of two molecules; dividing the volume by two; and finding the radius of a sphere for the resulting volume, or by the average density value of those given (i.e. 1.585 g cm$^{-3}$ for sucrose) as follows: the molecular weight is 342.3 g mol$^{-1}$, giving 358.622 Å$^3$ per molecule, and then finding the radius required for a sphere of this volume.

[n] From reference 169. Results from diffusion in $H_2O$ at 25°C and linear extrapolation to infinite dilution of concentration–diffusion data of sodium tetraphenylporphyrinetetrasulfonate ($Na_4TPPS$) for concentrations from 0.5 to 4 mM, but note as in reference 169 that there is an expected non-linear trend to infinite dilution due to the association of $Na^+$ with $TPPS^{4-}$; the expected $D_0$ is that from literature (i.e. o). The Stokes radii are calculated for the linear extrapolation value for $D_0$ and the viscosity of $H_2O$[212] at 25°C.

[o] Reference 8; this reference has a range of self-diffusion data for a variety of ions, but see also references 9, 10 for some concentration dependence data for some alkali metal ions.

[p] Reference 186. Diffusion coefficients from this reference are for infinite dilution in $H_2O$ at 25°C.

[q] Measurements made by Scott A. Willis where $H_2O$ was the solvent and the measurement was made at 37°C (unpublished results). The value is a result of linear extrapolation to infinite dilution of concentration–diffusion data of sodium tetraphenylporphyrinetetrasulfonate ($Na_4TPPS$) for concentrations from 0.5–4 mM, but note as in reference 169 for the results at 25°C that there is an expected non-linear trend to infinite dilution due to the association of $Na^+$ with $TPPS^{4-}$, and so this could be an underestimation. The Stokes radii are calculated for the linear extrapolation value for $D_0$ and the viscosity of $H_2O$[212] at 37°C. At higher temperature, $Na^+$ could be expected to have lower hydration due to a reduction in residence times as for $Li^+$ (see text). Also note that for these diffusion measurements of $Na^+$, this is also affected by binding with the counter-ion (see n and reference 169).

[r] Calculated using $D_0$ from o and the viscosity of $H_2O$[212] at 25°C.

[s] Reference 179, although, as noted in the text earlier in this section, the $r_{Hyd}$ values may only be qualitative due to the assumptions of the correction, and some of the $r_c$ values are estimates giving different values to elsewhere (see the earlier discussion in this section).

[t] Calculated using $D_0$ from p and the viscosity of $H_2O$[212] at 25°C.

[u] Reference 182; note that the value for $Cu^{2+}$ is an average due to the difficulty with this ion. This reference is useful for the description of numerous ion crystal radii and binding.

[v] Reference 184 at 30°C based on a calibration procedure for elution from a Sephadex G-10 size exclusion column, but there are assumptions, and the apparent dynamic hydration numbers in the reference do not correspond to the hydration numbers calculated in the table, that is, from f (see text). The values for multiple elutions were averaged.

[w] The values of $r_{Hyd}$ from reference 183 are for the primary (i.e. dynamic) hydration shell (i.e. from Equation 4.26). Note that this reference uses the same $r_c$ for all forms of orthophosphate.

[x] Not using the results for $r_{H,\,Stick}$ and $r_{H,\,Slip}$ from q as these are for 37°C.

[y] Using $r_{Hyd}$ from w only, as those in v are for 30°C.

[z] Using the $r_{Hyd}$ from s. The large numbers calculated further show that the results from s for $r_{Hyd}$ are mainly qualitative.

[aa] Summary of dynamic hydration numbers from various experimental methods and various models (given in integer value ranges to suit the format of the others in the table) from references 183, 186, 188 (but see also reference 185). This typically depends on the concentration and counter-ion (with the higher numbers for more dilute systems);[188] also note that the values calculated using the $r_{Hyd}$ from Equation 4.26 will always give the model hydration numbers in references 183, 186 simply because this is the model that they are based on, but the other experimental results from elsewhere are in good agreement with this model. The ranges highlight the difficulties.

symmetric, and so the average hydration based on the spherical molecule equivalent is not necessarily constant or an integer value.[185] Several references give the hydration numbers for ions or hydrated radius in solution,[179,182–188] but care needs to be taken when using this information to describe the hydrated radius of the moving ions as mentioned.

The Stokes radius of ions, with $r_c$ (i.e. bare ion) less than 0.25 nm (this is comparable to the calculation from Equation 4.25 using water as the solvent, giving the lower limit for the solute radius of 0.215 nm), is not a good measure of the size of moving ions in solution.[182] Ions with a hydrodynamic radius greater than 0.52 nm (calculated with stick conditions) may have an equivalent crystal radii, and so the Stokes–Einstein–Sutherland equation with stick conditions is valid for large ions (and they are effectively not hydrated).[179] There is also a broad minimum of $r_{Hyd}$ that exists for a small range of $r_c$, and for values outside of this range, the $r_{Hyd}$ increases;[179] this can be seen when plotting the $r_{Hyd}$ against the $r_c$ given in references 179, 183 (for reference 183, $r_{Hyd}$ can be found by adding the thickness of the hydration layer to the radius of the ion, e.g. see Equation 4.26). Generally, for larger values of the radius, $r_{Hyd}$ is expected to be equal to $r_c$ (i.e. the surface charge density is small, and there is lower energy for ion–dipole interaction, and so the ion is unhydrated),[179] and for smaller values of the $r_c$, there is increasing surface charge density, and so hydration becomes more significant with the possibility of further hydration shells giving larger $r_{Hyd}$.[179,180,183,185] An empirical procedure for estimating the hydrated radii from the Stokes 'stick' radius is given by Nightingale[179] and is based on the assumptions that the hydrated radius cannot go to zero as the Stokes 'stick' radius does and is limited to the diameter (not radius) of the water molecule, that $r_c$ values are known well, and that the tetraalkyl ammonium ions (except the tetramethyl ammonium ion) are unhydrated. A similar approach to estimating the corrected Stokes radii assumes that the tetraalkyl ammonium ions are minimally hydrated.[186] However, it is important to note that the 'bare' ion radius given by Nightingale[179] varies from those given elsewhere[182,183,185,186] for some ions (which is likely due to the data available at the time and/or the estimation method), but despite this and the other assumptions of these 'corrections' (i.e. the large tetraalkyl ammonium ions are probably hydrated, which would shift the corrections; see references 179, 190, 203), the results highlight the deviation of the radius in solution from the Stokes 'stick' radius (i.e. $r_{Hyd}$ is the more appropriate radius for ions in solution). It is also the deviation of the local viscosity around the ion from the viscosity of the continuum solvent (i.e. giving a different friction term for the Stokes–Einstein–Sutherland equation) that results in the deviation of the Stokes radii from $r_{Hyd}$.[188] The Hubbard–Onsager model[189,191–194] is one that modifies the friction coefficient of the Stokes–Einstein–Sutherland equation to include dielectric friction, but this may be limited to small ions.[193] Another model by Impey et al.[187] modifies the friction term to include the effects of non-uniform viscosity around the ion. In this model,[187] the change in viscosity is characterised by the rotational correlation times of the water near the ion or the residence times of the water molecules in the dynamic hydration layer. Some values for the correlation time of several ions and water are given in references 185, 186. Since the correlation times are related to the viscosity (whereby an increased viscosity is a longer correlation time),[1,187] the effect of structure-making/breaking ions is to change the local viscosity. The sign of the so-called 'B coefficient' of the Dole–Jones viscosity relationship[204] can be used to assign ions as either kosmotropes or chaotropes.[179,181] Structure-breaking ions reduce the average number of hydrogen bonds in the water around them,[186] which decreases the local viscosity and increases the mobility.[190] Values for B coefficients can be found in the literature (e.g. references 181, 186). Structure-making ions increase the average number of hydrogen bonds that the water molecules around the ions participate in.[186]

Marcus[182] and Ohtaki and Radnai[185] give similar summaries and results for ion–water distances and also dynamic hydration/coordination numbers for the hydration shells. Ohtaki and Radnai[185] also provide more description of the structure of the first and second (if any) hydration shells (i.e. coordination polyhedra for transition metals; otherwise, the hydration is distributed in the hydration sphere), the orientation (proton or oxygen direction) and tilt angle of the water molecules at the ion (if applicable), and also consideration for the dynamics of the hydrated ions and the residence times of the hydration sites, giving a reduced hydration number to reflect the dynamic system. Impey et al.[187] also give description of the orientation of the water molecules and residence times. Residence times (i.e. for water binding to the ion and water binding to the bulk water) are very important in describing the dynamic hydration number of ions (which is equal to or lower than the static hydration number).[185,187] The dynamic hydration number is related to the residence times and static hydration/coordination number through an exponential decay.[185,187,188] If the residence time in the hydration shell is very small compared to the residence time in the bulk solution, then the dynamic hydration number is zero (i.e. unhydrated).[185] With increasing residence time in the hydration shell, the dynamic hydration number approaches the static value.[185]

The first hydration shell of $Li^+$ and $Na^+$ are tightly bound.[185,187] $Li^+$ shows only a slight change in ion–solvent distances with increasing temperature but the residence times decrease, and so the hydration is less stable.[187] $r_{Hyd}$ may be estimated by finding the equivalent spherical radius for the total volume of the ion and hydration, which is the 'dynamic' volume of its first and second (if any) hydration shells added to the crystallographic volume (i.e. the volume of a sphere found using $r_c$), but this relies on knowing the exact number of waters of dynamic hydration or the static hydration/coordination number and residence times (some values can be found in the literature, e.g. references 182, 183, 185, 188). Models for the thermodynamic properties (e.g. Gibbs free energy, entropy) for hydration are given in references 183, 186, and despite large uncertainties (some of which come from the difficulty in describing the hydrated size of the ion[183,186]), models like this may still be able to group ions according to their behaviour as structure making/breaking.[183,186] The width of the primary hydration shell (in the dynamic case) ($\Delta r_{Hyd}$) used to calculate the thermodynamic properties, as given in references 183, 186, 205, but written so the dimensions are in Å, is

$$\Delta r_{Hyd} = \left( \frac{0.45 d_w^3 |z|}{r_c} + r_c^3 \right)^{1/3} - r_c \tag{4.26}$$

where $d_w$ is the diameter of a water molecule (2.76 Å is used in references 183, 186, 205; this is the mean O–O distance in ice, whereas for liquid water, it is 2.92 Å[188]), $z$ is the integer charge number of the ion (all values in the relationship are in units of angstrom except for the integer charge number), and the value of 0.45 Å is the result of a fitting parameter for describing the hydration thickness based on the electrostriction affecting the relative permittivity of the immediately surrounding water.[183,186,205] Hence, $r_{Hyd}$ could be estimated for an ion as $r_c + \Delta r_{Hyd}$ from Equation 4.26, which only requires $r_c$ and $z$, as opposed to using a 'corrected' Stokes radius (e.g. as in reference 179) or knowledge of residence times and coordination numbers, although the smallest ions (e.g. $H^+$, $Be^{2+}$, and $Al^{3+}$) might not be best described by this model due to crowding near them.[186] The primary hydration shell given by Equation 4.26 is assumed to be the hydration that moves with the ion (but note that there is variation from model to model).[186,188]

#### 4.4.2.3 Hydrated Radius by Experiment

In an attempt to measure the apparent dynamic hydration numbers of some ions, Kiriukhin and Collins[184] used a calibration procedure for the elution times of ions from a Sephadex G-10 size exclusion column at 30°C.[184] Their apparent dynamic hydration numbers are then related to the number of tightly bound water molecules, which explain the apparent molecular weight of ions.[184] Interestingly, the apparent dynamic hydration numbers do not match the hydration number, which can be calculated from their calibrated $r_{Hyd}$ (i.e. from the volume difference between the unhydrated ion using $r_c$ and the hydrated ion using their $r_{Hyd}$ values).[184] The experimental results[184] are either over or underestimated by Equation 4.26, but this could be due to the experiments being performed at moderate concentrations (i.e. 0.1 M solutions) where ion–ion effects might be important or due to the assumptions of the calibration curve, but equally due to the assumptions of Equation 4.26 (i.e. spherical ions, immobilised hydration layer etc.[183,186,205]) and also the measurement technique dependence of the hydration numbers (i.e. reports on different aspects of hydration) as noted by Marcus,[186] although Equation 4.26 may be taken to give reasonable values for the dynamic hydration of ions in dilute solution.[186,188]

Support for the use of the $r_{Hydrated}$ for small ions can be seen in the work on membranes templated with lyotropic liquid crystals (LLCs) for reverse osmosis and nanofiltration.[206–209] For a membrane templated with an inverse hexagonal LLC phase with a pore radius of 0.6 nm, there was low rejection of the small metal ions and small polyethylene glycol molecules but high rejection of larger dyes and large polyethylene glycol molecules.[207,208] When compared to a membrane templated with a normal bicontinuous cubic LLC phase, there was high rejection of the small metal ions and molecules (e.g. glucose, Section 7.2) but lower rejection for smaller molecules (e.g. ethylene glycol).[206,209] The pore radius of this membrane was calculated by plotting the rejection against the radii (only for the non-ionic solutes) and fitting with the Ferry equation[208,209] and was found to be 0.375 nm.[206,209] However, both of the membrane types based on the LLCs had ionic/charged pores,[206–209] but this was only briefly mentioned as a possible way to generate effectively narrower pore sizes seen by ions.[208] Actually, the ionic nature of the membrane was presumably why only the non-ionic solutes were used to calculate the pore sizes[206] with the Ferry equation.[210,211] In addition to steric hindrance/size exclusion effects for the transport of ions through membranes, there are also electrostatic and dielectric exclusion effects (e.g. rejection of an ion because of a fixed charge, surface accumulation of charge, membrane potential etc.).[200,212–215] For the aforementioned studies (i.e. references 206–209), the assumed radii of the metal ions were those given in reference 179 for $r_{Hyd}$, and the radii for glucose and sucrose were the Stokes 'stick' radii from reference 5 (for this reference, the Stokes radius may be calculated using the stated infinite dilution diffusion coefficients and the Stokes–Einstein–Sutherland equation with stick conditions, but note that the value for the Stokes radius for glucose should be 0.355 nm and not 0.365 nm, but this small difference is of no real consequence to the estimated membrane pore size). However, as noted previously, the $r_{Hyd}$ of the ions given by Nightingale[179] might need to be corrected due to the assumption that the tetraalkyl ammonium ions were unhydrated and due to the differences in the $r_c$ values used (but the results are still qualitative). This means that, although for the results of the membrane studies of references 206–209 it seems as though the ions (e.g. $Na^+$) might be similar in size or larger than glucose, in fact, it is the combination of both steric, dielectric, and electrostatic effects that exclude the ions from the membrane, and so the ions should be smaller than glucose but still hydrated. For example, a Nanomax-50 membrane with negative charge and pores of 0.45-nm radius, the retention for NaCl was 65%; for $Na_2SO_4$, 98%; and for glucose, 76.5%.[212] As can be seen, the charge, ions, and counter-ions affected the results (along with pH and ionic strength for the other ions studied),[212] and if steric exclusion is assumed to be the only exclusion then, for the hydrated ion size given in reference 179, the ion would be too small in this case, to explain the retention and how it varies (i.e. compared to glucose). However, at pH values where the membrane is neutral, the exclusion might be mostly steric based.[212] The pH dependence (and, hence, dependence on ion charge) of the retention of ions by membranes is clearly seen for phosphoric acid and its ions (i.e. $H_3PO_4$, $H_2PO_4^-$, $HPO_4^{2-}$ and $PO_4^{3-}$),[213,215] and the size of phosphate or polyphosphate molecules is expected to change with pH since the molecular species present in a sample depends on this (pKas for the phosphates can be found in the literature[216–218]).

## 4.5 NMR MEASUREMENT INTERFERENCE

The basic NMR techniques described in Section 4.2 provide remarkable sensitivity to various aspects of molecular motion, yet the drawback of such sensitivity is the higher susceptibility of the method to external perturbations. NMR diffusion measurements are deleteriously affected by factors influencing general NMR spectroscopy, including radiation damping, field inhomogeneities, and additional issues resulting from the introduction of pulsed magnetic field gradients. The latter primarily include Eddy currents, background gradients, and gradient pulse mismatch. Nevertheless, most of these error factors can either be minimised or corrected using appropriate pulse sequence techniques.

### 4.5.1 Eddy Currents

Eddy current induction is a natural property of any conductor placed in a varying magnetic field. The 'Eddy' currents induced in the conductor generate a magnetic field that opposes the changes of the external varying magnetic field.[219] In NMR diffusion experiments, Eddy currents are mostly expected to appear during the magnetic field gradient pulse build-up and drop-off as these are the primal moments of magnetic field variations. The Eddy currents oppose the magnetic field changes, and in NMR, they result in distortion of the intended gradient pulse shape or persisting background gradients (i.e. when the Eddy current lifetime is in the range of the pulse sequence repetition time). Consequently, it is impossible to acquire good NMR spectra until the Eddy currents have decayed.

The two most common ways to reduce Eddy currents are shielding of the gradient coils[220] and using pre-emphasis[221] in the gradient pulse control circuit, both being hardware modifications to the equipment employed in most modern spectrometers. Yet, in some cases, such corrections might be insufficient, and then pulse sequences can be modified to even further reduce the Eddy current effects.

As far as Eddy currents affect magnetic field gradients, which act on the transverse magnetisation, one can reduce the effects of gradient pulse tails by flipping the magnetisation away from the transverse plane during the evolution time (i.e. $\Delta$) and prior to the signal acquisition, thus providing a delay for Eddy current dissipation in both cases. Longitudinal placement of magnetisation during the evolution time is known to happen if the PGSTE sequence (Figure 4.12) is employed for NMR diffusion measurements. This is often used to reduce the decay during long magnetisation evolution periods[49] and retain a sufficient portion of the observable signal at acquisition time if the longitudinal relaxation in the sample is slower than the transversal (i.e. $T_2 < T_1$). Note that the diffusion encoding and decoding pulse lobes in Figure 4.12 might act differently on transverse magnetisation due to Eddy current–induced pulse tails. If long enough, the encoding pulse (the first gradient pulse) tail effect is truncated by the second $\pi/2$ pulse, which flips the magnetisation to the longitudinal direction, making it insensitive to magnetic field gradients, yet the decoding pulse tail persists through the acquisition period, thus producing additional irrelevant echo attenuation.

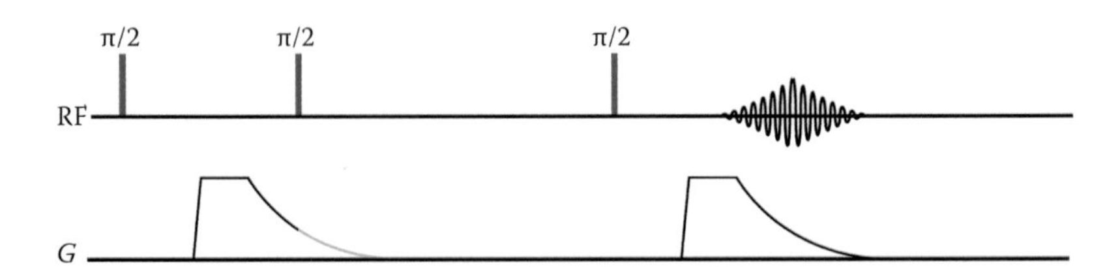


**FIGURE 4.12** PGSTE pulse sequence for diffusion measurements. Gradient pulses are shown to exhibit substantial pulse tails due to Eddy current effects (only end-slope tails are shown for brevity). The encoding pulse tail affects magnetisation only partially as at some time point, it turns to the longitudinal plane (i.e. pulse tail shown in grey).

The pulse mismatch, arising from the Eddy current–induced gradient pulse tails, can be overcome by delaying signal acquisition to the moment when the gradient pulse tail has completely decayed. This requires placing the spin magnetisation in the longitudinal direction again right after the decoding gradient pulse. Another refocussing pulse is then required to flip the magnetisation to the transverse plane before acquisition starts (Figure 4.13). The resulting pulse sequence,[222] therefore, employs a pre-acquisition delay with a longitudinal magnetisation or longitudinal Eddy current delay (LED) to similarly truncate the effects of the encoding and decoding gradients on the observed magnetisation, thus reducing the errors due to Eddy current–related gradient pulse mismatch.

Unfortunately, the LED sequence modification is only effective if added to the PGSTE sequence, which is known to provide the signal of only half an amplitude (not including the sample-dependent $T_1$ attenuation [Chapter 1]) of the one provided by the conventional PGSE sequence.[45] Yet, it is possible to produce Eddy current compensation in the PGSE-based pulse sequence using the fact that the gradient pulse switching produces the same Eddy current pattern during gradient rise and drop-off.[223,224] Therefore, by providing shorter gradient pulses, one can force the rise-time Eddy currents to cancel or, at least, to diminish the drop-off–time Eddy currents, thus reducing the gradient pulse tail. To sustain the desired echo diffusion weighting while shortening the gradient pulse duration, the latter should be performed by splitting the longer gradient pulses typical for the basic spin echo sequence into smaller parts and placing the shortest gradient pulses closer to the acquisition time. Double refocussed spin echo (or twice-refocussed spin echo [TRSE]) is proposed for this purpose,[223] employing two refocussing π pulses and a short gradient pulse right before the signal acquisition (Figure 4.14). The sequence consists of three evolution periods and four gradient lobes, providing sensitivity to motion. Compared to the basic single-refocussed spin echo experiment, this pulse sequence provides a better timing flexibility as the length of the last evolution period can be varied freely, allowing selection of the shorter gradient pulse duration, which provides maximum Eddy current reduction.

The cases above account for moderate Eddy current effects producing short-term decaying magnetic field distortions. In more severe cases, Eddy current effects

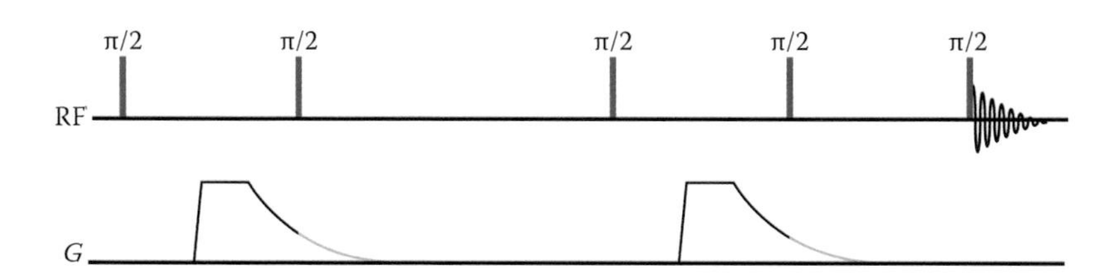


**FIGURE 4.13** PGSTE with LED pulse sequence for diffusion measurements. The fourth π/2 RF pulse flips the magnetisation to the longitudinal direction, where it is not affected by the gradient pulse tail. If introduced with appropriate timing, the longitudinal delay truncates the effects of the decoding gradient pulse in a same way that the second π/2 pulse and evolution period truncates the effects of the encoding gradient pulse tail (shown in grey), which results in matching gradient pulses.

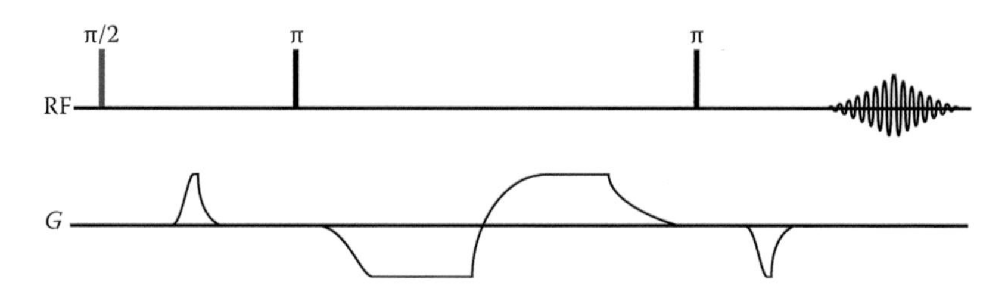


**FIGURE 4.14** Double refocussed spin echo sequence for diffusion measurements. Both front and end Eddy current–related gradient pulse tails are shown. Note that the long gradient pulses have larger pulse tails as the rise and drop-off–time Eddy currents do not counter each other but dissipate independently. Shorter gradient pulses use build-up–time Eddy currents to reduce drop-off–time Eddy currents, thus exhibiting much shorter pulse tails, allowing placing pulses closer to the acquisition time without the risk of distorting the echo signal by the gradient pulse tail.

may include long-term perturbations, which can hardly be overcome by introducing dissipation delays. These can either be represented as constant field inhomogeneities and treated accordingly or reduced by the use of the prepulsing technique.[11]

### 4.5.2 Field Inhomogeneities

Time-dependent Eddy current–related gradient pulse tails are not the only source of field inhomogeneities.[225] The spectrometer setup, the sample peculiarities, or both combined can cause time-independent field distortions, which persist throughout the pulse sequence and, thus, cannot be overcome by modifying pulse sequence timing. These field inhomogeneities are usually represented as additional constant (or background) gradients, which can vary throughout the sample volume. The prime method for correcting those is the use of shim coils,[226] which allow introducing minor, spatially distributed magnetic field corrections, which can counter the existing background gradients, thus providing a much more spatially homogenous field distribution. Nevertheless, some background fields, mostly the intrinsic sample fields or magnetic susceptibility borders, cannot be corrected for in this way and, thus, persist along with diffusion-sensitising gradient pulses. The effects of the two become co-dependent (due to cross terms; see references 61, 227), and as the exact distribution and value of the background gradients is unknown, molecular diffusion cannot be correctly estimated.

There are a number of methods in the field of pulse sequence design that help reduce the inhomogeneities effects. Such sequence modifications rely on the time-independent behaviour of field distortions. As these persist throughout the duration of the pulse sequence, one can modify the magnetisation behaviour in such a way that the errors induced in one part of the pulse sequence are corrected for in the other by the very same field inhomogeneities. Such magnetisation behaviour is achieved by introducing additional inversion pulses and changing the polarity of the diffusion-sensitising gradient pulses so that the desired diffusion weighting is retained, thus producing, in the simplest case (i.e. one inversion pulse added for winding and one for unwinding gradient pulse), the bipolar gradient pulse module for diffusion measurements (Figure 4.15) applicable to both PGSE (Figure 4.15b) and PGSTE (Figure 4.15d) pulse sequences. If appropriately timed, additional inversion π pulses allow diminishing of the cross terms to zero,[228,229] thus reducing the effect of the background gradients to a phase shift independent of the diffusion-weighting gradient value. This allows performing the NMR diffusion measurements in the ordinary manner regardless of the background gradient presence.

If shorter gradient pulses are desired (e.g. to minimise the Eddy current effects, as described above), it is possible to introduce any number of inversion pulses into the sequence. Such a technique actually results in a whole variety of multi-polar gradient pulse sequence modifications, allowing efficient cancellation of background gradients.[230,231]

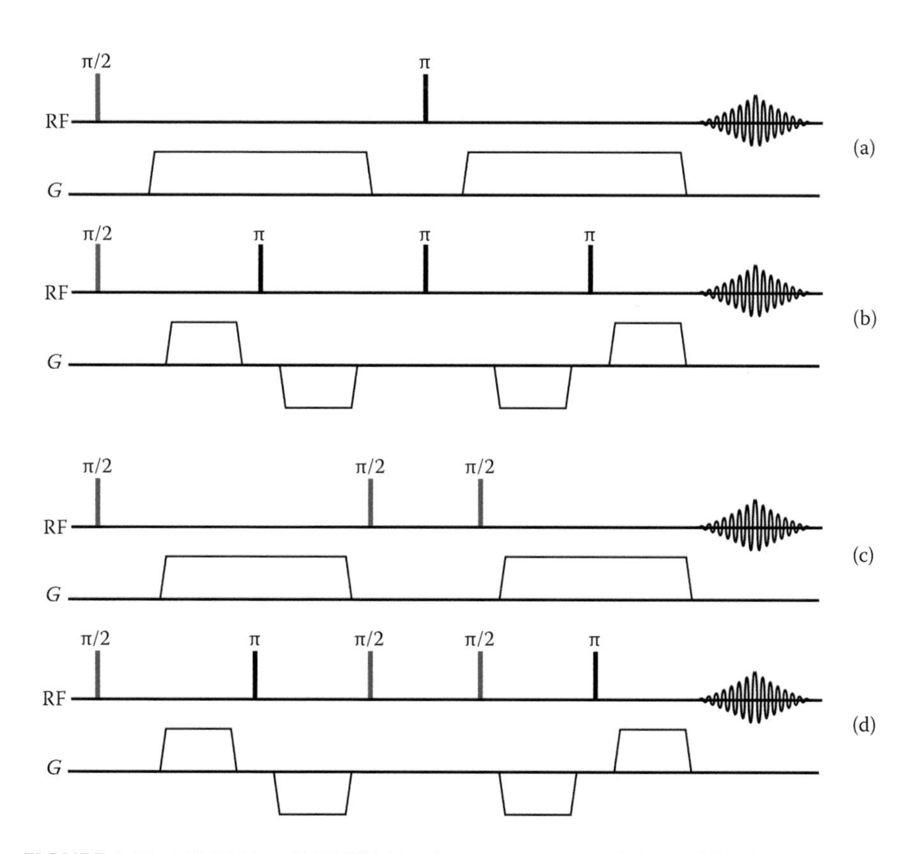


**FIGURE 4.15** PGSE (a) and PGSTE (c) pulse sequences, and their modifications, with additional inversion pulses and bipolar gradient pulses introduced (b and d). Such RF and gradient pulse placement allows separating the echo attenuations produced by diffusion weighting and background gradients. Note that the bipolar gradient with single inversion pulse is just one of the sequence options. More gradient pulses and inversion RF pulses can be introduced while retaining the achieved attenuation separation effect.

### 4.5.3 Solvent Suppression

In many drug development–related NMR experiments, the molecules of interest (e.g. drugs, proteins, and metabolites) are dissolved in water, which generates a huge signal due to its enormously higher (e.g. ~20,000 times) concentration than the biomolecules of interest (Figure 4.16 and Chapter 6) (e.g. references 232, 233). The efficient suppression of $^1H$ resonance of $H_2O$ is necessary when studying the diffusion and association of drug molecules or the drug–biomolecule interactions *in vitro* and *in vivo* using $^1H$ NMR (e.g. references 232–236). The most commonly used water signal–suppressed NMR diffusion experiment is a bipolar PGSTE sequence postfixed by a WATERGATE solvent signal suppression unit[237,238] (Figure 4.17) or excitation sculpting unit,[239] which is applicable to most samples unless some of the resonances are extremely close to the water resonance (Chapters 3 and 7).

For observing the close-to-water resonances, one can use a diffusion-based water suppression strategy (Chapter 3), in which one starts a standard PGSE or PGSTE NMR diffusion experiment with a relatively high diffusion measuring gradient strength (e.g. ~0.2 T $m^{-1}$) if the molecules of interest have a much higher molecular weight (e.g. ~1000) and, thus, a smaller diffusion coefficient than water (e.g. references 159, 240). An alternative is to have a standard PGSE or PGSTE experiment prefixed by a water-PRESS unit (Figure 4.18),[68,241] which can exceedingly selectively

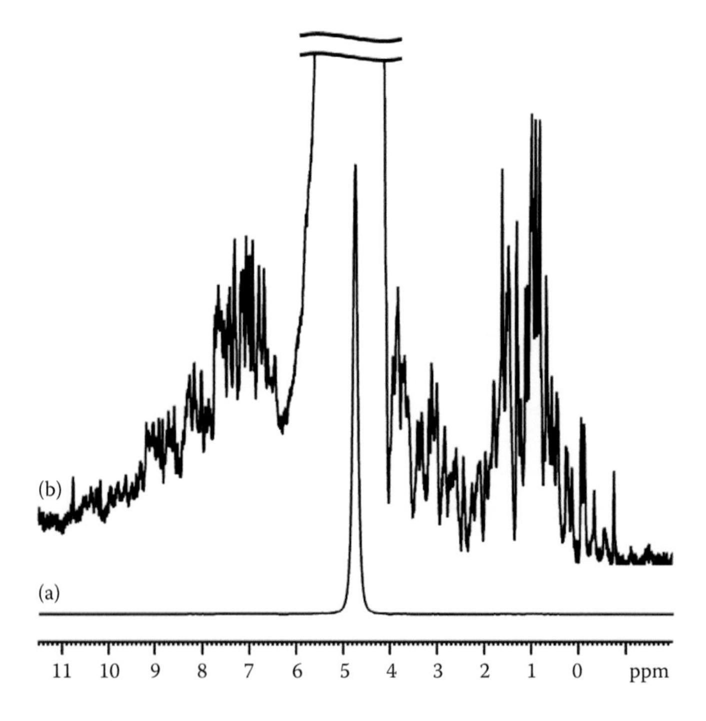


**FIGURE 4.16** (a) $^1H$ 500-MHz spectrum of 2-mM lysozyme in 90% $H_2O$ and 10% $D_2O$ using one scan and a very low receiver gain. (b) Same spectrum as (a) but with the vertical scale expanded approximately 900 times. (From Zheng, G. and W. S. Price, *Progress in Nuclear Magnetic Resonance Spectroscopy* 56, pp. 267–288, 2010.)

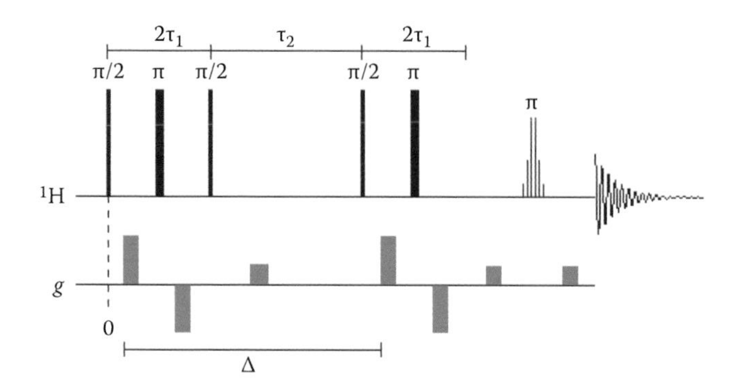


FIGURE 4.17 A bipolar PGSTE sequence post-fixed by a WATERGATE solvent signal suppression unit.

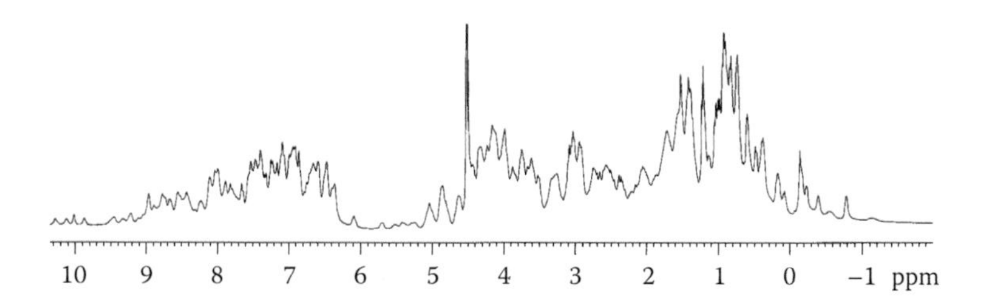


FIGURE 4.18 $^1$H 300-MHz spectrum of 10-mM lysozyme in 90% $H_2O$ and 10% $D_2O$ using the water-PRESS sequence. (From Price, W. S. and Y. Arata, *Journal of Magnetic Resonance, Series B* 112, pp. 190–192, 1996.)

suppress the water resonance due to its different $T_1$ relaxation behaviour to the resonances of interest (e.g. the $T_1$ relaxation time [Chapter 1] of water can be as short as ~15 ms when using a field strength of 18.8 T (i.e. $^1$H resonance frequency of 800 MHz) due to significant radiation damping effect (e.g. references 242, 243). The PURGE[244] pulse sequence can also be prefixed to a standard PGSE or PGSTE experiment for the conservation of the close-to-water resonances.

It is also quite common to work with samples with extremely low concentrations due to the low water solubility of newly synthesised molecules. The PGSTE–WATERGATE[47,245] sequence is well suited for studying low-concentration samples due to its highly efficient water signal suppression and, thus, capability of accumulation of a high number of transients/scans to achieve acceptable signal-to-noise ratio (Figure 4.19).

To probe the ligand–biomolecule interactions (Chapter 3), NMR diffusion experiments on the mixture of ligands and biomolecules are necessary. However, these experiments are often hampered not only by strong water signal, but also by intensive biomolecular signals. The PGSE–WATERGATE[156] sequence is then perfect for these experiments due to its double function of water and biomolecular signal suppression (Figure 4.20).

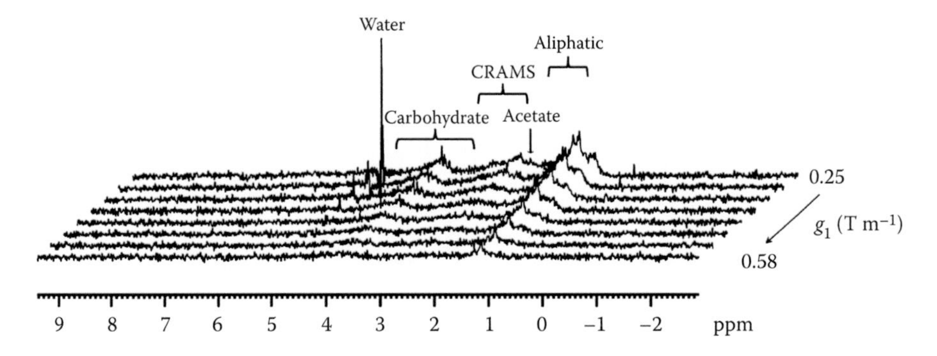


**FIGURE 4.19** $^1$H 500-MHz spectra of an unconcentrated pond water sample (with a dissolved organic matter concentration of ~7 μM) obtained using a modified PGSTE–WATERGATE method (number of scans = 16,384) at different $g_1$ values (i.e. the diffusion weighting gradient pulse). (From Zheng, G. and W. S. Price, *Environmental Science and Technology* 46, pp. 1675–1680, 2012.)

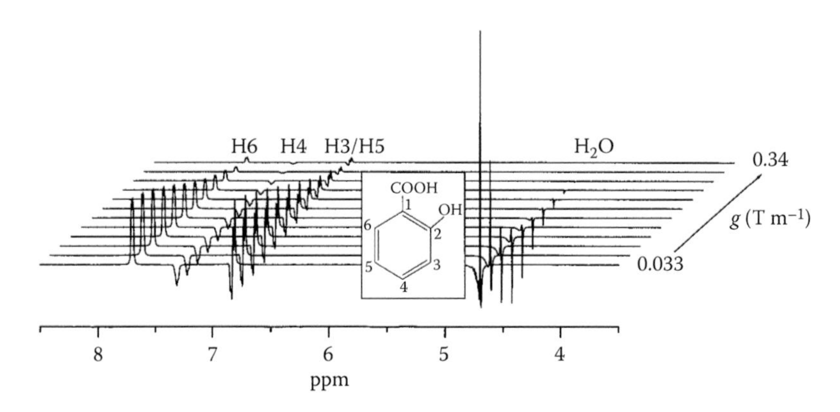


**FIGURE 4.20** $^1$H 500-MHz spectra of 80-mM salicylate and 0.5-mM BSA in 90% $H_2O$ and 10% $D_2O$ using the PGSE–WATERGATE sequence. (From Price, W. S. et al., *Magnetic Resonance in Chemistry* 40, pp. 391–395, 2002.)

### 4.5.4 Convection

Convective motion (or convective flow) is another hindrance for diffusion measurements. Convection is mainly related to the heat capacity of the solvent[66] (it is more likely to occur in solvents with low heat capacities, such as chloroform, even at low temperatures[246]), and so, in solvents with low heat capacities, temperature gradients can easily be created through sample heating from the magnetic pulsed field gradients.[247] Pulse sequences such as the double-stimulated echo can be used to compensate for convective flow (provided that it is laminar).[247] While unidirectional flow results in a net phase change of the signal, convection results in equal amounts of the sample moving in opposite directions, which leads to signal attenuation and, normally, to an increase in the diffusion coefficient.[14,52,247] Although unlike true diffusion, which depends on molecular size, all solutes are equally affected irrespective of size.[66] This behaviour mostly influences slower-diffusing molecules (such as macromolecules), which are more influenced by convective flow, since it results in a larger absolute change in the diffusion coefficient.

## 4.6 CONCLUDING REMARKS

It is impossible to cover all aspects of NMR diffusion measurements and applications to chemical systems. Therefore, only introductory basics are presented in this chapter, and the reader is strongly encouraged to consult any of the references listed in this chapter for additional information.

## REFERENCES


1. Levitt, M. H. 2008. *Spin Dynamics: Basics of Nuclear Magnetic Resonance*, 2nd ed. Chichester, United Kingdom: John Wiley & Sons, Ltd, 686 pp.
2. Tyrrell, H. J. V. and K. R. Harris. 1984. *Diffusion in Liquids: A Theoretical and Experimental Study*. London, United Kingdom: Butterworths, 464 pp.
3. Mills, R. 1965. The intra-diffusion and derived frictional coefficients for benzene and cyclohexane in their mixtures at 25°. *Journal of Physics and Chemistry* 69, p. 3116–3119.
4. Willis, S. A., G. R. Dennis, G. Zheng and W. S. Price. 2010. Hydrodynamic size and scaling relations for linear and 4-arm star PVAc studied using PGSE NMR. *Journal of Molecular Liquids* 156, p. 45–51.
5. Bowen, W. R., A. W. Mohammad and N. Hilal. 1997. Characterisation of nanofiltration membranes for predictive purposes: Use of salts, uncharged solutes, and atomic force microscopy. *Journal of Membrane Science* 126, p. 91–105.
6. Atkins, P. and J. de Paula. 2006. *Atkins' Physical Chemistry*, 8th ed. New York: W. H. Freeman, 1000 pp.
7. Price, W. S., F. Tsuchiya and Y. Arata. 1999. Lysozyme aggregation and solution properties studied using PGSE NMR diffusion measurements. *Journal of the American Chemical Society* 121, p. 11,503–11,512.
8. Li, Y.-H. and S. Gregory. 1974. Diffusion of ions in seawater and in deep-sea sediments. *Geochimica et Cosmochimica Acta* 88, p. 703–714.
9. Braun, B. M. and H. Weingaertner. 1988. Accurate self-diffusion coefficients of $Li^+$, $Na^+$, and $Cs^+$ ions in aqueous alkali metal halide solutions from NMR spin echo experiments. *Journal of Physics and Chemistry* 92, p. 1342–1346.
10. Mills, R. and J. W. Kennedy. 1953. The self-diffusion coefficients of iodide, potassium, and rubidium ions in aqueous solutions. *Journal of the American Chemical Society* 75, p. 5696–5701.
11. Price, W. S. 2009. *NMR Studies of Translational Motion: Principles and Applications*. New York: Cambridge University Press, 416 pp.
12. Kärger, J., C. Papadakis and F. Stallmach. 2004. Structure–mobility relations of molecular diffusion in interface systems. In *Molecules in Interaction with Surfaces and Interfaces*. Heidelberg, Germany: Springer, p. 127–162.
13. Sen, P. N. 2004. Time-dependent diffusion coefficient as a probe of geometry. *Concepts in Magnetic Resonance A* 23A, p. 1–21.
14. Price, W. S. 1997. Pulsed field gradient nuclear magnetic resonance as a tool for studying translational diffusion: Part 1. Basic theory. *Concepts in Magnetic Resonance* 9, p. 299–336.
15. Grebenkov, D. S. 2007. NMR survey of reflected Brownian motion. *Reviews of Modern Physics* 79, p. 1077.

16. Cussler, E. L. 2009. *Diffusion: Mass Transfer in Fluid Systems*, 3rd ed. New York: Cambridge University Press, 647 pp.
17. Kirkaldy, J. S. 1957. Diffusion in multi-component metallic systems. *Canadian Journal of Physics* 35, p. 435–440.
18. Kirkaldy, J. S., R. J. Brigham and D. H. Weichert. 1965. Diffusion interactions in Cu–Zn–Sn as determined from infinite and finite diffusion couples. *Acta Metallurgica* 13, p. 907–915.
19. Tang, X.-C., X.-W. Song, P.-Z. Shen and D.-Z. Jia. 2005. Capacity intermittent titration technique (CITT): A novel technique for determination of $Li^+$ solid diffusion coefficient of $LiMn_2O_4$. *Electrochimica Acta* 50, p. 5581–5587.
20. Cahn, R. W., J. E. Evetts, J. Patterson, R. E. Somekh and C. Kenway Jackson. 1980. Direct measurement by secondary ion mass spectrometry of self-diffusion of boron in $Fe_{40}Ni_{40}B_{20}$ glass. *Journal of Materials Science* 15, p. 702–710.
21. Kingshott, P., S. McArthur, H. Thissen, D. G. Castner and H. J. Griesser. 2002. Ultra-sensitive probing of the protein resistance of PEG surfaces by secondary ion mass spectrometry. *Biomaterials* 23, p. 4775–4785.
22. Heitjans, P. and J. Kärger, 2005. *Diffusion in Condensed Matter: Methods, Materials, Models*, 2nd ed. Berlin, Germany: Springer, 965 pp.
23. Suzuki, K., M. Adachi, and I. Kajizuka. 1994. Electron microprobe observations of Pb diffusion in metamorphosed detrital monazites. *Earth and Planetary Science Letters* 128, p. 391–405.
24. Danyluk, S., G. E. McGuire, K. M. Koliwad and M. G. Yang. 1975. Diffusion studies in Cr–Pt thin films using Auger electron spectroscopy. *Thin Solid Films* 25, p. 483–489.
25. Hall, P. M., J. M. Morabito, and J. M. Poate. 1976. Diffusion mechanisms in the Pd/Au thin film system and the correlation of resistivity changes with Auger electron spectroscopy and Rutherford backscattering profiles. *Thin Solid Films* 33, p. 107–134.
26. Stokes, R. H. 1950. An improved diaphragm cell for diffusion studies, and some tests of the method. *Journal of the American Chemical Society* 72, p. 763–767.
27. Mills, R. 1973. Self-diffusion in normal and heavy water in the range 1 to 45°. *Journal of Physics and Chemistry* 77, p. 685–688.
28. Minton, A. P. 1989. Analytical centrifugation with preparative ultra-centrifuges. *Analytical Biochemistry* 176, p. 209–216.
29. Muramatsu, N. and A. P. Minton. 1988. An automated method for rapid determination of diffusion coefficients via measurements of boundary spreading. *Analytical Biochemistry* 168, p. 345–351.
30. Ouano, A. 1972. Diffusion in liquid systems: I. A simple and fast method of measuring diffusion constants. *Industrial and Engineering Chemistry Fundamentals* 11, p. 268–271.
31. Levich, V. G. and D. B. Spalding. 1962. *Physicochemical Hydrodynamics*. Englewood Cliffs, New Jersey: Prentice-Hall.
32. Visser, A. W. G. and M. Hink. 1999. New perspectives of fluorescence correlation spectroscopy. *Journal of Fluorescence* 9, p. 81–87.
33. Philippe, L., C. Sammon, S. B. Lyon and J. Yarwood. 2004. An FTIR/ATR *in situ* study of sorption and transport in corrosion-protective organic coatings: 1. Water sorption and the role of inhibitor anions. *Progress in Organic Coatings* 49, p. 302–314.
34. Hiss, T. G. and E. L. Cussler. 1973. Diffusion in high-viscosity liquids. *AIChe Journal* 19, p. 698–703.
35. Gosting, L. J. and D. F. Akeley. 1952. A study of the diffusion of urea in water at 25° with the Gouy interference method. *Journal of the American Chemical Society* 74, p. 2058–2060.
36. Rard, J. A. and D. G. Miller. 1979. The mutual diffusion coefficients of NaCl–$H_2O$ and $CaCl_2$–$H_2O$ at 25°C from Rayleigh interferometry. *Journal of Solution Chemistry* 8, p. 701–716.

37. Caldwell, C. S., J. R. Hall and A. L. Babb. 1957. Zehnder interferometer for diffusion measurements in volatile liquid systems. *Review of Scientific Instruments* 28, p. 816–821.
38. Caldwell, C. S. and A. L. Babb. 1956. Diffusion in ideal binary liquid mixtures. *Journal of Physics and Chemistry* 60, p. 51–56.
39. Berne, B. J. 1976. *Dynamic Light Scattering: With Applications to Chemistry, Biology, and Physics*. Mineola, New York: Dover Publications, 384 pp.
40. Kuehner, D., C. Heyer, C. Rämsch et al. 1997. Interactions of lysozyme in concentrated electrolyte solutions from dynamic light-scattering measurements. *Biophysical Journal* 73, p. 3211–3224.
41. Wang, J. H., C. V. Robinson and I. S. Edelman. 1953. Self-diffusion and structure of liquid water: III. Measurement of the self-diffusion of liquid water with $H^2$, $H^3$, and $O^{18}$ as tracers. *Journal of the American Chemical Society* 75, p. 466–470.
42. Wang, J. H. 1952. Tracer diffusion in liquids: I. Diffusion of tracer amount of sodium ion in aqueous potassium chloride solutions. *Journal of the American Chemical Society* 74, p. 1182–1186.
43. Crist, B. 1991. Polymer self-diffusion measurements by small-angle neutron scattering. *Journal of Non-Crystalline Solids* 131–133: 709–714.
44. Boué, F., F. Lefaucheux, M. C. Robert and I. Rosenman. 1993. Small angle neutron scattering study of lysozyme solutions. *Journal of Non-Crystalline Solids* 133, p. 246–254.
45. Hahn, E. L. 1950. Spin echoes. *Physical Review* 80, p. 580–594.
46. Stejskal, E. O. and J. E. Tanner. 1965. Spin diffusion measurements: Spin echoes in the presence of a time-dependent field gradient. *Journal of Chemical Physics* 42, p. 288–292.
47. Zheng, G. and W. S. Price. 2012. Direct hydrodynamic radius measurement on dissolved organic matter in natural waters using diffusion NMR. *Environmental Science and Technology* 46, p. 1675–1680.
48. Keeler, J. 2010. *Understanding NMR Spectroscopy*, 2nd ed. Chichester, United Kingdom: Wiley, 562 pp.
49. Tanner, J. E. 1970. Use of the stimulated echo in NMR diffusion studies. *Journal of Chemical Physics* 52, p. 2523.
50. Torrey, H. C. 1956. Bloch equations with diffusion terms. *Physical Review* 104, p. 563–565.
51. Johnson Jr., C. S. 1999. Diffusion-ordered nuclear magnetic resonance spectroscopy: Principles and applications. *Progress in Nuclear Magnetic Resonance Spectroscopy* 34, p. 203–256.
52. Price, W. S. 1998. Pulsed field gradient nuclear magnetic resonance as a tool for studying translational diffusion: Part II. Experimental aspects. *Concepts in Magnetic Resonance* 10, p. 197–237.
53. Price, W. S. and P. W. Kuchel. 1991. Effect of non-rectangular field gradient pulses in the Stejskal and Tanner (diffusion) pulse sequence. *Journal of Magnetic Resonance* 94, p. 133–139.
54. Fleischer, G. 1985. The effect of polydispersity on measuring polymer self-diffusion with the NMR pulsed field gradient technique. *Polymer* 26, p. 1677–1682.
55. Callaghan, P. T. and D. N. Pinder. 1985. Influence of polydispersity on polymer self-diffusion measurements by pulsed field gradient nuclear magnetic resonance. *Macromolecules* 18, p. 373–379.
56. Callaghan, P. T. and D. N. Pinder. 1983. A pulsed field gradient NMR study of self-diffusion in a polydisperse polymer system: Dextran in water. *Macromolecules* 16, p. 968–973.
57. Willis, S. A., G. R. Dennis, G. Zheng and W. S. Price. 2013. Preparation and physical properties of a macroscopically aligned lyotropic hexagonal phase templated hydrogel. *Reactive and Functional Polymers* 73, p. 911–922.

58. Blinc, R., M. Burgar, M. Luzar et al. 1974. Anisotropy of self-diffusion in the smectic-A and smectic-C phases. *Physical Review Letters* 33, p. 1192.
59. Callaghan, P. T. 1984. Pulsed field gradient nuclear magnetic resonance as a probe of liquid-state molecular organisation. *Australian Journal of Physics* 37, p. 359–387.
60. Minati, L. and W. P. Węglarz. 2007. Physical foundations, models, and methods of diffusion magnetic resonance imaging of the brain: A review. *Concepts in Magnetic A* 30A, p. 278–307.
61. Mattiello, J., P. J. Basser and D. Lebihan. 1994. Analytical expressions for the b matrix in NMR diffusion imaging and spectroscopy. *Journal of Magnetic Resonance, Series A* 108, p. 131–141.
62. Basser, P. J. 1995. Inferring microstructural features and the physiological state of tissues from diffusion-weighted images. *NMR in Biomedicine* 8, p. 333–344.
63. Furó, I. and H. Jóhannesson. 1996. Accurate anisotropic water diffusion measurements in liquid crystals. *Journal of Magnetic Resonance, Series A* 119, p. 15–21.
64. Price, W. S., F. Tsuchiya and Y. Arata. 2001. Time dependence of aggregation in crystallising lysozyme solutions probed using NMR self-diffusion measurements. *Biophysical Journal* 80, p. 1585–1590.
65. Price, W. S. 2006. Protein association studied by NMR diffusometry. *Current Opinion in Colloid and Interface Science* 11, p. 19–23.
66. Price, W. S. 2000. NMR gradient methods in the study of proteins. *Annual Reports, Section C: Physical Chemistry* 96, p. 3–53.
67. Nesmelova, I. V. and V. D. Fedotov. 1998. Self-diffusion and self-association of lysozyme molecules in solution. *Biochimica et Biophysica Acta, Protein Structure and Molecular Enzymology* 1383, p. 311–316.
68. Price, W. S., K. Hayamizu and Y. Arata. 1997. Optimisation of the water-PRESS pulse sequence and its integration into pulse sequences for studying biological macromolecules. *Journal of Magnetic Resonance* 126, p. 256–265.
69. Marqusee, J. A. and J. M. Deutch. 1980. Concentration dependence of the self-diffusion coefficient. *Journal of Chemical Physics* 73, p. 5396–5397.
70. Willis, S. A., G. R. Dennis, G. Zheng and W. S. Price. 2010. Averaging effects in PGSE NMR attenuations observed in bimodal molecular weight PMMA solutions. *Macromolecules* 43, p. 7351–7356.
71. Willis, S. A., W. S. Price, I. K. Eriksson-Scott, G. Zheng and G. R. Dennis. 2012. Influence of polymer architecture on the averaging effects in PGSE NMR attenuations for bimodal solutions of linear and star poly(vinyl acetates). *Journal of Molecular Liquids* 167, p. 110–114.
72. Stait-Gardner, T., S. A. Willis, N. N. Yadav, G. Zheng and W. S. Price. 2009. NMR diffusion measurements of complex systems. In Chmelik, C., N. Kanellopoulos, J. Kärger, and D. Theodorou, eds., *Diffusion Fundamentals III*. Leipzig: Leipzig University Press, p. 183–204 (Invited chapter; also published by the publisher as a journal article: Stait-Gardner, T., S. A. Willis, N. N. Yadav, G. Zheng, and W. S. Price. 2009. NMR diffusion measurements of complex systems. In *Diffusion Fundamentals 11*, Special Issue, *Diffusion Fundamentals III*, p. 1–22).
73. Stokes, G. G. 1856. On the effect of internal friction of fluids on the motion of pendulums. *Transactions of the Cambridge Philosophical Society* 9, p. 8–106.
74. Sutherland, W. 1902. XVIII. Ionisation, ionic velocities, and atomic sizes. *Philosophical Magazine Series 6* 3, p. 161–177.
75. Sutherland, W. 1905. A dynamical theory of diffusion for non-electrolytes and the molecular mass of albumin. *Philosophical Magazine* 9, p. 781–785.
76. Einstein, A. 1956. *Investigations on the Theory of the Brownian Movement*. New York: Dover Publications, 132 pp.

77. Stilbs, P. 1981. Molecular self-diffusion coefficients in Fourier-transform nuclear magnetic resonance spectrometric analysis of complex mixtures. *Analytical Chemistry* 53, p. 2135–2137.
78. van Zijl, P. C. M., C. T. W. Moonen, P. Faustino et al. 1991. Complete separation of intracellular and extra-cellular information in NMR spectra of perfused cells by diffusion-weighted spectroscopy. *Proceedings of the National Academy of Science U.S.A.* 88, p. 3228.
79. Nicolay, K., K. P. J. Braun, R. A. D. Graaf, R. M. Dijkhuizen and M. J. Kruiskamp. 2001. Diffusion NMR spectroscopy. *NMR in Biomedicine* 14, p. 94.
80. Allison, S. A., M. Potter and J. A. McCammon. 1997. Modeling the electrophoresis of lysozyme: II. Inclusion of ion relaxation. *Biophysical Journal* 73, p. 133–140.
81. Asish Xavier, K. and R. C. Willson. 1998. Association and dissociation kinetics of anti–hen egg lysozyme monoclonal antibodies HyHEL-5 and HyHEL-10. *Biophysics Journal* 74, p. 2036–2045.
82. Li, M., A. Nadarajah and M. L. Pusey. 1995. Modeling the growth rates of tetragonal lysozyme crystals. *Journal of Crystal Growth* 156, p. 121–132.
83. Nohara, D., A. Mizutani and T. Sakai. 1999. Kinetic study on thermal denaturation of hen egg-white lysozyme involving precipitation. *Journal of Bioscience and Bioengineering* 87, p. 199–205.
84. Raeymaekers, H. H., H. Eisendrath, A. Verbeken, Y. Van and R. N. Muller. 1989. Nuclear magnetic relaxation dispersion in protein solutions as a probe for protein transformation: Example, the dimerisation of lysozyme. *Journal of Magnetic Resonance* 85, p. 421–425.
85. Retailleau, P., M. Ries-Kautt and A. Ducruix. 1997. No salting-in of lysozyme chloride observed at low ionic strength over a large range of pH. *Biophysics Journal* 73, p. 2156–2163.
86. Velev, O. D., E. W. Kaler and A. M. Lenhoff. 1998. Protein interactions in solution characterised by light and neutron scattering: Comparison of lysozyme and chymotrypsinogen. *Biophysics Journal* 75, p. 2682–2697.
87. Wilson, L. J., L. Adcock-Downey and M. L. Pusey. 1996. Monomer concentrations and dimerisation constants in crystallising lysozyme solutions by dialysis kinetics. *Biophysics Journal* 71, p. 2123–2129.
88. Han, J. and J. Herzfeld. 1993. Macromolecular diffusion in crowded solutions. *Biophysics Journal* 65, p. 1155–1161.
89. Walderhaug, H., O. Söderman and D. Topgaard. 2010. Self-diffusion in polymer systems studied by magnetic field gradient–spin echo NMR methods. *Progress in Nuclear Magnetic Resonance Spectroscopy* 56, p. 406–425.
90. Amsden, B. 1999. An obstruction-scaling model for diffusion in homogeneous hydrogels. *Macromolecules* 32, p. 874–879.
91. Amsden, B. 1998. Solute diffusion within hydrogels. Mechanisms and models. *Macromolecules* 31, p. 8382–8395.
92. Cukier, R. I. 1984. Diffusion of Brownian spheres in semi-dilute polymer solutions. *Macromolecules* 17, p. 252–255.
93. Ogston, A. G., B. N. Preston and J. D. Wells. 1973. On the transport of compact particles through solutions of chain polymers. *Proceedings of the Royal Society of London A* 333, p. 297–316.
94. Darwish, M. I. M., J. R. C. van der Maarel and P. L. J. Zitha. 2004. Ionic transport in polyelectrolyte gels: Model and NMR investigations. *Macromolecules* 37, p. 2307–2312.
95. Johansson, L., C. Elvingson and J. E. Loefroth. 1991. Diffusion and interaction in gels and solutions: 3. Theoretical results on the obstruction effect. *Macromolecules* 24, p. 6024–6029.
96. Schlick, S., J. Pilar, S. C. Kweon et al. 1995. Measurements of diffusion processes in HEMA–DEGMA hydrogels by ESR imaging. *Macromolecules* 28, p. 5780–5788.

97. Feil, H., Y. H. Bae, J. Feijen and S. W. Kim. 1991. Molecular separation by thermosensitive hydrogel membranes. *Journal of Membrane Science* 64, p. 283–294.
98. Gagnon, M.-A. and M. Lafleur. 2009. Self-diffusion and mutual diffusion of small molecules in high-set curdlan hydrogels studied by $^{31}$P NMR. *Journal of Physics and Chemistry B* 113, p. 9084–9091.
99. Söderman, O., P. Stilbs and W. S. Price. 2004. NMR studies of surfactants. *Concepts in Magnetic Resonance A* 23A, p. 121–135.
100. Palit, S. and A. Yethiraj. 2008. A new model system for diffusion NMR studies of concentrated monodisperse and bidisperse colloids. *Langmuir* 24, p. 3747–3751.
101. Jönsson, B., H. Wennerström, P. G. Nilsson and P. Linse. 1986. Self-diffusion of small molecules in colloidal systems. *Colloid and Polymer Science* 264, p. 77–88.
102. Jóhannesson, H. and B. Halle. 1996. Solvent diffusion in ordered macrofluids: A stochastic simulation study of the obstruction effect. *Journal of Chemical Physics* 104, p. 6807–6817.
103. Cooper, R. L., D. B. Chang, A. C. Young, C. J. Martin and B. Ancker-Johnson. 1974. Restricted diffusion in biophysical systems: Experiment. *Biophysics Journal* 14, p. 161–177.
104. Lindblom, G. and G. Orädd. 1994. NMR studies of translational diffusion in lyotropic liquid crystals and lipid membranes. *Progress in Nuclear Magnetic Resonance Spectroscopy* 26, p. 483–515.
105. Celebre, G., L. Coppola and G. A. Ranieri. 1992. Water self-diffusion in lyotropic systems by simulation of pulsed field gradient–spin echo nuclear magnetic resonance experiments. *Journal of Chemical Physics* 97, p. 7781–7785.
106. Celebre, G., L. Coppola, G. A. Ranieri and M. Terenzi. 1994. Analysis of the PFG-SE NMR experiments in lyotropic mesophases: The hexagonal case. *Molecular Crystals and Liquid Crystals Science* 238, p. 117–123.
107. Gaemers, S. and A. Bax. 2001. Morphology of three lyotropic liquid-crystalline biological NMR media studied by translational diffusion anisotropy. *Journal of the American Chemical Society* 123, p. 12,343–12,352.
108. Kleinschmidt, F., M. Hickl, K. Saalwachter, C. Schmidt and H. Finkelmann. 2005. Lamellar liquid single crystal hydrogels: Synthesis and investigation of anisotropic water diffusion and swelling. *Macromolecules* 38, p. 9772–9782.
109. Kleinschmidt, F. 2005. *Anomalous Diffusion in Anisotropic Media*. PhD thesis, Albert-Ludwigs-Universität, Freiburg, Germany.
110. Chidichimo, G., D. de Fazio, G. A. Ranieri and M. Terenzi. 1986. Water diffusion and phase transition investigation in lyotropic mesophases: An NMR study. *Molecular Crystals and Liquid Crystals Science* 135, p. 223–236.
111. Chidichimo, G., D. de Fazio, G. A. Ranieri and M. Terenzi. 1985. Self-diffusion of water in a lamellar lyotropic liquid crystal: A study by pulsed field gradient NMR. *Chemical Physics Letters* 117, p. 514–517.
112. Coppola, L., S. D. Gregorio, G. A. Ranieri and G. Rocca. 1991. Simulation of restricted self-diffusion. *Molecular Simulation* 7, p. 241–247.
113. Anderson, D. M. and H. Wennerstroem. 1990. Self-diffusion in bicontinuous cubic phases, L3 phases, and microemulsions. *Journal of Physics and Chemistry* 94, p. 8683–8694.
114. Geil, B., T. Feiweier, E. M. Pospiech et al. 2000. Relating structure and translational dynamics in aqueous dispersions of monoolein. *Chemistry and Physics of Liquids* 106, p. 115–126.
115. Feiweier, T., B. Geil, E.-M. Pospiech, F. Fujara and R. Winter. 2000. NMR study of translational and rotational dynamics in monoolein-water mesophases: Obstruction and hydration effects. *Physical Review E* 62, p. 8182–8194.
116. Jóhannesson, H., I. Furó and B. Halle. 1996. Orientational order and micelle size in the nematic phase of the cesium pentadecafluorooctanoate-water system from the anisotropic self-diffusion of water. *Physical Review E* 53, p. 4904.

117. Yethiraj, A., D. Capitani, N. E. Burlinson and E. E. Burnell. 2005. An NMR study of translational diffusion and structural anisotropy in magnetically alignable non-ionic surfactant mesophases. *Langmuir* 21, p. 3311–3321.
118. Dagdug, L., M.-V. Vazquez, A. M. Berezhkovskii, V. Y. Zitserman and S. M. Bezrukov. 2012. Diffusion in the presence of cylindrical obstacles arranged in a square lattice analysed with generalised Fick–Jacobs equation. *Journal of Chemistry and Physics* 136, p. 204106-5.
119. Moggridge, G. D., N. K. Lape, C. Yang and E. L. Cussler. 2003. Barrier films using flakes and reactive additives. *Progress in Organic Coatings* 46, p. 231–240.
120. Cussler, E. L., S. E. Hughes, W. J. Ward III and R. Aris. 1988. Barrier membranes. *Journal of Membrane Science* 38, p. 161–174.
121. White, J. D. and E. L. Cussler. 2006. Anisotropic transport in water-swollen, flake-filled membranes. *Journal of Membrane Science* 278, p. 225–231.
122. Lape, N. K., C. Yang and E. L. Cussler. 2002. Flake-filled reactive membranes. *Journal of Membrane Science* 209, p. 271–282.
123. Lape, N. K., E. E. Nuxoll and E. L. Cussler. 2004. Polydisperse flakes in barrier films. *Journal of Membrane Science* 236, p. 29–37.
124. Aris, R. 1986. On a problem in hindered diffusion. *Archive for Rational Mechanics and Analysis* 95, p. 83–91.
125. Brydges, W. T., S. T. Gulati and G. Baum. 1975. Permeability of glass ribbon–reinforced composites. *Journal of Materials Science* 10, p. 2044–2049.
126. Latour, L. L., P. P. Mitra, R. L. Kleinberg and C. H. Sotak. 1993. Time-dependent diffusion coefficient of fluids in porous media as a probe of surface-to-volume ratio. *Journal of Magnetic Resonance, Series A* 101, p. 342–346.
127. Geier, O., R. Q. Snurr, F. Stallmach and J. Kärger. 2004. Boundary effects of molecular diffusion in nanoporous materials: A pulsed field gradient nuclear magnetic resonance study. *Journal of Chemical Physics* 120, p. 367–373.
128. Mitra, P. P., P. N. Sen and L. M. Schwartz. 1993. Short-time behaviour of the diffusion coefficient as a geometrical probe of porous media. *Physical Review B* 47, p. 8565.
129. Mitra, P. P., P. N. Sen, L. M. Schwartz and P. le Doussal. 1992. Diffusion propagator as a probe of the structure of porous media. *Physical Review Letters* 68, p. 3555.
130. Hizi, U. 2000. Molecular diffusion in periodic porous media. *Journal of Applied Physics* 87, p. 1704.
131. Syková, E. and C. Nicholson. 2008. Diffusion in brain extracellular space. *Physiological Reviews* 88, p. 1277–1340.
132. Crick, F. 1970. Diffusion in embryogenesis. *Nature* 225, p. 420–422.
133. Tanner, J. E. 1978. Transient diffusion in a system partitioned by permeable barriers: Application to NMR measurements with a pulsed field gradient. *Journal of Chemical Physics* 69, p. 1748–1754.
134. Dudko, O. K., A. M. Berezhkovskii and G. H. Weiss. 2004. Diffusion in the presence of periodically spaced permeable membranes. *Journal of Chemical Physics* 121, p. 11,283–11,288.
135. Wästerby, P., G. Orädd and G. Lindblom. 2002. Anisotropic water diffusion in macroscopically oriented lipid bilayers studied by pulsed magnetic field gradient NMR. *Journal of Magnetic Resonance* 157, p. 156–159.
136. Kärger, J., H. Pfeifer and W. Heink. 1988. Principles and application of self-diffusion measurements by nuclear magnetic resonance. *Advances in Magnetic and Optical Resonance* 12, p. 1–89.
137. Markworth, A. J. 1993. The transverse thermal conductivity of a unidirectional fiber composite with fiber matrix debonding: A calculation based on effective-medium theory. *Journal of Materials Science Letters* 12, p. 1487–1489.

138. Manteufel, R. D. and N. E. Todreas. 1994. Analytic formulae for the effective conductivity of a square or hexagonal array of parallel tubes. *International Journal of Heat and Mass Transfer* 37, p. 647–657.
139. Rayleigh, R. S. 1892. LVI. On the influence of obstacles arranged in rectangular order upon the properties of a medium. *Philosophical Magazine Series 5* 34, p. 481–502.
140. Photinos, P. and A. Saupe. 1981. Calculation of the electric conductivity of the hexagonal lyotropic mesophase. *Journal of Chemical Physics* 75, p. 1313–1315.
141. Photinos, P. J. and A. Saupe. 1984. Calculations on the electric conductivity of a lyotropic mesophase with perforated lamellae. *Journal of Chemical Physics* 81, p. 563–566.
142. Zimmerman, S. B. and A. P. Minton. 1993. Macromolecular crowding: Biochemical, biophysical, and physiological consequences. *Annual Review of Biophysics and Biomolecular Structure* 22, p. 27–65.
143. Bernadó, P., J. G. de la Torre and M. Pons. 2004. Macromolecular crowding in biological systems: Hydrodynamics and NMR methods. *Journal of Molecular Recognition* 17, p. 397–407.
144. Tokuyama, M. and R. I. Cukier. 1982. Dynamics of diffusion-controlled reactions among stationary sinks: Scaling expansion approach. *Journal of Chemical Physics* 76, p. 6202–6214.
145. Wang, J. H. 1954. Theory of the self-diffusion of water in protein solutions: A new method for studying the hydration and shape of protein molecules. *Journal of the American Chemical Society* 76, p. 4755–4763.
146. Clark, M., E. Burnell, N. Chapman and J. Hinke. 1982. Water in barnacle muscle: IV. Factors contributing to reduced self-diffusion. *Biophysics Journal* 39, p. 289–299.
147. Tokuyama, M. and I. Oppenheim. 1994. Dynamics of hard-sphere suspensions. *Physical Review E* 50, p. 16–19.
148. Lekkerkerker, H. N. W. and J. K. G. Dhont. 1984. On the calculation of the self-diffusion coefficient of interacting Brownian particles. *Journal of Chemical Physics* 80, p. 5790–5792.
149. von Goldammer, E. and H. G. Hertz. 1970. Molecular motion and structure of aqueous mixtures with non-electrolytes as studied by nuclear magnetic relaxation methods. *Journal of Physics and Chemistry* 74, p. 3734–3755.
150. Lindman, B. and B. Brun. 1973. Translational motion in aqueous sodium N-octanoate solutions. *Journal of Colloid and Interface Science* 42, p. 388–399.
151. Lindman, B., P. Stilbs and M. E. Moseley. 1981. Fourier-transform NMR self-diffusion and microemulsion structure. *Journal of Colloid and Interface Science* 83, p. 569–582.
152. Stilbs, P. 1982. Fourier-transform NMR pulsed gradient spin echo (FT-PGSE) self-diffusion measurements of solubilisation equilibria in SDS solutions. *Journal of Colloid and Interface Science* 87, p. 385–394.
153. Stilbs, P. 1982. Micellar breakdown by short-chain alcohols: A multi-component FT-PGSE NMR self-diffusion study. *Journal of Colloid and Interface Science* 89, p. 547–554.
154. Stilbs, P. 1983. A comparative study of micellar solubilisation for combinations of surfactants and solubilisates using the Fourier-transform pulsed gradient spin echo NMR multi-component self-diffusion technique. *Journal of Colloid and Interface Science* 94, p. 463–469.
155. Kärger, J. 1985. NMR self-diffusion studies in heterogeneous systems. *Advances in Colloid and Interface Science* 23, p. 129–148.
156. Price, W. S., F. Elwinger, C. Vigouroux and P. Stilbs. 2002. PGSE–WATERGATE, a new tool for NMR diffusion-based studies of ligand–macromolecule binding. *Magnetic Resonance in Chemistry* 40, p. 391–395.
157. Price, W. S., F. Hallberg and P. Stilbs. 2007. A PGSE diffusion and electrophoretic NMR study of $Cs^+$ and $Na^+$ dynamics in aqueous crown ether systems. *Magnetic Resonance in Chemistry* 45, p. 152–156.

158. Fielding, L. 2000. Determination of association constants (Ka) from solution NMR data. *Tetrahedron* 56, p. 6151–6170.
159. Stilbs, P. 1987. Fourier-transform pulsed gradient spin echo studies of molecular diffusion. *Progress in Nuclear Magnetic Resonance Spectroscopy* 19, p. 1–45.
160. Nelson, D. L. and M. M. Cox. 2005. *Lehninger Principles of Biochemistry*, 4th ed. New York: W. H. Freeman and Company, 1158 pp.
161. Housecroft, C. E. and A. G. Sharpe. 2005. *Inorganic Chemistry*, 2nd ed. England, United Kingdom: Pearson Education, Ltd, 210 pp.
162. Bringuier, E. and A. Bourdon. 2007. Kinetic theory of colloid thermodiffusion. *Physica A* 385, p. 9–24.
163. Hansen, S. 2004. Translational friction coefficients for cylinders of arbitrary axial ratios estimated by Monte Carlo simulation. *Journal of Chemical Physics* 121, p. 9111–9115.
164. Ortega, A. and J. García de la Torre. 2003. Hydrodynamic properties of rod-like and disc-like particles in dilute solution. *Journal of Chemical Physics* 119, p. 9914.
165. García de la Torre, J. and V. A. Bloomfield. 1981. Hydrodynamic properties of complex, rigid, biological macromolecules: Theory and applications. *Quarterly Reviews of Biophysics* 14, p. 81–139.
166. Doi, M. and S. F. Edwards. 1986. *The Theory of Polymer Dynamics*. New York: Oxford University Press, 406 pp.
167. Tirado, M. M. and J. García de la Torre. 1979. Translational friction coefficients of rigid, symmetric top macromolecules: Application to circular cylinders. *Journal of Chemical Physics* 71, p. 2581–2587.
168. Tirado, M. M., C. L. Martinez and J. García de la Torre. 1984. Comparison of theories for the translational and rotational diffusion coefficients of rod-like macromolecules: Application to short DNA fragments. *Journal of Chemical Physics* 81, p. 2047–2052.
169. da Costa, V. C. P., A. C. F. Ribeiro, A. J. F. N. Sobral et al. 2012. Mutual and self-diffusion of charged porphyrines in aqueous solutions. *Journal of Chemical Thermodynamics* 47, p. 312–319.
170. Young, H. D. and R. A. Freedman, 2000. *Sears and Zemansky's University Physics with Modern Physics*, 10th ed. San Francisco, California: Addison Wesley Longman, Inc, 1550 pp.
171. Ravi, R. and D. Ben-Amotz. 1994. Translational and rotational dynamics in liquids: Comparison of experiment, kinetic theory, and hydrodynamics. *Chemical Physics* 183, p. 385–392.
172. García de la Torre, J. 2001. Hydration from hydrodynamics: General considerations and applications of bead modelling to globular proteins. *Biophysical Chemistry* 93, p. 159–170.
173. Zwanzig, R. and M. Bixon. 1970. Hydrodynamic theory of the velocity correlation function. *Physical Review A* 2, p. 2005–2012.
174. Evans, D. F., T. Tominaga and H. T. Davis. 1981. Tracer diffusion in polyatomic liquids. *Journal of Chemical Physics* 74, p. 1298–1305.
175. Gerstein, M., J. Tsai and M. Levitt. 1995. The volume of atoms on the protein surface: Calculated from simulation, using Voronoi polyhedra. *Journal of Molecular Biology* 249, p. 955–966.
176. Halle, B. and M. Davidovic. 2003. Biomolecular hydration: From water dynamics to hydrodynamics. *Proceedings of the National Academy of Science U.S.A.* 100, p. 12,135–12,140.
177. Zipper, P. and H. Durchschlag. 2002. Prediction of structural and hydrodynamic parameters of hydrated proteins by computer modelling based on the results from high-resolution techniques. *Physica A* 304, p. 283–293.
178. Schultz, S. G. and A. K. Solomon. 1961. Determination of the effective hydrodynamic radii of small molecules by viscometry. *Journal of General Physiology* 44, p. 1189–1199.

179. Nightingale, E. R. 1959. Phenomenological theory of ion solvation: Effective radii of hydrated ions. *Journal of Physics and Chemistry* 63, p. 1381–1387.
180. Collins, K. D. and M. W. Washabaugh. 1985. The Hofmeister effect and the behaviour of water at interfaces. *Quarterly Reviews of Biophysics* 18, p. 323–422.
181. Collins, K. D. 1997. Charge density–dependent strength of hydration and biological structure. *Biophysics Journal* 72, p. 65–76.
182. Marcus, Y. 1988. Ionic radii in aqueous solutions. *Chemical Reviews* 88, p. 1475–1498.
183. Marcus, Y. 1994. A simple empirical model describing the thermodynamics of hydration of ions of widely varying charges, sizes, and shapes. *Biophysical Chemistry* 51, p. 111–127.
184. Kiriukhin, M. Y. and K. D. Collins. 2002. Dynamic hydration numbers for biologically important ions. *Biophysical Chemistry* 99, p. 155–168.
185. Ohtaki, H. and T. Radnai. 1993. Structure and dynamics of hydrated ions. *Chemical Reviews* 93, p. 1157–1204.
186. Marcus, Y. 1997. *Ion Properties*. New York: Marcel Dekker, Inc, 272 pp.
187. Impey, R. W., P. A. Madden and I. R. McDonald. 1983. Hydration and mobility of ions in solution. *Journal of Physics and Chemistry* 87, p. 5071–5083.
188. Bockris, J. O. M. and A. K. N. Reddy. 1998. *Modern Electrochemistry 1: Ionics*, 2nd ed. New York: Plenum Press, 828 pp.
189. Evans, D. F., T. Tominaga, J. B. Hubbard and P. G. Wolynes. 1979. Ionic mobility: Theory meets experiment. *Journal of Physics and Chemistry* 83, p. 2669–2677.
190. Kay, R. L. and D. F. Evans. 1966. The effect of solvent structure on the mobility of symmetrical ions in aqueous solution. *Journal of Physics and Chemistry* 70, p. 2325–2335.
191. Hubbard, J. and L. Onsager. 1977. Dielectric dispersion and dielectric friction in electrolyte solutions: I. *Journal of Chemical Physics* 67, p. 4850–4857.
192. Kay, R. L. 1991. The current state of our understanding of ionic mobilities. *Pure and Applied Chemistry* 63, p. 1393–1399.
193. Ibuki, K. and M. Nakahara. 1987. Test of the Hubbard–Onsager dielectric friction theory of ion mobility in non-aqueous solvents: 1. Ion-size effect. *Journal of Physics and Chemistry* 91, p. 1864–1867.
194. Ibuki, K. and M. Nakahara. 1986. Interpretation of the 'Stokes radius' in terms of Hubbard–Onsager's dielectric friction theory. *Journal of Physics and Chemistry* 90, p. 6362–6365.
195. Cho, C. H., J. Urquidi, S. Singh and G. W. Robinson. 1999. Thermal offset viscosities of liquid $H_2O$, $D_2O$, and $T_2O$. *Journal of Physics and Chemistry B* 103, p. 1991–1994.
196. Pappenheimer, J. R., E. M. Renkin and L. M. Borrero. 1951. Filtration, diffusion, and molecular sieving through peripheral capillary membranes: A contribution to the pore theory of capillary permeability. *American Journal of Physiology* 167, p. 13–46.
197. McDonald, T. R. R. and C. A. Beevers. 1950. The crystal structure of α-D-glucose. *Acta Crystallographica* 3, p. 394–395.
198. McDonald, T. R. R. and C. A. Beevers. 1952. The crystal and molecular structure of α-glucose. *Acta Crystallographica* 5, p. 654–659.
199. Ferrier, W. G. 1960. The crystal structure of β-D-glucose. *Acta Crystallographica* 13, p. 678–679.
200. Krasilnikov, O. V., R. Z. Sabirov, V. I. Ternovsky, P. G. Merzliak and J. N. Muratkhodjaev. 1992. A simple method for the determination of the pore radius of ion channels in planar lipid bilayer membranes. *FEMS Microbiology Letters* 105, p. 93–100.
201. Beevers, C. A., T. R. R. McDonald, J. H. Robertson and F. Stern. 1952. The crystal structure of sucrose. *Acta Crystallographica* 5, p. 689–690.
202. Brown, G. M. and H. A. Levy. 1973. Further refinement of the structure of sucrose based on neutron-diffraction data. *Acta Crystallographica, Section B* 29, p. 790–797.

203. Desnoyers, J. E., G. E. Pelletier and C. Jolicoeur. 1965. Salting-in by quaternary ammonium salts. *Canadian Journal of Chemistry* 43, p. 3232–3237.
204. Jones, G. and M. Dole. 1929. The viscosity of aqueous solutions of strong electrolytes with special reference to barium chloride. *Journal of the American Chemical Society* 51, p. 2950–2964.
205. Marcus, Y. 1987. Thermodynamics of ion hydration and its interpretation in terms of a common model. *Pure and Applied Chemistry* 59, p. 1093–1101.
206. Zhou, M., P. R. Nemade, X. Lu et al. 2007. New type of membrane material for water desalination based on a cross-linked bicontinuous cubic lyotropic liquid-crystal assembly. *Journal of the American Chemical Society* 129, p. 9574–9575.
207. Zhou, M., T. J. Kidd, R. D. Noble and D. L. Gin. 2005. Supported lyotropic liquid-crystal polymer membranes: Promising materials for molecular size–selective aqueous nanofiltration. *Advanced Materials* 17, p. 1850–1853.
208. Gin, D. L., X. Lu, P. R. Nemade et al. 2006. Recent advances in the design of polymerisable lyotropic liquid-crystal assemblies for heterogeneous catalysis and selective separations. *Advanced Functional Materials* 16, p. 865–878.
209. Gin, D. L., J. E. Bara, R. D. Noble and B. J. Elliott. 2008. Polymerised lyotropic liquid-crystal assemblies for membrane applications. *Macromolecular Rapid Communications* 29, p. 367–389.
210. Ferry, J. D. 1936. Statistical evaluation of sieve constants in ultrafiltration. *Journal of General Physiology* 20, p. 95–104.
211. Nakao, S.-I. 1994. Determination of pore size and pore size distribution: 3. Filtration membranes. *Journal of Membrane Science* 96, p. 131–165.
212. Lanteri, Y., A. Szymczyk and P. Fievet. 2008. Influence of steric, electric, and dielectric effects on membrane potential. *Langmuir* 24, p. 7955–7962.
213. Abidi, A., N. Gherraf, S. Ladjel, M. Rabiller-Baudry and T. Bouchami. 2011. Effect of operating parameters on the selectivity of nanofiltration phosphates transfer through a Nanomax-50 membrane. *Arabian Journal of Chemistry* 20, doi:10.1016/j.arabjc.2011.04.014.
214. Yaroshchuk, A. E. 2000. Dielectric exclusion of ions from membranes. *Advances in Colloid and Interface Science* 85, p. 193–230.
215. Ballet, G. T., A. Hafiane and M. Dhahbi. 2007. Influence of operating conditions on the retention of phosphate in water by nanofiltration. *Journal of Membrane Science* 290, p. 164–172.
216. Lee, A. and G. M. Whitesides. 2010. Analysis of inorganic polyphosphates by capillary gel electrophoresis. *Analytical Chemistry* 82, p. 6838–6846.
217. Degani, C. and M. Halmann. 1966. Solvolysis of phosphoric acid esters: Hydrolysis of glucose 6-phosphate – Kinetic and tracer studies. *Journal of the American Chemical Society* 88, p. 4075–4082.
218. Aylward, G. and T. Findlay. 2002. *SI Chemical Data*, 5th ed. Singapore: John Wiley & Sons Australia, Ltd, 202 pp.
219. Burl, M. and I. R. Young. 2007. Eddy currents and their control. In *eMagRes: The Ultimate Online Resource for NMR and MRI*. United Kingdom: John Wiley & Sons, Ltd.
220. Mansfield, P. and B. Chapman. 1986. Active magnetic screening of gradient coils in NMR imaging. *Journal of Magnetic Resonance* 66, p. 573–576.
221. Majors, P. D., J. L. Blackley, S. A. Altobelli, A. Caprihan and E. Fukushima. 1990. Eddy current compensation by direct field detection and digital gradient modification. *Journal of Magnetic Resonance* 87, p. 548–553.
222. Gibbs, S. J. and C. S. Johnson Jr. 1991. A PFG NMR experiment for accurate diffusion and flow studies in the presence of Eddy currents. *Journal of Magnetic Resonance* 93, p. 395–402.

223. Reese, T. G., O. Heid, R. M. Weisskoff and V. J. Wedeen. 2003. Reduction of Eddy current–induced distortion in diffusion MRI using a twice-refocussed spin echo. *Magnetic Resonance in Medicine* 49, p. 177–182.
224. Heid, O. 2000. Eddy current–nulled diffusion weighting. *Proceedings of the 8th Annual Meeting of ISMRM* 8, p. 799.
225. Weiger, M. and T. Speck. 2007. Shimming for high-resolution NMR spectroscopy. In *eMagRes: The Ultimate Online Resource for NMR and MRI*. United Kingdom: John Wiley & Sons, Ltd.
226. Miner, V. W. and W. W. Conover. 2007. *Shimming of superconducting magnets*. In *eMagRes: The Ultimate Online Resource for NMR and MRI*. United Kingdom: John Wiley & Sons, Ltd.
227. Price, W. S., P. Stilbs, B. Jonsson and O. Soderman. 2001. Macroscopic background gradient and radiation damping effects on high-field PGSE NMR diffusion measurements. *Journal of Magnetic Resonance* 150, p. 49–56.
228. Cotts, R. M., M. J. R. Hoch, T. Sun and J. T. Markert. 1989. Pulsed field gradient stimulated echo methods for improved NMR diffusion measurements in heterogeneous systems. *Journal of Magnetic Resonance* 83, p. 252–266.
229. Karlicek, R. F. and I. J. Lowe. 1980. A modified pulsed gradient technique for measuring diffusion in the presence of large background gradients. *Journal of Magnetic Resonance* 37, p. 75–91.
230. Latour, L. L., L. M. Li and C. H. Sotak. 1993. Improved PFG stimulated echo method for the measurement of diffusion in inhomogeneous fields. *Journal of Magnetic Resonance, Series B* 101, p. 72–77.
231. Wider, G., V. Dotsch and K. Wuthrich. 1994. Self-compensating pulsed magnetic field gradients for short recovery times. *Journal of Magnetic Resonance, Series A* 108, p. 255–258.
232. Price, W. S. 1999. Water signal suppression in NMR spectroscopy. *Annual Reports on NMR Spectroscopy* 38, p. 289–354.
233. Zheng, G. and W. S. Price. 2010. Solvent signal suppression in NMR. *Progress in Nuclear Magnetic Resonance Spectroscopy* 56, p. 267–288.
234. Guéron, M. and P. Plateau. 2007. Water signal suppression in NMR of biomolecules. In *eMagRes: The Ultimate Online Resource for NMR and MRI*. United Kingdom: John Wiley & Sons, Ltd.
235. Moonen, C. T. W. and P. C. M. van Zijl. 2007. Water suppression in proton MRS of humans and animals. In *eMagRes: The Ultimate Online Resource for NMR and MRI*. United Kingdom: John Wiley & Sons, Ltd.
236. Liu, M. and X.-A. Mao. 1999. Solvent suppression methods in NMR spectroscopy. In *Encyclopedia of Spectroscopy and Spectrometry*. Oxford, United Kingdom: Elsevier, p. 2145–2152.
237. Piotto, M., V. Saudek and V. Sklenář. 1992. Gradient-tailored excitation for single-quantum NMR spectroscopy of aqueous solutions. *Journal of Biomolecular NMR* 2, p. 661–665.
238. Sklenar, V., M. Piotto, R. Leppik and V. Saudek. 1993. Gradient-tailored water suppression for $^{1}H$–$^{15}N$ HSQC experiments optimised to retain full sensitivity. *Journal of Magnetic Resonance, Series A* 102, p. 241–245.
239. Nguyen, B. D., X. Meng, K. J. Donovan and A. J. Shaka. 2007. SOGGY: Solvent-optimised double gradient spectroscopy for water suppression: A comparison with some existing techniques. *Journal of Magnetic Resonance* 184, p. 263–274.
240. van Zijl, P. C. M. and C. T. W. Moonen. 1990. Complete water suppression for solutions of large molecules based on diffusional differences between solute and solvent (DRYCLEAN). *Journal of Magnetic Resonance* 87, p. 18–25.

241. Price, W. S. and Y. Arata. 1996. The manipulation of water relaxation and water suppression in biological systems using the water-PRESS pulse sequence. *Journal of Magnetic Resonance, Series B* 112, p. 190–192.
242. Mao, X. A. and C. H. Ye. 1997. Understanding radiation damping in a simple way. *Concepts in Magnetic Resonance* 9, p. 173–187.
243. Warren, W. S. and W. Richter. 2007. Concentrated solution effects. In *eMagRes: The Ultimate Online Resource for NMR and MRI.* United Kingdom: John Wiley & Sons, Ltd.
244. Simpson, A. J. and S. A. Brown. 2005. Purge NMR: Effective and easy solvent suppression. *Journal of Magnetic Resonance* 175, p. 340–346.
245. Zheng, G., T. Stait-Gardner, P. G. Anil Kumar, A. M. Torres and W. S. Price. 2008. PGSTE–WATERGATE: An STE-based PGSE NMR sequence with excellent solvent suppression. *Journal of Magnetic Resonance* 191, p. 159–163.
246. Esturau, N., F. Sanchez-Ferrando, J. A. Gavin et al. 2001. The use of sample rotation for minimising convection effects in self-diffusion NMR measurements. *Journal of Magnetic Resonance* 153, p. 48–55.
247. Jerschow, A. and N. Müller. 1997. Suppression of convection artefacts in stimulated echo diffusion experiments: Double-stimulated echo experiments. *Journal of Magnetic Resonance* 125, p. 372–375.

# 5 Multi-Nuclear Magnetic Resonance Spectroscopy

*Jonathan A. Iggo and Konstantin V. Luzyanin*

## CONTENTS

5.1 Introduction ........ 178
5.1.1 Chemical Shift ........ 185
5.2 Acquisition of the NMR Spectrum of an 'Inorganic' Nuclide: For the First Time ........ 186
5.2.1 Spectrometer ........ 186
5.2.2 Test Sample ........ 187
5.2.3 Referencing ........ 188
5.2.4 Solvent and Temperature Considerations ........ 189
5.2.5 Running 'Off Lock' ........ 189
5.3 Choice of Experiment ........ 189
5.3.1 Direct Detection: INEPT ........ 189
5.3.2 Indirect Detection ........ 190
5.3.2.1 Instrumental Considerations ........ 191
5.3.2.2 HMQC and HSQC Approaches ........ 191
5.3.2.3 HMBC Spectroscopy ........ 196
5.3.3 Selective Observation and Insensitive (*S*) Spin Satellites ........ 197
5.4 Studying Dynamic Systems ........ 198
5.4.1 Simulation of VT Spectra ........ 198
5.4.2 Saturation Transfer Exchange Spectroscopy ........ 198
5.4.3 Exchange Spectroscopy Experiments ........ 199
5.5 More Experiment Options and Considerations ........ 200
5.5.1 Gradient-Accelerated HOESY Spectroscopy ........ 200
5.5.2 Pulsed Field Gradient Spin Echo Diffusion Spectroscopy ........ 201
5.5.3 Para-Hydrogen–Enhanced Spectroscopy ........ 203
5.5.4 Quadrupolar Nuclei ........ 203
5.6 Selected Examples for Measurements of Inorganic Nuclides ........ 203
5.6.1 Applications of $^{15}N$ NMR ........ 203
5.6.1.1 Properties of $^{15}N$ Isotope ........ 203
5.6.1.2 Chemical Shifts Range and Coupling Constants ........ 204
5.6.1.3 Examples of the Use of $^{15}N$ NMR ........ 204
5.6.2 Applications of $^{195}Pt$ NMR ........ 207
5.6.2.1 Properties of $^{195}Pt$ Isotope ........ 207
5.6.2.2 Chemical Shifts Range and Coupling Constants ........ 207
5.6.2.3 Examples of the Uses of $^{195}Pt$ NMR ........ 209

5.7 Measurement of Quadrupolar Nuclei ........ 214
5.7.1 General Considerations ........ 214
5.7.2 Applications of $^{61}$Ni NMR as a Typical Quadrupolar Nuclide ........ 216
5.7.2.1 Properties of $^{61}$Ni Isotope ........ 216
5.7.2.2 Chemical Shifts Range and Coupling Constants ........ 216
5.7.2.3 Examples of the Use of $^{61}$Ni NMR ........ 216
5.8 Summary ........ 219
References ........ 219
Further Readings ........ 224

## 5.1 INTRODUCTION

This chapter is aimed at a reader who has some experience in operating an nuclear magnetic resonance (NMR) spectrometer 'out of automation mode' to record $^{1}$H and $^{13}$C NMR spectra but is not an NMR specialist, and who now has a need to run the NMR spectra of 'inorganic' nuclei as heteroatoms either in organic compounds or in organometallic or inorganic molecules. Tables 5.1 and 5.2 list the types of information accessible via NMR spectroscopic studies of inorganic nuclei together with information not accessible and some limitations.

The power of NMR spectroscopy in the characterisation of unknown compounds follows from the sensitivity of the chemical shift to the chemical environment and

**TABLE 5.1**
**Information Available via Heteronuclear NMR Spectroscopy**

| Information Available, Advantages | |
|---|---|
| Number of types and number of each type of NMR active nuclide present | Via chemical shift, scalar coupling patterns, resonance integrals |
| Topology of the NMR active parts of the molecule | Via scalar couplings. Structure of (the NMR active parts of) molecules can be deduced with confidence |
| Technique is multi-nuclear and nuclide selective | Resonances from other nuclides do not appear in the spectra |
| Instantaneous excitation of spins is possible | Allows complex systems to be simplified using 2-D techniques |
| Relative disposition of NMR active groups | Via scalar coupling patterns and magnitude of *J*, correlation spectroscopies, and heteronuclear NOEs |
| Exchange rates and pathways between sites | Via VT, saturation transfer, and EXSY measurements |
| Reaction pathways | Via saturation transfer and labelling measurements |
| | Topology of the exchange and reaction pathways of the NMR active parts of the molecule can be obtained |
| Through-space interactions can be probed via Overhauser spectroscopy | Distance between NMR active nuclei |
| | Ion pairing and ion contacts (in combination with diffusion measurements) |

**TABLE 5.2**
**Information Not Available via Heteronuclear NMR Spectroscopy**

| Information Not Available, Limitations | |
|---|---|
| Must have useful NMR active nuclei present | Not available for all important transition metals and ligand donors |
| High-resolution spectra, in which all sites are resolved, are only obtained | In solution or in the gas phase, for spin-1/2 nuclides, in the slow exchange limit |
| Relatively low sensitivity | Strong samples are often needed |
| Relatively long NMR timescale, approximately $10^{-7}$ s | Limits routine detection to long-lived species and slow exchange rates |
| Quadrupolar nuclei and paramagnetism | Give broad lines and may cause extreme line broadening of the resonances of other nuclei |

the presence of scalar couplings between spins, which allows the topology of the spin system to be mapped. Most users will be familiar with the application of $^{1}$H and $^{13}$C{$^{1}$H}NMR spectroscopy to organic molecules, but many may have only passing or no experience of the use of NMR spectroscopy to study other nuclei. It is worth remembering that there is nothing special about proton NMR spectroscopy; $^{1}$H NMR spectra are simply those most commonly encountered. The rules for interpretation of the NMR spectrum of any nuclide are essentially the same, certainly as far as spin-1/2 nuclei are concerned; we are interested in the chemical shift and in what it tells us about the chemical environment of the nucleus, and in the coupling patterns caused by neighboring spins, which allow us to deduce the connectivity in the molecule under study. The numerical value of scalar couplings will be dependent on the pair of coupling nuclei, for example, couplings to $^{31}$P range from a few hertz to a few thousand hertz, depending on the coupling partner and the geometrical relationship of the coupling nuclei; however, the origin of and the rules governing the coupling patterns observed are exactly the same for all spin-1/2 nuclei (Table 5.3); there is nothing special about the interpretation of heteronuclear scalar coupling. For quadrupolar nuclei, that is, those with spins greater than 1/2, the situation is slightly different in as much as fast relaxation results in quadrupolar broadening of the spectral lines, which often obscures scalar couplings or prevents observation of the NMR spectrum entirely. Table 5.4 lists line width factors for quadrupolar nuclei, which serve as a guide; nuclides with small line width factors, for example, $^{2}$H, $^{6}$Li, $^{11}$B and $^{17}$O, should give well-resolved spectra in which scalar coupling may be observed; depending, of course, on the magnitude of the coupling, those with intermediate factors may be observable in symmetrical environments, for example, $^{59}$Co, but it is unlikely that scalar couplings will be observed, whereas those with very large line width factors may be unobservable, for example, Ir and Au.

The information from the heteronuclear NMR spectrum allows us to answer questions such as follows:

1. What is the chemical environment of the nuclide? (via the chemical shift)
2. What neighbours does the nuclide have? (via the scalar coupling)

**TABLE 5.3**
**Spin Properties of Selected Spin-1/2 Nuclei**

| Isotope | Natural Abundance ($\chi$/%) | Magnetogyric Ratio ($\gamma/10^7$ rad $s^{-1}$ $T^{-1}$) | Frequency Ratio[a] ($\Xi$%) | Reference Compound | Relative Receptivity,[b] $D^C$ |
|---|---|---|---|---|---|
| $^{1}H$ | 99.9885 | 26.752 2128 | 100.000 000[c] | $Me_4Si$ | $5.87 \times 10^3$ |
| $^{13}C$ | 1.07 | 6.728 284 | 25.145 020 | $Me_4Si$ | 1.00 |
| $^{15}N$ | 0.368 | −2.712 618 04 | 10.136 767 | $MeNO_2$ | $2.25 \times 10^{-2}$ |
| $^{19}F$ | 100.0 | 25.181 48 | 94.094 011 | $CCl_3F$ | $4.90 \times 10^3$ |
| $^{29}Si$ | 4.6832 | −5.3190 | 19.867 187 | $Me_4Si$ | 2.16 |
| $^{31}P$ | 100 | 10.8394 | 40.480 742 | 80% $H_3PO_4$ | $3.91 \times 10^2$ |
| $^{57}Fe$ | 2.119 | 0.868 0624 | 3.237 778 | $Fe(CO)_5$ | $4.25 \times 10^{-3}$ |
| $^{77}Se$ | 7.63 | 5.125 3857 | 19.071 513 | $Me_2Se$ | 3.15 |
| $^{89}Y$ | 100 | −1.316 2791 | 4.900 198 | $Y(NO_3)_3$ | 0.700 |
| $^{103}Rh$ | 100 | −0.8468 | 3.186 447[d,e] | $Rh(acac)_3$ | 0.186 |
| $^{109}Ag$ | 48.161 | −1.251 8634 | 4.653 533 | $AgNO_3$ | 0.290 |
| $^{113}Cd$[f] | 12.22 | −5.960 9155 | 22.193 175 | $Me_2Cd$ | 7.94 |

| | | | | | |
|---|---|---|---|---|---|
| $^{119}Sn$ | −8.59 | −10.0317 | 37.290 632 | $Me_4Sn$ | 26.6 |
| $^{125}Te$ | −7.07 | −8.510 8404 | 31.549 769 | $Me_2Te$ | 13.4 |
| $^{129}Xe$ | 26.44 | −7.452 103 | 27.810 186 | $XeOF_4$ | 33.6 |
| $^{171}Yb$ | 14.28 | 4.7288 | 17.499 306 | $Yb(C_5Me_5).THF_2$ | 7.89 |
| $^{183}W$ | 14.31 | 1.128 2403 | 4.166 387 | $Na_2WO_4$ | $6.31 \times 10^{-2}$ |
| $^{187}Os$ | 1.96 | 0.619 2895 | 2.282 331 | $OsO_4$ | $1.43 \times 10^{-3}$ |
| $^{195}Pt$ | 33.832 | 5.8385 | 21.496 784[d] | $Na_2PtCl_6$ | 20.7 |
| $^{199}Hg$ | 16.87 | 4.845 7916 | 17.910 822 | $Me_2Hg$[g] | 5.89 |
| $^{205}Tl$ | 70.476 | 15.692 1808 | 57.683 838 | $Tl(NO_3)_3$ | $8.36 \times 10^{2}$ |
| $^{207}Pb$ | 22.1 | 5.580 46 | 20.920 599 | $Me_4Pb$ | 11.8 |

*Source:* Adapted from Harris, R. K. et al., *Pure and Applied Chemistry* 73, pp. 1795–1818, 2001.

[a] Ratio, expressed as a percentage, of the resonance frequency of the reference to that of the protons of TMS at infinite dilution (in practice at $\phi = 1\%$) in $CDCl_3$.

[b] $D^C$ is the receptivity relative to that of $^{13}C$. See Harris, R. K. and B. E. Mann, eds. 1979. *NMR and the Periodic Table*. London, United Kingdom: Academic Press.

[c] By definition.

[d] The precise values 3.160 000 and 21.400 000 MHz have also been suggested as the references for $^{103}Rh$ and $^{195}Pt$, respectively.[1,2]

[e] Subject to considerable variation with temperature.

[f] Long-lived radioactive isotope.

[g] The high toxicity of this compound means that its use should be avoided and is strongly discouraged.

**TABLE 5.4**
**Spin Properties of Selected Quadrupolar Nuclei**

| Isotope | Spin | Natural Abundance ($\chi$/%) | Magnetogyric Ratio ($\gamma/10^7\ s^{-1}\ T^{-1}$) | Frequency Ratio[a] ($\Xi$%) | Reference Sample | Line-Width Factor[b] ($\lambda/fm^4$) | Relative Receptivity,[c] $D^C$ |
|---|---|---|---|---|---|---|---|
| $^{2}H$ | 1 | 0.0115 | 4.106 627 91 | 15.350 609 | $(CD_3)_4Si$ | 0.41 | $6.52 \times 10^{-3}$ |
| $^{6}Li$ | 1 | 7.59 | 3.937 1709 | 14.716 086 | LiCl | 0.033 | 3.79 |
| $^{7}Li$ | 3/2 | 92.41 | 10.397 7013 | 38.863 797 | LiCl | 21 | $1.59 \times 10^{3}$ |
| $^{9}Be$ | 3/2 | 100 | −3.759 666 | 14.051 813 | $BeSO_4$ | 37 | 81.5 |
| $^{10}B$ | 3 | 19.9 | 2.874 6786 | 10.743 658 | $BF_3{\cdot}Et_2O$ | 14 | 23.2 |
| $^{11}B$ | 3/2 | 80.1 | 8.584 7044 | 32.083 974 | $BF_3{\cdot}Et_2O$ | 22 | $7.77 \times 10^{2}$ |
| $^{14}N$ | 1 | 99.632 | 1.933 7792 | 7.226 317 | $CH_3NO_2$ | 21 | 5.90 |
| $^{17}O$ | 5/2 | 0.038 | −3.628 08 | 13.556 457 | $D_2O$ | 2.1 | $6.50 \times 10^{-2}$ |
| $^{21}Ne$ | 3/2 | 0.27 | −2.113 08 | 7.894 296 | Ne | 140 | $3.91 \times 10^{-2}$ |
| $^{23}Na$ | 3/2 | 100 | 7.080 8493 | 26.451 900 | NaCl | 140 | $5.45 \times 10^{2}$ |
| $^{25}Mg$ | 5/2 | 10.00 | −1.638 87 | 6.121 635 | $MgCl_2$ | 130 | 1.58 |
| $^{27}Al$ | 5/2 | 100 | 6.976 2715 | 26.056 859 | $Al(NO_3)_3$ | 69 | $1.22 \times 10^{3}$ |
| $^{33}S$ | 3/2 | 0.76 | 2.055 685 | 7.676 000 | $(NH_4)_2SO_4$ | 61 | 0.101 |
| $^{35}Cl$ | 3/2 | 75.78 | 2.624 198 | 9.797 909 | NaCl | 89 | 21.0 |
| $^{37}Cl$ | 3/2 | 24.22 | 2.184 368 | 8.155 725 | NaCl | 55 | 3.87 |
| $^{39}K$ | 3/2 | 93.2581 | 1.250 0608 | 4.666 373 | KCl | 46 | 2.79 |
| $^{43}Ca$ | 7/2 | 0.135 | −1.803 069 | 6.730 029 | $CaCl_2$ | 2.3 | $5.10 \times 10^{-2}$ |

| | | | | | | | |
|---|---|---|---|---|---|---|---|
| $^{45}Sc$ | 7/2 | 100 | 6.508 7973 | 24.291 747 | $Sc(NO_3)_3$ | 66 | $1.78 \times 10^3$ |
| $^{47}Ti$ | 5/2 | 7.44 | −1.5105 | 5.637 534 | $TiCl_4$ | 290 | 0.918 |
| $^{49}Ti$ | 7/2 | 5.41 | −1.510 95 | 5.639 037 | $TiCl_4$ | 83 | 1.20 |
| $^{51}V$ | 7/2 | 99.750 | 7.045 5117 | 26.302 948 | $VOCl_3$ | 3.7 | $2.25 \times 10^3$ |
| $^{53}Cr$ | 3/2 | 9.501 | −1.5152 | 5.652 496 | $K_2CrO_4$ | 300 | 0.507 |
| $^{55}Mn$ | 5/2 | 100 | 6.645 2546 | 24.789 218 | $KMnO_4$ | 350 | $1.05 \times 10^3$ |
| $^{59}Co$ | 7/2 | 100 | 6.332 | 23.727 074 | $K_3[Co(CN)_6]$ | 240 | $1.64 \times 10^3$ |
| $^{61}Ni$ | 3/2 | 1.1399 | −2.3948 | 8.936 051 | $Ni(CO)_4$ | 350 | 0.240 |
| $^{63}Cu$ | 3/2 | 69.17 | 7.111 7890 | 26.515 473 | $[Cu(CH_3CN)_4][ClO_4]$ | 650 | $3.82 \times 10^2$ |
| $^{65}Cu$ | 3/2 | 30.83 | 7.604 35 | 28.403 693 | $[Cu(CH_3CN)_4][ClO_4]$ | 550 | $2.08 \times 10^2$ |
| $^{67}Zn$ | 5/2 | 4.10 | 1.676 688 | 6.256 803 | $Zn(NO_3)_2$ | 72 | 0.692 |
| $^{71}Ga$ | 3/2 | 39.892 | 8.181 171 | 30.496 704 | $Ga(NO_3)_3$ | 150 | 3.35 × 102 |
| $^{73}Ge$ | 9/2 | 7.73 | −0.936 0303 | 3.488 315 | $(CH_3)_4Ge$ | 28 | 0.642 |
| $^{75}As$ | 3/2 | 100 | 4.596 163 | 17.122 614 | $NaAsF_6$ | 1300 | $1.49 \times 10^2$ |
| $^{81}Br$ | 3/2 | 49.31 | 7.249 776 | 27.006 518 | NaBr | 920 | $2.88 \times 10^2$ |
| $^{83}Kr$ | 9/2 | 11.49 | −1.033 10 | 3.847 600 | Kr | 50 | 1.28 |
| $^{87}Rb$[d] | 3/2 | 27.83 | 8.786 400 | 32.720 454 | RbCl | 240 | $2.90 \times 10^2$ |
| $^{87}Sr$ | 9/2 | 7.00 | −1.163 9376 | 4.333 822 | $SrCl_2$ | 83 | 1.12 |
| $^{91}Zr$ | 5/2 | 11.22 | −2.497 43 | 9.296 298 | $Zr(C_5H_5)_2Cl_2$ | 99 | 6.26 |
| $^{93}Nb$ | 9/2 | 100 | 6.5674 | 24.476 170 | $K[NbCl_6]$ | 76 | $2.87 \times 10^3$ |
| $^{95}Mo$ | 5/2 | 15.92 | −1.751 | 6.516 926 | $Na_2MoO_4$ | 1.5 | 3.06 |
| $^{99}Tc$[d] | 9/2 | – | 6.046 | 22.508 326 | $NH_4TcO_4$ | 12 | – |

*(continued)*

**TABLE 5.4 (Continued)**
**Spin Properties of Selected Quadrupolar Nuclei**

| Isotope | Spin | Natural Abundance ($\chi$/%) | Magnetogyric Ratio ($\gamma/10^7$ s$^{-1}$ T$^{-1}$) | Frequency Ratio[a] ($\Xi$%) | Reference Sample | Line-Width Factor[b] ($\lambda$/fm$^4$) | Relative Receptivity,[c] $D^C$ |
|---|---|---|---|---|---|---|---|
| $^{99}$Ru | 5/2 | 12.76 | −1.229 | 4.605 151 | $K_4[Ru(CN)_6]$ | 20 | 0.848 |
| $^{101}$Ru | 5/2 | 17.06 | −1.377 | 5.161 369 | $K_4[Ru(CN)_6]$ | 670 | 1.59 |
| $^{115}$In[d] | 9/2 | 95.71 | 5.8972 | 21.912 629 | $In(NO_3)_3$ | 490 | $1.98 \times 10^3$ |
| $^{121}$Sb | 5/2 | 57.21 | 6.4435 | 23.930 577 | $KSbCl_6$ | 410 | $5.48 \times 10^2$ |
| $^{127}$I | 5/2 | 100 | 5.389 573 | 20.007 486 | KI | 1600 | $5.60 \times 10^2$ |
| $^{131}$Xe | 3/2 | 21.18 | 2.209 076 | 8.243 921 | $XeOF_4$ | 170 | 3.50 |
| $^{133}$Cs | 7/2 | 100 | 3.533 2539 | 13.116 142 | $CsNO_3$ | 0.016 | $2.84 \times 10^2$ |
| $^{137}$Ba | 3/2 | 11.232 | 2.992 95 | 11.112 928 | $BaCl_2$ | 800 | 4.62 |
| $^{139}$La | 7/2 | 99.910 | 3.808 3318 | 14.125 641 | $LaCl_3$ | 54 | $3.56 \times 10^2$ |
| $^{181}$Ta | 7/2 | 99.988 | 3.2438 | 11.989 600 | $KTaCl_6$ | $1.4 \times 10^4$ | $2.20 \times 10^2$ |
| $^{187}$Re[d] | 5/2 | 62.60 | 6.1682 | 22.751 600 | $KReO_4$ | $1.4 \times 10^4$ | $5.26 \times 10^2$ |
| $^{209}$Bi | 9/2 | 100 | 4.3750 | 16.069 288 | $Bi(NO_3)_2$ | 200 | $8.48 \times 10^2$ |

*Source:* Adapted from Harris, R. K. et al., *Pure and Applied Chemistry* 73, pp. 1795–1818, 2001.

[a] Ratio of the resonance frequency of the reference to that of the protons of TMS at infinite dilution (in practice at $\phi = 1\%$) in $CDCl_3$.

[b] $l = (2I + 3)Q^2/I^2(2I - 1)$. The values are quoted, arbitrarily, to two significant figures.

[c] $D^C$ is the receptivity relative to that of $^{13}$C.

[d] Radioactive, with a long half-life.

3. What is the topology of the molecule? (via scalar couplings and correlation spectroscopy)
4. Is chemical exchange or ligand dynamics occurring? (via variable temperature [VT] or saturation transfer measurements)
5. Are these groups proximal in space? (via Overhauser spectroscopy)
6. Is a reaction occurring? (via, for example, *in situ* methods, isotopic labelling, para-hydrogen)

However, it must be remembered that instrumental and physical constraints might limit the feasibility of the proposed study. For example, signal to noise improves linearly with sample strength and as the square root of the number of transients is recorded. Depending on the spectrometer field and the nuclide under study, typical sample concentrations, in solution, of 0.1 to 400 mM are required to give spectra of good quality in reasonable (2–300 min) time; however, sample concentration may be limited by solubility or the available amount of the sample under study; 2-D spectra do not necessarily require higher sample concentrations, but experiment times may be considerably shortened for stronger samples. Degassed samples are usually preferred for polarisation experiments and those studying relaxation phenomena, and may be essential to prevent decomposition of the sample; however, degassed samples may require inconveniently long recovery delays. Conversely, fast relaxation caused by the presence of quadrupolar nuclei or paramagnetic species may destroy the desired signal.

### 5.1.1 Chemical Shift

Chemical shift ranges are, of course, specific to the nucleus under study and may seem unusual at first, for example, $^{195}Pt$ chemical shifts range more than 10,000 ppm. The shielding of a nucleus can be described by Equation 5.1, where $\sigma_d$ is the diamagnetic term and arises from the influence of core electrons and $\sigma_p$ is the paramagnetic term:

$$\sigma = \sigma_d + \sigma_p \tag{5.1}$$

The paramagnetic term, $\sigma_p$, gives large shielding effects and dominates the shielding for transition and p-block metals, resulting in large chemical shift ranges for these nuclides, of the order of $10^3$ to $10^4$ ppm. For transition metals, there are two principal contributions to $\sigma_p$: the effect of the ligands and contributions from mixing in of excited states.[3–5] Recent advances in density functional theory are enabling the precise calculation of metal chemical shifts.[6–10] However, (some) metal nuclei show large temperature and solvent effects; thus, for example, Carlton has recently determined the solvent and temperature effects on $\delta_{Rh}$ for a large number of square planar Rh(I) complexes, $[Rh(X)(PPh_3)_3]$ (X = Cl, $N_3$, NCO, NCS, $N(CN)_2$, $NCBPh_3$, $CNBPh_3$, CN), and derivatives containing CO, isocyanide, pyridine, $H_2$, and $O_2$.[11]

By contrast, the diamagnetic term is small and can effectively be ignored when considering differences in δ, except in the s-block, when $\sigma_p$ is approximately 0, and

$\sigma_d$ alone determines the shielding. The chemical shift range of s-block nuclides is small, of the order of a few tens of parts per million.

Chemical shift range is one of the issues that needs to be considered when planning and implementing NMR experiments to detect heavier elements. In this chapter, we discuss other experimental considerations in recording heteronuclear NMR spectra and illustrate some of the common heteronuclear NMR experiments, returning to chemical shift and coupling constant considerations toward the end, providing some representative examples of compounds containing common heteronuclei.

## 5.2 ACQUISITION OF THE NMR SPECTRUM OF AN 'INORGANIC' NUCLIDE: FOR THE FIRST TIME

The vast majority of researchers start their NMR career working with the most sensitive (non-radioactive) nuclide, $^{1}H$, moving on to the significantly less abundant (but still spin-1/2) nuclide, $^{13}C$. While there are clearly many similarities in the processes surrounding the acquisition of such data and data for more 'exotic' nuclides, there are also some differences that need proper consideration so as to achieve optimum results. In this section, we discuss some of these matters, which are of particular relevance when attempting data acquisition (for a particular nuclide) for the first time.

### 5.2.1 Spectrometer

These might seem obvious questions, but ask! Has anyone measured the nuclide of interest, X, on this spectrometer before? Can the probe be tuned to nuclide X? The electronics of many (most) modern spectrometers are broadbanded, that is, capable of measuring nuclide X for all X resonating above $^{15}N$; however, the probe may not be a broadband type, particularly if the spectrometer is in an environment where only $^{1}H$ and $^{13}C$ NMR spectra are routinely recorded, in which case it is likely that only a dual $^{1}H$ and $^{13}C$ probe is fitted. Ask and check the hardware yourself. If you are interested in low magnetogyric ratio ($\gamma$) nuclei, e.g. $^{103}Rh$ or $^{109}Ag$, specialist equipment will be required – it may be more sensible and cost effective to seek a collaboration with a specialist group.

Other questions to ask are as follows: Is the probe direct or inverse? The former is good for direct observation with or without insensitive nucleus enhancement by polarization transfer (INEPT) (see later) enhancement. The latter will give poor signal to noise in direct experiments since the sample does not fill the coil space, but is much preferred for indirect detection via, for example, an heteronuclear multiple-quantum coherence (HMQC) or an heteronuclear single-quantum coherence (HSQC) experiment. Some manufacturers now offer direct geometry broadband probes that afford close to inverse geometry proton sensitivity, for example, JEOL's 'Royal'™ probe.

Is the spectrometer/probe capable of VT operation? Dynamic samples need to be measured at low temperature to freeze out the dynamic processes (Chapter 2), quadrupolar nuclei are often best observed at high temperature – the increased molecular tumbling rates help to average out the quadrupolar, line broadening interaction.

Finally, what is the $^{1}H$ frequency? In common parlance, spectrometers are referred to as 300, 400, 500 etc. However, this does not usually correspond to the actual spectrometer $^{1}H$ frequency. For example, the default frequency of a Bruker spectrometer is 0.13 MHz above this, that is, 400.13 MHz, etc. Furthermore, the frequency may have been shifted ±0.07 MHz from the default value during installation to prevent interference with another, nearby spectrometer of the same nominal $^{1}H$ frequency. It is very unlikely that the protons of tetramethylsilane (TMS) will resonate at exactly the notional spectrometer frequency. Measure a proton spectrum of a standard sample containing TMS and note the absolute TMS frequency. Use the measured value (in the appropriate lock solvent) in calculating the observation frequency for other nuclides on the International Union of Pure and Applied Chemistry (IUPAC) scale (Tables 5.3 and 5.4). It is good practice to add TMS to the sample of interest (if it is compatible) and record a $^{1}H$ NMR spectrum as well as the desired spectrum of nuclide X.

### 5.2.2 Test Sample

It is rarely a good idea to use a precious, real sample when observing a nuclide for the first time; rather, choose a test sample of known chemical shift giving a strong signal. If the receptivity of the desired nucleus is so low that a suitable test sample for direct observation does not exist or a suitable sample containing the desired nuclide is not available or gives a weak signal, choose a sample of a nuclide with a similar resonance frequency and use that, for example, $GeMe_4$ can be used to set up a spectrometer for observation of nuclides in the range $^{57}Fe$ to $^{187}Os$, which includes $^{103}Rh$. Calculate the expected absolute frequency on the IUPAC scale (Tables 5.3 and 5.4) for the test sample versus the measured absolute proton frequency of TMS on your spectrometer. Note, however, that, although IUPAC has issued recommended reference frequencies for all nuclei, these are not in universal use, so the chemical shift given in the literature for the selected test compound may be on a 'different' parts-per-million scale and, hence, may not appear at the 'expected' position in the spectrum (or, indeed, at all) using the manufacturer's default settings; thus, the offset may need to be adjusted. Use a small pulse length and a moderate-power hard pulse – check what has been used before for nuclides of similar $\gamma$. If you would be using the probe's lock channel to observe deuterium, $^{2}H$, remember that this will likely be intended for low-power operation; attenuate the transmitter pulse to avoid damage to the probe.

Once you have found the signal of interest, measure the 90° (or $\pi/2$) transmitter pulse for X. If inverse detection experiments are to be used, measure the X decoupler pulse. The spectrometer manufacturer might have provided suitable pulse programmes and routines to do this. When measuring the transmitter pulse, you should observe a sinusoidal variation in intensity of the NMR resonance (Figure 5.1). If the intensity initially varies sinusoidally but then the sinusoidal envelope 'stretches' as the pulse length increases, you are not allowing a sufficient recovery delay between measurements (Chapter 3). Increase the recovery delay or add a relaxation agent such as $Cr(acac)_3$. If the intensity reaches a plateau as the pulse length increases, or varies randomly, then the probe is 'breaking down'. Reduce the pulse power to avoid damaging the probe and try again.

When measuring the decoupler pulse, make sure that the phase of the lines of the multiplet inverts as you go through the zero crossing point (Figure 5.1b). If it does not, then you are probably far off resonance for nucleus X. You need to check both the tuning of the probe and the spectral offset. Is your test sample's chemical shift

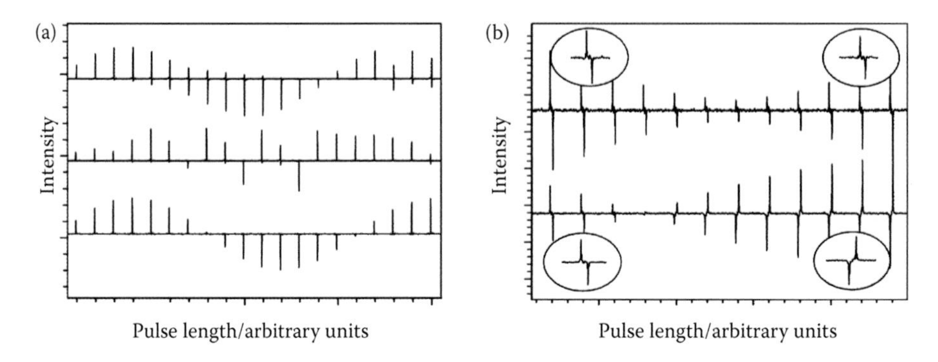


**FIGURE 5.1** (a) $^{119}Sn$ 90° transmitter pulse measurement. $SnMe_4$ in $CDCl_3$. Top: Signal 100-ppm off-resonance gives phase distortion as pulse length varies. Middle: Probe breakdown gives random intensity variations. Bottom: Correct. (b) $^{15}N$ 90° decoupler pulse measurement via $^{31}P$, using Bruker pulse programme DECP90F3, sample $[Pt(C_5H_5{}^{15}N)(PPh_3)Cl_2]$ in $d_8$-THF. Top: Decoupler off-resonance–anti-phase doublet does not invert, gives an incorrect pulse length. Bottom: Decoupler on-resonance, correct.

on the IUPAC scale or on some other? Is the manufacturer's default X frequency on the IUPAC scale or on some other? Has someone changed the default X frequency on your spectrometer?

### 5.2.3 Referencing

When referencing a heteronuclear NMR spectrum, there are two options. The first is to measure the spectrum of a compound with known chemical shift and use that as a secondary reference (Tables 5.3 and 5.4). Note that some IUPAC reference compounds differ from common practice, for example, $^{103}Rh$ is often referenced to $\Theta_{Rh}$ = 3.16 MHz when $\Theta_{TMS}$ = 100 MHz. For $^{15}N$, chemists usually reference to $MeNO_2$, whereas structural biologists use $NH_3$. Shifts on the $MeNO_2$ scale are approximately 380.5 ppm less positive than those on the $NH_3$ scale, for example, on the IUPAC scale, $\delta(^{15}NH_3) = -380.4$. The exact correction required depends on the resonance frequency of the sample.

Alternatively, reference the spectrum to the $^1H$ signal of TMS – this is the IUPAC-recommended method since this establishes a universal scale of chemical shifts.[12] The procedure is straightforward. Simply record a $^1H$ NMR spectrum of your sample immediately before you record the X spectrum. Do not change the lock frequency, and do not re-shim the sample between recording the proton and X spectra. If TMS is present in the sample, measure the absolute frequency of TMS = $\Theta_{TMS}$ directly; if not, use the solvent $^1H$ NMR resonance as a secondary reference to determine $\Theta_{TMS}$. Look up the reference frequency, $\Theta_X$ (against TMS = 100 MHz exactly), for the nuclide of interest (X) in Table 5.3 or 5.4. The zero on the parts-per-million scale for the nuclide X is then $\Theta_X \times \Theta_{TMS}/100$ MHz, exactly. Record the NMR spectrum of the sample of interest. This method avoids the use of some particularly noxious compounds, for example, $HgMe_2$, $Fe(CO)_5$, $Ni(CO)_4$, $PbMe_4$, etc. and has the added advantage that the $^1H$ NMR spectrum provides a quick check that the sample is correctly shimmed.

### 5.2.4 Solvent and Temperature Considerations

The chemical shifts of many heteronuclides, for example, $^{31}P$, $^{103}Rh$ and $^{195}Pt$, show large temperature, concentration, and/or solvent shifts. For example, $\delta(^{103}Rh)$ shows a temperature effect of approximately 0.5 ppm/K, whereas the $^{31}P$ chemical shift of phosphine oxides can vary by ±5 ppm on changing the solvent but are relatively insensitive to temperature, while $\delta(^{31}P)$ of phosphines show large temperature effects but may be relatively insensitive to the solvent. Obviously, the sample temperature and concentration, as well as the solvent, need to be given in reporting data.

### 5.2.5 Running 'Off Lock'

In heteronuclear studies, there is no need to use deuterated solvents to suppress unwanted solvent proton NMR resonances; it is, thus, possible, and common, to record heteronuclear NMR spectra in protio-solvent, for example, to allow monitoring a reaction without first removing the solvent and then dissolving the sample in deuterated solvent. There are a couple of issues with this approach that an inexperienced user must be aware of: (1) the locking system must be turned off, and the field-frequency sweep used to excite the solvent deuterium resonance during locking must be disabled; otherwise, the spectrometer will continue to sweep the magnetic field, searching for a lock during sample acquisition. Resonance frequency varies linearly with field – if the field is continuously changing, so too will be the frequency of the resonance of X, resulting in unusable spectra. (2) The sample will need to be shimmed on the solvent resonance. If the solvent contains a strong, well-separated singlet resonance, gradient shimming using a selective $^{1}H$ pulse is possible; otherwise, shimming on the free induction decay (FID) will be necessary.

## 5.3 CHOICE OF EXPERIMENT

The information above should allow successful acquisition of simple 1-D and 2-D spectra of heteronuclides; however, for many heteronuclei, the sensitivity is very low. This can be for two reasons: the nuclide under study has low receptivity and, in effect, gives an intrinsically weak signal, and/or the nuclide under study has low natural abundance. We pay attention to these points when describing some of the wide range of pulse sequences utilised for the direct or indirect detection of inorganic nuclei.

### 5.3.1 Direct Detection: INEPT

If the nucleus of interest ($S$) couples to a higher $\gamma$ nucleus ($I$), sensitivity enhancement should be considered. The best 1-D method for inorganic samples is the INEPT family of experiments since the sensitivity enhancement is greater than that of distortionless enhancement by polarization transfer (DEPT) (Section 7.1) and the spectral editing features of DEPT are not usually required.[13] In the authors' experience, refocussed INEPT, with or without decoupling (of the $I$ nucleus), is the best choice since this yields spectra similar in appearance to normal 1-D spectra – the number of coupled, neighboring spins is, thus, readily extracted from the spectrum. There is, however, a slight loss of signal intensity compared to the basic INEPT sequence due to the longer pulse sequence.[14,15] Note that integrals in polarisation transfer experiments are dependent on factors such as the efficiency of polarisation transfer and the number of $I$ spins to which the insensitive nuclide $S$ couples. Thus, integrals are not a good measure of the number of $S$ nuclei present.

### 5.3.2 Indirect Detection

The organic chemist will be familiar with inverse polarisation transfer experiments such as HSQC,[16] heteronuclear multiple bond correlation (HMBC) and HMQC[17–19] (Section 7.1), in which polarisation transfer from $^{1}H$ is used to obtain proton–carbon correlation spectra that reveal proton–carbon connectivities. These experiments can also be used to mitigate the problems of low signal intensity when acquiring heteronuclear spectra. If coupling to a nuclide such as $^{1}H$, $^{19}F$ or $^{31}P$ is present, polarisation transfer from the nucleus of higher resonance frequency and receptivity can be employed to 'boost' the signal from the insensitive nuclide. The principle disadvantages of these indirect detection experiments are that a three-channel spectrometer and probe is required if the sensitive spin is other than $^{1}H$ (and, possibly, $^{19}F$, depending on spectrometer configuration), and the number of (directly detected) sensitive spins cannot be determined directly from the multiplicity of the indirectly detected spin since this information is intentionally 'refocussed' by the pulse sequence. Where this information is required, the INEPT direct detection experiment can be used.

The sensitivity of any heteronuclear NMR experiment is related to the magnetogyric ratios of the starting and the detected spins and a relaxation term:

$$S/N \propto \gamma_{\text{start}}\gamma_{\text{detect}}^{3/2}[1-\exp\{-1/T_{1,\text{start}} \times RD\}] \tag{5.2}$$

where $\gamma_{\text{start}}$ and $\gamma_{\text{detect}}$ are the magnetogyric ratios of the initially excited and detector spins, respectively; $T_{1,\text{start}}$ is the spin–lattice relaxation time constant of the excited spin; and $RD$ is the experiment recycle time.

For HMQC- and HSQC-type experiments, $\gamma_{\text{start}} = \gamma_I$, $\gamma_{\text{detect}} = \gamma_I$; hence, the sensitivity enhancement is approximately $(\gamma_I/\gamma_S)^{5/2}$ (cf. for a direct polarisation transfer experiment such as INEPT, $\gamma_{\text{start}} = \gamma_I$, $\gamma_{\text{detect}} = \gamma_S$, the sensitivity enhancement is approximately $\gamma_I/\gamma_S$). Clearly, if detection of the resonance of the heteronuclide is the aim of the experiment, HMQC and HSQC should be considered and will be most effective when the detector nucleus has a high $\gamma$. 2-D correlation spectroscopy has revolutionised the NMR spectroscopy of low-$\gamma$ heteronuclei as a result of this sensitivity gain, allowing much shorter experiment times; for example, it is possible to record $^{31}P$-detected $^{103}Rh$ spectra in minutes in contrast to the direct detection of the same sample, which might take many hours.[3,20] Indirect detection, however, cannot compensate for a low natural abundance of the spin of interest since only the satellites from coupling to the metal are observed. Nonetheless, the sensitivity enhancement that can be achieved can be very large (Table 5.5). Additionally, the HSQC and HMQC experiments provide a correlation map linking coupled sensitive and insensitive spins.

**TABLE 5.5**
**Theoretical Enhancement Factors for Selected Spin-1/2 Nuclides**

| | | Enhancement Factor via Detection by | | | | |
|---|---|---|---|---|---|---|
| | | HXQC[a] | | | | INEPT |
| Nuclide | $\gamma$ ($10^7$ rad s$^{-1}$ T$^{-1}$) | $^1$H | $^{13}$C[b] | $^{19}$F | $^{31}$P | $^1$H |
| $^{15}$N | 2.71262 | 305 | 10 | 263 | 32 | 10 |
| $^{57}$Fe | 0.8680624 | 5273 | 167 | 4532 | 551 | 31 |
| $^{77}$Se | 5.1253857 | 62 | 2 | 54 | 7 | 5 |
| $^{89}$Y | 1.3162791 | 1862 | 59 | 1601 | 195 | 20 |
| $^{103}$Rh | 0.8468 | 5610 | 178 | 4822 | 586 | 32 |
| $^{109}$Ag | 1.2518634 | 2111 | 67 | 1815 | 221 | 21 |
| $^{113}$Cd | 5.9609155 | 43 | 1 | 37 | 4 | 4 |
| $^{119}$Sn | 10.0317 | 12 | <1 | 10 | 1 | 3 |
| $^{125}$Te | 8.5108404 | 18 | 1 | 15 | 2 | 3 |
| $^{171}$Yb | 4.7288 | 76 | 2 | 65 | 8 | 6 |
| $^{183}$W | 1.1282403 | 2738 | 87 | 2353 | 286 | 24 |
| $^{187}$Os | 0.6192895 | 12,265 | 389 | 10,543 | 1282 | 43 |
| $^{195}$Pt | 5.8385 | 45 | 1 | 39 | 5 | 5 |
| $^{199}$Hg | 4.8457916 | 72 | 2 | 62 | 7 | 6 |
| $^{205}$Tl | 15.6921808 | 4 | <1 | 3 | <1 | 2 |
| $^{207}$Pb | 5.58046 | 50 | 2 | 43 | 5 | 5 |

[a] HXQC = HMQC or HSQC.
[b] Assuming enrichment to 100%.

### 5.3.2.1 Instrumental Considerations

HXQC spectra using $^1$H as the detector (*I*) spin can be recorded on a conventional two-channel NMR spectrometer with a 'normal' broadband or a selective dual probehead; however, better signal to noise will be achieved using an inverse probehead due to the better filling factor.

For HXQC spectroscopy, in which both the detected *I* and indirect *S* nuclei are 'heteronuclei,' a third spectrometer channel is required. An inverse broadband, triple-resonance probehead, in which the inner coil is tuned to $^1$H, *I* and $^2$H lock, and the outer coil is broadbanded, is then the best option, particularly where a single *I* nuclide or $^1$H will be used for detection. Such probeheads are 'stock items' for some manufacturers.

### 5.3.2.2 HMQC and HSQC Approaches

There is a bewildering array of HMQC and HSQC pulse sequences. A detailed analysis of the workings of these sequences, and the consequences, is beyond the scope of this chapter; Keeler (http://www.keeler.ch.cam.ac.uk/lectures/index.html) and Mandal[21] have written excellent reviews that are highly recommended to a reader interested in detailed information on the workings of pulse sequences, phase cycling, gradient selection, etc. The distinction between these two families of experiments is that, in the first, magnetisation is stored as multiple quantum coherence during the evolution period, while in the second, it is stored as single quantum coherence. This impacts on the interactions that are active during the pulse sequence, on relaxation processes, and hence, on, for example, the signal strength, artefacts, splittings, and line widths in the final spectrum.

The HMQC sequence uses a small number of pulses, that is, is short, while the basic HSQC sequence uses 10 pulses. The HMQC sequence is, therefore, relatively less affected by pulse imperfections/mis-setting. Furthermore, storing the magnetisation as multiple quantum coherence, which has a relatively longer characteristic spin–spin relaxation time, should afford higher resolution. HMQC has, therefore, been favoured by most workers in the field since, in principle, it should give better signal to noise.

Other factors to bear in mind when choosing between these pulse sequences include passive coupling and exchange (Chapter 2), which may favour the use of the HSQC experiment. Passive coupling between $I$ spins modulates the multiple quantum coherence, resulting in splitting of the signals in the F1 ($S$ [insensitive] spin) dimension of an HMQC spectrum by $I$ spin–$I$ spin couplings; these splittings are absent from the HSQC spectrum (Figure 5.2a and b). The HMQC experiment is also more affected by exchange broadening because both the $I$ and $S$ spins are in the $xy$ plane; exchange is, of course, commonplace in transition metal chemistry. Decoupling of other spins may also be required (Figure 5.2c and d). Which experiment gives the 'better' spectrum will, therefore, depend both on the peculiarities of the system under study and the accuracy with which pulse lengths and delays are set.

#### *5.3.2.2.1 Gradient versus Phase Cycle Selection*

Field gradient pulses and phase cycling are alternative methods to select the desired coherence pathways. Phase cycling is a subtractive process – the sign of unwanted signals is alternated positive then negative by altering the phases of the pulses in the pulse sequence so that, overall, the unwanted signals cancel while the desired signal is cumulative. By contrast, gradient selection is not subtractive; thus, subtraction errors are absent from the spectra, giving much reduced $t_1$ noise. The application of a field gradient to the sample causes transverse magnetisation and other coherences to dephase. Subsequent application of an equal and opposite gradient pulse later in the pulse sequence causes an equal but opposite 'de-dephasing' of the magnetisation, producing a gradient echo (Chapter 4). The refocussing gradient can be adjusted so that only some coherences are refocussed, while unwanted signals continue to dephase and, hence, are 'lost'. Modern spectrometers should have routines for calculating correctly the required gradient ratios, which can differ between variants of an experiment.

For the basic HMQC sequence set $g1$, $g2$, and $g3$ to satisfy,

$$g1(\gamma_I + \gamma_S) - g2(-\gamma_I + \gamma_S) + g3(-\gamma) = 0 \qquad (5.3)$$

For HSQC set $g1$ and $g2$ to satisfy,

$$g1(\gamma_S) - g2(\gamma_I) = 0 \qquad (5.4)$$

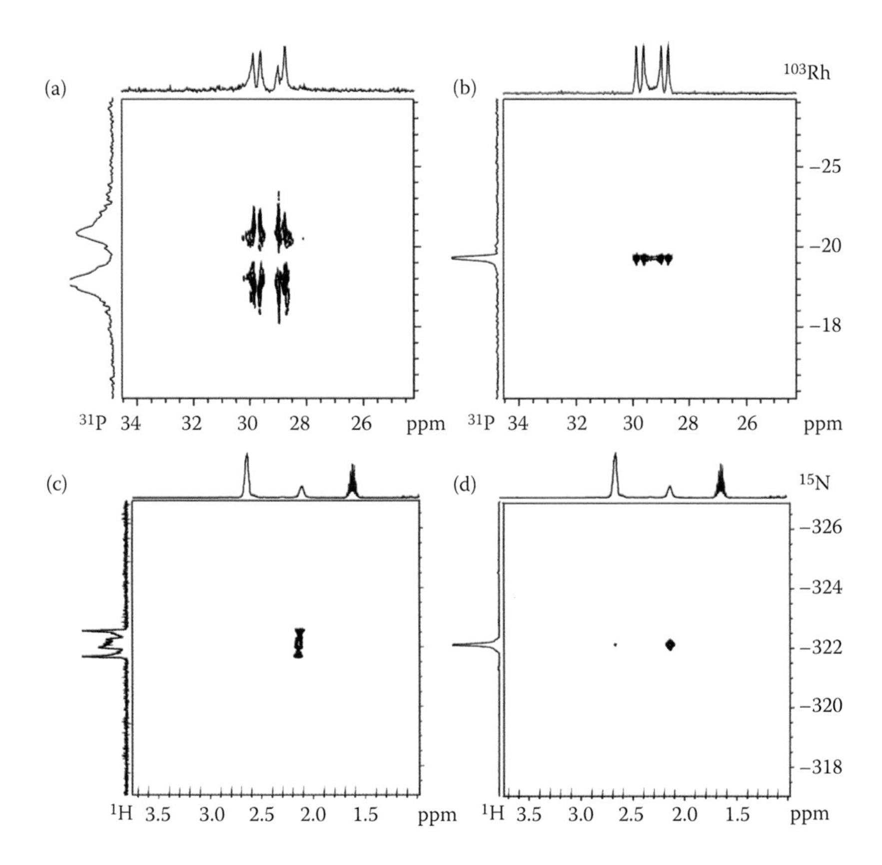


**FIGURE 5.2** Phase-sensitive $^{31}$P,$^{103}$Rh{$^{1}$H} HMQC (a) and HSQC (b) spectra of Wilkinson's catalyst, [Rh(PPh$_3$)$_3$Cl] (30 mg in 0.8 mL CDCl$_3$, 295 K), illustrating the splitting of the resonances in F1 ($^{103}$Rh) by passive couplings between the cis- and trans-phosphine ligands. Only the trans-PPh$_3$ region is shown. $^{1}$H,$^{15}$N{$^{15}$N} HSQC spectra of [(N$_3$(P(Bu$^i$)$_2$)$_3$] in CDCl$_3$ at 295 K recorded: without (c) and with (d) CHIRP60 adiabatic $^{31}$P decoupling pulse. Bruker pulse programme hsqcetf3gpsi2.

However, gradients cannot always refocus all the desired signal, for example, in the HMQC sequence, only one of the two coherence pathways is refocussed and half the desired magnetisation is lost, resulting in a lower intrinsic sensitivity compared with phase cycle selection. The principle disadvantage of phase cycle selection is that each phase cycle may contain 8, 16, or more steps, considerably extending the required experiment time, whereas gradient selection may require only 2 or 4 transients to be acquired to achieve good selection of the desired signal. On spectrometers equipped with gradients (most modern spectrometers are), the gradient-selected experiment is, therefore, now routinely used since it maximises throughput, even though it may not always be the optimal experiment.

In practice, the advantage that gradient acceleration offers in HMQC/HSQC spectroscopy will be different for different *S* spins[22]; remember that only one coherence pathway is selected in gradient-accelerated HMQC, that is, half the signal intensity is discarded, resulting in a loss of sensitivity. Paradoxically, gradient selection may be the better choice where the *S* nuclide is less than 100% abundant since breakthrough of signals from molecules containing NMR silent '*S*' nuclei is much better suppressed. Furthermore, sensitivity-improved gradient experiments are available; however, the increased length of the pulse sequences allows for additional dephasing of the desired magnetisation with some loss of signal. Note also that the sensitivity improvement is often restricted to a particular type of spin system (normally, one *I* spin and one *S*). The principal advantage of gradient selection is that it is possible to acquire HMQC and HSQC spectra using only one, two, or four scan(s) per $t_1$ increment; there is no need to complete a large number of steps determined by the phase cycle, resulting in considerable saving of spectrometer time.

Phase cycle, rather than gradient, selection retains both coherence pathways, in principle, increasing the signal intensity. Furthermore, many *S* nuclides have high natural abundance ($^{89}$Yb, $^{103}$Rh, $^{109}$Ag etc.); suppression of signals from molecules containing NMR silent '*S*' nuclei is, therefore, not such an issue. If the intensity of the desired signal is the limiting factor, then phase cycle selection may be preferred. If phase cycling must be used and *S* is appreciably less than 100% abundant, then a bilinear rotation decoupling (BIRD) sequence[23] to purge signals from molecules containing NMR silent '*S*' nuclei should be considered.

Overall, S/N may be better in the gradient-accelerated experiment because of the better rejection of unwanted signals, even though the desired signal is weaker. Paradoxically, therefore, it may be useful to use gradient acceleration to suppress the noise in the spectra if the indirectly observed spin has low natural abundance.

##### *5.3.2.2.2 Spin Systems Involving Coupling to Multiple Insensitive Spins*

Unlike $^{13}$C or $^{15}$N, the natural abundance of many metal nuclides is high and can be 100%; this introduces specific problems in recording HMQC spectra since degenerate insensitive nuclei may be present.[24] Examples include transition-metal cluster compounds that contain edge- and/or face-bridging carbonyl ligands of, for example, rhodium or platinum, and bimetallic and cluster compounds with phosphorus-containing hetero-ligands. It is important to appreciate that such spin systems will not behave as 'normal' $^1$H–$^{13}$C or $^1$H–$^{15}$N systems and will give spurious or no correlations if experimental methodology is simply carried over 'organic' systems. One solution is to modify the phase cycling such that all coherences will appear and then follow Ruegger and Moskau's rules to identify those correlations corresponding to single insensitive spin flips.[25] The disadvantage of this approach is the plethora of coherences detected. An alternative approach is to modify the preparation delay, $d_2$, such that only coherences arising from single metal spin flips are detected.[26–28] Podkorytov has analysed two such situations likely to be commonly encountered: (1) the detector nucleus couples to more than one insensitive spin in an approximately equal fashion, for example, edge bridging, face bridging, or interstitial ligands; and (2) the detector spin couples very differently to several metal spins through one, two, or more bond couplings. An example of the first is the face-bridging carbonyl group in $[Rh_6(CO)_{15}L]$, (L = $PR_3$ [R = alkyl, aryl], $P(OPh)_3$, NCMe, *I* etc.) (Figure 5.3).[29] Podkorytov showed that the maximum intensity of correlations arising from single metal spin flips occurs when a preparation delay of $d_2 = 1/(5J)$ rather than the

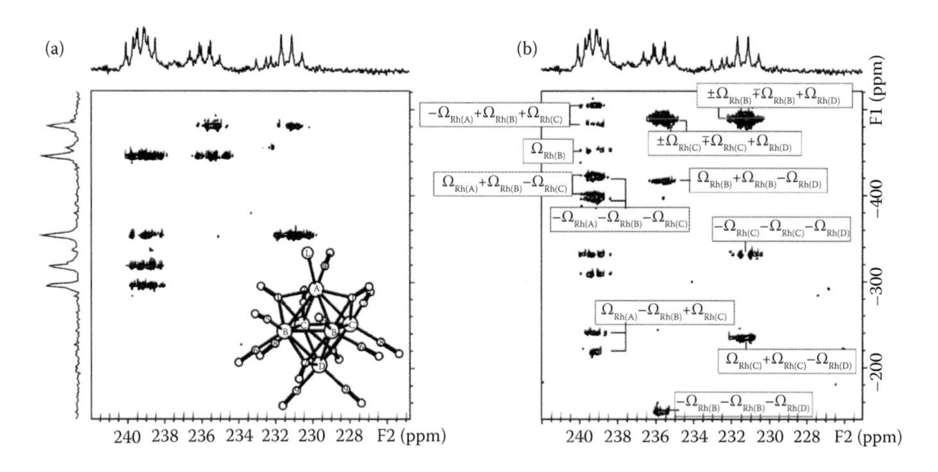


**FIGURE 5.3** The inverse detected HMQC $^{13}C\{^{103}Rh\}$ spectra of $[Rh_6(CO)_{15}\{P(4\text{-}F\text{-}C_6H_4)_3\}]$ recorded using preparation delays, $d_2$, of $1/(2J)$ (a) and $1/(5J)$ (b), $J = 28$ Hz.

conventional delay of $d_2 = 1/(2J)$ is used. Use of the latter results in close-to-zero intensities of these cross peaks and the observation of a large number of 'spurious' correlations resulting from multiple metal spin flips (Figure 5.3).

The effects described above are expected whenever the detector nucleus is coupled to several metal spins that can act as a unit. Edge-bridging carbonyls, hydrides, or diorganophosphides, and interstitial carbides and phosphides are also frequently encountered in transition-metal carbonyl clusters and other polymetallic compounds.[26] The above analysis is readily generalised, and it is found that the 'correct' signals for an $IS_n$ group are maximal at

$$d_2 = \frac{1}{\pi J} arctg \frac{1}{\sqrt{n-1}} \tag{5.5}$$

An example of the second type of spin system analysed by Podkorytov include the dimeric complexes $[Rh_2\{\mu_2\text{-MTPA}\}_4(PPh_3)]$ (MTPA = $(PhC(OMe)(CF_3)CO_2)$)[27] and $[Rh_2(\mu_2\text{-}C_7H_{15}CO_2)_4(PPh_3)]$, for which $^1J$(Rh-P) = 95.5 and $^2J$(Rh-P) = 21.6 Hz, and 96.6 and 33.6 Hz, respectively. In the case of $[Rh_2\{\mu_2\text{-MPTA})(CF_3)CO_2\}_4(PPh_3)]$, multiple metal spin flips are an inconvenience; Rh(A) and Rh(B) can be observed, albeit in separate experiments and with reduced intensity for Rh(B), using the conventional preparation delay, $d_2 = 1/(2J)$. However, for $[Rh_2(\mu_2\text{-}C_7H_{15}CO_2)_4(PPh_3)]$, in which $^1J$ is nearly an odd integral multiple of $^2J$,$^1J/^2J = 2.9$, the intensity of the correlation due to Rh(B) is close to zero if $d_2 = 1/(2\ ^2J)$ is used (Figure 5.4). Using a 'conventional' preparation delay, $d_2 = 1/(2\ ^1J)$, the correlation due to Rh(A) is seen with good sensitivity (Figure 5.4a); a delay of $1/(2\ ^2J)$, conventionally appropriate for the detection of Rh(B), gives only weak correlations (Figure 5.4b); and an 'unconventional' delay of 7.7 ms, derived from a simulation to determine the optimum value of $d_2$,[28] allows cross peaks from both Rh(A) and Rh(B) to be observed with good sensitivity (Figure 5.4c).

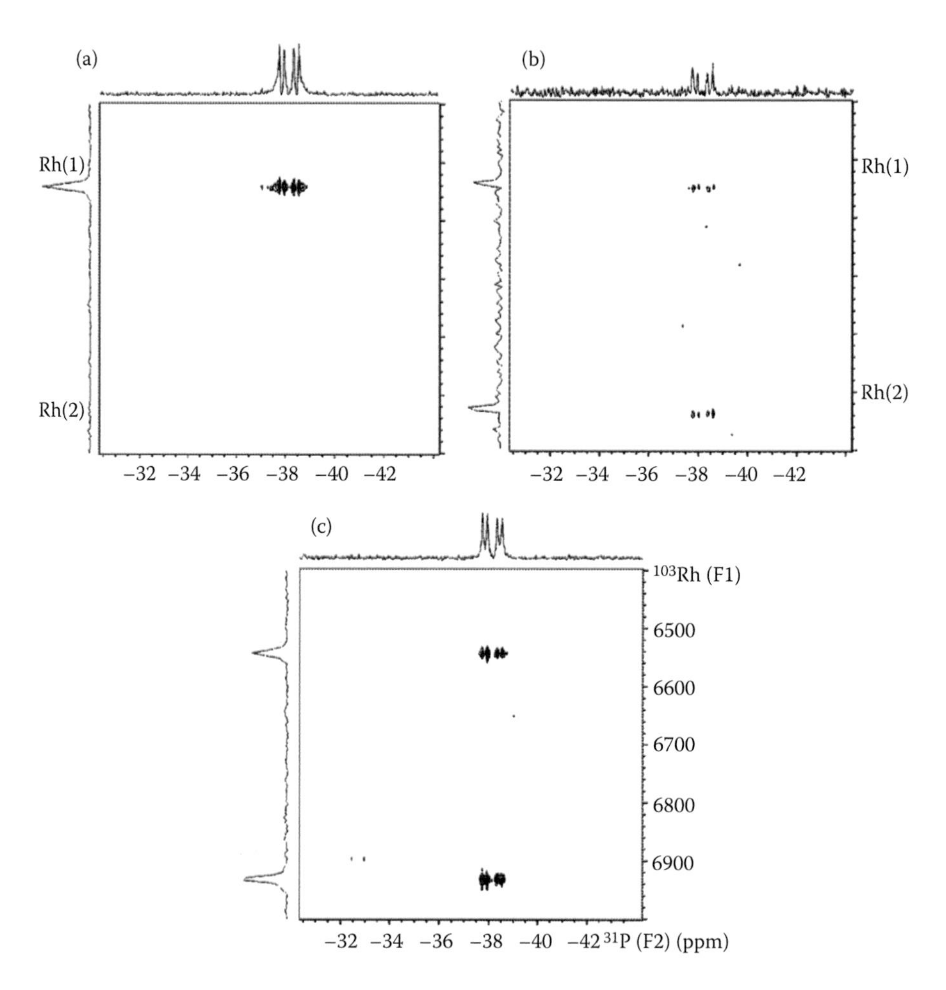


**FIGURE 5.4** The $^{31}P$–$^{103}Rh$ HMQC NMR spectrum of $[Rh_2(\mu_2\text{-}C_7H_{15}CO_2)_4(PPh_3)]$. (a) Using a 'conventional' preparation delay, $d_2 = 1/(2J_1) = 5.2$ ms gives the correlation due to Rh(A). (b) A delay of $1/(2J_2) = 14.9$ ms gives very weak correlations to Rh(A) and Rh(B). (c) A delay of 7.7 ms allows cross peaks from both Rh(A) and Rh(B) to be observed with good sensitivity.

### 5.3.2.3 HMBC Spectroscopy

The HMBC sequence is, in effect, an HMQC sequence preceded by a low-pass filter and was originally designed to eliminate magnetisation arising from one-bond couplings in $^1H,^{13}C$ HMQC spectroscopy. In principle, this is unlikely to be required in heteronuclear studies, in which HMQC is applied to the indirect detection of insensitive nuclei since there is unlikely to be a need to distinguish between one-bond and long-range couplings. On the other hand, the longer HMBC sequence and additional delays introduced by the low-pass filter allow for further dephasing/relaxation of the desired magnetisation, reducing the sensitivity of the HMBC experiment.

### 5.3.3 Selective Observation and Insensitive (*S*) Spin Satellites

It may be necessary to detect signals arising from only part of the NMR spectrum of X or of those spins (in the spectrum) of nuclide 1 (*I*) that are coupled to nuclide 2 (*S* = X). For example, it may be necessary for chemical reasons to record the $^{1}$H NMR spectrum of a metal-hydride in protio- rather than deutero-solvent, with associated issues of dynamic range. It may also be that the exact value of the coupling *J*(*IS*) is required for a polarisation transfer experiment but cannot be measured directly from the spectrum of *I* due, for example, to a low natural abundance of *S*. The simplest method to achieve the former is to take advantage of the digital filtering implemented on all modern spectrometers and reduce the observation window to include just the resonances of interest. However, these may not be well separated from other, stronger resonances, in which case selective excitation of the desired spectral region using a shaped pulse, or delays alternating with nutations for tailored excitation (DANTE) excitation, or solvent suppression may be used. None of these methods use spin–spin coupling to edit the spectrum, and so they can be applied to any resonance.[30]

Selective observation of the *S* spin multiplet can be achieved via a 1-D HMQC (or HMBC or HSQC) experiment (Figure 5.5).

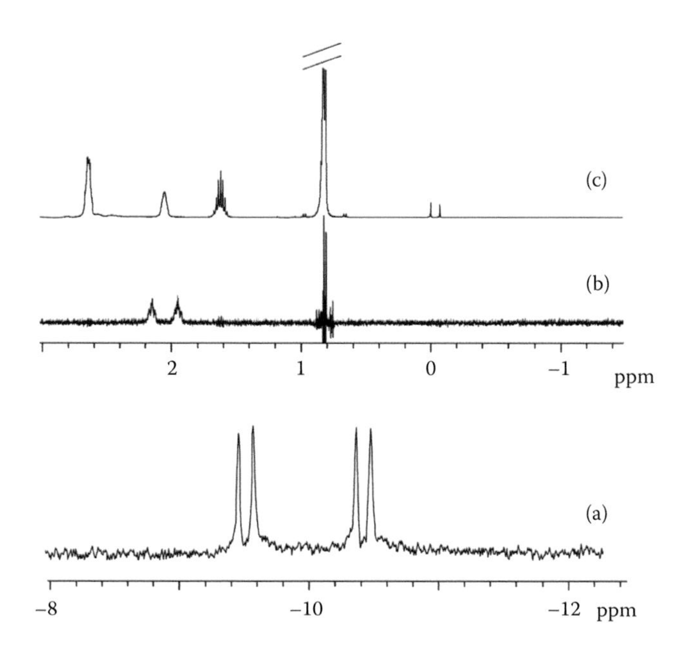


**FIGURE 5.5** (a) $^{1}$H,$^{31}$P 1-D HMQC phase cycle selected NMR spectrum of [PdH(LL)(solv)]$^{+}$, (LL = 1,2-bis((di-tert-butylphosphino)methyl)benzene) recorded in protio-MeOH at 295 K. Note: Resonances not coupled to $^{31}$P are suppressed, allowing the metal hydride resonance to be seen clearly in protio solvent. $^{1}$H NMR spectrum of [$N_3$(P(Bu$^{i}$)$_2$)$_3$] in $CDCl_3$ at 295 K. (b) $^{1}$H,$^{15}$N 1-D HMQC gradient-selected spectrum. Note that $^{15}$N satellites are readily observed at natural abundance. (c) Normal 1-D spectrum.

## 5.4 STUDYING DYNAMIC SYSTEMS

The use of VT measurement to study dynamic systems allows exchange pathways to be determined and is well documented (Chapter 2). Exchange rate constants and, hence, the thermodynamic parameters for the exchange process can be obtained by simulation and analysis of spectra recorded with VT. For quantitative work, the sample concentration must be known accurately, good shimming at each temperature is required, and the VT unit must be correctly calibrated – how to do this will be described in the manufacturers' manuals or see, for example, Berger and Braun.[31]

The recorded and actual temperatures of the sample should be within 1 to 2 K. This systematic error should be born in mind when reporting, for example, reaction rates and thermodynamic parameters.

### 5.4.1 Simulation of VT Spectra

Band-shape analysis can be a powerful mechanistic tool and gives 'pseudo–first-order rate constants', from which activation parameters can be derived via Arrhenius and/or Eyring plots; remember, however, that a good kinetic model for the proposed dynamic process is also required to relate these rate constants to the chemical processes actually happening (Chapter 2). It is then possible to distinguish between proposed exchange/reaction mechanisms. Near the slow- and fast-exchange limits, factors other than exchange, especially shim quality, contribute significantly to the line width; rate constants determined near these limits will be inaccurate. Another factor that must be borne in mind is temperature dependence of the chemical shift. Although it is possible to fit both the chemical shifts and the rate constants simultaneously, the two are interdependent, and extra care will be required. Coupling constants are less temperature dependent and should be fixed at the values measured from the slow-exchange limiting spectrum. Small amounts of impurities may provide additional exchange pathways having a large effect on reaction rates, and minor undetected species may affect the behaviour of the major species, resulting in spectra that are difficult to interpret. If more than one dynamic process is suspected, then it is a difference in rate constants and not a difference in coalescence temperatures that prove the case.

Several packages are available in addition to the instrument manufacturers' offerings, for example, gNMR (http://www.home.cc.umanitoba.ca/~budzelaa/gNMR/gNMR.html).

### 5.4.2 Saturation Transfer Exchange Spectroscopy

If the dynamic process is too slow to allow study by conventional VT methods, then saturation transfer (Chapter 3) can be used to study the exchange.[32] In these experiments, the resonance of one species is selectively excited. If the rate of exchange process is comparable to or faster than the NMR relaxation rate, then this excitation will be carried to each site visited by the exchanging species, perturbing the resonance of each site visited (Figure 5.6) and allowing both the exchange pathway(s) and the rate(s) of the exchange to be determined. The required selective excitation can be achieved either by a DANTE (Chapter 2) pulse sequence or by a selective

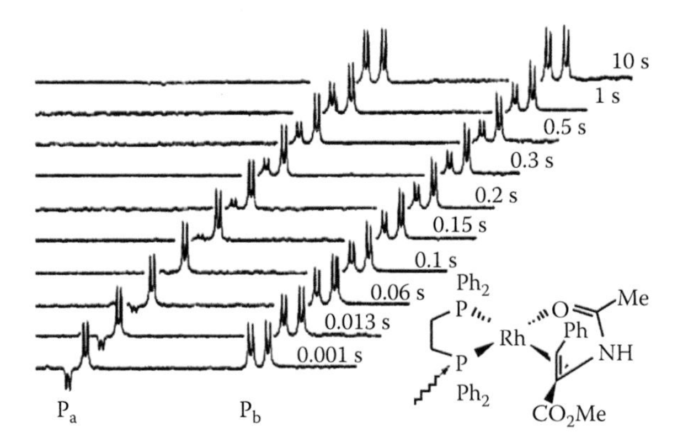


**FIGURE 5.6** Saturation transfer experiment; the downfield branch of the $P_a$ doublet of **1** was inverted using a DANTE selective excitation pulse. The exchange of the magnetisation was followed as a function of mixing time. $T_1$ was found to be 1.26 s and $k_{exchange}$ = 2.85 $s^{-1}$ at 300 K. (From Brown, J. M., P. A. Chaloner and G. A. Morris, 1987. *Journal of the Chemical Society, Perkin Transactions* 1583. Reproduced by permission of The Royal Society of Chemistry.)

pulse; both methods work equally well, although on modern spectrometers, the latter may be simpler to implement, although less selective. Account must be taken of relaxation as well as exchange in analysing the intensity variation in the spectra. For simple two-site exchanges, a quasi spin–lattice relaxation time, $T_1$, is obtained by plotting $\ln\left(\Sigma(I_o) - \Sigma I_{vd}\right)$ versus vd, the variable delay time, where $\Sigma(I_o)$ is the summed equilibrium intensity of the irradiated resonance and its exchange partner and $\Sigma(I_{vd})$ is the summed intensity at time vd. $T_1$ is the inverse of the slope of the line. The rate constant for exchange is then obtained by plotting $\ln(I_{Ir} - I_{Ex})$ versus vd, where $I_{Ir}$ is the intensity of the irradiated resonance and $I_{Ex}$ is that of the resonance of the exchanging site at time vd. The slope of the line is $(2k + 1/T_1)$. The procedure assumes that the spin–lattice relaxation time is identical at the two sites. See Akitt and Mann for more details.[33]

### 5.4.3 Exchange Spectroscopy Experiments

Dynamic processes can also be studied by exchange spectroscopy (EXSY), a 2-D method (Chapter 2) that can detect chemical exchange and fluxional processes before line broadening occurs and allows complex exchange networks, for example, CO fluxionality in a metal carbonyl cluster, to be readily mapped (Figure 5.7).[34,35] EXSY uses the same pulse sequence as nuclear overhauser effect spectroscopy (NOESY); however, NOESY and EXSY peaks are easily distinguished – EXSY peaks are very strong and have the same sign (positive vs. negative) as the diagonal peaks, whereas NOESY peaks are weak, and, except for very large molecules, will have the opposite sign to the diagonal peaks. If several exchange processes, with different rates, are occurring, several values of $t_{mix}$, the mixing time, will be needed (Chapter 2).

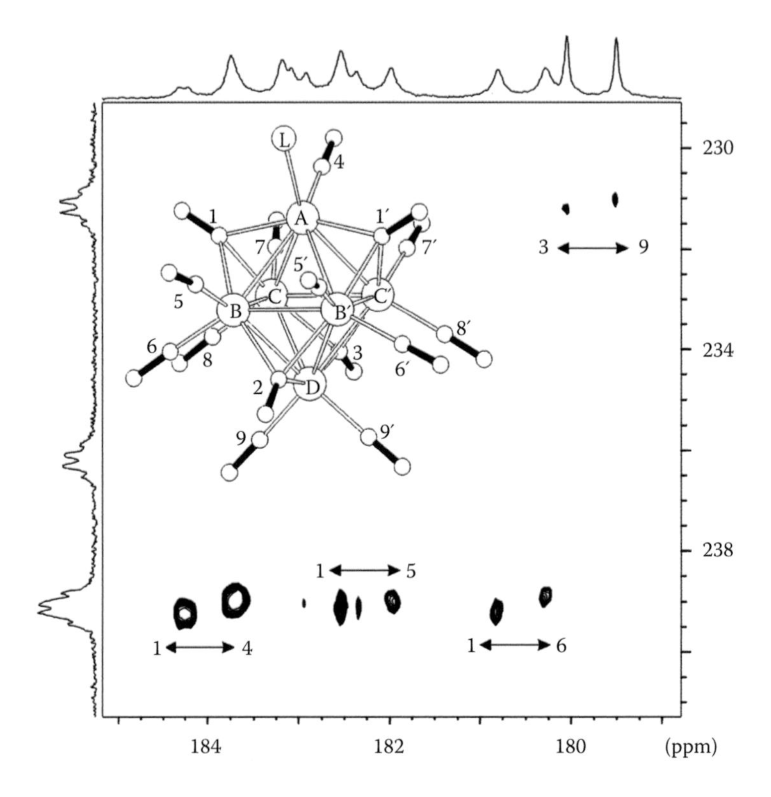


**FIGURE 5.7** $^{13}C$ EXSY spectrum showing stereospecific exchange correlations between bridging and terminal carbonyls in $[Rh_6(CO)_{15}\{P(4\text{-}(MeO)C_6H_4)_3\}]$ in $CDCl_3$ at 274 K; $t_{mix} = 0.02$ s.

## 5.5 MORE EXPERIMENT OPTIONS AND CONSIDERATIONS

There are a number of further experimental approaches that may be applicable under specific circumstances. A selection of these is now outlined.

### 5.5.1 Gradient-Accelerated HOESY Spectroscopy

HOESY spectroscopy is the heteronuclear equivalent of NOESY[36] and is useful in determining the relative dispositions of, for example, the cation and anion in solutions of organometallic salts, especially in systems where both the anion and cation contain NMR-active nuclei of high receptivity (e.g. $^{1}H$ or $^{19}F$) and several magnetically distinct nuclei are present in each ion so that different anion–cation orientations can be distinguished. Several excellent reviews have appeared recently.[37–40] Quantitative analysis requires a reference distance that will be an intra-molecular distance in the anion or cation, although the contacts of interest will be inter-ionic. If accurate inter-ionic distances are to be obtained, the reference nuclei must have the same correlation time as the pair of nuclei of interest; thus, the correlation times must also be checked. The anion–cation distances obtained will be average values reflecting the transient nature of, for example, an ion pair. In practice, the distances obtained are usually accurate to about 10%.[41]

### 5.5.2 Pulsed Field Gradient Spin Echo Diffusion Spectroscopy

Pulsed field gradient spin echo (PFGSE) NMR diffusion methods (Chapter 4) can provide complementary information which, when combined with HOESY measurements, can allow the structure of a self-assembled aggregate, for example, between the anion and catalyst–substrate cationic complex[38,39] or between a larger aggregate involving, for example, hydrogen bonding between the catalyst, co-catalyst, and substrate[42] to be established. PFGSE routinely gives diffusion constants accurate to approximately 2%.[36,37]

A detailed description of the factors relating to diffusion measurements is provided in Chapter 4, and you are advised to consult that chapter if you require more than the brief coverage included here.

Pregosin recommends the stimulated echo version of the PFGSE experiment since $T_1$, rather than $T_2$, is the effective relaxation path during the diffusion period, $\Delta$; less relaxation affords a better signal-to-noise ratio. In the PFGSE experiment, gradient pulses are used first to defocus, and then to refocus, the magnetisation. If, during the delay, $\Delta$, between the gradient pulses, the molecules diffuse from their positions after the first gradient pulse, the second pulse will not completely refocus the magnetisation, and there will be a decrease in the intensity of the NMR signal according to Equation 5.6, where $G$ is the gradient strength, $\delta$ is its length, and $\Delta$ is the diffusion delay. $D$ is the diffusion coefficient, and $\gamma$ is the magnetogyric ratio.

$$\ln\left(\frac{I}{I_0}\right) = -(\gamma_X \delta)^2 G^2 \left(\Delta - \frac{\delta}{3}\right) D \qquad (5.6)$$

Experimentally, it is important to consider the effect of temperature variation across the sample length and the solvent viscosity. Temperature gradients cause convection currents, which produce much larger displacements of the nuclei than self-diffusion. Several procedures have been suggested to minimise this effect, including the use of a high flow rate for the temperature control gas, the use of Shigemi tubes (i.e. very narrow bore tubes), the use of a 5-mm tube inside a 10-mm tube to provide a thermal jacket for the sample, and the use of inserts to disrupt convection currents.[43] The last is particularly easy to implement in modern 5-mm probes. Low solvent viscosity adds to the problem of convection when using solvents such as dichloromethane or acetone; in this case, a pulse sequence incorporating convection compensation must be used.[44] The absence of convection should be checked for; convection will be seen as a change in apparent diffusion coefficient as $\Delta$ increases. Do not spin the sample; the use of a heavy ceramic spinner is also beneficial to suppress vibrational effects. Spectrometer manufacturers provide basic data analysis tools; more advanced analysis tool packs are available from several specialist NMR groups, for example, the diffusion ordered spectroscopy (DOSY) Toolbox from Nilsson.[45] The hydrodynamic radius, $R_H$, can be used to estimate the size of the species/aggregate and is obtained from $D$ by substitution in the Stokes–Einstein equation:

$$D = k_B T / n.\pi.\eta.R_H \qquad (5.7)$$

where η is the viscosity and $R_H$ is the hydrodynamic radius. $n$ is usually taken to be 6; however, a value of 4 has also been used when the solute and the solvent have similar molecular size. (Note: In Section 4.4.1, Equation 5.7 [Equation 4.23 in Chapter 4] is referred to as the Stokes–Einstein–Sutherland equation and has $\chi$ rather than $n$ in the denominator, although the meaning is the same. $D$ is presented as $D_o$, the diffusion coefficient at infinite dilution.)

PFGSE diffusion measurements in conjunction with $^1H$–$^{19}F$ HOESY data have been used by, for example, Macchioni to establish both the existence of ion pairing and the location of the anion with respect to the bound substrate in catalyst–substrate cationic complex. Diffusion NMR measurements were first performed for solutions of precursor, substrate, and catalyst–substrate complex **1a** (Figure 5.8) at two different concentrations. Aggregation (i.e. ion pairing) in the solution of the catalyst–substrate complex was demonstrated by an increase in the hydrodynamic volume of both the catalyst–substrate complex and the anion with concentration, and the volume of both the catalyst–substrate complex and of the catalyst–substrate complex anion–ion pair obtained.

The diffusion NMR measurements indicated that only free ions and ion pairs were present in the solution, which allowed complications such as contacts arising from ion quadruples to be excluded in the interpretation of the $^{19}F$,$^1H$-HOESY spectrum. Note, however, that correlations due to contact with free substrate will be present due to the transient nature of the catalyst–substrate complex (ligand exchange)

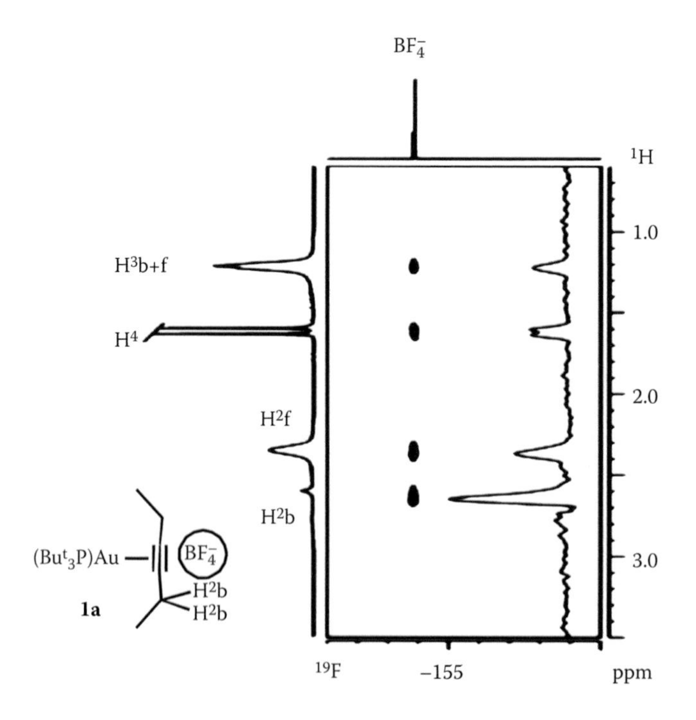


**FIGURE 5.8** $^1H$–$^{19}F$ HOESY spectrum of **1a**[$BF_4$]. B, bound; f, free hexyne. (From Ciancaleoni, G., L. Belpassi, F. Tarantelli, D. Zuccaccia and A. Macchioni, 2013. *Dalton Transactions* 42, pp. 4122–4131. Reproduced by permission of The Royal Society of Chemistry.)

and that the intensities of the nuclear overhauser effect (NOE) inter-ionic contacts must be normalised for the number of magnetically equivalent nuclei. Macchioni concluded that the most favourable anion position is close to the triple bond, halfway between the two quaternary carbon atoms (Figure 5.8).[46]

### 5.5.3 Para-Hydrogen–Enhanced Spectroscopy

The use of polarisation transfer from para-hydrogen has been demonstrated to further enhance the signal of heteronuclei in indirect detection experiments such as HMQC. Indeed, with suitable modification of the pulse sequences, the para-hydrogen effect can be used to enhance the signals in many NMR experiments, for example, correlation spectroscopy (COSY), HMQC, HSQC, EXSY and Overhauser spectroscopy.[47]

### 5.5.4 Quadrupolar Nuclei

Polarisation transfer methods can also be used to enhance the recording of spectra of quadrupolar nuclei; however, the indirect detection of the NMR spectrum of a quadrupolar nucleus presents problems due to efficient quadrupolar relaxation, which results in dephasing of the desired resonances, and the presence of the manifold of quadrupolar energy levels. Nevertheless, with appropriate modification of the pulse sequences, HMQC spectra can be observed.[48–50] For example, optimised experiments using the quadrupolar nucleus $^{6}Li$ as detector (Chapter 2) have been described.[51] For more information about the observation quadrupolar nuclei, see the examples in Section 5.7.

## 5.6 SELECTED EXAMPLES FOR MEASUREMENTS OF INORGANIC NUCLIDES

Measurement of $^{1}H$, $^{13}C$ and $^{31}P$ is rather routine as indicated above. In this chapter, attention is focussed on providing examples of the measurement of less common nuclei, for instance, $^{15}N$[52] and $^{195}Pt$[53], and the detection of typical quadrupolar nuclei, such as $^{27}Al$ and $^{51}Ni$. The goal is to illustrate the range of application of NMR measurements based on those nuclei and to show what kind of chemical information one can extract. We start by describing the essential properties of the nuclei that will justify the use of one or another technique, discussing the chemical shifts, and coupling constants, followed by several modern examples for the application of the NMR techniques. Whenever possible, additional information on what other nuclei can be detected using the same approach is provided.

### 5.6.1 Applications of $^{15}N$ NMR

#### 5.6.1.1 Properties of $^{15}N$ Isotope

Nitrogen has two NMR-measurable nuclei, for example, $^{14}N$ and $^{15}N$. While $^{14}N$ is much more sensitive as compared to $^{15}N$, its spectra are typically very broad due to quadrupolar interactions, which precludes the analysis of the fine structure in $^{14}N$ NMR spectra, in particular, when two or more non-symmetrically substituted atoms are present in the molecule. This accounts for the chemists' preference to observe $^{15}N$ instead of $^{14}N$ despite the low sensitivity of $^{15}N$.

The $^{15}N$ nucleus possesses a spin equal to 1/2 and a low and negative magnetogyric ratio (–2.71261804 rad $s^{-1}$ $T^{-1}$) and, consequently, a low Larmor frequency (30.423 MHz at 7.05 T). Furthermore, it has very low natural abundance (0.368%) and, consequently, low sensitivity ($2.25 \times 10^{-2}$). Hence, when measuring $^{15}N$ NMR, one should take into account two considerations.[52] First, the direct 1-D detection of $^{15}N$ will be much less efficient than corresponding $^{1}H$ and $^{13}C$ measurements (due to its low natural abundance). Furthermore, the negative magnetogyric ratio results in a negative NOE effect (intensity drops by ~3.94 times when compared with the direct observation without NOE); inverse gated decoupling should be used in direct 1-D NMR measurements. INEPT-based techniques should be used whenever possible since the signal increase due to the INEPT enhancement is approximately 9.87 times for the $^{1}H$–$^{15}N$ pair. The disadvantage of using of these methods is that they do not allow the observation of non-protonated nitrogen. The same approach can be used for the measurement of $^{29}Si$, $^{103}Rh$, $^{109}Ag$, and $^{199}Sn$, which all exhibit negative NOE enhancement. Second, among available 2D NMR methods, one should prefer those based on indirect detection, that is, $^{1}H,^{15}N$-HSQC, or $^{1}H,^{15}N$-HMBC. To obtain a short-range heteronuclear correlation (i.e. $^{1}H,^{15}N$-HSQC), the NH protons should not be involved in fast exchange processes and must not be replaced with deuterons from the solvent. Hence, this precludes the use of $D_2O$ and deuterated alcohols as solvents since these exchange easily with the protons in the sample. Deuterated dimethylsulphoxide or mixed $D_2O/H_2O$ (1:9) is typically used for biological samples containing compounds (e.g. amino acids) that are insoluble in, or whose structure can be irreversibly affected by, common organic solvents. The measurement of long-range correlations (i.e. $^{1}H,^{15}N$-HMBC) requires very sharp proton signals with a practical line-width limit of approximately 3 Hz; however, sharp NH resonances are not required in this case.

#### 5.6.1.2 Chemical Shifts Range and Coupling Constants

$^{15}N$ yields narrow signals (the line width for neat $NH_3$ is 0.2 Hz) that spread over an approximately 900-ppm chemical shift range (Table 5.6). Heteronuclear couplings between $^{15}N$ and $^{1}H$ are of the order of 90 Hz for one-bond couplings, and smaller two-, three- and four-bond couplings are also frequently observed. Sometimes, very weak satellite signals are detected in the $^{1}H$ NMR spectrum in case of NH resonances and can be used to confirm the assignment of NH signals.

#### 5.6.1.3 Examples of the Use of $^{15}N$ NMR

Since nitrogen is one of the major components of biological systems, it is not surprising that the major application of the $^{15}N$ NMR lies in the field of structure elucidation of biomolecules (Chapter 3).[52,54,55] Several comprehensive reviews covering the application of different $^{15}N$ NMR techniques are available.[52,56,57] Here, we provide several representative examples of $^{15}N$ measurements.

Marek et al. used $^{15}N$ NMR for the discrimination of a series of N7- and N9-substituted purine derivatives (Figure 5.9a).[54] Their chemical shifts were measured at the natural abundance level of the $^{15}N$ isotope via direct 1-D observation (Figure 5.9b).

**TABLE 5.6**
**$^{15}N$ Chemical Shifts for Exemplar Organic Compounds**

| Compound | Chemical Shift Range Related to $MeNO_2$, ppm |
|---|---|
| Nitroso compounds | −480 to −540 |
| Nitrites | −180 to −210 |
| Nitrosamines | +150 to −180 (NO) |
| | +150 to +120 ($NR_2$) |
| Azo compounds | −130 to −170 |
| Nitroalkanes | 0 to −40 |
| Nitroarenes | +20 to 0 |
| Imines, oximes | +170 to −30 |
| Nitrones | +100 to +80 |
| Azines | +120 to −40 |
| Azides | +130 to +110 (RN*N*N) |
| | +270 to +250 (RNN*N*) |
| | +320 to +290 (R*N*NN) |
| Nitriles | +140 to +110 |
| Nitrile N-oxides | +180 to +150 |
| Cyanamides | +200 to +170 (CN) |
| | +390 to +330 ($NR_2$) |
| Guanidines | +210 to +190 (NR) |
| | +330 to +310 ($NR_2$) |
| Isocyanides | +220 to +170 |
| Amides | +280 to +230 |
| Amine N-oxides | +280 to +250 |
| Hydrazines | +330 to +270 |
| Aryl amines | +330 to +270 |
| Isocyanates | +360 to +320 |
| Alkyl amines | +390 to +300 |

*Source:* M. Witanowski, G. A. Webb (eds.), *Nitrogen NMR*, Plenum Press. London and New York, 1973; M. Witanowski, G. A. Webb, *Nitrogen NMR Spectroscopy, Annual Reports on NMR Spectroscopy*, Vol. 5A, pp. 395–464 (ed. E. F. Mooney), Academic Press, London and New York, 1972; R. L. Lichter, *$^{15}N$ Nuclear Magnetic Resonance, Determination of Organic Structures by Physical Methods*, Vol. 4, pp. 195–232 (eds. F. C. Nachod and J. J. Zuckerman), Academic Press. New York and London, 1971.

At the same time, the correct assignment was performed using gradient-enhanced $^{1}H,^{15}N$-HSQC/HMQC and $^{1}H,^{15}N$-HMBC/SQHMBC techniques (Figure 5.9c).

In another study, Larina and Milata studied the structure and tautomerism of the nitro-derivatives of benzotriazole, prepared by nitration of the corresponding benzotriazoles or by methylation or cyclisation of appropriate nitro-1,2-phenylenediamines[58] (Figure 5.10) using a combination of $^{1}H$, $^{13}C$, $^{15}N$ and 2-D NMR

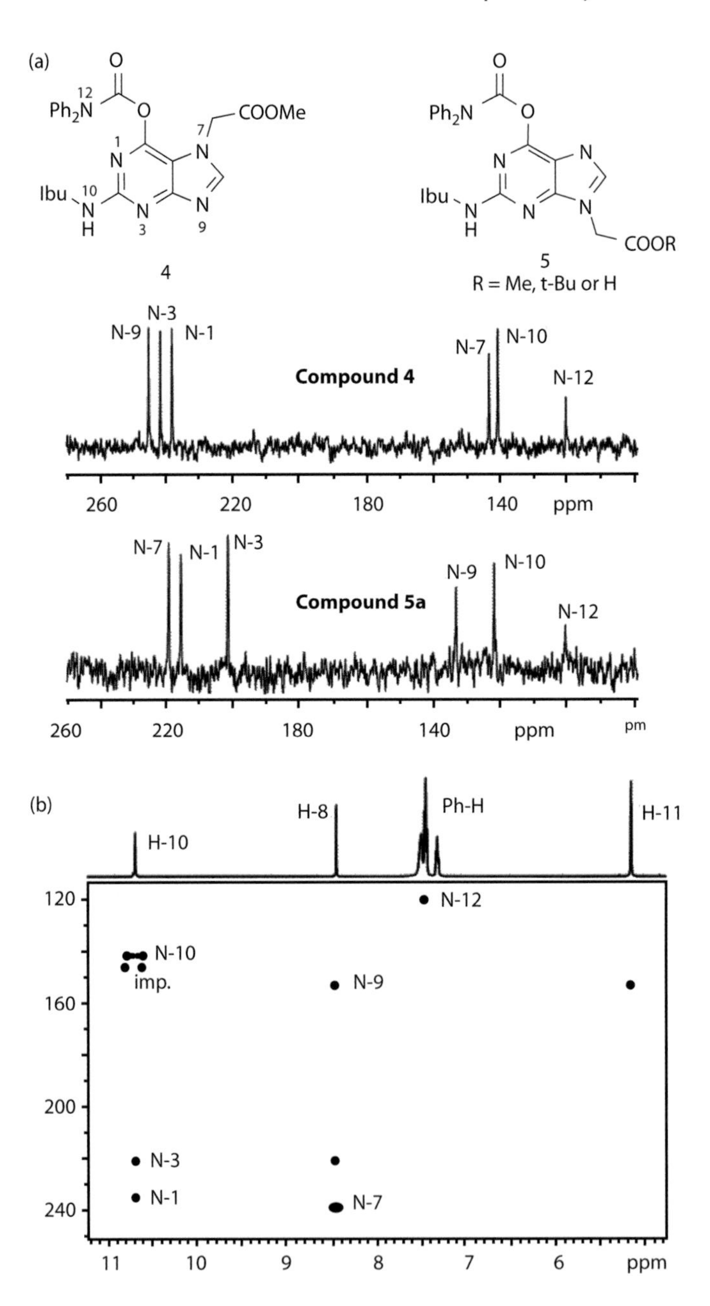


**FIGURE 5.9** $^{15}$N NMR spectra of two representative compounds (**4** and **5a**) of a series of $^{7}$N- and $^{9}$N-substituted purine derivatives studied in ref. 54, measured via direct $^{15}$N observation (a) and gradient-enhanced $^{1}$H,$^{15}$N-SQHMBC spectrum (b) of **5a**. (Marek, R., J. Brus, J. Tousek, L. Kovacs and D. Hockova. *Magnetic Resonance in Chemistry.* 2002. 40. pp. 353–360. Copyright Wiley-VCH Verlag GmbH & Co. KGaA. Reproduced with permission.)

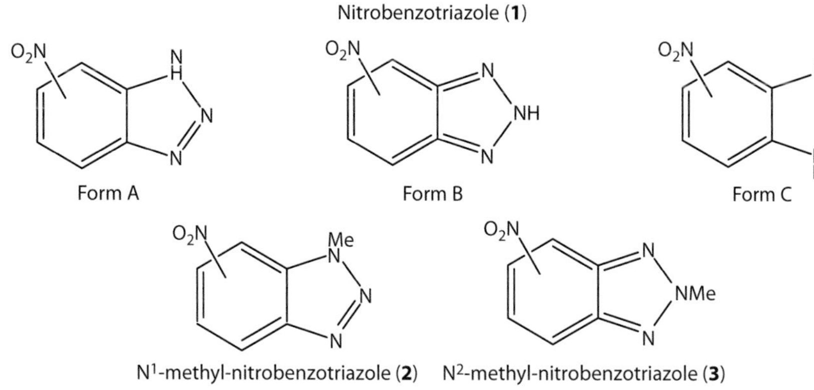

Nitrobenzotriazole (**1**)

Form A Form B Form C

$N^1$-methyl-nitrobenzotriazole (**2**) $N^2$-methyl-nitrobenzotriazole (**3**)

**FIGURE 5.10** Structures of nitrobenzotriazole (**1**) and its methylated derivatives (**2** and **3**) studied in ref. 58.

spectroscopy and quantum chemistry calculations. The experimental and calculated screening constant values indicated that N-unsubstituted benzotriazoles undergo a prototropic exchange only between forms A and C and not through B, and the position of a nitro group at the phenylene fragment of benzimidazole cycle does not influence the tautomeric equilibrium significantly but has impact only on the screening constants of the magnetic active nuclei of the heterocycle. The authors concluded that $^{15}N$ NMR spectroscopy, along with quantum–chemical calculations, is a convenient approach in the examination of tautomeric processes.

## 5.6.2 Applications of $^{195}Pt$ NMR

### 5.6.2.1 Properties of $^{195}Pt$ Isotope

Of the five naturally occurring platinum isotopes ($^{192}Pt$, $^{194}Pt$, $^{195}Pt$, $^{196}Pt$, $^{198}Pt$), only $^{195}Pt$ is NMR active (Table 5.3). It exhibits favourable properties for NMR measurements, that is, a spin quantum number of 1/2, and moderate magnetogyric ratio (5.7686107 rad $s^{-1}$ $T^{-1}$) and Larmor frequency (64.5 MHz at 7.05 T). Furthermore, significant natural abundance (33.8%) and sensitivity ($9.94 \times 10^{-3}$) guarantee the widespread use of $^{195}Pt$ NMR measurements for structural characterisation of platinum-containing molecules.[53,59] Both direct and indirect measurement through correlation with abundant, higher $\gamma$ nuclei (e.g. $^{1}H$,$^{31}P$-based HSQC and HMBC) are typically employed in the $^{195}Pt$ NMR spectroscopy; however, the sensitivity gain from such experiments is relatively small, except for $^{1}H$, the value being in the connectivity information obtained. All known oxidation states of platinum (0, I, II, III, IV) are diamagnetic and produce narrow resonances in the NMR.

### 5.6.2.2 Chemical Shifts Range and Coupling Constants

$^{195}Pt$ yields narrow signals (line width for 1.2 M $Na_2PtCl_6$ in $D_2O$ used as reference compound is 5.4 Hz) that occur over a very wide chemical shift range (~13,000 ppm). The exact position in $^{195}Pt$ NMR is influenced by the metal oxidation state, ligand type and its coordination mode, and the stereochemistry around the metal.

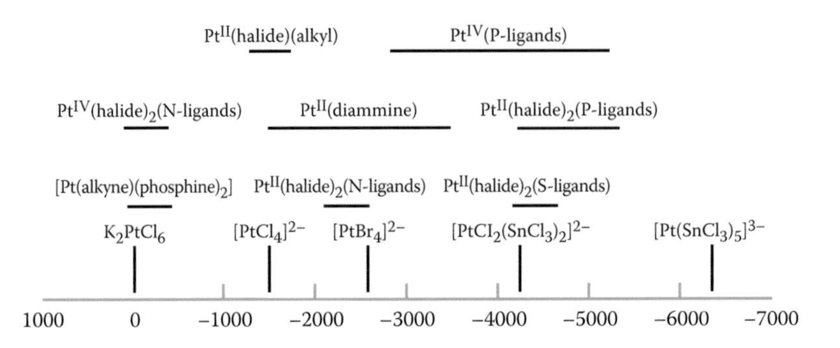


**FIGURE 5.11** Selected chemical shifts for $^{195}$Pt.

Approximate ranges for Pt(0), Pt(II), Pt(III), and Pt(IV) species are indicated in Figure 5.11. The calculated shift for Pt(0) is −10,427 ppm,[60] while removal of electrons, that is, an increase in the oxidation state, leads to de-shielding and results in the high-frequency shift. Despite the fact that, typically, Pt(II) species resonate at a higher field compared to Pt(IV) counterparts, a considerable overlap between the chemical shift ranges of the different oxidation states is commonly observed since the shift is strongly influenced by the ligands.

In referencing platinum NMR spectra, notice must be taken of (1) the large temperature dependence of a chemical shift (e.g. using broadband $^1$H decoupling can cause a temperature drift of a signal), (2) interaction of the platinum compounds with solvent or a substance used as an internal reference, (3) uncertainties due to isotopomers, and (4) the existence of two referencing schemes for this nucleus.[61] $^{195}$Pt chemical shifts are sometimes expressed relative to a standard sample, $[PtCl_6]^{2-}$ (however, this is not very convenient due to the limited solubility of Pt(IV) chlorides) or $[PtCl_4]^{2-}$; in $D_2O$; however, other compounds can also be used provided that their chemical shifts are known.[53] Alternatively, $^{195}$Pt shifts may be reported relative to the absolute frequency of 21.400 000 MHz (Table 5.3).

Platinum shows a wide variety of couplings with other nuclei; coupling to $^{195}$Pt is normally visible in $^1$H, $^{13}$C, $^{15}$N, $^{19}$F, and $^{31}$P spectra as $^{195}$Pt satellites on either side of the NMR signal. Analysis of the coupling constant can provide valuable information on the interaction between the metal center and the ligand, including the trans-influence of the ligand and the geometry of the complex. In general, the spin–spin interactions vary over several orders of magnitude, for example, $^1J(^1H,^{195}Pt) > 1$ kHz, $^1J(^{31}P,^{195}Pt) > 2$ kHz, and $^1J(^{119}Sn,^{195}Pt) > 20$ kHz.[53] Two-bond couplings to protons are typically between 25 and 90 Hz, one-bond $^{195}$Pt–$^{15}$N couplings are in the region of 160 to 390 Hz, and couplings to $^{31}$P are around 1300 to 4000 Hz for one bond and 30 Hz for two bond. The platinum coupling to $^{119}$Sn is especially large and can be over 33,000 Hz. Couplings to $^{195}$Pt are observable in both 1-D and 2-D measurements. $^nJ(^{195}Pt,^1H)$ interactions (where $n$ = 1–5) can be readily detected using $^1$H signals (Figure 5.12),[62] and HMQC experiments using $^{31}$P as the high magnetogyric nucleus for polarisation transfer has been employed to study platinum–phosphine clusters.[25]

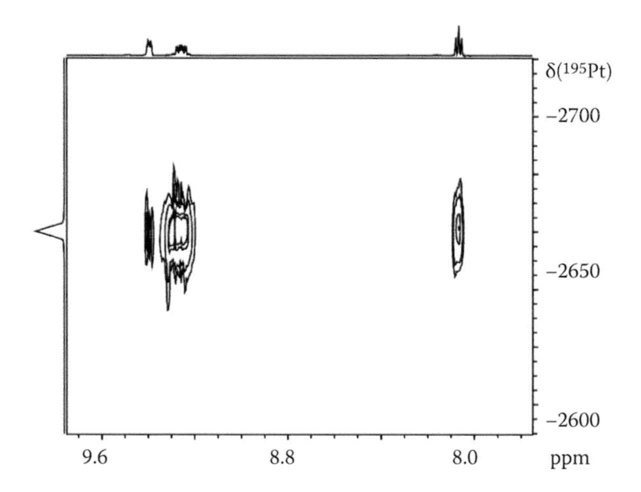


**FIGURE 5.12** Expansion of the $^1H,^{195}Pt$ HMQC spectrum of the Pt–bipyrimidine complex shown, displaying correlation via $^3J(^1H,^{195}Pt)$, $^4J(^1H,^{195}Pt)$ and $^5J(^1H,^{195}Pt)$ between the aromatic protons and the platinum center. (Gudat, D., A. Dogan, W. Kaim and A. Klein. *Magnetic Resonance in Chemistry*. 2004. 42. p. 781. Copyright Wiley-VCH Verlag GmbH & Co. KGaA. Reproduced with permission.)

### 5.6.2.3 Examples of the Uses of $^{195}$Pt NMR

Two extensive reviews are available with detailed analysis of applications of $^{195}$Pt NMR.[53,59] Herein, we describe some representative studies demonstrating the application of different NMR methods in $^{195}$Pt NMR spectroscopy.

Kukushkin et al. described the enantiomeric discrimination of anti-tumor active cis- and trans-isomers of platinum(IV) carboxamide complexes $[PtCl_4\{NH{=}C(OH)R\}_2]$ (Me, Et) and the identification of those in cis-$[PtCl_4\{NH{=}C(OH)Me\}_2]_{0.66}$ · cis-$[PtCl_4\{NH{=}C(OH)Et\}_2]_{0.33}$.[63] A small (30°) flip angle and the 'aring' pulse sequence were used to suppress acoustic ringing in direct observation of the $^{195}$Pt NMR spectrum of the trans-complex $[PtCl_4\{NH{=}C(OH)Et\}_2]$, which showed a resonance at 36 ppm due to the platinum center, while that of the cis-species occurs at a higher field of −202 ppm. The same authors reported[64] the use of $^{195}$Pt for the identification of the oxidation state in novel Pt(IV)– and Pt(II)–oxadiazole complexes $\left[PtCl_4\left\{N^a = C(R)ON = CC_6^bH_2R_3^1\right\}\right]^{a-b}$ (R = Me, Et, $CH_2Ph$; $R^1$ = Me, OMe); Pt(IV) chemical shifts were in the range of −170 to 20, and Pt(II) were around −2200 ppm.

$^{195}$Pt NMR measurements in solutions have also been used to estimate first-order rate constants in cis–trans isomerisation of some square-planar platinum(II) nitroimidazole complexes.[65] The kinetics of the isomerisation process in solution was obtained from the rate of disappearance of the $^{195}$Pt resonance of the trans-isomer and the reaction mechanism found to be associative. Values for the chemical shift of $^{195}$Pt and coupling constants (Table 5.7) have also been used to study the effect of solvent coordination and ligand variation on the cis- and trans-configurations in square planar $[Pt(DMSO)_2(Ar)_2]$ and allowed the assignment of the configurational isomers and their isomerisation mechanisms.[66,67]

**TABLE 5.7**
**Selected Experimental and Calculated $\delta(^{195}Pt)$ and Coupling Constants for the Complexes $[Pt(DMSO)_2(Ar)_2]$**

| Complex | $^3J(^{13}CH_3,^{195}Pt)$ DMSO/Hz | $^1J(^{13}C(1),^{195}Pt)$/Hz | $\delta(^{195}Pt)$/ppm | Conformation Based on NMR Data | Calculated $^1J(^{13}C(1),^{195}Pt)$/Hz |
|---|---|---|---|---|---|
| $[Pt(DMSO)_2(Ph)_2]$ | 14.9 | 994.8 | –4217 | cis | cis: 1018.3<br>trans: 538.7 |
| $[Pt(DMSO)_2(2\text{-}Tol)_2]$ (anti) | 13.8, 15.5 | 1022.3 | –4157 | cis | cis, anti: 1042.9<br>trans, anti: 548.1 |
| $[Pt(DMSO)_2(2\text{-}Tol)_2]$ (syn) | 14.1, 16.0 | 1021.3 | –4165 | cis | cis, syn: 1029.2<br>trans, syn: 536.1 |
| $[Pt(DMSO)_2(3\text{-}Tol)_2]$ | 14.8 | 945.5 | –4220 | cis | cis, anti: 1017.3<br>trans, anti: 536.1 |
| $[Pt(DMSO)_2(4\text{-}Tol)_2]$ | 14.9 | 1002.6 | –4212 | cis | cis: 1027.5<br>trans: 541.3 |
| $[Pt(DMSO)_2(Xyl)_2]$ | 28.8 | 602.9 | –4146 | trans | cis: 1139.6<br>trans: 541.2 |
| $[Pt(DMSO)_2(Mes)_2]$ | 28.6 | 605.0 | –4148 | trans | cis: 1130.4<br>trans: 546.0 |
| $[Pt(DMSO)_2(Me_5Ph)_2]$ | 28.5 | 617.5 | –4089 | trans | – |

*Source:* Kerrison, S. J. S. and P. J. Sadler, *Inorganica Chimica Acta* 104, p. 197, 1985; Klein, A. et al., *Organometallics* 24, p. 4125, 2005.

Beni et al. have identified a unique hydride complex containing a triangular cluster $[Pt_3(\mu\text{-}CO)(PCy_3)_3(Ph_3Sn)H]$.[68] Proton-coupled and decoupled $^{195}Pt$ NMR measurements (Figure 5.13) were performed to confirm the presence of the hydride ligand. The deduced $^{1}J(^{1}H,^{195}Pt)$ coupling constant of 1018 Hz suggested that the hydride ligand was coplanar with the metal core.

Abel et al. used 2-D exchange spectroscopy (EXSY) (Chapter 2) to study the kinetics associated with pyramidal sulphur conversion in the $[PtIMe_3(MeSCH_2CH_2SEt)]$ complex (Figures 5.14 and 5.15).[69] The authors were able to find nine distinct rate constants associated with the extensive fluxionality in the platinum(IV) complex.

Chiral Pt(II) complexes have also been used as chiral derivatising agents for the enantio-discrimination of unsaturated compounds and the assignment of their absolute configuration (R or S).[70,71] It was shown that $^{195}Pt$ NMR was helpful in detecting the inequivalences of the diastereoisomeric mixtures arising from the complexation of the metal to the unsaturated carbons.[70,71]

$^{195}Pt$ and $^{205}Tl$ NMR was used to establish the structure of some newly prepared porphyrin–thallium–platinum intermetallic complexes. Very strong one-bond $^{195}Pt$–$^{205}Tl$ spin–spin couplings, 47.8 and 48.3 kHz for two different porphyrin complexes (Figure 5.16), confirmed the existence of a 'naked' metal–metal bond.[72]

$^{195}Pt$ NMR is an important technique in bio- and medicinal chemistry, in particular, in the development of platinum-containing anti-tumor drugs and in the evaluation of the mechanism of action. For example, for the widely used drugs cisplatin and carboplatin, it is believed that anti-tumor action results primarily from the strong binding of the platinum core to DNA. Many kinds of DNA adducts have been identified by NMR spectroscopy[73,74] and time-course 1-D $^{195}Pt$ NMR experiments, observing changes in chemical shifts and intensities of the resonances used in binding studies.[75–78] For example, it was established that the preferred binding site of cisplatin

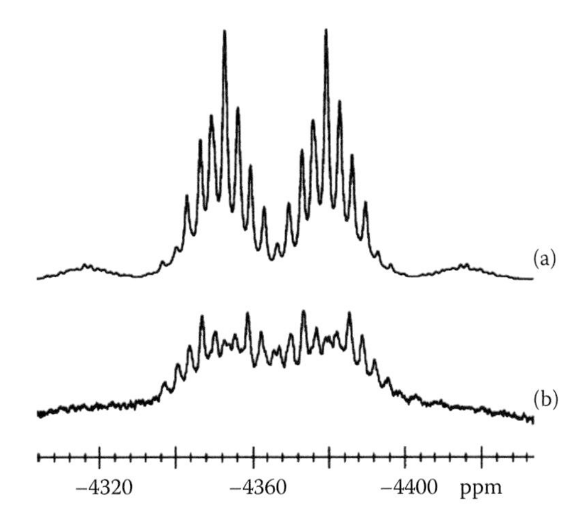


**FIGURE 5.13** Hydride indication from $^{1}H$ coupled (a) and $^{1}H$ decoupled (b) $^{195}Pt$ NMR spectra of $[Pt_3(\mu\text{-}CO)(PCy_3)_3(Ph_3Sn)H]$ in $CD_2Cl_2$. (Reprinted from *Inorganic Chemistry Communications*, 7, Béni, Z., R. Scopelliti and R. Roulet, pp. 935–937, Copyright 2004, with permission from Elsevier.)

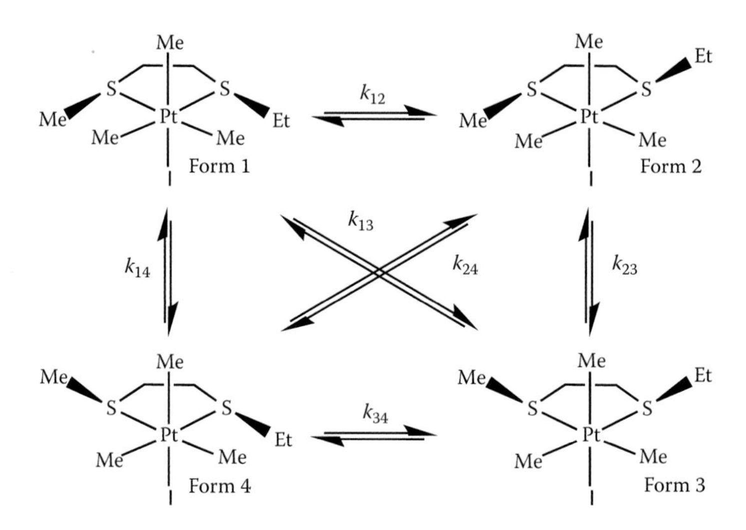


**FIGURE 5.14** The inter-conversion between isomers of [PtIMe$_3$(MeSCH$_2$CH$_2$SEt)]. (From Abel, E. W., I. Moss, K. G. Orrell, V. Sik and D. Stephenson, 1987. *Journal of the Chemical Society, Dalton Transactions*, pp. 2695–2701. Reproduced by permission of The Royal Society of Chemistry.)

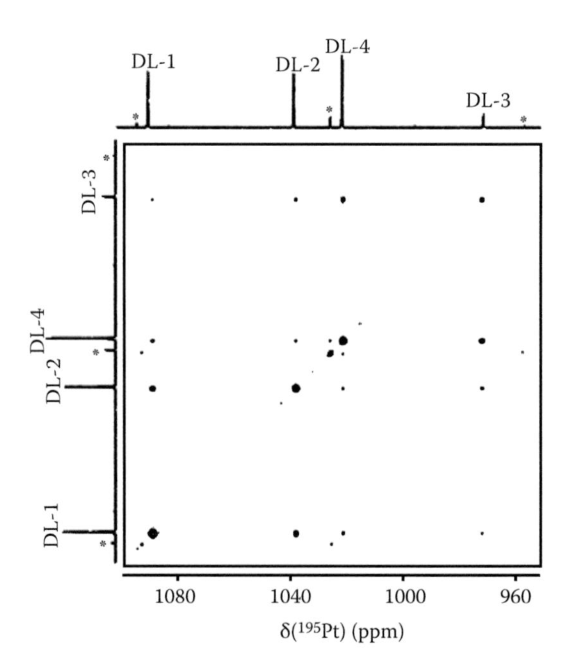


**FIGURE 5.15** $^{195}$Pt 2-D exchange spectrum of [PtIMe$_3$(MeSCH$_2$CH$_2$SEt)] (CDCl$_3$ solution, 263 K; $t_{mix}$ = 0.08 s). Signals (marked *) are due to a small amount of [PtIMe$_3$(MeSCH$_2$CH$_2$SMe)]. (From Abel, E. W., I. Moss, K. G. Orrell, V. Sik and D. Stephenson, 1987. *Journal of the Chemical Society, Dalton Transactions*, pp. 2695–2701. Reproduced by permission of The Royal Society of Chemistry.)

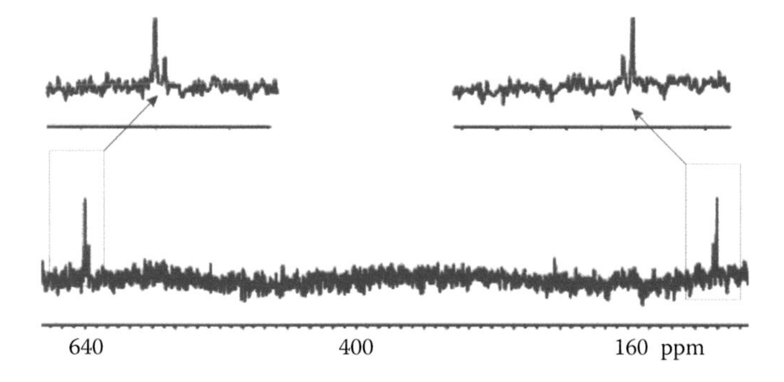


**FIGURE 5.16** $^{195}$Pt NMR spectrum of $[(NC)_5Pt–Tl(tpp)]^{2-}$ at 298 K (lower trace) and expanded doublet signals (upper trace). $J(^{195}Pt–^{205}Tl) \approx 48$ kHz. (Reprinted from *Inorganica Chimica Acta*, 357, Ma, G., M. Maliarik, L. Sun and J. Glaser, 2004, pp. 4073–4077. Copyright 2004, with permission from Elsevier.)

to DNA is the N7 of guanine, and binding involves successive displacement of $Cl^-$ by the purine N, thus leading to low-frequency shifts in the $^{195}$Pt NMR spectrum.[79] Furthermore, formation and closure of adducts during the course of DNA binding reactions was confirmed by monitoring the 1-D $^{195}$Pt NMR spectrum and suggested that the biological inactivity of the transplatin isomer can be explained by its selective trapping *in vivo* with glutathione.[80–83]

An extended study on the toxicity and DNA groove binding of platinum–nitrogen complexes has been carried out.[84–86] The binding of several dipyrazolylmethane complexes with guanosine and adenosine was characterised using $^{195}$Pt NMR. Binding of guanosine and adenosine to a dinuclear platinum species caused an upfield shift in the $^{195}$Pt resonance from 2338 to 2467 and 2486 ppm, respectively (Figure 5.17).

In another study, the local disposition kinetics of carboplatin after subcutaneous injection in rats was monitored using $^{195}$Pt NMR.[84] *In vitro* measurements were

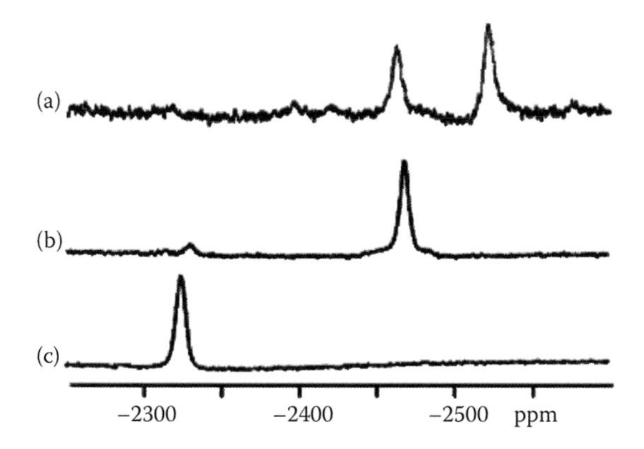


**FIGURE 5.17** $^{195}$Pt NMR spectra of di-Pt (inset) after reaction with adenosine for 1 week (a), di-Pt after 24-h reaction with guanosine (b), and di-Pt (c). (From Wheate, N. J., B. J. Evison, A. J. Herlt, D. R. Phillips and J. G. Collins, 2003. *Dalton Transactions*, pp. 3486–3492. Reproduced by permission of The Royal Society of Chemistry.)

performed in different solvents containing potassium tetrachloroplatinate(II), carboplatin, and cisplatin, which showed resonances at $\delta(^{195}Pt)$ of −1623, −1705 and −2060 ppm, respectively; the $T_1$ relaxation time of carboplatin was found to be around 100 ms. Subsequent *in vivo* measurements undertaken for carboplatin showed a broad resonance at $\delta = -1715$, thus confirming the local disposition kinetics.

## 5.7 MEASUREMENT OF QUADRUPOLAR NUCLEI

As noted earlier, NMR detection of quadrupolar nuclei is complicated by the very efficient spin relaxation such nuclei experience. Nevertheless, under appropriate conditions, it is possible to derive useful information from such NMR studies as described in the following.

### 5.7.1 General Considerations

The principal difficulty in NMR spectroscopic studies of quadrupolar nuclei is that the spin–lattice relaxation time ($T_1$) is typically very short, resulting in significant broadening of the lines. Since $T_1$ is controlled by two factors – the electric quadrupole moment ($Q$) and the presence of electric field gradients across the nucleus – both components should be considered during the detection of the quadrupolar nuclei.

If a nucleus possesses a large quadrupole moment, efficient relaxation by molecular motion will occur, giving very broad lines. For example, observation of the iodide anion in water by $^{125}$I NMR ($I = 5/2$, $Q = 0.6$) gives $\nu_{1/2}$ of 1800 Hz, while all other iodo compounds, being less symmetric, produce lines so broad that they cannot be detected in the NMR spectra. For nuclei with small quadrupolar moments, for example, $^{2}$H ($I = 1$, $Q = 0.0028$) and $^{6}$Li ($I = 1$, $Q = 0.00046$), which have the smallest electric quadrupole moments of all quadrupole nuclides, give narrow lines facilitating detection, and often, resolved $J$ coupling is observed.

Broadening due to the presence of an electric field gradient across the nucleus is least for molecules with high symmetry (e.g. tetrahedral and octahedral species in which the quadrupolar nucleus in question is at the center of symmetry); hence, highly symmetrical species can give sharp lines and resolved $J$ multiplets. For example, the $^{14}$N NMR spectrum of $NH_4^+$ ($T_1 > 50$ s) shows well-resolved lines (line width < 0.1 Hz), while that of the unsymmetrical MeCN shows very broad lines (line width > 50 Hz) as a result of the very fast relaxation of the $^{14}$N spins ($T_1 = 0.022$ s) (Figure 5.18).[87]

Another characteristic example is illustrated in Figure 5.19,[88] which shows the NMR spectra of some trimethyl phosphate complexes of aluminium, $^{27}$Al ($I = 5/2$, $Q = 0.15\ 10^{24}$ cm$^{-1}$). The octahedral aluminum complex with six trimethylphosphate ligands gives a sharp multiplet due to $J_{P-Al}$ in both the $^{27}$Al and $^{31}$P spectra. However, replacement of one phosphate ligand with water results in the desymmetrisation of the species and, consequently, produces a broad signal because there is now an electric field gradient at the aluminum.

Similar considerations/effects are valid for the detection of nearly all quadrupolar nuclei. As an example, we provide in the following the achievement regarding the observation of $^{61}$Ni NMR, one of the most attractive metal nuclei for NMR study due to its catalytic properties.

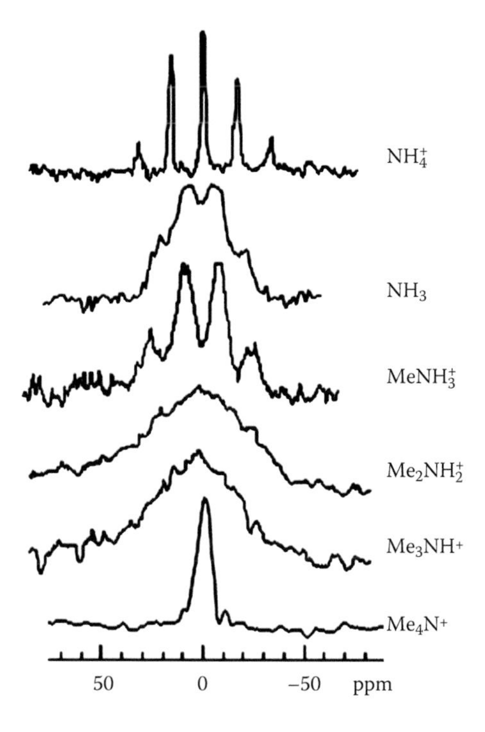


**FIGURE 5.18** $^{14}N$ NMR spectra ($I = 1$, $Q = 0.016$) of representative nitrogen compounds, illustrating the influence of symmetry on the broadening of the NMR signals. All are on the same horizontal scale, but chemical shifts are widely different. (Reprinted with permission from Ogg, R. A. and J. D. Ray. 1957. *Journal of Chemical Physics* 26, pp. 1339–1340. Copyright 1957, American Institute of Physics.)

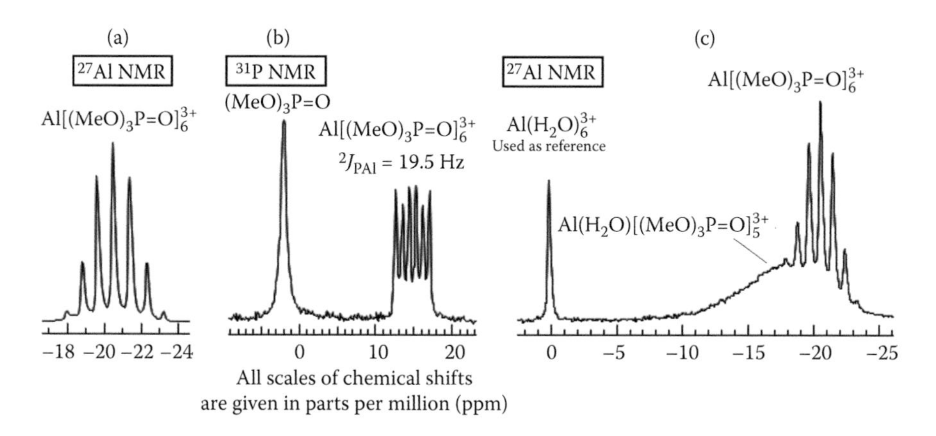


**FIGURE 5.19** $^{27}Al$ (22.63 MHz) and $^{31}P$ (24.288 MHz) NMR spectra of a solution, $Al(ClO_4)_3$, containing excess trimethyl phosphate in nitromethane (a and b) and aqueous nitromethane (c). (Reprinted with permission from Delpuech, J.-J., M. R. Khaddar, A. A. Peguy and P. R. Rubini. 1975. *Journal of the American Chemical Society* 97, pp. 3373–3379. Copyright 1975 American Chemical Society.)

## 5.7.2 Applications of $^{61}$Ni NMR as a Typical Quadrupolar Nuclide

### 5.7.2.1 Properties of $^{61}$Ni Isotope

$^{61}$Ni is the only magnetically active isotope of nickel and is a typical quadrupolar nucleus – spin-3/2, with a quadrupole moment of $16.2 \times 10^{-30}$ m$^2$. The latter leads to extremely broadened $^{61}$Ni NMR resonances even for slightly unsymmetric nickel complexes. It has low natural abundance (1.14%) and sensitivity of measurement comparable to $^{13}$C (0.240). The low resonance frequency (26.8 MHz, with $^1$H frequency of 300 MHz) often leads to severe ringing effects and rolling baselines. Furthermore, owing to the fast relaxation, indirect methods of detection can no longer be applied.[89]

Nonetheless, $^{61}$Ni NMR has been used successfully to study diamagnetic Ni(0) complexes. Unfortunately, Ni(II), the most common oxidation state of nickel, is paramagnetic in the tetrahedral form; hence, it cannot be observed on a high-resolution NMR spectrometer. However, the square-planar form is diamagnetic, and $^{61}$Ni NMR spectra can be measured.

It is worth mentioning that the solution NMR spectroscopy of another catalytically important metal, that is, $^{105}$Pd NMR (spin-5/2, with a quadrupole moment of $66.0 \times 10^{-30}$ m$^2$) is nearly unknown, the main reasons being the high quadrupole moment and the low sensitivity of the nuclide.[90]

### 5.7.2.2 Chemical Shifts Range and Coupling Constants

$^{61}$Ni yields broad signals (line width for $Ni(CO)_4$; 80% in $C_6D_6$ used as reference compound is 7 Hz) that spread over an approximately 1200-ppm–wide chemical shift range. The signal width decreases with the symmetry of the environment, allowing the use of $^{61}$Ni for symmetrical complexes.

### 5.7.2.3 Examples of the Use of $^{61}$Ni NMR

Examples of $^{61}$Ni NMR spectroscopy are still scarce[89,91–95] due to the aforementioned difficulties in the detection of this nucleus. Two representative studies are given in the following.

Hao et al. established the ranges of $^{61}$Ni NMR chemical shifts, coupling constants, relaxation times, and natural line widths using natural-abundance $^{61}$Ni NMR measurements.[92] Several $d^{10}$ symmetrical tetrahedral $NiL_4$ complexes (where L is a phosphine or a phosphite ligand) were prepared and studied. In these symmetrical complexes, the electric field gradient at $^{61}$Ni is minimal, and quadrupolar line broadening is correspondingly small, resulting in $^{61}$Ni line widths that are sufficiently narrow to permit the observation of spin–spin coupling interactions with $^{61}$Ni (see Table 5.8 and Figure 5.20, which shows the $^{61}$Ni NMR spectrum of $Ni(PF_3)_4$ – one-bond $^{61}$Ni–$^{31}$P and two-bond $^{61}$Ni–$^{19}$F couplings are resolved).

Behringer et al. report an extensive $^{61}$Ni NMR spectroscopic study of di- and tricarbonylnickel(0) phosphine complexes that were recorded using solenoid glass sample tubes in a special solenoid probehead to maximise probe efficiency.[89] Representative spectra are presented in Figure 5.21, and values of chemical shifts and coupling constants are given in Table 5.9.

$^1J(^{61}\mathrm{Ni},^{31}\mathrm{P})$ coupling constants can be obtained alternatively by measuring the $^{31}$P NMR spectrum of a saturated tetrahydrofuran (THF) solution of the complex

**TABLE 5.8**
**$^{61}$Ni NMR Data for $d^{10}$ Symmetrical Tetrahedral $NiL_4$ Complexes**

| Compound | $\delta(^{61}Ni)^a$/ppm | $J(^{61}Ni–^{31}P)$ | $\varpi_{1/2}{}^b$/Hz | Solvent |
|---|---|---|---|---|
| $Ni(PF_3)_4{}^c$ | −929.1 | 482 | 20 (±5) | THF |
| $Ni[P(OMe)_3]_4$ | −742.2 | 398 | 80 (±10) | $d_6$-benzene |
| $Ni[P(OEt)_3]_4$ | −703.0 | 405 | 90 (±10) | $d_6$-benzene |
| $Ni(CO)_4$ | 0 | NA | 4 (±0.5) | Neat |
| $Ni[PCl_3]_4$ | 266.9 | 450 | 20 (±5) | $PCl_3$ |

[a] Recorded at 298 K. Chemical shifts referenced are following IUPAC guidelines (see p. 188).
[b] Line width.
[c] $J(^{61}Ni–^{19}F)$ = 28 Hz.

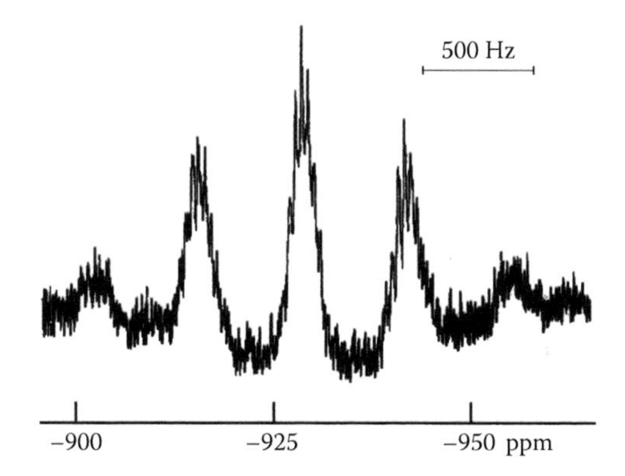


**FIGURE 5.20** Natural-abundance $^{61}$Ni NMR spectrum of $Ni(PF_3)_4$ recorded at 35.727 MHz (pulse width, 50 μs; 32.000 scans) in THF solution, showing coupling between $^{61}$Ni and $^{31}$P and $^{19}$F. (Reprinted from *Journal of Magnetic Resonance*, 46, Hao, N., M. J. McGlinchey, B. G. Sayer and G. J. Schrobilgen, 1982, pp. 158–162. Copyright 1982, with permission from Elsevier.)

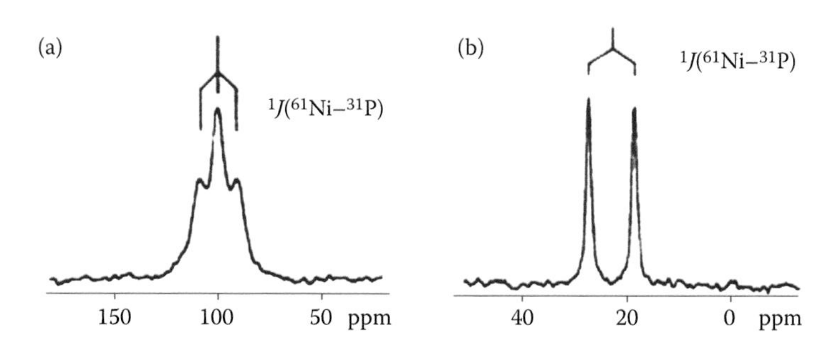


**FIGURE 5.21** 26.8-MHz $^{61}$Ni NMR spectrum of $(CO)_2Ni(PMe_3)_2$ (a) and $(CO)_3Ni(PPh_3)$ (b) in THF. $^1J(^{61}Ni,^{31}P)$ = 250 and 236 Hz, respectively. (Behringer, K. D. and J. Blumel: *Journal of Magnetic Resonance*. 1995. 33. pp. 729–733. Copyright Wiley-VCH Verlag GmbH & Co. KGaA. Reproduced with permission.)

**TABLE 5.9**
**$^{61}$Ni NMR Chemical Shifts, Signal Half-Widths, and Coupling Constants for Tricarbonylnickel(0) Complexes Relative to External $Ni(CO)_4$ in THF Solution**

| Compound | $\delta(^{61}Ni)$/ppm | $J(^{61}Ni–^{31}P)$/Hz | $\omega_{1/2}$/Hz |
|---|---|---|---|
| $Ni(CO)_3(PMe_3)$ | 25.8 | 230 | 140 |
| $Ni(CO)_3$(dppe) | −2.1 | 210 | ND |
| $Ni(CO)_3$(dppb) | 6.7 | 235 | 131 |
| $Ni(CO)_3(PPh_2H)$ | 5.0 | 234 | 90 |
| $Ni(CO)_3(P(OPh)_3)$ | −129.5 | 351 | ND |
| $Ni(CO)_3(PBu_3)$ | −16.2 | 196 | ND |
| $Ni(CO)_3(PPh_2Cl)$ | 48.5 | 236 | ND |
| $Ni(CO)_3(PPh_2Et)$ | −0.1 | 236 | 131 |
| $Ni(CO)_3(PCy_2H)$ | −23.5 | 230 | 55 |
| $Ni(CO)_3(PPh_3)$ | 24.2 | 236 | 16 |
| $Ni(CO)_3(PCy_3)$ | −11.2 | 211 | 55 |

*Source:* Behringer, K. D. and J. Blumel: *Journal of Magnetic Resonance*. 1995. 33. pp. 729–733.
*Note:* Ph = $C_6H_5$; Cy = $C_6H_{11}$; dppe = $Ph_2PC_2H_4PPh_2$; dppb = $Ph_2P(CH_2)_4PPh_2$.

(Figure 5.22). The central $^{31}$P signal is flanked by four nickel satellites with equal intensity, as is expected for a $^{31}$P NMR nucleus, coupled to a spin-3/2 $^{61}$Ni nucleus.

$^{61}$Ni chemical shifts for both di- and tricarbonylnickel complexes are strongly temperature dependent. For example, for $(CO)_3Ni(PCy_3)$, the $^{61}$Ni chemical shift changes from −11.2 to 3.7 on raising the measurement temperature from 298 to 330 K.[89]

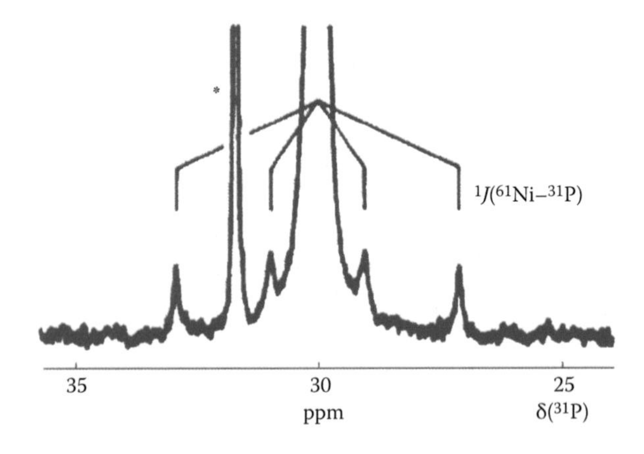


**FIGURE 5.22** 121.5-MHz $^{31}$P NMR spectrum of a saturated THF solution of $(CO)_3Ni(PPh_3)$. * denotes the $^{31}$P NMR signal of trace $(CO)_2Ni(PPh_3)_2$. (Behringer, K. D. and J. Blumel: *Journal of Magnetic Resonance*. 1995. 33. pp. 729–733. Copyright Wiley-VCH Verlag GmbH & Co. KGaA. Reproduced with permission.)

## 5.8 SUMMARY

In this chapter, we attempted to provide the reader with an introduction to the techniques available for the measurement of heteronuclei, illustrating the possibilities with examples from less common nuclei such as $^{15}N$ or $^{195}Pt$. It is important to note that, despite the fact that the majority of NMR structural elucidation methods were originally developed for and based on $^{1}H$ and $^{13}C$ measurement, observation of other nuclei, requiring only slight modifications to the pulse sequences and timings, can provide equally valuable or even more important information that frequently cannot easily be gathered using alternative techniques. For instance, $^{195}Pt$ NMR spectroscopy can shed light on a binding of platinum compounds to biomolecules, such as proteins and amino acids, the information of which can only be obtained from, for example, X-ray diffraction, if it is possible to crystallise platinum–protein adduct. Furthermore, $^{15}N$ NMR spectroscopy can be employed to study mutations in DNA and to evaluate the interaction of drugs with nucleobases. Finally, driven by the modern discoveries in the field of catalysis and materials sciences, we may expect that measurement of the quadrupolar nuclei, such as $^{51}Ni$ or $^{23}Al$, will become of high significance.

## REFERENCES

1. Anderson, S. J., J. R. Barnes, P. L. Goggin and R. J. Goodfellow. 1978. Nuclear magnetic resonance studies of some dimethyl sulphide and dimethyl telluride complexes of rhodium (III). *Journal of Chemical Resonance* S, pp. 286–287.
2. Goggin, P. L., R. Goodfellow and F. J. S. Reed. 1974. Preparation and identification by proton nuclear magnetic resonance and proton $^{195}Pt$ INDOR and vibrational spectroscopy of some platinum(II) complexes containing dimethyl sulphide as a bridging ligand. *Journal of the Chemical Society, Dalton Transactions*, pp. 576–585.
3. Ernsting, J. M., S. Gaemers and C. J. Elsevier. 2004. $^{103}Rh$ NMR spectroscopy and its application to rhodium chemistry. *Magnetic Resonance in Chemistry* 42, pp. 721–736.
4. von Philipsborn, W. 1999. Probing organometallic structure and reactivity by transition metal NMR spectroscopy. *Chemical Society Reviews* 28, pp. 95–105.
5. Mason, J. 1987. Patterns of nuclear magnetic shielding of transition-metal nuclei. *Chemical Reviews* 87, pp. 1299–1312.
6. Wrackmeyer, B. and M. Herberhold. 2006. NMR parameters of the tetrahedrane [$^{2}Fe$ (CO)(6)(mu-SNH)], studied by experiment and DFT. *Structural Chemistry* 17, pp. 79–83.
7. Buhl, M., S. Grigoleit, H. Kabrede and F. T. Mauschick. 2005. Simulation of $^{59}Co$ NMR chemical shifts in aqueous solution. *Chemistry: A European Journal* 12, pp. 477–488.
8. Autschbach, J. and S. H. Zheng. 2006. Density functional computations of $^{99}Ru$ chemical shifts: Relativistic effects, influence of the density functional, and study of solvent effects on fac-$[Ru(CO)_3I_3]^-$. *Magnetic Resonance in Chemistry* 44, pp. 989–1007.
9. Bagno, A., G. Casella and G. Saielli. 2006. Relativistic DFT calculation of $^{119}Sn$ chemical shifts and coupling constants in tin compounds. *Journal of Chemical Theory and Computation* 2, pp. 37–46.
10. Hratchian, H. P. and M. C. Milletti. 2005. First principles determination of $^{99}Ru$ chemical shifts using moderately sized basis sets. *Journal of Molecular Structure: THEOCHEM* 724, pp. 45–52.
11. Carlton, L. 2004. $^{103}$Rhodium NMR of $[Rh(X)(PPh_3)_3]$ [X = Cl, $N_3^-$, NCO, NCS, $N(CN)_2$, $NCBPh_3$, $CNBPh_3$, CN] and derivatives containing CO, isocyanide, pyridine, $^{2}H$ and $^{2}O$: Ligand, solvent and temperature effects. *Magnetic Resonance in Chemistry* 42, pp. 760–768.

12. Harris, R. K., E. D. Becker, S. M. C. de Menezes, R. Goodfellow and P. Granger. 2001. NMR nomenclature: Nuclear spin properties and conventions for chemical shifts – IUPAC recommendations 2001. *Pure and Applied Chemistry* 73, pp. 1795–1818.
13. Morris, G. A. and R. Freeman. 1979. Enhancement of nuclear magnetic resonance signals by polarisation transfer. *Journal of the American Chemical Society* 101, pp. 760–762.
14. Sorensen, O. W. and R. R. Ernst. 1983. Elimination of spectral distortion in polarisation transfer experiments: Improvements and comparison of techniques. *Journal of Magnetic Resonance* 51, pp. 477–489.
15. Burum, D. P. and R. R. Ernst. 1980. Net polarisation transfer via a *J*-ordered state for signal enhancement of low-sensitivity nuclei. *Journal of Magnetic Resonance* 39, pp. 163–168.
16. Bodenhausen, G. and D. J. Ruben. 1980. Natural abundance $^{15}$N NMR by enhanced heteronuclear spectroscopy. *Chemical Physics Letters* 69, pp. 185–189.
17. Muller, L. 1979. Sensitivity-enhanced detection of weak nuclei using heteronuclear multiple quantum coherence. *Journal of the American Chemical Society* 101, pp. 4481–4484.
18. Bax, A. and S. Subramanian. 1986. Sensitivity-enhanced two-dimensional heteronuclear shift correlation NMR spectroscopy. *Journal of Magnetic Resonance* 67, pp. 565–569.
19. Bax, A., R. H. Griffey and B. L. Hawkins. 1983. Correlation of proton and $^{15}$N chemical shifts by multiple quantum NMR. *Journal of Magnetic Resonance* 55, pp. 301–315.
20. Lopez-Ortiz, F. and R. J. Carbajo. 1998. Applications of polarisation transfer and indirect detection NMR spectroscopic methods based on $^{31}$P in organic and organometallic chemistry. *Current Organic Chemistry* 2, pp. 97–130.
21. Mandal, P. K. and A. Majumdar. 2004. A comprehensive discussion of HSQC and HMQC pulse sequences. *Concepts in Magnetic Resonance A* 20A, pp. 1–23.
22. Reynolds, W. F. and R. G. Enriquez. 2001. Gradient-selected versus phase-cycled HMBC and HSQC: Pros and cons. *Magnetic Resonance in Chemistry* 39, pp. 531–538.
23. Garbow, J. R., D. P. Weitekamp and A. Pines. 1982. Bilinear rotation decoupling of homonuclear scalar interactions. *Chemical Physics Letters* 93, pp. 504–509.
24. Nanz, D. and W. Vonphilipsborn. 1991. Transition-metal NMR spectroscopy: 15. Coherence pathways and inverse spectroscopy of $I_nS_m$ spin systems. *Journal of Magnetic Resonance* 92, pp. 560–571.
25. Ruegger, H. and D. Moskau. 1991. Platinum–phosphine cluster chemistry: New potentials From $^{195}$Pt–$^{31}$P($^{1}$H) triple-resonance spectroscopy. *Magnetic Resonance in Chemistry* 29, pp. S11–S15.
26. Heaton, B. T., J. A. Iggo, I. S. Podkorytov, D. J. Smawfield, S. P. Tunik and R. Whyman. 1999. Observation of triple-quantum effects in the HMQC spectra of substituted derivatives of $^{6}$Rh(CO)(16). *Journal of the Chemical Society, Dalton Transactions*, pp. 1917–1919.
27. Magiera, D., W. Baumann, I. S. Podkorytov, J. Omelanczuk and H. Duddeck. 2002. Stable $Rh_2(MTPA)_4$ phosphane adducts: The first example of P-chirality recognition by $^{103}$Rh NMR spectroscopy. *European Journal of Inorganic Chemistry*, pp. 3253–3257.
28. Heaton, B. T., J. A. Iggo, I. S. Podkorytov and S. P. Tunik. 2004. Application of HMQC spectroscopy to the observation of the metal resonance in polymetallic compounds: The effect of multiple-metal spin transitions. *Magnetic Resonance in Chemistry* 42, pp. 769–775.
29. Farrar, D. H., E. V. Grachova, A. Lough, C. Patirana, A. J. Poe and S. P. Tunik. 2001. Ligand effects on the structures of $^{6}$Rh(CO)(15)L clusters. *Journal of the Chemical Society, Dalton Transactions*, pp. 2015–2019.
30. Boudot, D., D. Canet, J. Brondeau and J. C. Boubel. 1989. DANTE-Z: A new approach for accurate frequency–selectivity using hard pulses. *Journal of Magnetic Resonance* 83, pp. 428–439.

31. Berger, S. and S. Braun. 2004. *200 and More NMR Experiments: A Practical Course*. Weinheim, Germany: Wiley-VCH, 854 pp.
32. Bengtsson, L. A., B. T. Heaton, J. A. Iggo, C. Jacob, G. L. Monks, J. Ratnam and A. K. Smith. 1994. Kinetics of ligand exchange in rhodium-(I) and rhodium-(III) complexes from magnetisation transfer measurements: The crystal structure of $[Rh(NH_3)(PPh_3)_3]ClO_4$. *Journal of the Chemical Society, Dalton Transactions*, pp. 1857–1865.
33. Akitt, J. W. and B. E. Mann. 2000. *NMR and Chemistry*. Cheltenham, United Kingdom: Stanley Thornes, pp. 198–203.
34. Jeener, J., B. H. Meier, P. Bachmann and R. R. Ernst. 1979. Investigation of exchange processes by two-dimensional NMR spectroscopy. *Journal of Chemical Physics* 71, pp. 4546–4553.
35. Perrin, C. L. and T. J. Dwyer. 1990. Application of two-dimensional NMR to kinetics of chemical exchange. *Chemical Reviews* 90, pp. 935–967.
36. Bauer, W. 1996. Pulsed field gradient 'inverse' HOESY applied to the isotope pairs $^{1}H$,$^{31}P$ and $^{1}H$,$^{7}Li$. *Magnetic Resonance in Chemistry* 34, pp. 532–537.
37. Pregosin, P. S., P. G. A. Kumar and I. Fernandez. 2005. Pulsed gradient spin echo (PGSE) diffusion and $^{1}H$,$^{19}F$ heteronuclear Overhauser spectroscopy (HOESY) NMR methods in inorganic and organometallic chemistry: Something old and something new. *Chemical Reviews* 105, pp. 2977–2998.
38. Macchioni, A. 2005. Ion pairing in transition-metal organometallic chemistry. *Chemical Reviews* 105, pp. 2039–2073.
39. Binotti, B., G. Bellachioma, G. Cardaci, A. Macchioni, C. Zuccaccia, E. Foresti and P. Sabatino. 2002. Intra-molecular and inter-ionic structural studies of novel olefin palladium(II) and platinum(II) complexes containing poly-(pyrazol-1-yl)borate and methane ligands: X-ray structures of palladium five-coordinate complexes. *Organometallics* 21, pp. 346–354.
40. Binotti, B., A. Macchioni, C. Zuccaccia and D. Zuccaccia. 2002. Application of NOE and PGSE NMR methodologies to investigate non-covalent intimate inorganic adducts in solution. *Comments on Inorganic Chemistry* 23, pp. 417–450.
41. Neuhaus, D. and M. Williamson. 1989. *The Nuclear Overhauser Effect in Structural and Conformational Analysis*. Weinheim, Germany: VCH, 522 pp.
42. Tang, W., S. Johnston, J. A. Iggo, N. G. Berry, M. Phelan, L. Lian, J. Bacsa and J. Xiao. 2013. Cooperative catalysis through non-covalent interactions. *Angewandte Chemie International Edition* 52, pp. 1668–1672.
43. Martínez-Viviente, E. and P. S. Pregosin. 2003. Low-temperature $^{1}H$-, $^{19}F$-, and $^{31}P$-PGSE diffusion measurements: Applications to cationic alcohol complexes. *Helvetica Chimica Acta* 86, pp. 2364–2378.
44. Jerschow, A. and N. Müller. 1997. Suppression of convection artefacts in stimulated-echo diffusion experiments. Double–stimulated-echo experiments. *Journal of Magnetic Resonance* 125, pp. 372–375.
45. Nilsson, M. 2009. The DOSY Toolbox: A new tool for processing PFG NMR diffusion data. *Journal of Magnetic Resonance* 200, pp. 296–302.
46. Ciancaleoni, G., L. Belpassi, F. Tarantelli, D. Zuccaccia and A. Macchioni. 2013. A combined NMR/DFT study on the ion-pair structure of ((PR2R2)-R-1)Au(eta(2)-3-hexyne) BF4 complexes. *Dalton Transactions* 42, pp. 4122–4131.
47. Blazina, D., S. B. Duckett, J. P. Dunne and C. Godard. 2004. Applications of the para-hydrogen phenomenon in inorganic chemistry. *Dalton Transactions*, pp. 2601–2609.
48. Amoureux, J. P., J. Trebosc, J. Wiench and M. Pruski. 2007. HMQC and refocussed-INEPT experiments involving half-integer quadrupolar nuclei in solids. *Journal of Magnetic Resonance* 184, pp. 1–14.
49. Gudat, D., U. Fischbeck, F. Tabellion, M. Billen and F. Preuss. 2002. Multi-nuclear magnetic resonance study of some imidovanadium complexes. *Magnetic Resonance in Chemistry* 40, pp. 139–146.

50. Nanz, D. and W. Vonphilipsborn. 1992. Indirect detection of spin-1 nuclei: Application and product-operator description of inverse correlation experiments with $I_nS_m$ spin systems. *Journal of Magnetic Resonance* 100, pp. 243–255.
51. Xiang, B. S., M. D. Winemiller, T. F. Briggs, D. J. Fuller and D. B. Collum. 2001. Optimising HMQC for $IS_n$ spin systems. *Magnetic Resonance in Chemistry* 39, pp. 137–140.
52. von Philipsborn, W. and R. Müller. 1986. $^{15}$N NMR spectroscopy: New methods and applications. *Angewandte Chemie International Edition* 25, pp. 383–413.
53. Still, B. M., P. G. Anil Kumar, J. R. Aldrich-Wright and W. S. Price. 2007. $^{195}$Pt NMR: Theory and application. *Chemical Society Reviews* 36, pp. 665–686.
54. Marek, R., J. Brus, J. Tousek, L. Kovacs and D. Hockova. 2002. $^{7}$N- and $^{9}$N-substituted purine derivatives: A $^{15}$N NMR study. *Magnetic Resonance in Chemistry* 40, pp. 353–360.
55. Marek, R. and V. Sklenar. 2005. NMR studies of purines. *Annual Reports on NMR Spectroscopy* 54, pp. 201–242.
56. Marek, R. and A. Lycka. 2002. $^{15}$N NMR spectroscopy in structural analysis. *Current Organic Chemistry* 6, p. 35.
57. Witanowski, M., L. Stefaniak and G. A. Webb. 1993. Nitrogen NMR spectroscopy. *Annual Reports on NMR Spectroscopy* 25: p. 1.
58. Larina, L. I. and V. Milata. 2009. $^{1}$H, $^{13}$C, and $^{15}$N NMR spectroscopy and tautomerism of nitrobenzotriazoles. *Magnetic Resonance in Chemistry* 47, pp. 142–148.
59. Pregosin, P. S. 1982. $^{195}$Platinium NMR. *Coordination Chemistry Reviews* 44, pp. 247–291.
60. Ismail, I. M. and P. J. Sadler. 1983. *ACS Symposium Series*, pp. 171–190.
61. Pregosin, P. S. 1991. *Transition Metal Nuclear Magnetic Resonance*. Amsterdam, The Netherlands: Elsevier, 351 pp.
62. Gudat, D., A. Dogan, W. Kaim and A. Klein. 2004. Multi-nuclear NMR study of some organo-platinum complexes containing multifunctional azines as chelating ligands. *Magnetic Resonance in Chemistry* 42, pp. 781–787.
63. Luzyanin, K. V., M. Haukka, N. A. Bokach, M. L. Kuznetsov, V. Y. Kukushkin and A. J. L. Pombeiro. 2002. Platinum(IV)-mediated hydrolysis of nitriles giving metal-bound iminols. *Journal of the Chemical Society, Dalton Transactions*, pp. 1882–1887.
64. Bokach, N. A., A. V. Khripoun, V. Y. Kukushkin, M. Haukka and A. J. L. Pombeiro. 2003. A route to 1,2,4-oxadiazoles and their complexes via platinum-mediated 1,3-dipolar cycloaddition of nitrile oxides to organonitriles. *Inorganic Chemistry* 42, pp. 896–993.
65. Macdonald, F. M. and P. J. Sadler. 1991. Studies of the cis–trans isomerism of some square-planar platinum(II) nitroimidazole complexes. *Polyhedron* 10, pp. 1443–1448.
66. Kerrison, S. J. S. and P. J. Sadler. 1985. $^{195}$Pt NMR studies of platinum(II) dimethylsulphoxide complexes. *Inorganica Chimica Acta* 104, pp. 197–201.
67. Klein, A., T. Schurr, A. Knoedler, D. Gudat, K. W. Klinkhammer, V. K. Jain, S. Zalis and W. Kaim. 2005. Multiple isomerism (cis/trans, syn/anti) in $[(dmso)_2Pt(aryl)_2]$ complexes: A combined structural, spectroscopic, and theoretical investigation. *Organometallics* 24, pp. 4125–4131.
68. Béni, Z., R. Scopelliti and R. Roulet. 2004. Oxidative addition of $Ph_3SnH$ to $Pt_3(\mu\text{-}CO)_3(PR_3)_3$ (R = $PCy_3$, $PtBu_3$). *Inorganic Chemistry Communications* 7, pp. 935–937.
69. Abel, E. W., I. Moss, K. G. Orrell, V. Sik and D. Stephenson. 1987. A two-dimensional nuclear magnetic resonance exchange study of the fluxionality of $[PtIMe_3(MeSCH_2CH_2SEt)$. *Dalton Transactions*, pp. 2695–2701.
70. Uccello-Barretta, G., R. Bernardini, F. Balzano and P. Salvadori. 2002. Overall view of the use of chiral platinum(II) complexes as chiral derivatising agents (CDAs) for the enantiodiscrimination of unsaturated compounds by $^{195}$Pt NMR. *Chirality* 14, pp. 484–489.

71. Uccello-Barretta, G., R. Bernardini, F. Balzano and P. Salvadori. 2001. $[PtCl_3(C_2H_4)]^-$ $[AmH]^+$ complexes containing chiral secondary amines: Use as chiral derivatising agents for the enantiodiscrimination of unsaturated compounds by $^{195}$Pt NMR spectroscopy and NMR stereochemical investigation. *Journal of Organic Chemistry* 66, pp. 123–129.
72. Ma, G., M. Maliarik, L. Sun and J. Glaser. 2004. Novel porphyrin–thallium–platinum complex with 'naked' metal–metal bond: Multi-nuclear NMR characterisation of [(tpp) Tl–Pt(CN)5]2− and [(thpp)Tl–Pt(CN)5]2− in solution. *Inorganica Chimica Acta* 357, pp. 4073–4077.
73. Berners-Price, S. J. and P. J. Sadler. 1996. Coordination chemistry of metallodrugs: Insights into biological speciation from NMR spectroscopy. *Coordination Chemistry Reviews* 151, pp. 1–40.
74. Rosenberg, B., P. J. O'Dwyer, J. P. Stevenson, A. Eastman, S. W. Johnson, D. B. Zamble, S. J. Lippard, Y. Chen, Z. Guo and P. J. Sadler, eds. 1999. *Cisplatin: Chemistry and Biochemistry of a Leading Anti-Cancer Drug*. Chichester, United Kingdom: Wiley, 563 pp.
75. Bancroft, D. P., C. A. Lepre and S. J. Lippard. 1990. $^{195}$Pt NMR kinetic and mechanistic studies of cis-diamminedichloroplatinum and trans-diamminedichloroplatinum(II) binding to DNA. *Journal of the American Chemical Society* 112, pp. 6860–6871.
76. Wu, L., B. E. Schwederski and D. W. Margerum. 1990. Stepwise hydrolysis kinetics of tetrachloroplatinate(II) in base. *Inorganic Chemistry* 29, pp. 3578–3584.
77. Appleton, T. G., J. R. Hall, S. F. Ralph and C. S. M. Thompson. 1989. NMR study of acid-base equilibria and other reactions of ammineplatinum complexes with aqua and hydroxo ligands. *Inorganic Chemistry* 28, pp. 1989–1993.
78. Berners-Price, S. J. and P. W. Kuchel. 1990. Reaction of cis-$[PtCl_2(NH_3)_2]$ and trans-$[PtCl_2(NH_3)_2]$ with reduced glutathione studied by $^1$H[$^1$H], $^{13}$C [$^1$H], $^{195}$Pt[$^1$H], and $^{15}$N[$^1$H] DEPT NMR. *Journal of Inorganice Biochemistry* 38, pp. 305–326.
79. Barton, J. K. and S. J. Lippard. 1980. *Heavy Metal Interactions with Nucleic Acids*, in Metal Ions in Biology, ed. T. G. Spiro, John Wiley: New York, pp. 31–113.
80. Oehlsen, M. E., Y. Qu and N. Farrell. 2003. Reaction of polynuclear platinum antitumor compounds with reduced glutathione studied by multi-nuclear ($^1$H,$^1$H–$^{15}$N gradient heteronuclear single-quantum coherence, and $^{195}$Pt) NMR spectroscopy. *Inorganic Chemistry* 42, pp. 5498–5506.
81. Berners-Price, S. J., U. Frey, J. D. Ranford and P. J. Sadler. 1993. Stereospecific hydrogen bonding in mononucleotide adducts of platinum anti-cancer complexes in aqueous solution. *Journal of the American Chemical Society* 115, pp. 8649–8659.
82. Clore, G. M. and A. M. Gronenborn. 1982. Kinetic and structural studies on the intermediates formed in the reactions of 5′-adenosine monophosphate and 5′-guanosine monophosphate with cis-dichlorodiammineplatinum(II) using $^1$H and $^{195}$Pt magnetic resonance spectroscopy. *Journal of the American Chemical Society* 104, pp. 1369–1375.
83. Jansen, B. A. J., J. M. Perez, A. Pizarro, C. Alonso, J. Reedijk and C. Navarro-Ranninger. 2001. Sterically hindered cisplatin derivatives with multiple carboxylate auxiliary arms: Synthesis and reactions with guanosine-5′-monophosphate and plasmid DNA. *Journal of Inorganic Biochemistry* 85, pp. 229–235.
84. Wheate, N. J., L. K. Webster, C. R. Brodie and J. G. Collins. 2000. Synthesis, DNA binding, and cytotoxicity of isohelical DNA groove binding platinum complexes. *Anti-Cancer Drug Design* 15, pp. 313–322.
85. Wheate, N. J. and J. G. Collins. 2000. A $^1$H NMR study of the oligonucleotide binding of $[(en)Pt(mu\text{-}dpzm)_2Pt(en)]Cl_4$. *Journal of Inorganic Biochemistry* 78, pp. 313–320.
86. Wheate, N. J., B. J. Evison, A. J. Herlt, D. R. Phillips and J. G. Collins. 2003. DNA binding of the anti-cancer platinum complex [trans-{JPt$(NH_3)_2$Cl}$_2$mu-dpzm]$^{2+}$. *Dalton Transactions*, pp. 3486–3492.

87. Ogg, R.A. and J. D. Ray. 1957. Quadrupole relaxation and structures in nitrogen magnetic resonances of ammonia and ammonium salts. *Journal of Chemical Physics* 26, pp. 1339–1340.
88. Delpuech, J. J., M. R. Khaddar, A. A. Peguy and P. R. Rubini. 1975. Octahedral and tetrahedral solvates of aluminum cation: Study of exchange of free and bound organophosphorus ligands by nuclear magnetic resonance spectroscopy. *Journal of the American Chemical Society* 97, pp. 3373–3379.
89. Behringer, K. D. and J. Blumel. 1995. $^{61}$Ni NMR spectroscopy of dicarbonylnickel and tricarbonylnickel complexes. *Magnetic Resonance in Chemistry* 33, pp. 729–733.
90. Fedotov, M. A. and V. A. Likholobov. 1984. First direct observation of $^{105}$Pd NMR in solution. *Bulletin of the Academy of Sciences of the USSR Division of Chemical Science* 33, pp. 1751–1751.
91. Avent, A. G., F. G. N. Cloke, J. P. Day, E. A. Seddon, K. R. Seddon and S. M. Smedley. 1988. Tetrakis(trimethylphosphine)nickel(0). *Journal of Organometallic Chemistry* 341, pp. 535–541.
92. Hao, N., M. J. McGlinchey, B. G. Sayer and G. J. Schrobilgen. 1982. A $^{61}$Ni NMR study of some $d^{10}$ nickel complexes. *Journal of Magnetic Resonance* 46, pp. 158–162.
93. Schumann, H., M. Meissner and H. J. Kroth. 1980. NMR studies on organoelement(IVb) phosphines: 6. $^{61}$Ni NMR spectroscopic investigations on organoelement(IVb)-substituted complexes of tetracarbonylnickel. *Zeitschrift für Naturforschung B, A Journal of Chemical Sciences* 35, pp. 639–641.
94. Benn, R. and A. Rufinska. 1988. Indirect two-dimensional heteronuclear NMR spectroscopy of low-gamma metal nuclei (M = $^{183}$W, $^{57}$Fe, $^{103}$Rh, $^{61}$Ni). *Magnetic Resonance Chemistry* 26, pp. 895–902.
95. Benn, R. and A. Rufinska. 1986. High-resolution metal NMR spectroscopy of organometallic compounds. *Angewandte Chemie International Edition* 25, pp. 861–881.

## FURTHER READINGS

Derome, A. E. 1987. *Modern NMR Techniques for Chemistry Research.* Oxford, United Kingdom: Pergamon Press, 208 pp.

Freeman, R. 1988. *Spin Choreography: Basic Steps in High-Resolution NMR.* Oxford, United Kingdom: Oxford University Press, 404 pp.

gNMR. Available at http://www.home.cc.umanitoba.ca/~budzelaa/gNMR/gNMR.html.

Harris, R. K. and B. E. Mann, ed. 1979. *NMR and the Periodic Table.* London, United Kingdom: Academic Press, 459 pp.

# 6 NMR and Complex Mixtures

*Cassey McRae*

## CONTENTS

6.1 Introduction ........................................ 226
 6.1.1 Metabonomics ........................................ 226
 6.1.2 Lipidomics ........................................ 227
 6.1.3 Batch Processing ........................................ 227
6.2 Sample Preparation ........................................ 228
 6.2.1 Concentration References ........................................ 230
6.3 NMR Experimental Approaches and Considerations ........................................ 230
 6.3.1 Water Attenuation ........................................ 230
 6.3.2 $T_2$ Relaxation Editing: Detecting Low–Molecular-Weight Molecules ........................................ 235
 6.3.3 Diffusion Editing: Measuring High–Molecular-Weight Molecules ..... 238
 6.3.4 JRES Spectroscopy: Reducing Signal Overlap ........................................ 239
 6.3.5 Other Methods ........................................ 243
 6.3.6 Approaches to Solid-State Studies ........................................ 243
 6.3.7 X-Nucleus Detection ........................................ 244
6.4 Data Treatment for Statistical Analysis ........................................ 246
 6.4.1 Spectral Processing ........................................ 246
 6.4.2 Data Reduction ........................................ 247
 6.4.3 Data Normalisation ........................................ 248
 6.4.4 Scaling and Mean Centering ........................................ 249
6.5 Pattern Recognition ........................................ 249
 6.5.1 Multivariate Statistical Analysis ........................................ 249
  6.5.1.1 Interpretation of Scores and Loadings Plots ........................................ 249
 6.5.2 Data Analysis in Batch Processing ........................................ 252
  6.5.2.1 Observation Level ........................................ 253
  6.5.2.2 Batch Level ........................................ 253
6.6 Case Studies ........................................ 254
 6.6.1 Study 1: Application to Plasma ........................................ 254
 6.6.2 Study 2: Application to Saliva ........................................ 255
 6.6.3 Study 3: Batch Analysis ........................................ 256
6.7 Summary ........................................ 260
References ........................................ 261

## 6.1 INTRODUCTION

Nuclear magnetic resonance (NMR) has been applied to the analysis of mixtures (complex systems) for many years; however, advances in instrumentation and (just as importantly) the development of appropriate statistical tools to extract information from NMR data sets have seen this application area grow significantly since the 1990s. In this chapter, we consider three complex mixture 'themes': the metabolome, the lipidome, and batch processes. The importance of such studies is considered together with an insight to the practicalities of working within these themes. The statistical tools employed to analyse data arising from these studies are also mentioned, but only in brief.

### 6.1.1 METABONOMICS

The entire collection of metabolites in a cell, organ, or organism is called the 'metabolome'. The metabolome represents the end points of biological processes and so permits us to understand the phenotype of genetic or environmental influences (e.g. diet) on an organism. The metabolome may be analysed through NMR studies of biological fluids or tissue. By measuring the changes in the levels of metabolites (the 'metabolic profile') in cells or biofluids, we can learn about how a stimulus affects metabolism, which might be important for understanding the toxicology of drugs or for identifying biomarkers of disease. A metabolic profile can, therefore, be seen as a potential 'fingerprint' for disease etc. Some illustrative applications are provided in Section 6.6.

The term 'metabonomics' was coined by Jeremy Nicholson's research group at Imperial College London in 1999 and is defined as 'the quantitative measurement of the dynamic multi-parametric metabolic response of living systems to pathophysiological stimuli or genetic modification'.[1] It stems from a series of other 'omics' techniques, which, together, form the 'omics cascade' (Figure 6.1).

It should be noted that the term 'metabolomics' is often used instead of 'metabonomics', the former being defined as 'the comprehensive and quantitative analysis of all the metabolites in the metabolome'. Despite the difference in definition, the terms tend to be used interchangeably.

Metabonomics is a non-targeted approach, that is, it aims to measure and compare levels of all classes of metabolite simultaneously in a biological sample. This

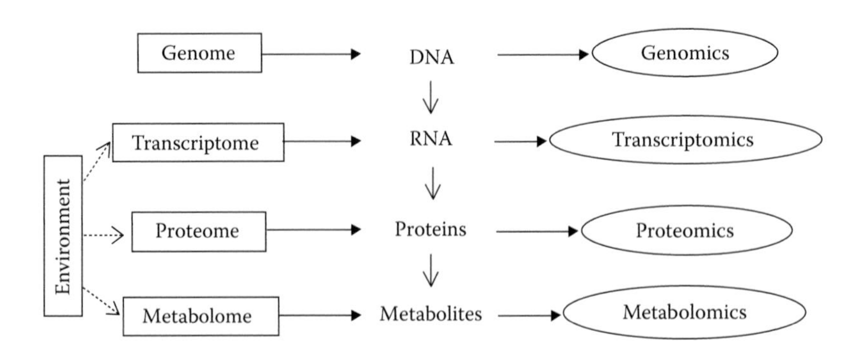


**FIGURE 6.1** The 'omics cascade'.

is often talked about as 'non–hypothesis-drive' research. If some idea of which classes of metabolites are related to the pathophysiological change of interest, then targeted analysis can be applied, whereby only that class of metabolites is quantified. Targeted analysis typically requires a longer sample preparation than non-targeted studies to separate out the metabolites of interest from the rest of the metabolites in the sample (cf. lipidomics described in the following).

Mass spectrometry (MS) is a popular platform for metabonomics studies, with advantages and disadvantages when compared with the NMR approach; these are not discussed here but are rehearsed in almost all of the articles referenced in this chapter.

Once NMR spectra have been collected for all of the samples, the data sets need to be compared to search for differences. This is where data reduction and pattern recognition methods are required; this is a multivariate problem as there are generally many more molecules being detected than there are samples measured. These statistical methods are common across all three themes referred to above and will be discussed in brief in Section 6.5.

### 6.1.2 Lipidomics

The 'lipidome' is the complete collection of lipids within a cell, tissue, or organism and is a subset of the metabolome along with sugars, amino acids/peptides, and nucleotides. Lipidomics has branched off relatively recently as its own field, which is concerned with identifying changes in lipid metabolism and lipid-mediated signaling processes.[2]

While technically, lipids can be measured in metabonomics studies, lipidomics became a field in its own right because of the uniqueness and functional specificity of lipids in comparison with other metabolites.

$^{1}$H-NMR spectroscopy has its place in lipidomics, particularly in tissue analysis, and $^{31}$P-NMR is important for the analysis of phospholipids (Section 6.3.7). To measure lipids specifically, they must first be extracted from tissue or biofluid samples (Section 6.2) and then separated into lipid classes using an appropriate separation technique. However, it is possible to analyse phospholipids with $^{31}$P-NMR in whole plasma with the addition of detergents as an alternative to perform lipid extractions (Section 6.3.7). Furthermore, two research groups have developed $^{1}$H-NMR line shape fitting techniques to quantify very low–density lipoprotein (VLDL), low-density lipoprotein (LDL), and high-density lipoprotein fractions in whole plasma spectra,[3,4] with the latter reporting absolute concentrations of triglycerides, phospholipids, total cholesterol, free cholesterol, and esterified cholesterol.[4]

### 6.1.3 Batch Processing

An abundance of processes in industries such as chemical, pharmaceutical, biotechnical, and semi-conductor manufacturers are of the 'batch type'. Batch processes are not continuous – they have a finite duration, and often, process variables have dependencies on batch progress. Trajectories of process variables over the timescale of the process display these dependencies and indicate when variables deviate from their target trajectories. This is important for identifying errors in the process, which might affect the end-product quality of a single batch or a sequence of batches. Batch processes often exhibit batch-to-batch variation, and so examining trajectories can also highlight causes of such between-batch variation and allow 'good batches' to be separated from 'bad batches'. Aside from being able to predict batch quality, it is also desirable to be able to predict batch maturity, and this may be possible by monitoring a variable whose numerical value changes dependently on process duration. Furthermore, it may be of interest to understand the relationship between all variables at a given time in a process, as well as the underlying properties that dominate the process. The batch process trajectory is analogous to a metabolic profile in metabonomics in that it acts as a fingerprint for each batch. Assuming an adequate abundance of NMR-detectable nuclei, industrial batch processes can be analysed by NMR spectroscopy in the same approach as applied to biofluids.

## 6.2 SAMPLE PREPARATION

The application of NMR to complex mixture analysis is only limited by any restrictions in the quantity of 'material' available. In Table 6.1, some indication is provided of the procedures required to prepare samples for NMR-based metabonomics and lipidomics studies; limited sample preparation would be expected for batch process investigations. Samples are best stored at −80°C until required.

When required, the samples are defrosted at room temperature and are usually centrifuged to remove any precipitate or cellular content. It may also be diluted before transferring to the NMR tube. A source of deuterium is required for locking, so deuterated solvents such as deuterium oxide ($D_2O$) is mixed with the sample, most commonly in a 1:1 or 2:1 sample to solvent; this ratio and the final volume used will depend on the amount of material and on the hardware available.

For samples like urine, which could vary significantly in pH, it is necessary to add a buffer, such as phosphate buffer (made up using $D_2O$) to regulate the pH of the samples.

Tissue samples can be analysed in-tact using high-resolution magic angle spinning (MAS) NMR (Section 6.3.6), or they can be analysed in solution state by extracting the metabolites and re-dissolving them in solvent (Table 6.1).

Preparation of intact tissue samples for MAS should be conducted on ice to avoid tissue degradation and should be as consistent as possible to avoid inter-sample variation. Typically, the frozen tissue sample is first cut to the size required for the rotor in the MAS probe (Section 6.3.6) and then washed with $D_2O$ or 0.9% saline in $D_2O$ to remove any blood. The sample can then be transferred to the MAS rotor and should fit without needing to be squeezed in. The rotor is then filled with $D_2O$ or 0.9% saline in $D_2O$ to displace the air and provide a field-frequency lock.

Lipid extraction techniques have been well documented. Traditional methods such as those of Folch[5] and Bligh and Dyer[6] (a modification of the Folch method) use a mixture of chloroform–methanol (2:1 by volume). Briefly, in these methods, tissue samples are homogenised with chloroform–methanol, shaken, centrifuged, and washed with water before separating the lipids into the chloroform layer (Table 6.1). Other methods for lipid extraction exist, and due to the diversity in lipid classes

**TABLE 6.1**
**Sample Preparation Methods for Different Types of Sample in Metabonomics Studies**

| Sample Type | Preparation for NMR | Optional Preparation Steps |
|---|---|---|
| Blood plasma/serum or similar biofluid | • Centrifuge.<br>• Mix in 1:1 ratio with $D_2O$ (optionally 0.17% [w/v] $D_2O$:TSP mixture). | Precipitation of proteins into an organic solvent. |
| Urine and other solution samples whose pH is expected to fluctuate significantly | • Centrifuge.<br>• Mix urine with phosphate buffer (pH 7.4) in a 2:1 ratio. | Removal of urea using urease treatment (for urine). |
| Tissue – intact | For MAS:<br>• Rinse tissue sample to remove any remaining blood (if necessary) using $D_2O$ or 0.9% saline in $D_2O$.<br>• Cut tissue to appropriate size and transfer to MAS rotor.<br>• Add $D_2O$ or 0.9% saline in $D_2O$ to rotor to displace air and provide frequency lock. | |
| Tissue – polar metabolites | • Grind or homogenise frozen tissue in organic solvent or perchloric acid.<br>• Centrifuge and remove supernatant. In case of perchloric acid, neutralise supernatant to pH 7.4 with 2-M $K_2CO_3$ and leave on ice for 30 min to precipitate salts.<br>• Remove solvent from supernatant (e.g. lyophilise).<br>• Re-suspend in $D_2O$ or other deuterated polar solvent. | |
| Tissue – lipophilic metabolites | • Grind or homogenise frozen tissue in methanol (4 mL/g of sample) and water (0.85 mL/g).<br>• Gradually add chloroform (4 mL/g) and water (2 mL/g), vortexing in between.<br>• Leave on ice for 15 min and centrifuge to obtain separate layers.<br>• Remove solvents off each layer (e.g. under stream of nitrogen gas).<br>• Re-suspend top layer (aqueous layer – polar metabolites) in $D_2O$ or other deuterated polar solvent.<br>• Re-suspend bottom layer (organic layer – lipophilic metabolites) in 2:1 $CDCl_3$:$CD_3OD$. | |

*Source:* Urine sample preparation and tissue extraction methods are from reference 7.
*Note:* $CDCl_3$, chloroform-d; $CD_3OD$, methanol-$d_4$; $D_2O$, deuterium oxide; $K_2CO_3$, potassium carbonate; MAS, magic angle spinning; and TSP, 3-(trimethylsilyl) propionic-2,2,3,3-d4 acid.

and their different polarities, no extraction method should be employed to obtain all lipid classes; often, it is necessary to perform several different types of extractions. Generally, it is advisable to extract lipids immediately after removal of the tissue from living organisms to avoid degradation. Otherwise, the tissue should be frozen and stored appropriately. Lipid peroxidation can also occur, and accordingly, some methods promote performing extractions at very low temperatures or under anaerobic conditions.

### 6.2.1 Concentration References

If one is interested in measuring concentrations of metabolites, rather than just comparing levels between samples, it may be beneficial to add a concentration reference to the sample. Since biofluid samples are aqueous, a polar compound is required, such as 3-(trimethylsilyl) propionic-2,2,3,3-d4 acid (TSP). TSP can be mixed with $D_2O$ in a 0.17% (w/v) ratio before adding to the biofluid sample. It produces a sharp singlet at 0.00 ppm, and since its concentration in the sample is known, the intensity of its signal can be used to calculate the concentrations of any metabolites of interest in the sample. 2,2-dimethyl-2-silapentane-5-sulfonate sodium salt (DSS) is another commonly used concentration reference for biological samples. Caution must be taken when using TSP and DSS in this way if the biological sample in question contains the protein albumin (as is the case for blood plasma). As TSP and DSS are charged molecules, they bind to albumin, and so if a $T_2$ editing pulse sequence is employed for NMR data collection (Section 6.3.2), then any bound reference compounds will not contribute to the signal measured in the final spectrum, and thus, they will no longer serve as an accurate concentration reference. As of yet, no suitable alternative to TSP and DSS has been established, although a similar concentration reference, 4,4-dimethyl-4-silapentane-1-ammonium trifluoroacetate (DSA) has been suggested.[8] Samples can be spiked with a chemical of known concentration, which can act as an internal reference. However, it is essential that the samples do not contain the reference naturally and that the reference signal(s) do not overlap with any metabolite signals.

## 6.3 NMR EXPERIMENTAL APPROACHES AND CONSIDERATIONS

### 6.3.1 Water Attenuation

If one were to analyse a biofluid such as blood plasma using a standard (one) 'pulse-and-acquire' sequence, its spectrum would appear rather like that shown in Figure 6.2 (Chapter 3, Sections 4.5.3 and 7.3). This spectrum is dominated by a very large solvent signal resulting from the high water content of biofluids. This solvent signal has to be attenuated before signals for the low-concentration metabolites can be seen. Some of the approaches to achieve this are indicated in Table 6.2 (Chapter 3). The simplest approach is to simply apply a (pre-) saturating radio frequency (RF) pulse (Chapter 3); however, application of the 1-D nuclear Overhauser spectroscopy (NOESY) experiment, in which a low-power (saturating) RF pulse is applied both during the pre-acquisition delay and the mixing time, is popular due to its success in

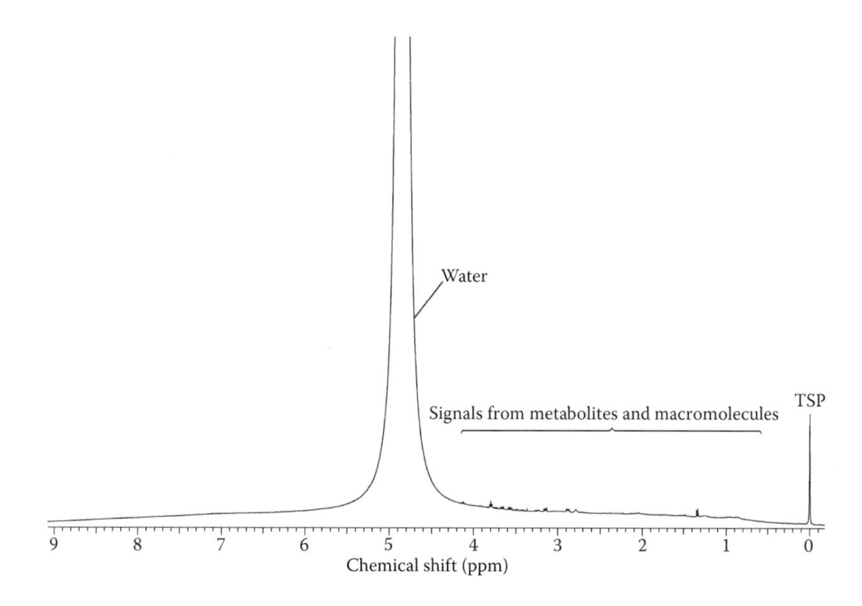


**FIGURE 6.2** 500-MHz $^1$H-NMR (one) 'pulse-and-acquire' spectrum of human blood plasma. The spectrum is dominated by the very large water signal even though the sample has been diluted by the addition of 50% v/v $D_2O$.

removing the water resonance 'hump'. However, a disadvantage of the 1-D NOESY–presat is that 90° (or π/2) pulses are used, and this might not be optimal for signal averaging or spectrum integration (Chapter 3). A pulse sequence called 'flip-angle adjustable 1-D NOESY' (FLIPSY) was devised to allow 'the best of both worlds', a variable pulse width as in the presat experiment, but with the superior water suppression of the 1-D NOESY–presat. Despite its advantages, FLIPSY is not routinely seen in metabonomics investigations, and it is usually more common to see either the presat or the 1-D NOESY–presat experiment in use. The length of the pre-saturation period is a compromise between effective suppression and extended experimental time, and typically, 1 to 3 s periods are used. Similarly, the strength of the irradiation must be selected so as to avoid affecting the intensities of resonances close to the water frequency, but without reducing the saturation of the water too much (Chapter 3). The main disadvantage of methods that use water irradiation is that the signals of any protons that exchange with water will also be suppressed through saturation transfer (Chapter 3 and Section 7.3). Thus, in cases where chemical exchange (Chapter 2) might be an issue, or also for more dilute samples where water suppression using pre-saturation might not be sufficient, it might be more appropriate to use, for example, sequences that employ pulsed field gradients to destroy the water resonance. Water suppression by gradient-tailored excitation (WATERGATE) and excitation sculpting are common examples of such experiments employing pulsed field gradients (Chapters 3 and 4, Section 7.3). Recent variations include MEshcher–GArwood (MEGA) and solvent-optimised double gradient spectroscopy (SOGGY),

**TABLE 6.2**
**Methods for Attenuation of the Large Water Signal that Dominates $^1$H-NMR Spectra of Biological Samples**

| | Method/Pulse Sequence | Pros/Cons |
|---|---|---|
| Suppression using weak RF radiation | Presat | Quick experiment time and variable pulse width, thus allowing selection of the Ernst angle for improved signal to noise and, hence, more accurate integrations. Prone to residual signals from solvent located at the outer edge of the RF coil, leading to loss of metabolites in those areas and/or baseline distortion. Also leads to suppression of signals in chemical exchange with water. |
| | 1-D NOESY–presat | Better suppression of residual solvent signal 'hump' than presat sequence. Maintains a flat baseline and narrows signals. Pulse sequence is based on 90° pulses so no variable pulse width option for improving integral accuracy. Leads to suppression of signals in chemical exchange with water. |
| | FLIPSY | Combines the advantages of both the presat and 1-D NOESY pulse sequences into one sequence, that is, allows selection of pulse width to improve signal integral accuracy, as in the presat experiment, but has the superior water signal 'hump' suppression of the 1-D NOESY pulse sequence. |
| Gradient-based methods | WATERGATE (plus variants) | Unlike RF saturation methods, does not attenuate resonances close to the water resonance or in chemical exchange with water. Provides flatter baselines and fewer distortions. |
| | Excitation sculpting | Superior phasing and less baseline distortion compared to WATERGATE. |
| | MEGA | Based on the excitation sculpting motif, with its superior water suppression, but with the advantage that the signals outside the water frequency band (the signals of interest) are refocussed by a single pulse (a spin echo) rather than two refocussing pulses, speeding up analysis time. |
| | SOGGY | Based on the excitation sculpting motif, but with the hard 180° pulse replaced with a simple phase-alternating composite pulse. The use of a composite pulse allows compensation for RF inhomogeneity to be achieved over a range of resonance offsets, resulting in optimised solute sensitivity and solvent suppression. |

(*continued*)

**TABLE 6.2 (Continued)**
**Methods for Attenuation of the Large Water Signal that Dominates $^1$H-NMR Spectra of Biological Samples**

| | Method/Pulse Sequence | Pros/Cons |
|---|---|---|
| Combinations of weak RF radiation and gradients | WET (plus variants) | A $T_1$ and $B_1$-independent method for solvent suppression, which is useful for cases where inhomogeneities in $T_1$ and $B_1$ (such as *in vivo* applications) cause spatially inhomogeneous water suppression. |
| Non-NMR methods | Freeze dry sample to evaporate off water, followed by re-suspension in deuterated solvent | Time consuming in comparison to using suppression techniques and potential for sample loss. |

*Note:* $B_1$, the applied radio-frequency field strength; RF, radio-frequency; and $T_1$, bulk magnetisation of nuclear spins.

both reported to improve upon the excitation sculpting method. All of these methods are more effective at attenuating water at the outer edges of the RF coil and tend to give flatter baselines than presat methods.

A disadvantage of the above gradient methods is that the signals of interest are stored in the transverse plane while the solvent is being suppressed, and so signal intensities may be affected by $T_2$ relaxation and diffusion in the time period between gradient pulses. Furthermore, evolution of homonuclear scalar coupling limits the selectivity of the selective (solvent-removing) pulse. Apart from the presat method, water suppression enhanced through $T_1$ effects (WET; Chapter 3) and several variants keep the interesting magnetisation along the $z$ axis while the solvent is suppressed. WET was originally developed as a $T_1$- and $B_1$-independent method for solvent suppression for *in vivo* samples, which suffer from inhomogeneities in $T_1$ and $B_1$ and can, thus, be prone to spatially inhomogeneous water suppression.

Biological fluids are highly complex mixtures, containing a vast number of molecules with a range of different structures. Once sufficient solvent suppression has been achieved, the $^1$H-NMR signals from these molecules become prominent in the spectrum. A 1-D NOESY–presat spectrum of blood plasma is shown in Figure 6.3 and can be compared to the spectrum of plasma without water suppression in Figure 6.2. This spectrum looks very complicated, with a vast number of signals from low–molecular-weight components such as amino acids, sugars such as lactate and glucose, and ketones, as well as signals from high–molecular-weight components such as proteins, lipoproteins, and phospholipids. The larger molecules produce very broad and intense signals due to their faster relaxation times (Chapters 2 and 3). These contributions can completely cover up weaker signals or distort the baseline around weaker and narrower signals from the low–molecular-weight components. Thus, if one is only interested in measuring low–molecular-weight

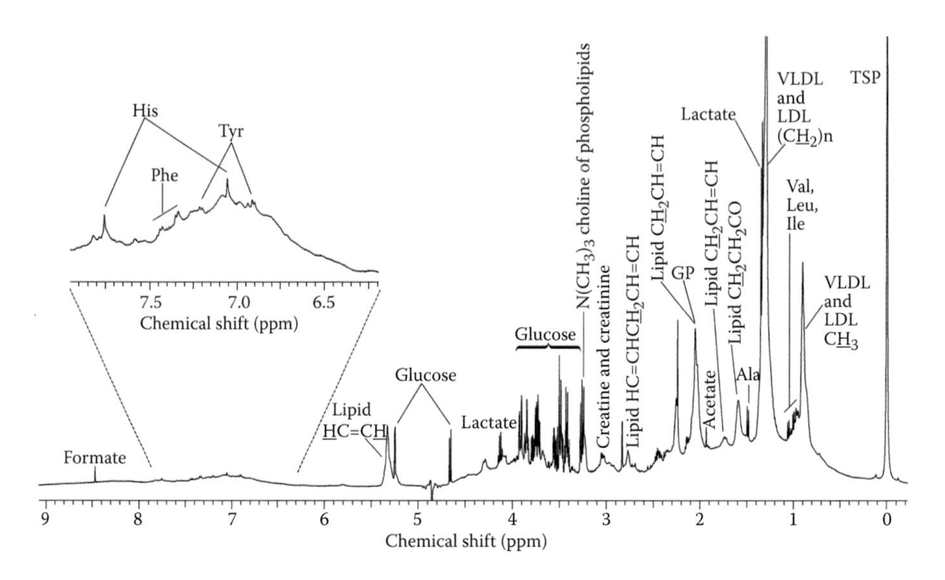


**FIGURE 6.3** 500-MHz $^1$H-NMR 1-D NOESY–presat spectrum of human blood plasma. Both low– and high–molecular-weight components are detected by the 1-D NOESY experiment, and so the broad signals of heavy molecules, such as proteins, are overlapping with the lower–molecular-weight components (metabolites). Ala, alanine; GP, glycoprotein N-acetyl groups; His, histidine; Ile, isoleucine; LDL, low-density lipoproteins; Leu, leucine; Phe, phenylalanine; TSP, 3-(trimethylsilyl) propionic-2,2,3,3-d4 acid; Tyr, tyrosine; Val, valine; and VLDL, very low–density lipoprotein.

metabolites, as is often the case in metabonomics, then steps must be taken to remove the signals of the heavy molecules. Alternatively, if it is the heavier molecules that are of interest, such as proteins or lipoproteins, then the signals from the overlapping low–molecular-weight molecules must be removed to obtain accurate integrals. There are two options to achieve this: one is to extract the metabolites of interest from the whole sample and then analyse the components (Section 6.2). Disadvantages of this method are that it can be very time consuming and that there is also a risk of loss of the sample during the extraction processes, thus giving inaccurate concentrations of the extracted components. An alternative method, which allows the entire sample to remain intact, is to edit the spectrum according to the transverse relaxation times ($T_2$; Section 6.3.2) or the diffusion coefficients (Chapter 4 and Section 6.3.3) of the molecules.

The presat and 1-D NOESY–presat experiments provide adequate water suppression for samples such as urine, which has a very high water content, yielding informative spectra. Unlike plasma, urine does not contain high–molecular-weight compounds such as proteins, and so the 1-D NOESY–presat and presat pulse sequences are usually sufficient for measuring metabolites and are most commonly adopted for metabonomics investigations of urine.

Table 6.3 summarises the different types of NMR experiments that can be performed on biological samples and which types of samples each is most appropriate for.

**TABLE 6.3**
**Appropriate NMR Experiments for Different Types of Complex Biological Fluids**

| Pulse Sequence | Types of Samples |
|---|---|
| One pulse sequence | • Lipid extracts from tissues since these do not require pre-saturation of the water signal |
| Presat (Section 6.3.1) | • Urine and other biofluids not containing high–molecular-weight components, such as proteins<br>• Blood plasma/serum and any other biofluid with a mixture of high–molecular-weight (proteins/lipoproteins) and low–molecular-weight (metabolites) chemicals, where the aim is to measure all species simultaneously |
| 1-D NOESY–presat (Section 6.3.1) | • Urine and other biofluids not containing high–molecular-weight components, such as proteins<br>• Blood plasma/serum and any other biofluid with a mixture of high–molecular-weight (proteins/lipoproteins) and low–molecular-weight (metabolites) chemicals, where the aim is to measure all species simultaneously |
| CPMG–presat (Section 6.3.2) | • Blood plasma/serum and any other biofluid with a mixture of high–molecular-weight (proteins/lipoproteins) and low–molecular-weight (metabolites) chemicals, where the low–molecular-weight species are of interest |
| Diffusion editing (Section 6.3.3) | • Blood plasma/serum and any other biofluid with a mixture of high–molecular-weight (proteins/lipoproteins) and low–molecular-weight (metabolites) chemicals, where the high–molecular-weight species are of interest |
| JRES (Section 6.3.4) | • Blood plasma/serum and any other biofluid with a complex mixture of species, where an overlap of signals is an issue |

*Note:* CPMG, Carr–Purcell–Meiboom–Gill; JRES, J-resolved spectroscopy; and 1-D NOESY–presat, 1-D nuclear Overhauser effect spectroscopy–presat.

### 6.3.2 $T_2$ Relaxation Editing: Detecting Low–Molecular-Weight Molecules

Nuclear spin relaxation is dependent on molecular motion, and as a result, larger molecules, which tumble more slowly, relax quicker than small, more rapidly tumbling molecules (Chapter 3). This property can be exploited in NMR to filter out the signals of unwanted molecules from a spectrum. This is useful in metabonomics because the signals of heavy proteins and lipoproteins can be removed, producing a spectrum with only the signals of the low–molecular-weight metabolites of interest. The most commonly used pulse sequence for achieving this is the Carr–Purcell–Meiboom–Gill (CPMG) pulse sequence, which is based on the spin echo experiment (Chapters 3 and 4) (Figure 6.4).

The spin echo pulse sequence has traditionally been used to measure $T_2$ for nuclei. However, it can also be used to obtain spectra free from the effects of field inhomogeneities because spins that have separated out due to differences in local magnetic

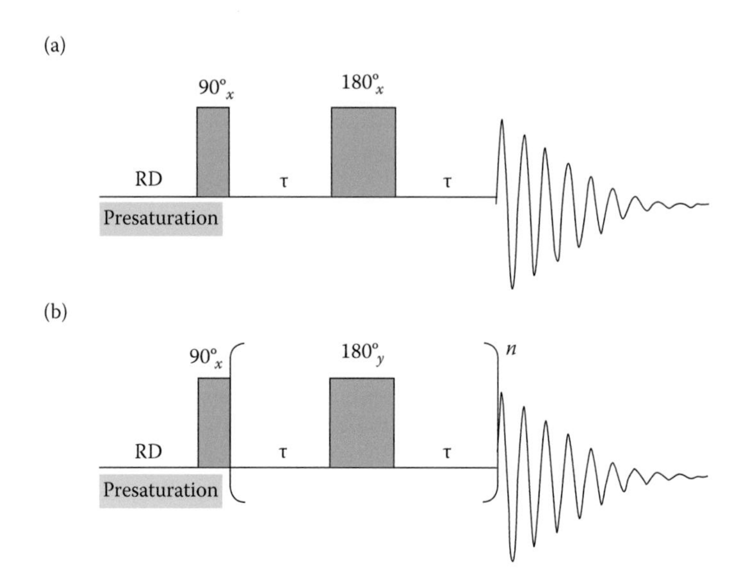


**FIGURE 6.4** Spin echo experiments. (a) The basic spin echo pulse sequence, traditionally used to measure $T_2$ relaxation times. (b) The CPMG pulse sequence, adapted from the spin echo sequence to include a $180°_y$ pulse instead of a $180°_x$ pulse to eliminate errors in the pulse width, and the repeating $[\tau - 180° - \tau]^n$ unit so that the parameter '*n*' can be increased instead of τ to reduce the amount of spin diffusion. The CPMG experiement is a popular method for $T_2$ relaxation editing spectra of complex biological fluids to remove the signals of high–molecular-weight components and enhance those of low-molecular-weight components. *n* is a parameter denoting the number of repeats of the $[\tau - 180° - \tau]$ unit, RD is recycle delay, and τ is the time delay.

field following $[90° - \tau]$ are refocussed by flipping them with the 180° pulse and allowing the same time delay, τ. The refocussing of the chemical shifts is called a 'spin echo'. While the chemical shifts are refocussed following the $[180° - \tau]$ step, the *J* couplings between homonuclear spins are not, and this phenomenon can be exploited, such as in the J-resolved (JRES) experiment (Section 6.3.4). The original spin echo experiment was modified to the CPMG experiment by the addition of the repeating $[\tau - 180° - \tau]^n$ unit so that the parameter '*n*' could be increased instead of τ to avoid having to use longer times where the effects of spin diffusion on the local magnetic field come into play (these are not refocussed). Furthermore, the $180°_x$ pulse was changed to a $180°_y$ pulse to eliminate any errors in the pulse width becoming cumulative; instead, they are cancelled out on every even number of repeats.[9]

The CPMG pulse sequence is widely used in metabonomics to obtain spectra containing only signals from low–molecular-weight metabolites, with those of the heavier molecules, such as proteins, filtered out. It works because, during the echo time, the spins of larger molecules relax on the timescale of the refocussing time, such that at the time of acquisition, no signal is detected. In contrast, the spins from slower-relaxing small molecules do not fully dephase during the echo time and produce a signal. Therefore, the CPMG experiment can be used as a $T_2$ relaxation editing technique (Figure 6.5).

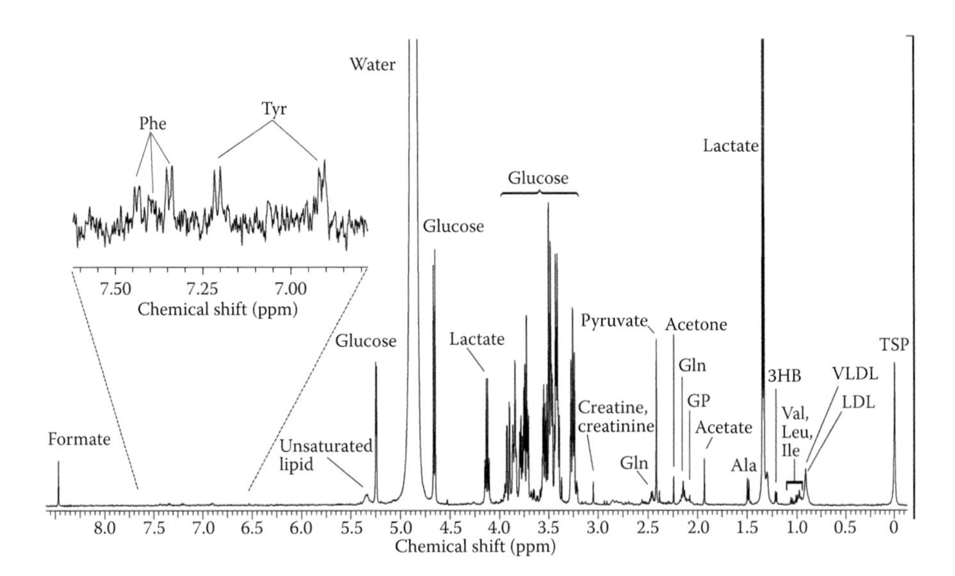


**FIGURE 6.5** 500-MHz $^1$H-NMR CPMG spectrum of human blood plasma; a pre-saturation pulse was included during the RD. 3HB, 3-hydroxybutyrate; Ala, alanine; Gln, glutamine; GP, glycoprotein N-acetyl groups; Ile, isoleucine; LDL, low-density lipoproteins; Leu, leucine; Phe, phenylalanine; TSP, 3-(trimethylsilyl) propionic-2,2,3,3-d4 acid; Tyr, tyrosine; Val, valine; and VLDL, very low–density lipoprotein.

In comparison to the 1-D NOESY–presat spectrum of blood plasma (Figure 6.3), in Figure 6.5, it is clear that the metabolite signals are no longer completely hidden or overlapped by the broad signals from the macromolecules. Instead, the metabolite signals are clear and resolved to the baseline, allowing for easier identification and more accurate integration. Note that some broad signals are still present due to $CH_3$-R protons in LDL and VLDL and the acetyl groups of N-acetylated sugars of glycoproteins. The appearance of signals from these larger molecules and their intensities depend on the echo time used in the pulse sequence. Thus, care must be taken when choosing the parameters. The delay time $\tau$ must be long enough to allow large molecules to fully relax so that they cannot produce detectable signals following the 180° pulse; however, not so long that relaxation of the smaller molecules begins to come into play and affect their signal intensities. Figure 6.6 shows a set of CPMG spectra of a biological fluid run at increasing echo times of 50, 150, and 450 ms in comparison to a presat spectrum, with no $T_2$ relaxation editing. At an echo time of 50 ms (Figure 6.6b), the signals of the macromolecules have not been fully suppressed, and the low–molecular-weight metabolite signals of interest exhibit phase distortions. As the total echo time increases, the intensities of the macromolecule signals decrease. The signals of the low–molecular-weight compounds, on the other hand, remain more similar as they are not given enough time to fully relax between the pulses. However, if echo times of over 450 ms were to be used, the signals of the low–molecular-weight metabolites would start to become attenuated too, giving an inaccurate reflection of the amounts of metabolites present. Furthermore, since nuclei relax at different rates depending on their chemical environment, some

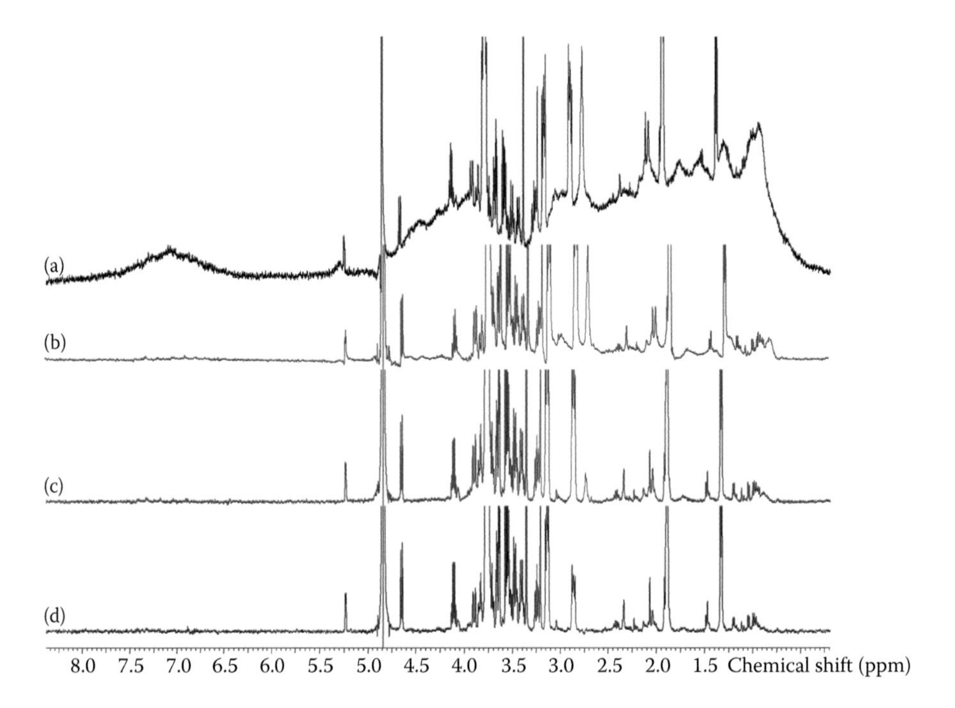


**FIGURE 6.6** The effect of total echo time on the suppression of macromolecule signals using the CPMG experiment. The figure shows four 500-MHz $^1$H-NMR spectra acquired on the same biological fluid, a presat spectrum, with no $T_2$ relaxation editing (a), and three CPMG spectra with total echo times of 50 ms (b), 150 ms (c), and 450 ms (d).

metabolite signals will be attenuated more than others, leading to false trends. It is worth experimenting with different values and observing how they affect the intensities of the metabolite signals, while ensuring full attenuation of the signals of macromolecules, rather like the experiments shown in Figure 6.6. Of course, providing all spectra for a set of samples is acquired using the same parameters; the signal intensities will be directly comparable across samples despite any attenuation due to relaxation. However, if quantitative analysis is being performed, it is important to find the most appropriate $\tau$ to use so that signal intensities correspond to the total amounts of metabolites. Delays of 1.5 ms have been used (with a total echo time of 450 ms), although smaller values such as 0.4 ms (with a total echo time of 64 ms) have also been successfully employed. The relaxation delay must be long enough to ensure full relaxation of all molecules in the sample (usually 2 s is sufficient), and during this delay, selective irradiation can be applied to suppress the water signal (Chapter 3).

### 6.3.3 Diffusion Editing: Measuring High–Molecular-Weight Molecules

The previous section described the CPMG experiment, which edits spectra according to $T_2$ relaxation times. It is also possible to edit on the basis of translational diffusion coefficients ($D$). A detailed description of the consequences of diffusion for the NMR signal and of the spin echo–based pulse sequences commonly employed

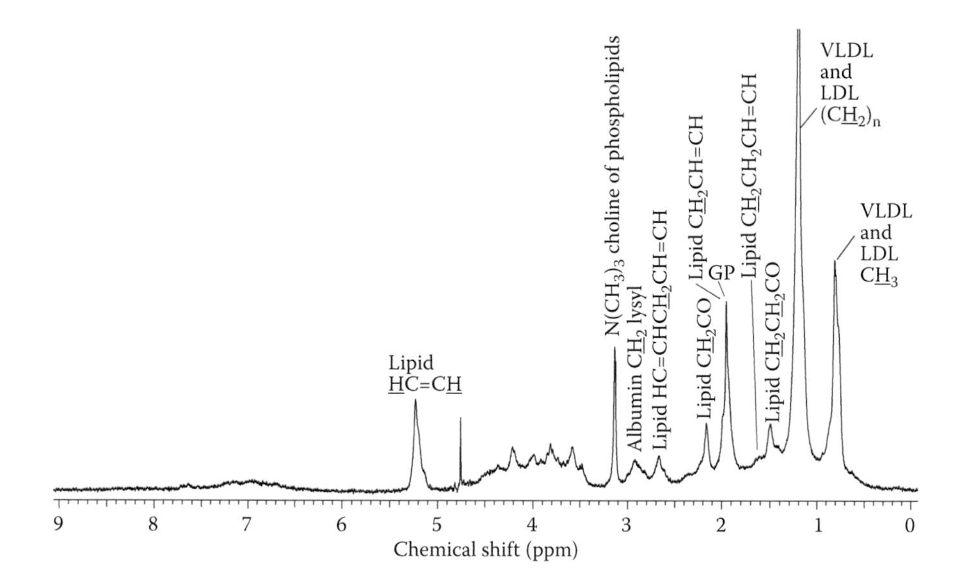


**FIGURE 6.7** 500-MHz $^1$H-NMR diffusion-edited spectrum of human blood plasma. GP, glycoprotein N-acetyl groups; LDL, low-density lipoproteins; and VLDL, very low–density lipoprotein.

is provided in Chapter 4. Following some calibration of diffusion delay times and gradient pulse duration, it is possible to use the sequences described in Chapter 4 to select slowly moving molecules in preference to rapidly moving molecules, thus selecting high molecular weights in preference to low–molecular-weight species. This approach has been used to good effect in lipidomics (and proteomics) studies; it may be used in a difference approach (i.e. subtracting the 'diffusion' spectrum of high–molecular-weight species from the spectrum recorded in the absence of diffusion editing results in a spectrum containing signals for low molecular weight alone) in metabonomics, but this is not very common.

A diffusion-edited spectrum of blood plasma is shown in Figure 6.7. Compared with the 1-D NOESY–presat spectrum in Figure 6.3, the peaks from the low–molecular-weight components have been greatly attenuated, while the signals of the macromolecules remain unaffected. Signals can be observed from lipoproteins through the methyl protons in the fatty acid acyl chains, phospholipids through the protons in the $N^+CH_3$ choline head groups, glycoproteins through protons in the N-acetyl groups, and also albumin through its lysyl groups.

### 6.3.4 JRES Spectroscopy: Reducing Signal Overlap

While the CPMG pulse sequence is excellent for removing broad macromolecule signals to reveal only signals from metabolites, there is still considerable peak overlap due to the fact that many metabolites give rise to complex multiplets with associated wide frequency range. This overlap and over-representation of metabolites can still make biomarker identification difficult. The 2-D JRES NMR experiment

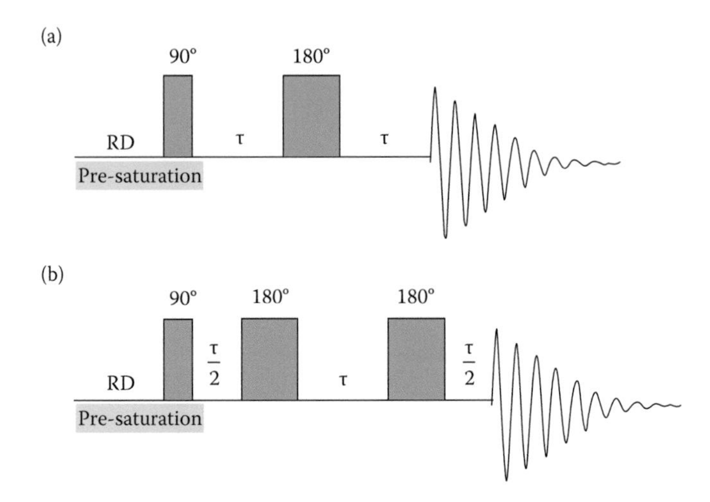


**FIGURE 6.8** 2-D JRES experiments. (a) The most basic 2-D JRES pulse sequence, which is based on the spin echo experiment, with water suppression during the relaxation delay (RD) (pre-saturation is shown here, although any method can be used). The pulse sequence is repeated for incremental values of 2τ to generate the second dimension. (b) A modified double spin echo version of the basic 2-D JRES experiment offers better removal of strong coupling artefacts and has become a recommended method for analysing complex biological fluids.[10] τ is a time delay.

is gaining popularity in the metabonomics community as an approach to reduce signal degeneracy. The most basic JRES pulse sequence is shown in Figure 6.8a. It is based on the spin echo pulse sequence; however, this is a 2-D experiment in which the time delay (2τ) is incremented to produce the $t_1$ dimension. While chemical shifts are refocussed (during the echo period) by the end of the second τ delay, the homonuclear *J* couplings continue to evolve during the entirety of τ – 180 – τ. As a result, the $F_1$ axis of the resulting 2-D JRES spectrum (the dimension resulting from the τ variable) shows *J* coupling constants. During the acquisition period (i.e. $t_2$), evolution of the signals due to both chemical shift and *J* coupling is active, and therefore, the $F_2$ axis appears as a normal 1-D (spin echo) spectrum. In the resulting 2-D spectrum, the proton multiplets appear at a 45° angle (Figure 6.9a); removing this 'tilt' enables the chemical shift and the *J* coupling information to be resolved into two separate dimensions (Figure 6.9b). The spectrum can be projected along $F_2$ to produce a 1-D spectrum where all signals appear as singlets regardless of their original multiplicity (Figure 6.9b), thus reducing overlap. This is demonstrated in Figure 6.10, which shows a projection of $F_2$ from a tilted 2-D JRES spectrum of follicular fluid (the biofluid surrounding eggs in ovaries). The *J* coupling information can be observed along the vertical axis of the 2-D JRES where a contour spot is present for each multiplet component at each chemical shift. For example, the lactate quartet at 4.12 ppm produces four contours along the vertical axis, one for each peak of the quartet. Because of the tilting, the horizontal axis only contains chemical shift information, and hence, the projected spectrum only contains singlets. Figure 6.11 shows this projected JRES spectrum compared to the CPMG spectrum of the same

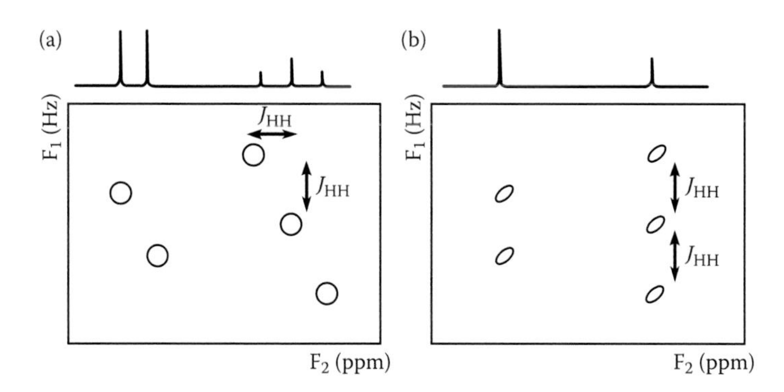


**FIGURE 6.9** The effect of tilting a 2-D JRES spectrum by 45°. (a) shows the 2-D JRES spectrum prior to tilting with the $F_2$ projection above. If the spectrum is tilted by 45°, as in (b), the *J* coupling between signals in the $F_2$ dimension is lost, and all signals appear as singlets in the $F_2$ 1-D projection (shown above). (Ludwig, C. and M. R. Viant: Two-dimensional J-resolved NMR spectroscopy: Review of a key methodology in the metabolomics toolbox. *Phytochemical Analysis.* 2010, 21, pp. 22–32. Copyright Wiley-VCH Verlag GmbH & Co. KGaA. Reproduced with permission.)

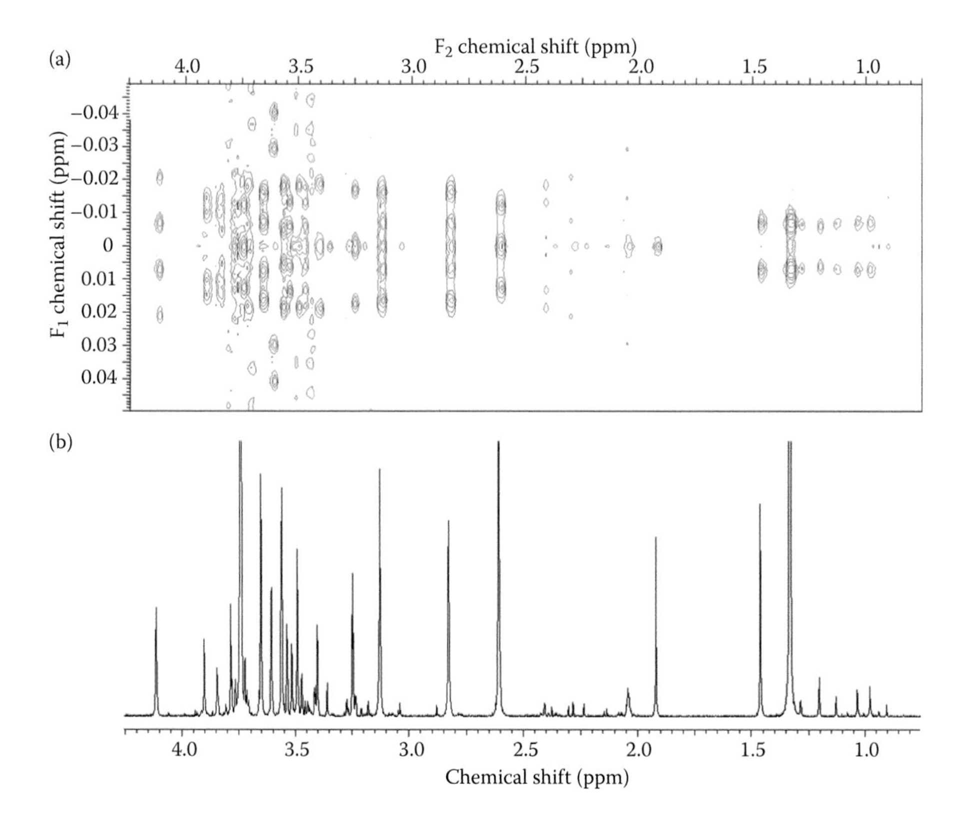


**FIGURE 6.10** 500-MHz [1]H-NMR 2-D JRES spectrum of human follicular fluid (a) and its 1-D $F_2$ projection (b) after tilting and symmetrising.

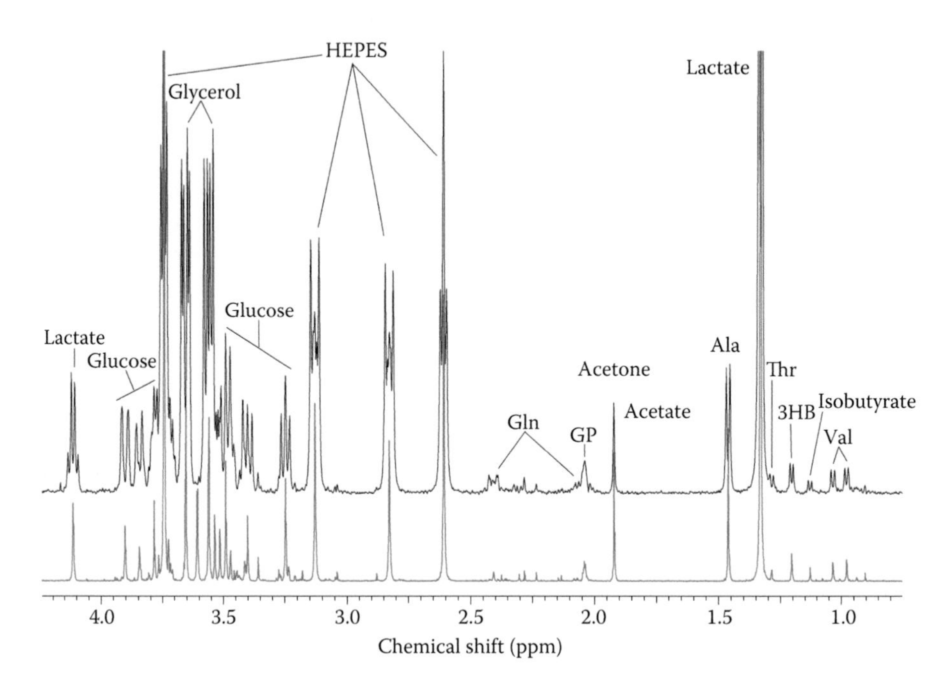


**FIGURE 6.11** The 1-D $F_2$ projection of the 500-MHz $^1$H-NMR 2-D JRES spectrum of human follicular fluid shown in Figure 6.10b (below, grey) and a 500-MHz $^1$H-NMR CPMG spectrum of the same sample for comparison (above, black). 3HB, 3-hydroxybutyrate; Ala, alanine; Gln, glutamine; GP, glycoprotein N-acetyl groups; HEPES, (hydroxyethyl) piperazine-1-ethanesulfonic acid; Thr, threonine; and Val, valine.

sample. There is a clear reduction in overlap between signals in the projected JRES spectrum, particularly in the glucose region (~3.24–3.92 ppm), making identification of individual signals easier and enabling them to be interrogated independently in any subsequent statistical analyses. This is particularly valuable in this example, given that the glucose signals overlap with intense glycerol and 4-(2-hydroxyethyl) piperazine-1-ethanesulfonic acid (HEPES) signals, which may dominate statistical models and potentially lead to false interpretations about glucose.

The loss of multiplicity may be disadvantageous for signal assignment; however, the *J* coupling information is available and can be accessed from the intact 2-D spectrum. This, of course, comes at the price of a much greater data file size compared to a 1-D spectrum. Furthermore, the total acquisition time to collect a set of JRES spectra is much greater than that for collecting a set of 1-D spectra. Reducing the acquisition time for JRES experiments is another matter for further investigation, and approaches, so far, have not produced results of a sufficiently high standard for metabonomics.

It should be noted that a source of artefacts in JRES spectra is strong coupling between protons, leading to signals in the 2-D spectrum, which do not appear at the correct chemical shift and so produce extra signals in the projected spectrum after tilting. A double spin echo method (Figure 6.8b) has been devised to suppress such signals and has been recommended for metabonomics studies.[10]

### 6.3.5 Other Methods

It can often be the case in metabonomics studies that it is necessary to acquire several different types of spectra (e.g. 1-D NOESY and CPMG) on the same sample to extract as much information as possible for the data. This can be very time consuming, particularly for large population cohorts. Therefore, methods have been devised for producing 'virtual CPMG' spectra by performing mathematical transformations on 1-D 'pulse-and-acquire' NMR data to suppress resonances from broad macromolecules, thus producing spectra with appearances similar to that of CPMG spectra.[11] In theory, this would allow for the collection of only one set of 1-D 'pulse-and-acquire' spectra in cases where time is restricted. Furthermore, this would be useful in cases where samples have been discarded and only 'pulse-and-acquire' data for them exist, although the creators affirm that these methods were not indented to completely replace CPMG experiments where it is possible to perform them. The two methods devised are called 'derivative-based relaxation-edited spectroscopy' (D-RESY) and 'gradient-based relaxation-edited spectroscopy' (G-RESY). In D-RESY, the first derivative of the Fourier-transformed 1-D ('pulse-and-acquire') NMR spectrum is calculated, whereas in G-RESY, Gaussian shaping is performed on the free induction decays (FIDs) of the spectra prior to Fourier transformation. Both methods have been shown to produce spectra with similar appearances to CPMG spectra but with an improved signal to noise and identify the same trends and biomarkers following data reduction and pattern recognition procedures.

2-D NMR spectroscopy has not commonly been collected for statistical analyses in metabonomic and lipidomic studies due to its complexity and the large time and computer memory requirements compared to 1-D NMR data, although some applications are beginning to appear.

### 6.3.6 Approaches to Solid-State Studies

NMR-based metabonomic studies of tissue-specific or cellular processes often utilise tissue extracts as the medium for investigation and have provided a wealth of information. However, studying tissue extracts comes with the loss of information on tissue compartmentalisation and localisation of metabolites, and which metabolites can be analysed is dictated by the solvents used for extraction. It is possible to measure metabolite levels in intact tissue samples using solid-state NMR, provided that the usual precautions are taken to produce narrow line widths (i.e. that MAS is employed). MAS NMR has been increasingly used in metabonomics studies, predominantly for the study of tumors and neurological tissues and, most recently, for studies where whole organisms (mice) are spun. Despite these advances, however, solution-state NMR currently remains the most popular technique for metabonomics.

MAS requires purpose-built rotors in which to spin the sample, which are positioned in a high-resolution MAS probe head at the magic angle (54.7°). Samples are typically spun at 4 to 6 kHz, and this is to displace spinning side bands outside the spectral range of interest, rather than to reduce line broadening. Spinning any faster than this begins to compromise sample integrity. Standard rotors allow sample volumes of 65 μL; however, spacers can be inserted, allowing smaller sample volumes, down to 12 μL. This is particularly useful for ensuring that the sample only contains the cells of interest. The use of spacers has become commonplace because they compact the tissue into a tight spherical shape, which improves field homogeneity and, hence, peak line shape. It is recommended to perform analyses with the temperature of the probe head 10 °C or lower to avoid tissue degradation.

The experimental approaches described above can be readily adapted for use in solid-state studies. Modern MAS probes now also allow detection of heteronuclei such as $^{31}P$, $^{13}C$, $^{15}N$ and $^{19}F$, and so homo- and heteronuclear 2-D NMR experiments are also possible. Just as in solution-state NMR, water suppression techniques (pre-saturation or excitation sculpting, etc.) are required to reduce the size of the water signal, and signal intensities can be edited based on NMR relaxation times and molecular diffusion coefficients. JRES experiments can also be used. The same parameters for these can be used, and acquisition times are the same. An example of a MAS NMR–CPMG spectrum of an intact liver tissue is shown in Figure 6.12a. For comparison, a typical $^{1}H$-NMR spectrum of a lipid extract (Figure 6.12b) and a presat $^{1}H$-NMR spectrum of an aqueous extract (Figure 6.12c) from liver tissue are also shown. The MAS spectrum of the intact sample is of comparable quality to those of the extracts and contains signals from water-soluble metabolites, such as sugars, amino acids, ketones, and osmolytes, as well as signals from various lipid groups. Thus, MAS allows simultaneous detection of lipid and water-soluble metabolites, whereas solution-state NMR requires the analysis of separate extracts.

### 6.3.7 X-Nucleus Detection

Most NMR-based metabonomics studies utilise $^{1}H$ observation due to sensitivity considerations. Carbon-13 NMR-based metabonomics may become more viable with the increasing use of cryoprobes to offset the problem of low natural abundance of this nucleus. Hyperpolarisation offers a route to utilise low-sensitivity nuclei. In this approach, the sample is mixed with a polarisation transfer agent and then frozen. The polarisation transfer agent is a species with an unpaired electron, which is excited from a ground spin state to a higher spin state. The difference in energies of electron spin states is much greater than that for nuclei spin states, and so microwave radiation is required to excite the electron, rather than RF radiation. While frozen, magnetisation from the transfer agent is transferred onto the sample, building up a huge population difference between the ground and excited nuclear spin states. The sample is then defrosted, and NMR is performed as normal. As soon as the sample is defrosted, the population difference will start to reduce through longitudinal relaxation, and this is a disadvantage of hyperpolarisation. $^{13}C$ nuclei have relatively long $T_1$ relaxation times and so has been successfully studied using this technique.

$^{31}P$ is another nucleus that is of interest in metabonomics and, particularly, in lipidomics, for measuring phospholipids. As there are fewer $^{31}P$ nuclei present in biological samples, $^{31}P$-NMR can offer a more practical alternative to measure phospholipids, with far simpler spectra. However, due to the large complexes formed by the phospholipids in the aqueous biological source and $^{31}P$ chemical shift anisotropy, the $^{31}P$ signals of untreated biofluids (such as blood plasma) are generally quite broad and

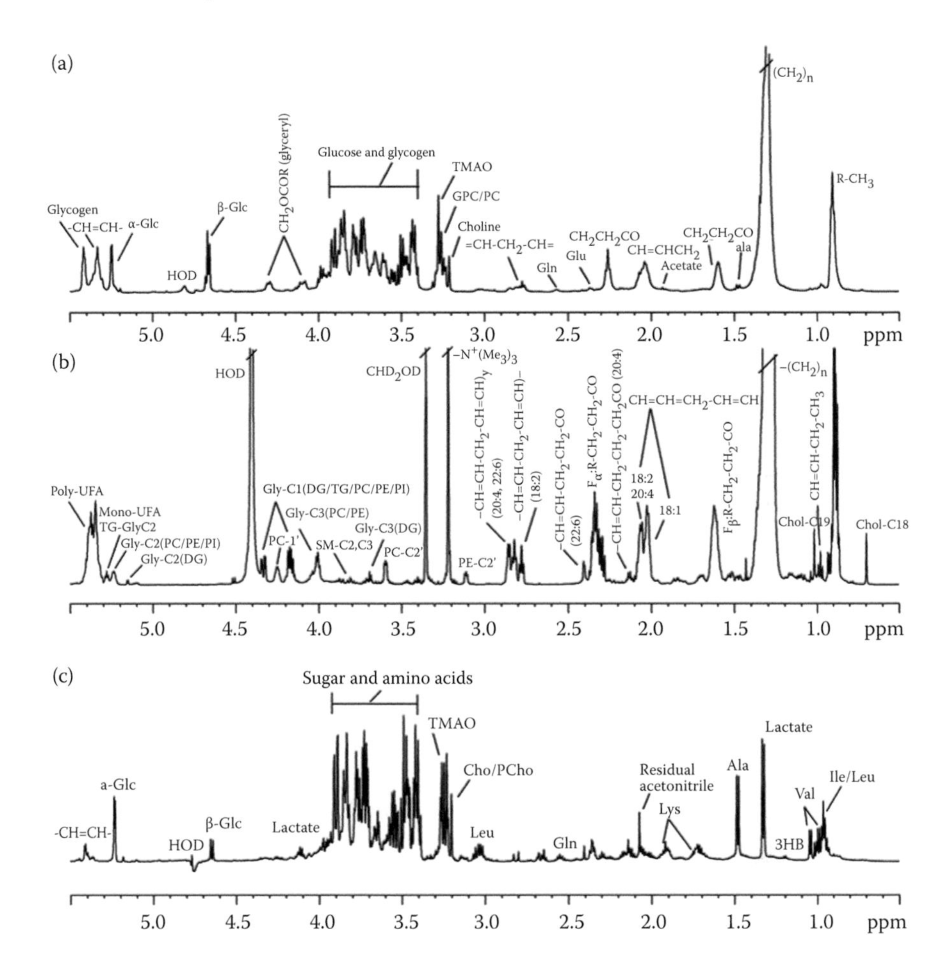

**FIGURE 6.12** 600-MHz $^1$H-NMR spectra of mouse liver tissue. (a) High-resolution MAS NMR–CPMG spectrum of intact liver tissue. (b) $^1$H-NMR spectrum of a lipid extract from liver tissue. (c) A presat $^1$H-NMR spectrum of an aqueous extract from liver tissue. Ala, alanine; 3HB, 3-hydroxybutyrate; Cho, choline; Chol, cholesterol; DG, diglyceride; Glc, glucose; Gln, glutamine; Glu, glutamate; Gly, glycerol; GPC, glycerophosphorylcholine; Ile, isoleucine; Leu, leucine; Lys, lysine; PC, phosphatidylcholine; PCho, phosphocholine; PE, phosphatidylethanolamine; PI, phosphatidylinositol; SM, sphingomyelin; TG, triglyceride; TMAO, trimethylamine-N-oxide; UFA, unsaturated fatty acid; and Val, valine. (Reprinted by permission from Macmillan Publishers Ltd., *Nature Protocols*, ref. 12, copyright 2010. Reprinted with permission from Coen, M., E. M. Lenz, J. K. Nicholson, I. D. Wilson, F. Pognan and J. C. Lindon. An integrated metabonomic investigation of acetaminophen toxicity in the mouse using NMR spectroscopy. *Chemical Research in Toxicology* 16, pp. 295–303. Copyright 2003 American Chemical Society.)

partially overlap, making quantification of the individual phospholipid groups non-trivial. One solution to this problem is to extract the lipids from the biological sample into organic solvents prior to $^{31}$P-NMR. This results in narrower line widths due to the disaggregation of the phospholipids. However, multi-step extraction methods are very time consuming, and there is a risk of loss of phospholipids through incomplete

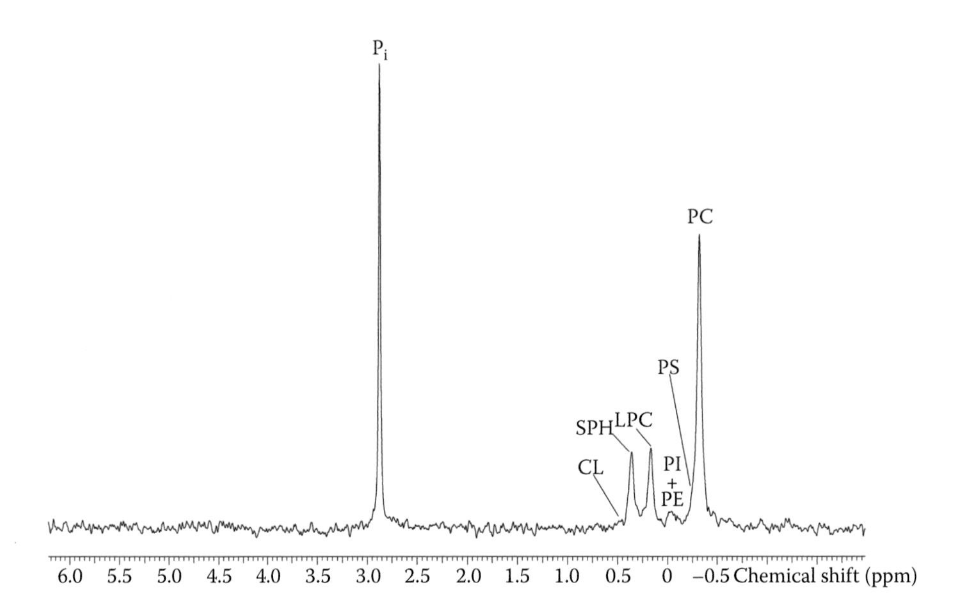


**FIGURE 6.13** 202.46-MHz $^{31}$P-NMR spectrum of human blood plasma treated with Triton X-100. CL, cardiolipin; LPC, lysophosphatidylcholine; PC, phosphatidylcholine; PE, phosphatidylethanolamine; PI, phosphatidylinositol; $P_i$, inorganic phosphate; PS, phosphatidylserine; and SPH, sphingomyelin.

extraction. Another alternative is to add detergents such as sodium cholate and Triton X-100 to biofluid samples to disperse the phospholipids into mixed micelles (which are smaller than the phospholipid aggregates in the natural biofluid), resulting in a more efficient averaging of the $^{31}$P chemical shift anisotropy. An example of a $^{31}$P-NMR spectrum of blood plasma treated with Triton X-100 is shown in Figure 6.13. It contains signals from phosphatidylcholine (PC), phosphatidylethanolamine (PE), phosphatidylinositol (PI), phosphatidylserine (PS), lysophosphatidylcholine (LPC), sphingomyelin (SPH), cardiolipin (CL), and inorganic phosphate ($P_i$).

## 6.4 DATA TREATMENT FOR STATISTICAL ANALYSIS

### 6.4.1 Spectral Processing

Raw NMR data should be processed (apodised, zero filled, baseline corrected) in the usual manner, and the resultant spectra (chemical shift) should be referenced. Referencing has traditionally been performed to the TSP signal. However, due to protein binding, the TSP signal can be prone to broadening and shifting. Therefore, metabolite signals that have minimal chemical shift variability are now favoured as reference compounds; for example, the formate singlet at approximately 8.45 ppm or the anomeric proton doublet of α-glucose (Section 7.3) at approximately 5.23 ppm is commonly used.

After the spectra have been identically processed, the signal intensities are calculated and exported as a table of numerical data, which is interpreted using pattern

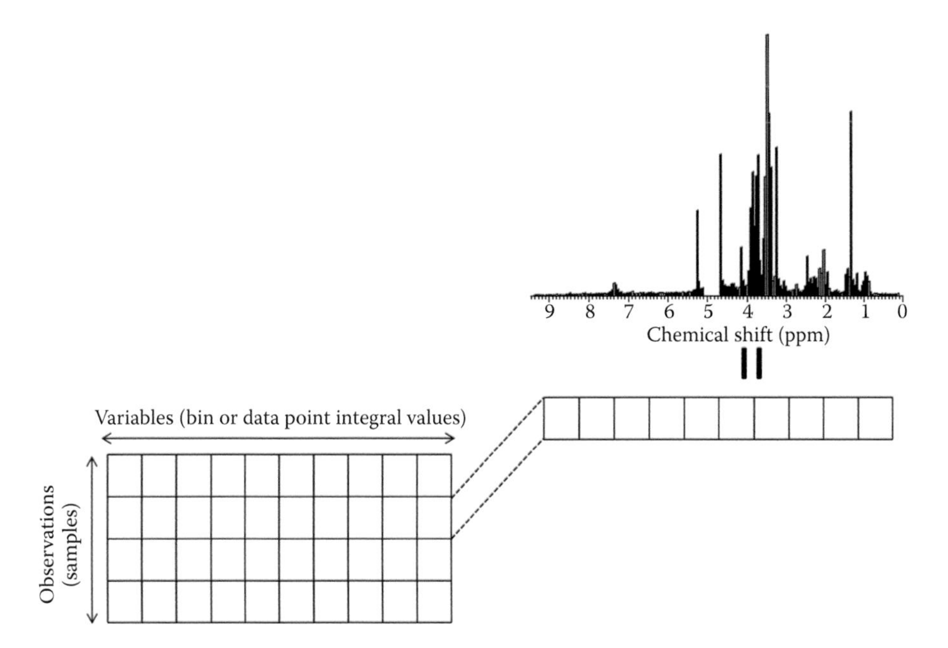


**FIGURE 6.14** A set of NMR spectra collected for $n$ samples can be thought of as a matrix of $k$ variables by $n$ observations. If full resolution data are being used, $k$ is the number of data points that the NMR data were acquired into. If the data are reduced, for example, by binning, then $k$ is the number of bins used. Each row of the matrix is an individual spectrum, and hence, a spectrum can be thought of as a vector of integral values. Each column of the matrix is a vector of the numerical values for a single variable (data point or bin region) over all samples.

recognition methods. An NMR spectrum is made up of a series of data points, each of which is associated with an intensity value. Thus, an NMR spectrum can be thought of as a vector of intensity values, and a set of spectra makes a matrix of intensity values (Figure 6.14). When we perform pattern recognition techniques on these data, we are comparing the magnitudes of the intensity values between experimental groups (e.g. samples from controls versus individuals with disease) to see if they differ and, hence, distinguish between our groups. Typically, statistical analyses are used to identify such trends. As we want to simultaneously compare all metabolites across groups (disease/control) at once, multivariate statistical analyses are required. Appropriate methods are discussed in brief in the following, with application to batch processes being emphasised.

### 6.4.2 Data Reduction

For either uni- or multivariate analysis, the NMR data can be used in its full resolution form or in a reduced form. A popular method for reducing the data is 'bucket integration' or 'binning'. In this process, a spectrum is divided into portions called 'buckets' or 'bins', and the intensity of the signal(s) that fall into each bin is calculated. This produces a smaller number of intensity values across the entire spectrum.

The advantage of reducing the data is that it provides much smaller data sets (~200 buckets/bins compared to ~64 K data points), which are easier to handle and store, and allow for faster computational times. Another important advantage of binning is that it eliminates the effects of signal shifting, which can occur due to changes in pH, ionic strength, and other inter-molecular interactions. When comparing full resolution data, which exhibit signal shifting, the data points do not line up, and the differences in intensity of the same data point between samples will influence the pattern recognition analysis. In some cases, such artefacts might even detract from real trends of biological relevance. The disadvantage of reducing data is the loss of resolution, which makes interpreting signals more difficult, and the loss of the fine structure of the spectrum. The width of the bin is optional, and many different methods are available for choosing the positions of the bins.[14–19] Early methods used fixed bin widths of 0.04 ppm. Over the years, as computational abilities have progressed, smaller bin widths have been encouraged, with bins as small as 0.01 and 0.005 ppm being used. The bin size to be used should be decided based on how much peak shifting your spectra exhibit; bin widths should not be smaller than the biggest signal shift in your spectrum. The correct bin size to use is a trade-off between minimum loss of resolution and maximum capture of signal shifting. A drawback to using bins of a fixed width is that, if peak shifts occur at the edge of a bin, the size of the adjacent bins will be altered according to where the peaks fall, resulting in spurious trends in the data. Furthermore, binning with fixed bin widths is more likely to result in more than one signal being pooled into one bin. For example, a small, but biologically influential, signal may not be flagged by pattern recognition analysis if it is in the same bin as a much larger (non-influential) signal. Recently, more sophisticated binning methods have become available in an attempt to overcome these problems. These include algorithms that allow variable bin widths across the spectra and exhibit peak picking methods that allow one signal to be represented in its own bin. Some NMR processing packages that perform binning, such as ACD/Labs (Canada), may offer variable bin widths and, again, work by identifying whether a bin boundary has been placed at a peak maximum and will stretch or shrink the boundary within the limits of a certain percentage change until it reaches a minimum.

### 6.4.3 Data Normalisation

Inter-sample variance due to differences in dilution (particularly in the case of urine); technical differences such as magnetic field inhomogeneities, the spectrometer used, and the number of scans; and differences in NMR line shape, must be accounted for so that they do not mask biologically meaningful changes in metabolite concentration. This step is called 'normalisation'. This is a row operation that acts on each spectrum in the data matrix (Figure 6.14). There are a number of methods for normalisation, the most widely used of which is normalisation to constant sum (CS) where each bin (or data point) is divided by the total spectral integral. Alternatively, if absolute concentrations are required, rather than relative concentrations, the integrals can be divided by those of a reference signal, such as TSP. Alternative methods are, however, available, and these should be considered alongside fluid (or sample matrix) type.[20–22]

### 6.4.4 Scaling and Mean Centering

NMR data require scaling prior to statistical analysis to account for the different numerical ranges that exist across the set of integrals. Even small changes in large signals will dominate over any changes in weaker or intermediate signals in statistical models, yet it may be the changes in the smaller signals that are important. Scaling is a column operation that acts on each NMR region (bin or data point) across all spectra in the data matrix (Figure 6.14). There are a number of scaling methods; Pareto scaling is commonly used with NMR data for plasma samples in which metabolite concentrations are essentially similar.

In addition to scaling, the data are usually mean-centered by subtracting the column mean from each variable. This is done so that, in pattern recognition analysis, deviating samples will be classed as those different to the mean.

## 6.5 PATTERN RECOGNITION

### 6.5.1 Multivariate Statistical Analysis

Multivariate statistical analyses are either unsupervised or supervised,[23–27] where the term 'supervised' refers to *a priori* knowledge of any grouping or classification in the samples. In unsupervised methods (Table 6.4), models are designed simply to find any intrinsic clustering within the data. The largest sources of variation in a data set will be highlighted by the model, regardless of whether they are of relevance to the grouping or classification of your samples. The most commonly used unsupervised technique in metabonomics is principal components analysis (PCA). PCA is generally carried out to perform an initial 'look–see' on the data and to identify any obvious outliers, problems with the data, and evidence of trends. Sometimes, trends of interest, such as between control and disease groups, are not initially picked up by unsupervised analyses, and either the confounding sources of variation must be removed to reveal the trend of interest or a supervised method must be applied instead.

Supervised methods (see Table 6.4), such as projections to latent structures by means of partial least squares (PLS), use class information about each sample (e.g. disease or control) when building the model and work to find the variation in the data that accounts for any differences between the classes. Because they are supervised, these methods must be validated to rule out over-fitting of a model. Often, supervised methods are applied even after a trend of biological interest has been identified using unsupervised methods because it allows a validatable model to be established. These models can also be used to predict the class membership of new unknown samples, hence providing a diagnostic tool.

#### 6.5.1.1 Interpretation of Scores and Loadings Plots

The output from the statistical analyses referred to above is commonly displayed as a scores plot or a loadings plot. Figure 6.15a shows a schematic of a PCA scores plot for a binned NMR data set containing two classes of observations–disease and control. The axes are the score values (t[1]) for two PCs of interest. Usually, the first

**TABLE 6.4**
**Most Commonly Applied Unsupervised and Supervised Multivariate Statistical Techniques and Correlation Analyses in NMR-Based Metabonomics/Lipidomics**

| | |
|---|---|
| | **Unsupervised Multivariate Statistical Analyses** |
| PCA | Displays intrinsic clustering within data. Good for an initial 'look–see' of the data and for identifying outliers. |
| | **Supervised Multivariate Statistical Analyses** |
| PLS | Identifies variation in a predictor matrix, which accounts for a response matrix. Can be used to predict response for new unknown observations. |
| PLS-DA | Identifies variation in a predictor matrix, which accounts for discrimination between two or more groups of observations. Can be used to predict class membership of new unknown observations. |
| Orthogonal projections to latent structures (OPLS) | Identifies variation in a predictor matrix, which accounts for a response matrix, while isolating and removing any orthogonal variation. Easier to interpret than PLS. Can be used to predict response for new unknown observations, identify sources of orthogonal variation, and filter out specific sources of variation from a data set. |
| OPLS-DA | Identifies variation in a predictor matrix, which accounts for discrimination between two or more groups of observations, while isolating and removing any orthogonal variation. Easier to interpret than PLS-DA. Can be used to predict class membership of new unknown observations, identify sources of orthogonal variation, and filter out specific sources of variation from a data set. |
| O2PLS | Identifies variation that is correlated in two data sets, as well as variation that is orthogonal for each. Can be used to predict one data set from another and identify sources of orthogonal variation in both data sets. |
| OnPLS | Identifies variation that is correlated in more than two data sets, as well as variation that is orthogonal for each. Can be used to predict one data set from another and identify sources of orthogonal variation in all data sets. |
| | **Correlation** |
| Statistical total correlation spectroscopy (STOCSY) | Measures correlations between all signals in an NMR spectrum or correlations between a single signal and the others in the spectrum to aid peak assignment. Can also be used to assign discriminating signals in a loadings plot and is informative of influential metabolic pathways. |
| STOCSY editing (E-STOCSY) | Edits a spectrum so as to remove the signals of drug molecules while leaving other signals intact. This avoids the need to cut out signals using dark regions, which risks simultaneous removal of potentially biologically relevant signals. |
| Iterative STOCSY (I-STOCSY) | Produces a network to show inter- and intra-metabolite relationships based on correlations. |

(*continued*)

**TABLE 6.4 (Continued)**
**Most Commonly Applied Unsupervised and Supervised Multivariate Statistical Techniques and Correlation Analyses in NMR-Based Metabonomics/Lipidomics**

| | |
|---|---|
| | **Correlation** |
| STOCSY scaling (STOCSY$^S$) | Edits a spectrum so as to remove unwanted signals (e.g. glucose signals) while leaving other signals intact. This avoids the need to cut out signals using dark regions, which risks simultaneous removal of potentially biologically relevant signals. Does not require a threshold value as in E-STOCSY and allows removed regions of the spectrum following editing to be analysed. |
| Statistical heterospectroscopy (SHY) | Identifies correlations between two data sets collected on the same individuals, for example, NMR data of two different biofluid samples from the same individuals or NMR versus MS data of the same samples. Aids peak assignment across analytical technique/biofluid and indicates metabolic pathways based on correlations between metabolites. |

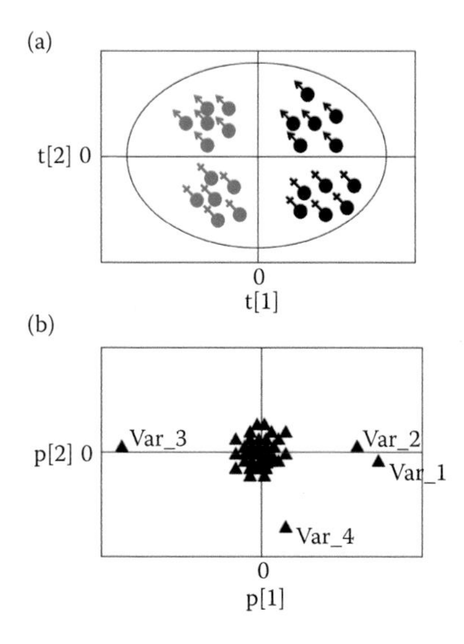


**FIGURE 6.15** Schematic PCA scores (a) and loadings (b) plots for an imaginary data set containing NMR data for diseased (black symbols) and control (grey symbols) individuals. The individuals are a mixture of males and females (denoted by the ♂ and ♀ symbols, respectively). The separation between diseased and control individuals is explained by the first PC, and the separation between sexes is accounted for by the second PC. The loadings plot shows which variables are influential in the grouping of the observations in the scores plot. In this diagram, the variables (which would be NMR spectral regions since this is an NMR data set) are called Var_1...Var_*k* for simplicity; however, in real-life cases, the variables would be labelled as the center of the bin for binned NMR data or the data point for full resolution data.

two PCs are plotted against each other to begin with. Inside the plot, each point represents an individual observation or a spectrum. These are plotted according to their score values for the two PCs being plotted. The first two PCs account for the two largest sources of variation in the data. In the example shown in Figure 6.15a, the largest source of variation between the observations is disease, and so we see a separation of control and disease samples along the PC 1 (t[1]) direction. The second largest source of variation is the differences due to sex in the observations, and so along the PC 2 direction, we see a separation of male and female observations. If the separation of interest is not observed in either PC 1 or 2, further PCs may be plotted to investigate whether this variation is captured later.

To interpret which variables are responsible for causing any separations in the scores plot, loadings plots can be produced for the same PCs. These plots show the scatter of the variables according to their loadings (p[i]) values for each PC on the axis. Since the loadings calculate how much each variable 'loads' into a PC, the loadings reveal how influential each variable is along the directions of each PC. So for example, the scores plot in Figure 6.15a showed a separation between disease and control observations along PC 1. Thus, in the loadings plot (Figure 6.15b), we look along the direction of PC 1 to see which variables are most influential along this component. Because the data are mean-centered and most spectral regions are not strongly influential in the PCs, the average clump of data points lies at zero. We see two variables influential at high p[1] and one variable influential at low p[1]. Where signals span more than one bin or multiple signals exist for the same metabolite, all relevant bins would be expected to show the same influence in the loadings plot. The direction of change in the variables between groups of observations can be determined from the loadings plot. In the example in Figure 6.15, the disease samples are at high t[1] in the scores plot, and the control samples are at low t[1]. Thus, variables 1 and 2, which are found at high p[1] in the corresponding loadings plot, must be at higher levels in the disease observations than the controls.

### 6.5.2 Data Analysis in Batch Processing

Batch process data form a ($N \times K \times J$) 3-D matrix made up of numerical values for $K$ variables at $J$ time points measured over $N$ batches. This differs from typical metabonomic and lipidomic data by the additional time dimension. Because the data are multivariate, multivariate statistical analyses are required to examine correlations between the variables, reveal properties of the batch process, and enable detection of batch deviations. However, to perform analyses such as PCA and PLS, the 3-D matrix must first be unfolded into a 2-D matrix, preserving one of the dimensions. This can be thought of as cutting the 3-D matrix into slices and then ordering the slices appropriately. For example, if the $N \times K \times J$ matrix is unfolded to preserve the direction of the variables, then a 2-D matrix is generated with $J$ columns and $N \times K$ rows, one row for each batch at each time point or 'observation' (i.e. observation $nj$ = batch $n$, time $j$). If the $N \times K \times J$ matrix is unfolded to preserve the batch direction, a 2-D matrix is produced with $N$ rows and $K \times J$

columns. Batch processing is typically performed at two levels – the observation level and the batch level.

#### 6.5.2.1 Observation Level

At the observation level, we are interested in modelling the development of 'good batches' so that this can be used to monitor new, evolving batches and detect faults as they arise. PCA can be performed to examine how individual observations relate to each other, or PLS can be performed using local batch time (or a similar representative index of batch maturity) as the dummy response matrix to enable batch phase to be understood and to allow its prediction. This works because PLS regresses a response matrix (in this case, local batch time) against a predictor matrix (the unfolded batch data matrix) and identifies the variation (if any) in the predictor matrix, which can be used to explain the response matrix, as well as builds a model that can be used to predict the response variables for new observations based on their predictor variables. At the observation level, PCA and PLS models should be built on data representing 'good' batch processes, and prior to this, the 3-D batch data matrix is unfolded to preserve the direction of the variables. Examination of scores plots for either PCA or PLS, with the observations from one batch chronologically joined up with a line, will show the trajectories of the batches in scores space and will give an idea of how similar the batches are. The corresponding loadings plots show how the variables correlate with batch maturity. Aside from scores and loadings plots, another type of chart that can be generated in batch processing is a 'control chart' (Section 6.6.3), which can be used to monitor new, developing batches and identify when they deviate from the average trajectory. In these plots, the score values for a component (e.g. PCA or PLS component 1) are plotted against the local batch time for all batches. The average trajectory of scores over time across all batches is calculated, and new batches are monitored according to their compliance with this average, which is taken as being the 'normal developmental trace'. Upper and lower limits of ±3 standard deviations are also computed as control limits. Corresponding 'contribution plots' show which variables are responsible for a deviation in the monitored trajectory from the average. Contribution plots show the changes in variables between two observations, in this case, the new batch's score at a given time point and that of the average trace, although contribution plots can also be produced to show differences between any two batches or groups of batches at one time point. The score contribution values are calculated by multiplying the difference in numerical value between the variables by the component loading value ($p$) of that variable.

#### 6.5.2.2 Batch Level

At the batch level, one whole batch model is produced using all available data, which might include data on initial conditions for the process, the batch evolution data (i.e. the batch matrix used at the observation level), and data on the final results and quality characteristics. For the batch evolution data, either the original batch data matrix can be used or the scores matrix from the observation level. If the original variables are used, then the 3-D batch matrix must be unfolded, this time, to preserve the batch direction. The initial and evolution data are combined into one matrix, which is then regressed against the final result/quality characteristic matrix using PLS. The resulting model can be used to predict the final conditions of new batches and, hence, predict whether batches will be 'good' or 'bad'. Furthermore, the model provides information on how the final results are influenced by the initial conditions and batch evolution.

Alternatively, a PCA model can be built from the scores of the observation-level components (PLS or PCA) or from the unfolded original variables to explore groupings/ trends in batches based on data over all time points. In other words, the batch level compares overall trajectories of batches, consisting of the initial conditions, the evolution, and the final result data, so each batch is represented by a single coordinate in the scores plot (Section 6.6.3).

## 6.6 CASE STUDIES

### 6.6.1 Study 1: Application to Plasma

Our first example is taken from a study investigating the metabolic pathways of diabetes.[28] CPMG $^1$H-NMR spectra were acquired for blood plasma samples from people with type 1 diabetes while treated with insulin and, again, during insulin deprivation, and from non-diabetic controls. The spectra were binned into 0.003-ppm bins, normalised to CS and mean-centered before PCA was performed. The water regions of the spectra were also removed prior to PCA, along with the glucose regions, to allow differentiation of spectra without the dominant influence of the strong glucose signals. The resulting PCA scores and loadings plots are shown in Figure 6.16. Notice that the loadings are represented in the form of a 1-D plot, with the loadings values on the $y$ axis and the variable numbers on the $x$ axis. Thus, the 1-D version of the loadings plot resembles the appearance of an NMR spectrum, which makes identification of influential signals easier. The scores plot shows a clear separation between the diabetic patients deprived of insulin and the insulin-treated diabetic patients and the controls. The insulin-treated diabetic patients and the control samples are clustered together, indicating that the metabolic profile of a diabetic patient treated with insulin is similar to that of a non-diabetic individual. However, without the insulin, the metabolic pathways of diabetic individuals become altered. To understand which metabolic alterations occur, we look at the loadings plot. Because the differentiation of insulin-deprived diabetics occurs along PC 1, we look at the loadings for PC 1. It can be seen that ketone signals (3HB, acetoacetate and acetone) and fatty acid signals are strongly influential in the deviation of the insulin-deprived diabetic samples. Because the ketone signals are influential along the $+[p_1]$ axis, this indicates that insulin-deprived diabetic patients are characterised with higher plasma levels of ketones. Thus, by the same logic, fatty acid levels are lower in these patients. Ketogenesis occurs when the body is deprived of insulin and its cells are unable to take up glucose from the blood. Instead, the body begins to break down fatty acids for energy, producing ketones as bi-products. Thus, the PCA scores and loadings plots beautifully demonstrate the presence of ketogenesis in the individuals deprived of insulin.

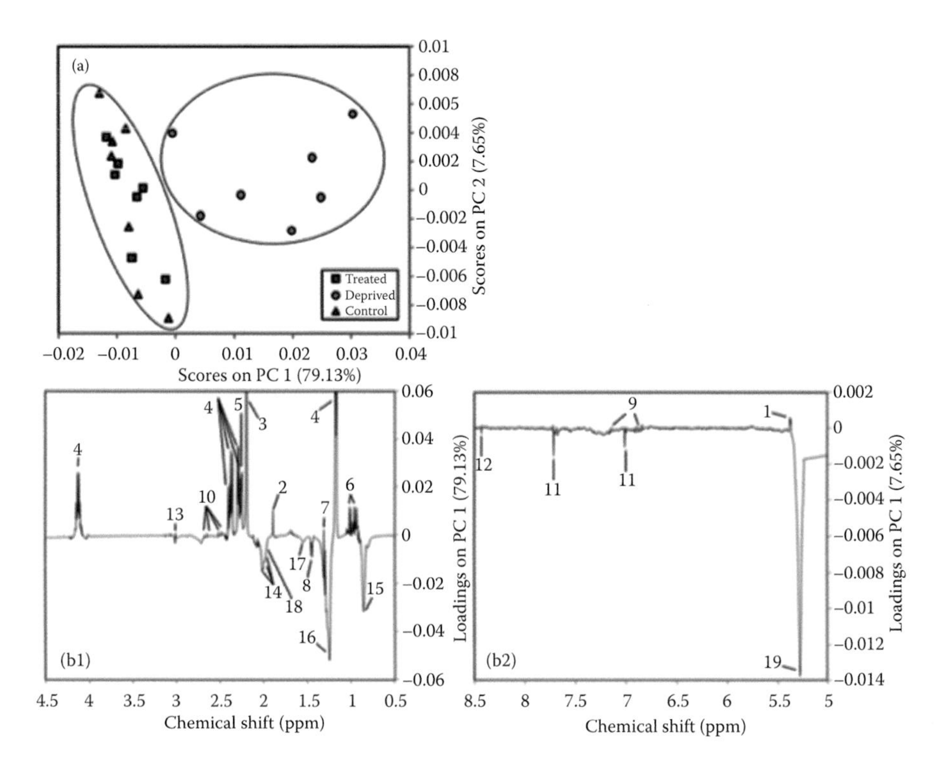


**FIGURE 6.16** PCA scores (a) and PC 1 1-D loadings (b1 and b2) plots from $^1$H-NMR spectra of plasma from insulin-deprived diabetics (circles), insulin-treated diabetics (squares), and controls (triangles). The scores plot shows a separation between the insulin-deprived diabetic patients and the insulin-treated diabetic patients and the controls. (1) allantoin; (2) acetate; (3) acetone; (4) 3-hydroxybutyrate; (5) acetoacetate; (6) valine; (7) lactate; (8) alanine; (9) tyrosine; (10) citrate; (11) histidine; (12) formate; (13) creatinine; (14) glycoprotein N-acetyl groups; (15) lipid:$CH_3$; (16) lipid:$CH_2$; (17) lipid:$CH_2CH_2$–CO; (18) lipid:$CH_2$–C = C; and (19) lipid:fatty acyl groups=CH. (Reprinted from Lanza, I. R. et al., *PLoS ONE* 5, e10538. doi:10.1371/journal.pone.0010538, 2010. With permission.)

### 6.6.2 Study 2: Application to Saliva

In our second example,[29] $^1$H-NMR spectra were acquired using a single 90° pulse experiment with water pre-saturation on saliva samples collected from the same individuals in the morning and at night. The spectra were reduced into 0.0074-ppm bins, normalised to the TSP signal, and mean-centered prior to PCA. The water signal was also removed prior to PCA. The resulting PCA scores plot is shown in Figure 6.17. Some separation due to sampling time can be seen along PC 1; however, there is an overlap between the two groups, and the separation is not well defined. Therefore, projections to latent structures by means of PLS-discriminant analysis (PLS-DA) was performed on the saliva samples, using sampling time as the response variable. The PLS-DA scores plot is shown in Figure 6.18a, and this now shows a much more defined separation between the morning and night samples. The corresponding

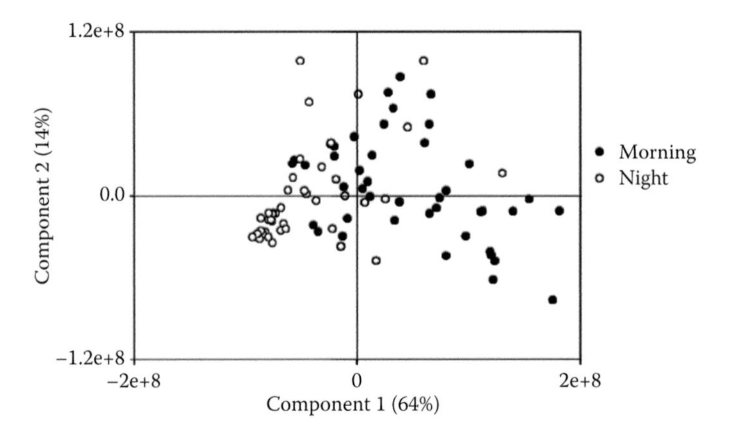


**FIGURE 6.17** PCA scores plot from $^1$H-NMR spectral data of human saliva samples collected in the morning (black circles) and at night (white circles). Some discrimination between the sampling times can be made out in PC 1; however, there is an overlap between the two classes. (Reprinted with permission from Bertram, H. C. et al. Potential of human saliva for nuclear magnetic resonance–based metabolomics and for health-related biomarker identification. *Analytical Chemistry* 81, pp. 9188–9193. Copyright 2009 American Chemical Society.)

regression coefficient plot for PLS-DA component 1 (Figure 6.18b) shows that morning samples have higher intensities of signals due to alanine, propionate, choline, and trimethylamine-N-oxide (TMAO) and lower intensities of signals due to methanol and N-acetyl groups than the night samples. The authors of this study suggested that increased propionate in the morning indicated microorganism activity in the oral cavity – alanine was likely to be produced in the highly metabolically active salivary glands, methanol was derived from cigarette smoke (smokers were included in the study), which would be elevated on evening, and N-acetyl sugars were likely derived from bacterial enzyme activity. Because it is a supervised method, the PLS-DA model was validated using a method called 'cross validation', whereby each sample is left out one at a time and a PLS-DA model is built from the remaining samples' data. The second model is then used to predict the class membership of the excluded sample. In this case, 78 of the 92 samples were correctly classified, giving a predictive ability of 85%.

### 6.6.3 Study 3: Batch Analysis

This final example demonstrates the application of NMR spectroscopy and batch processing to investigate biological dynamic processes, in this case,[30] the time-dependent metabolic response to liver toxin, α-naphthylisothiocyanate (ANIT). $^1$H-NMR presat spectra were acquired on urine samples from control and ANIT-dosed rats 24 to 0 h pre-dose, and 0 to 8, 8 to 24, 24 to 32, 32 to 48, 48 to 72, 72 to 96, 96 to 120, 120 to 144 and 144 to 168 h post-dose. The spectral data were reduced to 0.04-ppm bins, the water regions were removed, and the data CS was normalised.

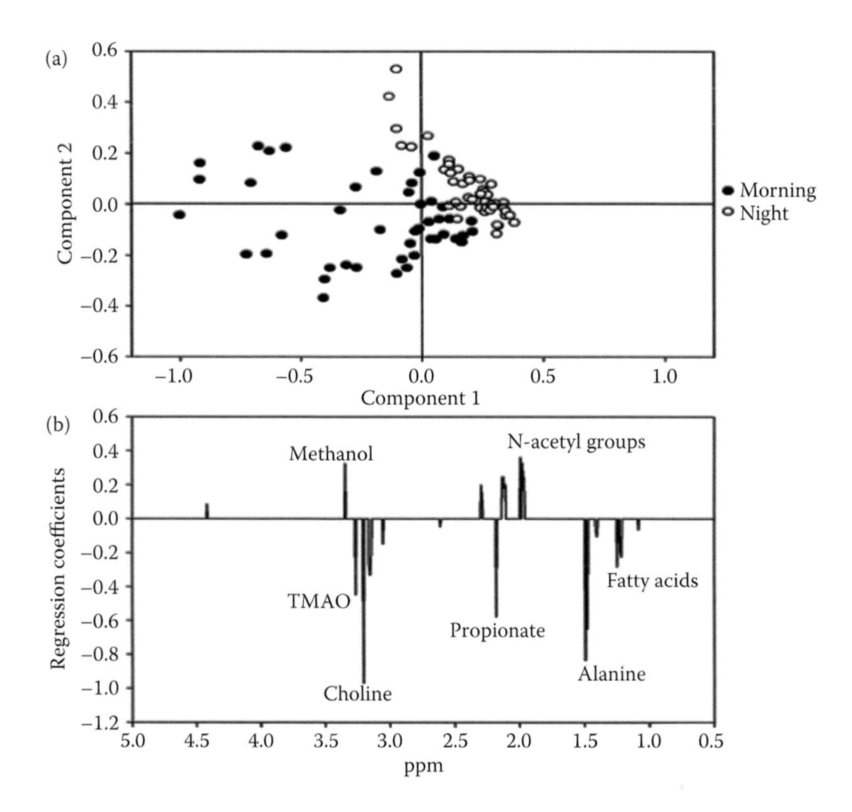


**FIGURE 6.18** PLS-DA scores (a) and component 1 regression coefficients (b) plots from [1]H-NMR spectral data of human saliva samples collected in the morning (black circles) and at night (white circles). The samples can be differentiated on the basis of sampling time, and the regression coefficients reveal that alanine, propionate, choline, trimethylamine-N-oxide (TMAO), methanol, and N-acetyl groups are influential in the separation. (Reprinted with permission from Bertram, H. C. et al. Potential of human saliva for nuclear magnetic resonance–based metabolomics and for health-related biomarker identification. *Analytical Chemistry* 81, pp. 9188–9193. Copyright 2009 American Chemical Society.)

A 3-D matrix was generated, consisting of the individual rats (batches), the NMR data (variables), and the urine collection time points.

First, batch processing was performed at the observation level. At this level, a PLS model built only on the data from the control rats was produced to allow characterisation and monitoring of 'normal' urine over time. The 3-D matrix was unfolded so that the NMR data for each rat at each time point were regressed against the time variable vector. The PLS model explains how the NMR data vary with respect to time, and so this information can be used to generate trajectories of the data for each mouse. The average trajectory for 'normal' urine was obtained, along with control limits of ±2 and ±3 standard deviations from the mean. These can be seen on the control charts in Figure 6.19; the thick black lines are the average trajectories, the dotted lines are the ±2 standard deviation limits, and the dashed

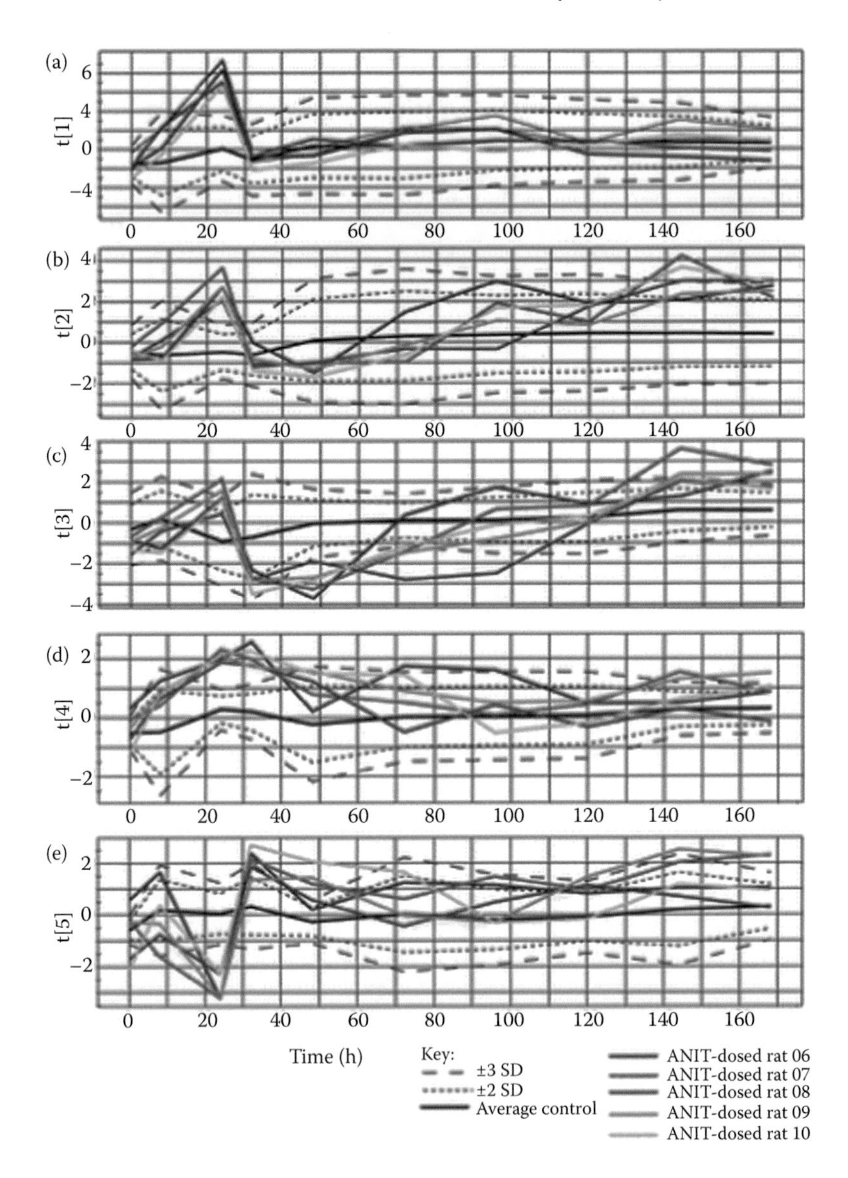


**FIGURE 6.19** Observation-level batch processing control charts for urine $^1$H-NMR data from control and ANIT-dosed rats. The plots are produced by mapping the PLS components from the ANIT-dosed rats onto the PLS components for the control rats, over all sampling time points. Plots were produced for the first five PLS components, whose corresponding scores are t[1] (a), t[2] (b), t[3] (c), t[4] (d) and t[5] (e). The thick black lines are the average trajectories for the control rats, the dotted lines are the ±2 standard deviation control limits, and the dashed lines are the ±3 standard deviation control limits. The other lines represent the trajectories of the ANIT-dosed rats. (From Azmi, J. et al. 2002. Metabolic trajectory characterisation of xenobiotic-induced hepatotoxic lesions using statistical batch processing of NMR data. *Analyst* 127, pp. 271–276. Reproduced by permission of The Royal Society of Chemistry.)

lines are the ±3 standard deviation limits. The scores for each component generated from PLS of the NMR spectral data of the ANIT-dosed rats were then mapped individually onto the control model to determine how the treated rats deviated from 'normality'. The control charts in Figure 6.19 show the ANIT-dosed rat trajectories for the first five PLS components plotted along with the average control trajectory and control limits. In the first component (t[1]) (Figure 6.19a), the treated rats initially begin to deviate from the controls 0 to 8 h post-dose, peaking at the 8- to 24-h post-dose time point. This deviation, in fact, lies outside of the uppermost control limit. Following this time point, the dosed rat trajectories become more similar to that of the average control and are within the ±2 standard deviation limits, indicating some recovery from the initial response to ANIT. Contribution plots at the 8- to 24-h post-dose time point indicated that the deviation in treated rats was caused by increased levels of glucose and lower levels of citrate, 2-oxoglutarate, and succinate in the urine. Since ANIT is a liver toxin, the authors suggested that the high levels of glucose could indicate affected glycogen storage in the liver due to the onset of liver damage.

In the second component (t[2]) (Figure 6.19b), the deviation at 8 to 24 h post-dose is still evident, but there is also deviation later on in the trajectory, at the post-dose time points 96 to 120, 120 to 144 and 144 to 168 h. Contributions plots revealed that elevations in urinary succinate and acetate occurred at these time points in the ANIT-treated rats compared to the controls. The lower components (Figure 6.19c–e) also revealed further differences in the ANIT rats compared to the controls at various time points, providing further information about the more subtle (PLS/PCA components account for less and less variation the lower they become) metabolic responses to ANIT and the dynamics of these responses. Thus, this example demonstrates the power of batch processing in elucidating the mechanisms of dynamic processes. Furthermore, the control charts for the lower components also demonstrate the power of batch processing in identifying individual batches whose trajectories deviate from those of the main group. For example, in the control plot for t[2] (Figure 6.19b), rats 6 and 8 both show abnormal responses to ANIT toxicity 72 to 96 h post-dose. Rat 6 had higher levels of succinate than the other treated mice, and Rat 8 deviated due to poor suppression of the water signal in the NMR spectrum. Thus, whether biologically relevant or not, batch processing is very sensitive to deviations in individual batches. In industrial applications, this would reveal 'bad batches'.

Finally, at the batch level, a PCA model was built using the observation-level PLS components as the variables to compare the overall metabolic responses of the rats to ANIT-induced toxicity. Thus, one single coordinate is present in the scores plot for each rat over all time points, including data on the pre-dose condition, the initial onset of the toxin, the response to the toxin, and the recovery. This differs from the observation level where several coordinates are produced for the same rat for each separate time point. The resulting scores plot is shown in Figure 6.20. A clear separation is achieved between control and ANIT-treated rats, supporting the findings from the observation level that control rats and ANIT-treated rats have different trajectories. Furthermore, this model can be used to predict ANIT-induced toxicity in other mice.

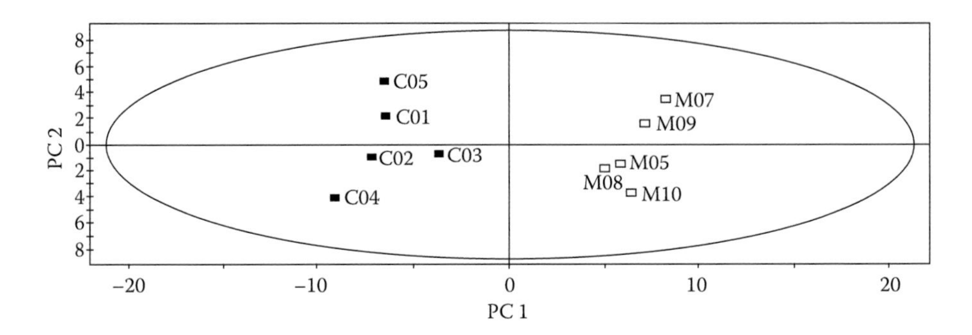


**FIGURE 6.20** Batch-level processing PCA scores plot built from the PLS coordinates from the observation level. The plot shows a separation between the control and the ANIT-treated rats based on the overall metabolic response of each animal. Thus, each rat is depicted as a single coordinate representing all time points. (From Azmi, J. et al. 2002. Metabolic trajectory characterisation of xenobiotic-induced hepatotoxic lesions using statistical batch processing of NMR data. *Analyst* 127, pp. 271–276. Reproduced by permission of The Royal Society of Chemistry.)

## 6.7 SUMMARY

We have shown here that NMR can be applied to investigate the response of biological systems to pathophysiological stimuli by measuring the levels of metabolites and lipids in biofluids and tissues. A range of NMR experiments can be employed for metabonomic and lipidomic studies, including presat and 1-D NOESY–presat experiments to simultaneously examine all species in biological samples while suppressing the large water signal; CPMG experiments to examine only low–molecular-weight species; JRES experiments to reduce the overlap of these same signals; DOSY experiments to examine only high–molecular-weight species; and MAS to study solid biological samples, such as tissues. A range of statistical techniques exist to explore NMR data and determine trends in the metabolic profiles, which can lead to understanding biological processes and biomarker identification. These techniques can also be applied to study batch processes in the industry, allowing characteristic trajectories of good batches to be determined, batch errors to be identified as they occur, and new batches to be predicted as 'good' or 'bad'. Care must be taken when designing investigations to ensure that samples are well collected, representative, and statistically significant, and NMR data should be collected and processed carefully and consistently to avoid producing spurious trends. While methods for NMR analysis, processing, and modelling in metabonomics, lipidomics, and batch processing have come a long way in the past 20 years, there are still limitations and challenges that have yet to be addressed, such as the low sensitivity of NMR in comparison with techniques such as MS and the ongoing task of completely assigning all metabolite NMR signals across all biological media.[31–33] However, metabonomics, lipidomics, and batch processing have certainly become very prevalent and fruitful applications of NMR, which will continue to thrive with further advancements in NMR technology and growing opportunities to explore new biological domains.

## REFERENCES

1. Nicholson, J. K., J. C. Lindon and E. Holmes. 1999. 'Metabonomics': Understanding the metabolic responses of living systems to pathophysiological stimuli via multivariate statistical analysis of biological NMR spectroscopic data. *Xenobiotica* 29, p. 1181–1189.
2. Spener, F., M. Lagarde, A. Géloën and M. Record. 2003. What is lipidomics? *European Journal of Lipid Science and Technology* 105, p. 481–482.
3. Otvos, J. D., E. J. Jeyarajah, D. W. Bennett and R. M. Krauss. 1992. Development of a proton nuclear magnetic resonance spectroscopic method for determining plasma lipoprotein concentrations and subspecies distributions from a single, rapid Measurement. *Clinical Chemistry* 38, p. 1632–1638.
4. Ala-Korpela, M., A. Korhonen, J. Keisala et al. 1994. $^{1}$H-NMR–based absolute quantitation of human lipoproteins and their lipid contents directly from plasma. *Journal of Lipid Research* 35, p. 2292–2304.
5. Folch, J., M. Lees and S. G. H. Sloane. 1957. A simple method for the isolation and purification of total lipides from animal tissues. *Journal of Biological Chemistry* 226, p. 497–509.
6. Bligh, E. G. and W. J. Dyer. 1959. A rapid method of total lipid extraction and purification. *Canadian Journal of Physiology and Biochemistry* 37, p. 911–917.
7. Beckonert, O., H. C. Keun, T. M. Ebbels et al. 2007. Metabolic profiling, metabolomics, and metabonomic procedures for NMR spectroscopy of urine, plasma, serum, and tissue extracts. *Nature Protocols* 2, p. 2692–2703.
8. Alum, M. F., P. A. Shaw, B. C. Sweatman, B. K. Ubhi, J. N. Haselden and S. C. Connor. 2008. 4,4-dimethyl-4-silapentane-1-ammonium trifluoroacetate (DSA), a promising universal internal standard for NMR-based metabolic profiling studies of biofluids, including blood plasma and serum. *Metabolomics* 4, p. 122–127.
9. Claridge, T. W. D. 1999. *High-Resolution NMR Techniques in Organic Chemistry*. Oxford, United Kingdom: Elsevier, p. 33.
10. Ludwig, C. and M. R. Viant. 2010. Two-dimensional J-resolved NMR spectroscopy: Review of a key methodology in the metabolomics toolbox. *Phytochemical Analysis* 21, p. 22–32.
11. Maher, A. D., D. Crockford, H. Toft et al. 2008. Optimisation of human plasma $^{1}$H-NMR spectroscopic data processing for high-throughput metabolic phenotyping studies and detection of insulin resistance related to type 2 diabetes. *Analytical Chemistry* 80, p. 7354–7362.
12. Beckonert, O., M. Coen, H. C. Keun et al. 2010. High-resolution magic angle spinning NMR spectroscopy for metabolic profiling of intact tissues. *Nature Protocols* 5, p. 1019–1032.
13. Coen, M., E. M. Lenz, J. K. Nicholson, I. D. Wilson, F. Pognan and J. C. Lindon. 2003. An integrated metabonomic investigation of acetaminophen toxicity in the mouse using NMR spectroscopy. *Chemical Research in Toxicology* 16, p. 295–303.
14. Veselkov, K. A., J. C. Lindon, T. M. D. Ebbels et al. 2009. Recursive segment-wise peak alignment of biological H-1 NMR spectra for improved metabolic biomarker recovery. *Analytical Chemistry* 81, p. 56–66.
15. Savorani, F., G. Tomasi and S. B. Engelsen. 2010. icoshift: A versatile tool for the rapid alignment of 1-D NMR spectra. *Journal of Magnetic Resonance* 202, p. 190–202.
16. Davis, R. A., A. J. Charlton, J. Godward, S. A. Jones, M. Harrison and J. C. Wilson. 2007. Adaptive binning: An improved binning method for metabolomics data using the undecimated wavelet transform. *Chemometrics and Intelligent Laboratory Systems* 85, p. 144–154.
17. de Meyer, T., D. Sinnaeve, B. van Gasse et al. 2008. NMR-based characterisation of metabolic alterations in hypertension using an adaptive, intelligent binning algorithm. *Analytical Chemistry* 80, p. 3783–3790.

18. Anderson, P. E., D. A. Mahle, T. E. Doom, N. V. Reo, N. J. DelRaso and M. L. Raymer. 2011. Dynamic adaptive binning: An improved quantification technique for NMR spectroscopic data. *Metabolomics* 7, p. 179–190.
19. Anderson, P. E., N. V. Reo, N. J. DelRaso, T. E. Doom and M. L. Raymer. 2008. Gaussian binning: A new kernel-based method for processing NMR spectroscopic data for metabolomics. *Metabolomics* 4, p. 261–272.
20. Dieterle, F., A. Ross, G. Schlotterbeck and H. Senn. 2006. Probabilistic quotient normalisation as robust method to account for dilution of complex biological mixtures. Application in H-1 NMR metabonomics. *Analytical Chemistry* 78, p. 4281–4290.
21. Torgrip, R. J. O., K. M. Aberg, E. Alm, I. Schuppe-Koistinen and J. Lindberg. 2008. A note on normalisation of biofluid 1-D H-1 NMR data. *Metabolomics* 4, p. 114–121.
22. Dong, J. Y., K.-K. Cheng, J. J. Xu, Z. Chen and J. L. Griffin. 2011. Group aggregating normalisation method for the pre-processing of NMR-based metabolomic data. *Chemometrics and Intelligent Laboratory Systems* 108, p. 123–132.
23. Parsons, H. M., C. Ludwig, U. L. Gunther and M. R. Viant. 2007. Improved classification accuracy in 1- and 2-dimensional NMR metabolomics data using the variance stabilising generalised logarithm transformation. *BMC Bioinformatics* 8, p. 16.
24. Westerhuis, J. A., H. C. J. Hoefsloot, S. Smit et al. 2008. Assessment of PLSDA cross validation. *Metabolomics* 4, p. 81–89.
25. Lofstedt, T. and J. Trygg. 2011. OnPLS: A novel multi-block method for the modelling of predictive and orthogonal variation. *Journal of Chemotherapy* 25, p. 441–455.
26. Bylesjö, M., M. Rantalainen, O. Cloarec, J. K. Nicholson, E. Holmes and J. Trygg. 2006. OPLS discriminant analysis: Combining the strengths of PLS-DA and SIMCA classification. *Journal of Chemotherapy* 20, p. 341–351.
27. Cloarec, O., M.-E. Dumas, A. Craig et al. 2005. Statistical total correlation spectroscopy: An exploratory approach for latent biomarker identification from metabolic H-1 NMR data sets. *Analytical Chemistry* 77, p. 1282–1289.
28. Lanza, I. R., S. Zhang, L. E. Ward, H. Karakelides, D. Raftery and S. Nair. 2010. Quantitative metabolomics by $^{1}$H-NMR and LC-MS/MS confirms altered metabolic pathways in diabetes. *PLoS ONE* 5, p. e10538.
29. Bertram, H. C., N. Eggers and N. Eller. 2009. Potential of human saliva for nuclear magnetic resonance–based metabolomics and for health-related biomarker identification. *Analytical Chemistry* 81, p. 9188–9193.
30. Azmi, J., J. L. Griffin, H. Antti et al. 2002. Metabolic trajectory characterisation of xenobiotic-induced hepatotoxic lesions using statistical batch processing of NMR data. *Analyst* 127, p. 271–276.
31. Wishart, D. S., C. Knox, A. C. Guo et al. 2009. HMDB: A knowledge base for the human metabolome. *Nucleic Acids Research* 37, S1, p. D603–D610.
32. Cui, Q., I. A. Lewis, A. D. Hegeman et al. 2008. Metabolite identification via the Madison Metabolomics Consortium database. *Nature Biotechnology* 26, p. 162–164.
33. Ludwig, C., J. M. Easton, A. Lodi et al. 2012. Birmingham Metabolite Library: A publicly accessible database of 1-D $^{1}$H and 2-D $^{1}$H J-resolved NMR spectra of authentic metabolite standards (BML–NMR). *Metabolomics* 8, p. 8–18.

# 7 Selected Applications of NMR Spectroscopy

*Robert Brkljača, Sylvia Urban,*
*Kristian Hollingsworth, W. Bruce Turnbull*
*and John A. Parkinson*

## CONTENTS

7.1 Natural Products ........ 264
*Robert Brkljača and Sylvia Urban*
7.1.1 Introduction ........ 264
7.1.2 1-D NMR Experiments ........ 267
7.1.2.1 1-D NOE and 1-D TOCSY Experiments ........ 267
7.1.2.2 DEPT and APT ........ 269
7.1.3 2-D NMR Experiments ........ 270
7.1.3.1 Homonuclear Approaches ........ 270
7.1.3.2 Heteronuclear Approaches ........ 271
7.1.4 Hyphenation: HPLC-NMR ........ 278
7.1.4.1 Application to Organism Profiling and Structure Elucidation ........ 279
7.1.5 Summary ........ 281
7.2 Carbohydrates ........ 282
*Kristian Hollingsworth and W. Bruce Turnbull*
7.2.1 Introduction ........ 282
7.2.2 Sequence Determination by NMR ........ 282
7.2.3 Conformational Determination ........ 286
7.2.3.1 Monosaccharide Conformation from Coupling Constants ........ 286
7.2.3.2 Conformations of Oligosaccharides ........ 288
7.2.4 Protein–Carbohydrate Interactions ........ 290
7.2.4.1 Identification of Binding Interfaces ........ 290
7.2.4.2 Oligosaccharide Conformation When Bound ........ 293
7.2.4.3 Relative Orientations of Protein and Carbohydrate ........ 294
7.2.5 Summary ........ 295
7.3 Nucleic Acids ........ 296
*John A. Parkinson*
7.3.1 Introduction ........ 296
7.3.2 Sample Selection ........ 296

7.3.2.1 Sequence Design ... 296
7.3.2.2 Length, Shape and Complementarity ... 299
7.3.3 Sample Handling and Preparation for NMR Studies ... 300
7.3.3.1 Handling DNA ... 300
7.3.3.2 Sample Preparation for NMR Studies ... 300
7.3.3.3 Choice of Solvent ... 301
7.3.4 Data Acquisition and Processing ... 301
7.3.4.1 Samples in 90% $H_2O$/10% $D_2O$ ... 302
7.3.4.2 Samples in 99% $D_2O$ ... 307
7.3.4.3 2-D NMR Data Processing ... 312
7.3.4.4 Temperature Considerations ... 313
7.3.5 Assessment of Ligand Binding Using 1-D NMR ... 316
7.3.6 Resonance Assignment ... 317
7.3.7 Summary ... 322
References ... 323
Further Readings ... 328

## 7.1 NATURAL PRODUCTS

Robert Brkljača and Sylvia Urban

### 7.1.1 Introduction

For the last few centuries, natural products have played a pivotal role as a source of new pharmaceutical agents.[1] During the period 1981 to 2010, it has been documented that more than 40% of new chemical drug entities released were either a natural product or derived from a natural product.[2] While the total number of new drugs derived from a natural source decreased to only 12.2% in 1997, this figure soared to a new high of 50% in 2010.[2] This trend illustrates that, despite the efforts made in the combinatorial chemistry area, where potential new drug entities are systematically synthesised, natural products continue to play an important part in the future of drug discovery. This is especially important as new diseases emerge and the resistance to drugs, such as antibiotics, escalates.[3]

Many areas of the natural environment, including the marine and microbial environments, remain understudied. It is from these environments that natural product researchers are focussing their efforts to discover further new bioactive secondary metabolites possessing unique pharmacophores with pharmaceutical potential.[4–6]

In a natural product study, marine or terrestrial organisms are, first, solvent-extracted and then typically screened for potential biological activity. This can be achieved by screening the crude extract against a selected bioassay or, alternatively, against a variety of assays. If biological activity is detected, the crude extract is fractionated in an effort to isolate and then determine the structure of the active component(s) via a bioassay-directed approach. Any bioactive compounds identified are then further evaluated. The isolation and identification exercise can typically take a few short weeks or, in extreme cases, it can take years. Two important examples of anti-cancer natural products that took years to resolve include ecteinascidin 743 and taxol (Figure 7.1).

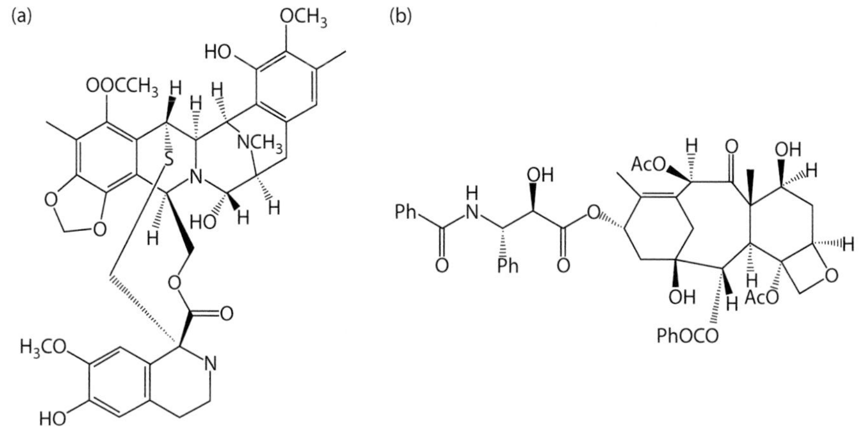

**FIGURE 7.1** Structure of the anti-tumour agents ecteinascidin 743 (a) and taxol (paclitaxel) (b).

The tetrahydroisoquinoline alkaloid, ecteinascidin 743, is a potent anti-tumour natural product that was isolated from the Caribbean tunicate/ascidian *Ecteinascidia turbinata*.[7] In 1970, it had been established that the crude extract derived from this ascidian displayed potent anti-tumour activity,[8] but it was not until 1987 that the isolation and complete structure elucidation of the component responsible for this activity could be reported.[7] Trabectedin (also known as ecteinascidin 743 or ET-743) is sold by Zeltia and Johnson and Johnson under the brand name Yondelis™. It is approved for use in Europe, Russia, and South Korea for the treatment of advanced soft tissue sarcoma. In the case of the anti-cancer natural product taxol, which was isolated from the Pacific yew tree *Taxus brevifolia*, the compound was isolated in its pure form in 1967, but its complete structure was not reported until 1971.[9] When it was developed commercially by Bristol-Myers Squibb (BMS), the generic name was changed to paclitaxel™, and the BMS compound is sold under the trademark Taxol™. A newer formulation, in which paclitaxel is bound to albumin, is sold under the trademark Abraxane™. Taxol (paclitaxel) is used to treat a range of cancers, such as lung, ovarian, and breast cancers. Owing to a combination of the complexity in these structures, the low quantities isolated, and in the case of ET-743, the difficulty encountered with the isolation, it took many years and a tremendous effort to identify these important natural product drugs. Particularly, in the case of taxol (paclitaxel), it was the advances made in NMR spectroscopic techniques that enabled its structure to be finally elucidated.

The typical process for the isolation and identification of new bioactive natural products in a marine or terrestrial organism is summarised in Figure 7.2. Once an organism is collected and identified, a crude extract is generated by solvent extraction and evaluated for bioactivity against a panel of biological assays (e.g. anti-microbial and anti-tumour activities). Selection of an organism can be either random or targeted (e.g. a plant with known ethnopharmacological precedence or a

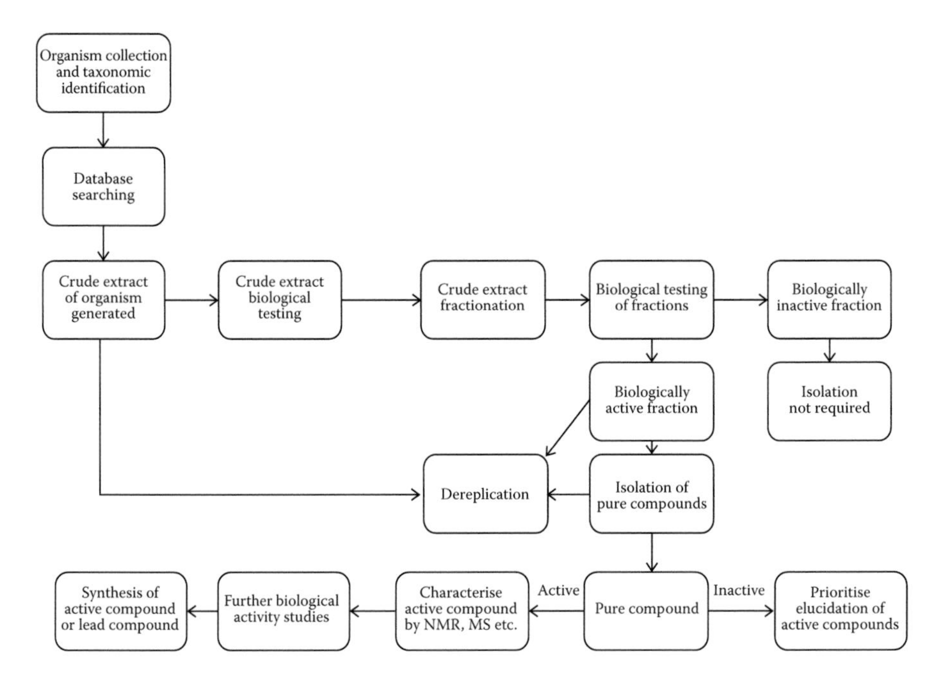


**FIGURE 7.2** General process for the discovery of new bioactive natural products.

marine organism known for its chemical defense strategies). If biological activity is observed, the crude extract is further fractionated usually by column chromatography. In a bioassay-guided isolation, resulting fractions are tested against the particular biological assay(s) to confirm which fraction(s) contains the active component(s) of interest. Once a bioactive fraction(s) has been identified, each compound present is further fractionated and purified using high-performance liquid chromatography (HPLC). When a pure compound has been obtained, the structure of the natural product is elucidated and characterised using a variety of methodologies. These include infrared spectrometry, ultraviolet-visible (UV-VIS) spectroscopy, mass spectrometry (MS), X-ray diffraction, and most importantly, nuclear magnetic resonance (NMR) spectroscopy. The use of NMR spectroscopy for the structure determination of an unidentified natural product provides the greatest and most useful structure connectivity information. MS is used to confirm the molecular formula and weight. Once a natural product's structure has been elucidated, the compound can then be subjected to further biological evaluation, and if necessary/feasible, a total synthesis can be formulated to provide larger quantities for further studies. These further studies would include *in vivo* biological activity studies and, possibly, the determination of the mechanism of action, particularly for new bioactive natural products.

Since the isolation process can be lengthy and require considerable effort, it is important to recognise the presence of potential new secondary metabolites at the earliest possible stage. The process known as 'dereplication' is vital to avoid the time-consuming process of isolating known and/or nuisance compounds. Combining both chemical and biological screening provides the quickest method to identify new lead compounds. Simple biological assays can be carried out to identify activity in crude extracts or fractions. These active extracts or fractions can then be chemically screened/profiled using techniques such as HPLC, UV-VIS spectroscopy, and NMR spectroscopy, or alternatively using hyphenated spectroscopic techniques such as HPLC-NMR or HPLC-MS.

Natural product structure elucidation studies have become increasingly easier over the last three to four decades due to the major advances that have occurred in a variety of NMR techniques. This is particularly evident when the introduction of higher magnetic field strengths and improved pulse sequences that became available is considered.[10] There are many NMR experiments available for carrying out natural product structure elucidations. These can be broadly categorised into two main groups: the 1-D or the 2-D NMR experiments. Following an illustration of the utility of these two classes of experiment, we describe a high-throughput, hyphenated approach, HPLC-NMR, which is proving advantageous in the search for, and characterisation of, new natural products.

### 7.1.2 1-D NMR Experiments

For natural product structure elucidations, $^{1}H$ and $^{13}C$ are the most important nuclei, and several 1-D pulse sequences (some of which are discussed in the following) used for their observation are of fundamental importance. Together, they provide a detailed account of the number of protons and their associated coupling to other protons, the chemical environment, and an estimate of the number of carbons present. Experiments such as the 1-D single-irradiation (or steady-state; Chapter 3) nuclear Overhauser enhancement (1-D NOE) and the distortionless enhancement by polarisation transfer (DEPT) NMR approach play a pivotal role in natural product characterisation.

#### 7.1.2.1 1-D NOE and 1-D TOCSY Experiments

The steady-state NOE experiment involves the selected irradiation (or saturation, Chapter 3) of a proton, which then relaxes via other protons that are close in space.[11] If the outcome of this experiment is subtracted from a control (off-resonance) experiment, the resultant difference spectrum will display a peak for the originally saturated resonance, which (for relatively low–molecular-weight species) will have a phase opposite to that for resonances of protons with which it shares a close spatial proximity. Typically, proton nuclei of up to 5 to 6 Å away will be detected via this through-space NMR experiment.[12] The strength of an observed NOE is greatly dependent on the combination of the molecular weight of the molecule and the saturation time.[13] If the combination of these parameters is not optimised, no NOE will be observed (Chapter 3). The 1-D NOE is favoured over the 2-D nuclear overhauser effect spectroscopy (NOESY) (transient NOE equivalent) for low– and medium–molecular-weight compounds primarily due to differing time demands.

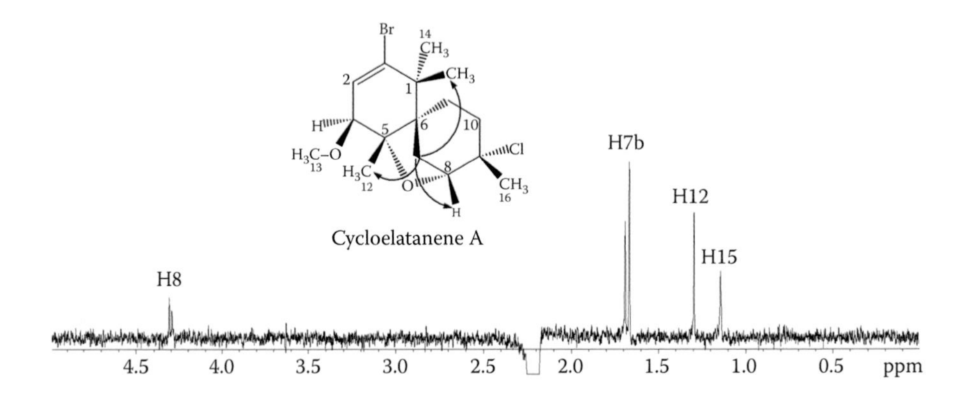


**FIGURE 7.3** 1-D NOE difference NMR spectrum of cycloelatanene A, illustrating the irradiation of the proton resonance at $\delta_H$ 2.21 (H7a).

A 1-D NOE NMR experiment was utilised to determine the relative configuration of the natural product cycloelatanene A, which was isolated from the Australian marine red alga *Laurencia elata*.[14] The $^1H$ and $^{13}C$ NMR experiments confirmed this compound to be a $C_{16}$ chamigrene, and the relative configuration was deduced through a series of 1-D NOE NMR experiments. Figure 7.3 illustrates the NOE difference spectrum for cycloelatanene A where the proton resonance at $\delta_H$ 2.21 (H7a) was irradiated. This irradiation resulted in four NOE enhancements being observed to the proton resonances at positions 7b, 8, 12 and 15 which concluded that these protons were all situated on the same face of the molecule.

The 1-D NOE NMR experiment only provides information on the protons connected through space. In contrast, the 1-D total correlation spectroscopy (1-D TOCSY) NMR experiment uses variable spin-locking times to allow for the determination of proton couplings along a compound chain.[15] As the mixing time is changed, the polarisation energy is transferred through the bonds to other protons. Signals that appear with a shorter spin-locking time are closer (through-bonds) to the excited proton, and correspondingly, those appearing when longer spin-lock times are employed are further away from the selected proton. This NMR experiment is particularly useful for defining individual spin systems within a compound.

The natural product furospinosulin-1 was first isolated from the marine sponge *Ircinia spinosula*[16] and recently re-isolated by the marine and terrestrial natural product (MATNAP) research group from *Dactylospongia* sp.[17] Figure 7.4 shows the 1-D TOCSY spectra obtained for furospinosulin-1 using spin-lock times of 30, 70, 150 and 200 ms. The proton at $\delta_H$ 2.45 (H1) was selected, and by comparing the peak heights in each spectrum, it was possible to determine which protons were closer or further away in the covalently linked spin system.

This 1-D TOCSY NMR experiment indicated that the H2 and H3 protons were closest to the H1 proton and that the H18 and H3′ protons were further away in terms of bonding from the H1 proton. The 2-D variant of the experiment provides a wealth of information, is not as easily interpreted as the 1-D experiment.

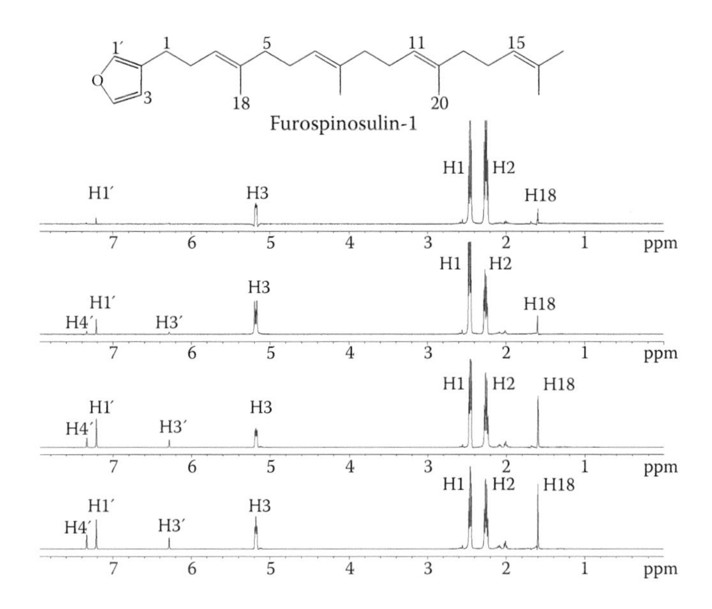


**FIGURE 7.4** 1-D TOCSY NMR spectra (spin-lock times [from top] of 30, 70, 150 and 200 ms) of furospinosulin-1.

#### 7.1.2.2 DEPT and APT

The DEPT NMR experiment is a direct-detect heteronuclear experiment (Chapter 5), in this case, $^{13}C$ detect, which provides information of the number of protons attached to the carbon nucleus. There are several versions of this experiment, which lead to slightly different data, but all of these reflect the attached-proton status of the $^{13}C$ atom. The DEPT45 NMR experiment shows all CH (methine), $CH_2$ (methylene), and $CH_3$ (methyl) carbons having the same (positive) phase.[11] Comparing the DEPT45 to the normal-broadband, proton-decoupled $^{13}C$ NMR spectrum, thus, enables quaternary carbons to be identified. The DEPT90 experiment is more selective and only displays CH (methine) carbons.[11] The DEPT135 shows all CH (methine), $CH_2$ (methylene), and $CH_3$ (methyl) carbons but with phase modulation. The CH (methine) and $CH_3$ (methyl) carbons appear with the same (positive) phase, and $CH_2$ (methylene) carbons appear with a 180° (negative) phase shift.[11] Combining a DEPT90 and DEPT135 NMR allows for all carbon-substitution types to be assigned. Another experiment similar to DEPT is the attached proton test (APT) experiment. The APT NMR experiment results in quaternary (C) and methylene ($CH_2$) carbons being positive and methyl ($CH_3$) and methine (CH) carbons being negative in their phasing.[18] While these experiments still have their uses, they have been made almost redundant with the advent of the indirect-detection heteronuclear experiments (see in the following section).

Figure 7.5 shows the DEPT $^{13}C$ NMR spectrum as well as the APT NMR spectrum of the natural product furospinosulin-1. Comparison of the three different DEPT NMR spectra allows for each of the carbon multiplicities to be determined. Analysis of the APT will also allow for multiplicities to be determined; however, unlike in the DEPT NMR experiment, the quaternary carbons are observed.

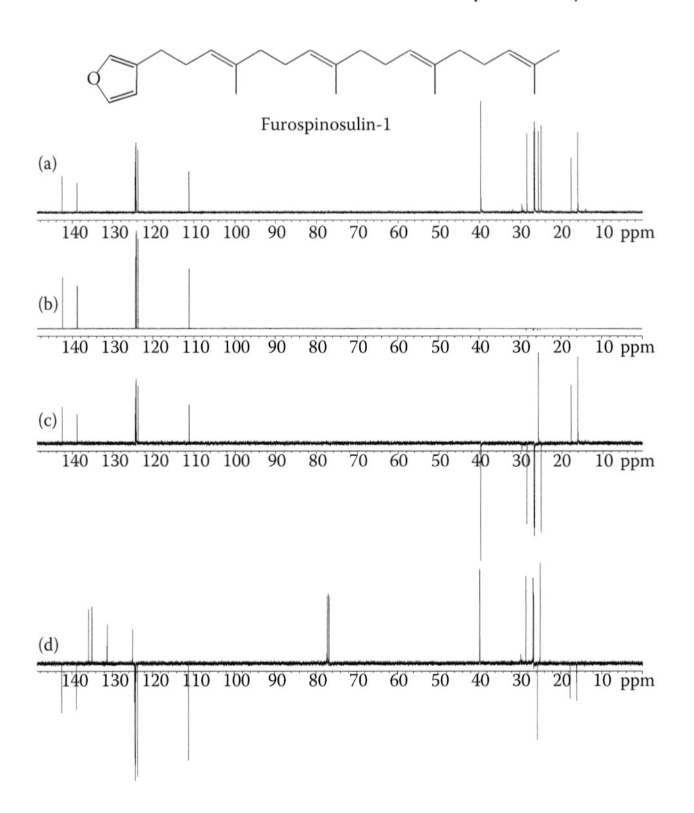


**FIGURE 7.5** DEPT 45 (a), DEPT 90 (b), DEPT 135 (c) and APT (d) NMR spectra of furospinosulin-1, displaying selective carbon multiplicities.

### 7.1.3 2-D NMR Experiments

In this section, we illustrate the application of 2-D NMR techniques to natural product characterisation. These are divided into two types. The homonuclear 2-D NMR experiments typically utilised include the correlation spectroscopy (COSY), NOESY, rotating frame Overhauser effect spectroscopy (ROESY), and TOCSY experiments. The heteronuclear 2-D NMR experiments (proton and carbon nuclei here) frequently implemented include the heteronuclear single quantum coherence (HSQC) and heteronuclear multiple quantum correlation (HMBC) NMR experiments. Most 2-D experiments come with non-gradient and gradient options (usually indicated by a 'g' in the name for the pulse sequence). The use of gradients typically increases the signal-to-noise as well as the quality of the spectra, and for this reason, it is very uncommon for gradient (Chapter 4) versions of the 2-D NMR experiments not to be used.

#### 7.1.3.1 Homonuclear Approaches

##### *7.1.3.1.1 Correlation Spectroscopy*

The COSY NMR experiment provides information on which protons are immediate (bonding) neighbours, allowing for the spin systems of the natural product to be determined.[13] While coupling constants and splitting patterns observed in the 1-D $^1$H NMR spectrum can be used to determine the number and geometric relationship of neighbouring protons, in complex systems, sometimes, this cannot be achieved. In these situations, 2-D COSY and related experiments are useful. The double-quantum–filtered COSY (DQFCOSY) NMR experiment, like the COSY NMR experiment, provides cross-peaks, indicating proton–proton coupling. However, unlike the COSY NMR experiment, the DQFCOSY is able to eliminate any peaks that do not have a double-quantum transition (i.e. singlets); this phase-sensitive spectrum has higher resolution and clearer-looking spectra, but the longer pulse sequence and quantum filter employed results in a loss of sensitivity.[19]

To illustrate the ability of the DQFCOSY to simplify and improve resolution, the natural product furospinosulin-1 was subjected to a gCOSY and a gDQFCOSY experiment. Figure 7.6 shows an expansion of the gCOSY and gDQFCOSY NMR experiments. It can be seen that the gCOSY NMR spectrum has regions where clear contours cannot be easily distinguished as well as the presence of long-range couplings. The gDQFCOSY NMR spectrum shows better resolution and a clearer-looking spectrum. In this NMR experiment, the long-range gCOSY correlations are not present.

##### *7.1.3.1.2 NOESY and ROESY*

The 2-D NOESY experiment provides the same type of information as the 1-D NOE spectrum (Chapter 3) except that all through-space connectivities within a molecule can be explored in a single experiment. Once again, proton–proton through-space relationships up to a distance of approximately 5 to 6 Å will be detected.

The compound 2,5,6-trimethoxy-9-phenyl-$^1$H-phenalen-1-one has been previously reported as being synthetically derived from a natural product.[20,21] We have recently isolated this compound for the first time, arising as a naturally occurring pigment from the Australian plant *Haemodorum* sp.[22] In the structure elucidation of this natural product, the point of attachment of the aromatic ring to either position 7 or 9 needed to be confirmed. The NOESY spectrum shown in Figure 7.7 clearly revealed a cross-peak between the aromatic proton at position 7 ($\delta_H$ 8.50) and the methoxy protons at position 6 ($\delta_H$ 4.10), demonstrating that the aromatic moiety must be attached at position 9. In this way, the NOESY spectrum could unequivocally confirm the structure of the natural pigment.

Just as the single-irradiation 1-D NOE NMR experiment requires the combination of the molecular weight and mixing time to be optimised to observe an NOE, the same criteria applies to the 2-D NOESY experiment. For intermediate-sized compounds or those that do not show strong NOESY correlations, the ROESY experiment is appropriate.[15,23] While the ROESY NMR experiment may provide more information for intermediate-sized molecules compared to NOESY, the ROESY experiment is susceptible to also showing COSY- and TOCSY-type correlations, which adds complexity to the spectrum.[11]

#### 7.1.3.2 Heteronuclear Approaches

##### *7.1.3.2.1 HSQC and HMBC*

Earlier heteronuclear 2-D NMR experiments (pre-1990s) utilised the X nucleus, in this case, the carbon, as the observation frequency. These experiments were known

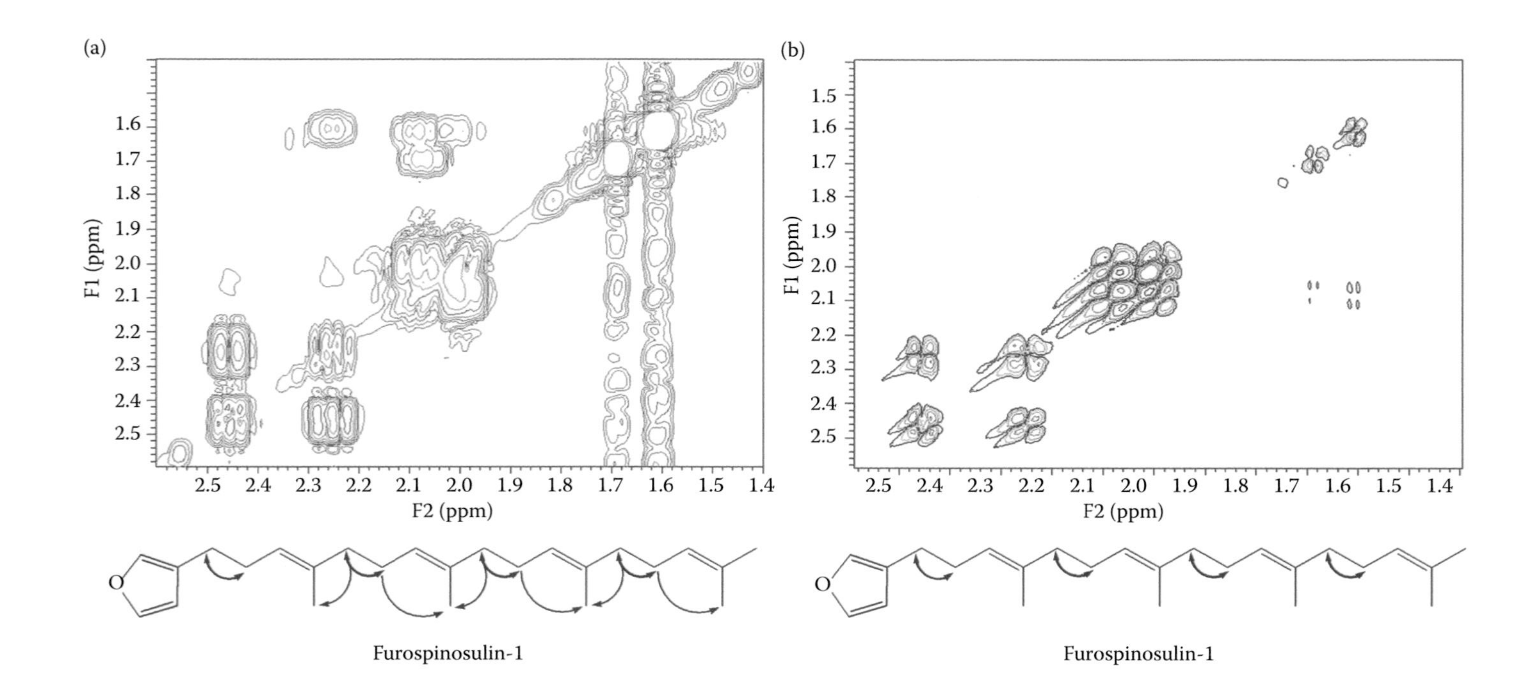


**FIGURE 7.6** Expansion of gCOSY (a) and gDQFCOSY (b) NMR spectra, showing a selection of the observed $^{1}H$–$^{1}H$ scalar ($J$) coupling connectivities in furospinosulin-1.

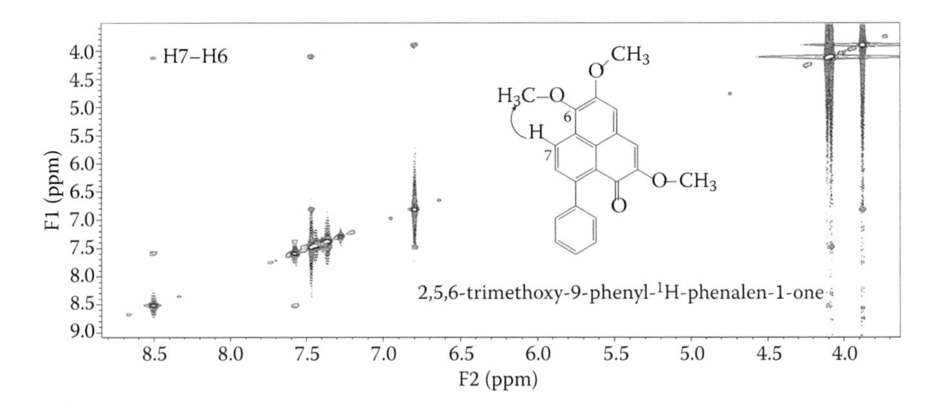


**FIGURE 7.7** 2-D NOESY NMR spectrum of 2,5,6-trimethoxy-9-phenyl-$^1$H-phenalen-1-one.

as the 'heteronuclear shift correlation' (HETCOR) and 'correlation through long-range coupling' (COLOC) approaches. The sensitivity of these experiments was dictated by that of the X nucleus and, thus, were not high. Subsequently, experiments have been implemented that employ the proton as the observation nucleus and carbon as the indirectly detected nucleus, therefore substantially increasing the sensitivity of these experiments. Thus, the indirect-detect equivalent of the HETCOR is the heteronuclear multiple quantum correlation (HMQC) NMR experiment, although the single-quantum equivalent (HSQC) is more frequently employed and the indirect equivalent of COLOC is the HMBC. These indirect-detection experiments are much more time efficient, and this is an imperative in the field of natural products, where frequently low to sub-milligram quantities are typically isolated.

In the past, assigning each proton to a specific carbon was almost exclusively done on the basis of the chemical shifts, and as such, sometimes, assignments were interchangeable or incorrectly applied. The HSQC NMR experiment has removed this ambiguity and determines which protons are directly attached to which carbon.[18] The most basic HSQC NMR experiment will only correlate protons and carbons that are bonded to one another, and therefore, quaternary carbons do not appear in a HSQC NMR experiment. There are variants of the HSQC NMR experiment that offer spectra of improved quality with increased information. For example, the use of gradients along with adiabatic pulse versions during experiments offers increased sensitivity.[24–26] There is also a version of the HSQC that provides carbon multiplicity information. In this experiment, methylene ($CH_2$) multiplicities are displayed in the negative phase, and methine/methyl ($CH/CH_3$) multiplicities are displayed in the positive phase.[11]

While a COSY NMR experiment allows for the spin systems to be determined, the HMBC NMR experiment is the most important of the NMR experiments for natural product structure determination as it allows for the carbon skeleton of a natural product to be pieced together. This is achieved by formulating structure fragments on the basis of all the NMR (both 1-D and 2-D) information gathered and then forming connections of these structure fragments via through-bond connectivities observed in the HMBC NMR experiment. While the HSQC NMR experiment displays direct one-bond proton to carbon connectivities, the HMBC NMR experiment

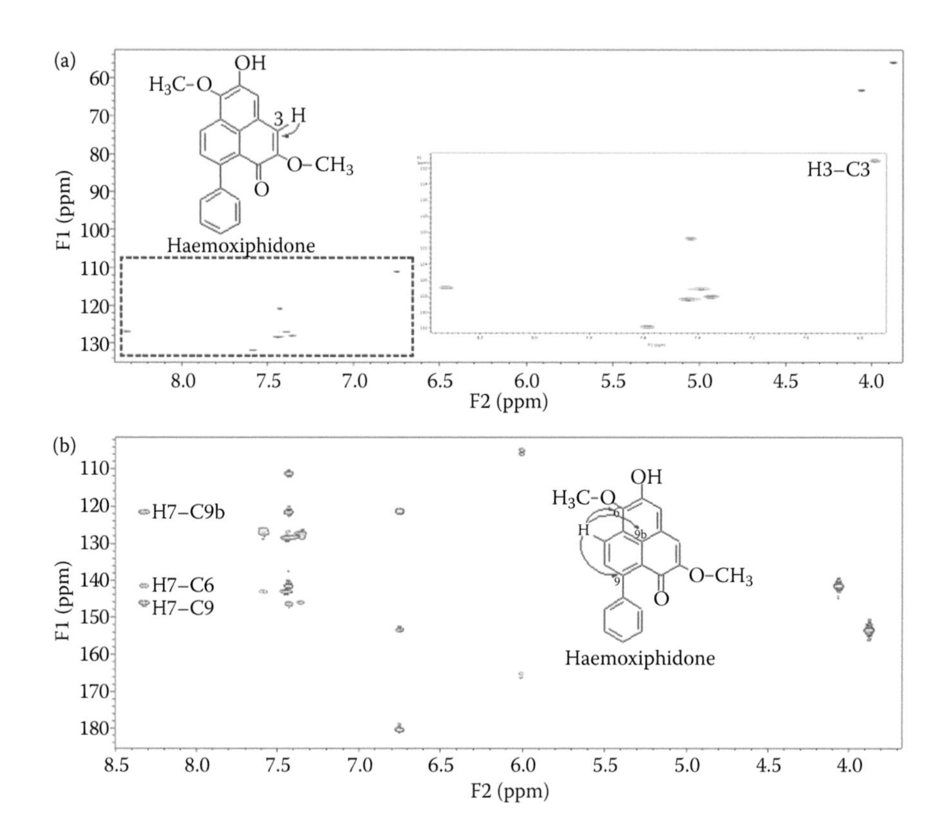


**FIGURE 7.8** Sections of the 2-D gHSQCAD (a) and gHMBCAD (b) NMR spectra of haemoxiphidone.

provides proton to carbon connectivity usually up to two to three bonds removed[27] (in aromatic systems, four-bond correlations are also encountered). Unlike the HSQC NMR experiment, quaternary carbons are observed in the HMBC NMR experiment, thereby providing vital structure connectivity information.

The gradient heteronuclear single quantum coherence adiabatic (gHSQCAD) and gradient heteronuclear multiple-bond correlation adiabatic (gHMBCAD) NMR spectra of haemoxiphidone, a pigment isolated from the Australian plant *Haemodorum simulans*, are illustrated in Figure 7.8.[28] All the correlations observed in the gHSQCAD NMR spectrum were in the positive phase, indicating that the multiplicities of the carbons were either CH (methines) or $CH_3$ (methyls).

In the structure elucidation of this pigment, it was uncertain whether the aromatic moiety was attached at position 7 or 9 of the compound. Analysis of the HMBC NMR spectrum supported a position-9 attachment, which was also confirmed by 1-D NOE NMR experiments.[28]

### *7.1.3.2.2 Coupled HSQC*

In many natural products, it is common for proton signals to overlap, particularly, signals for those protons having complex splitting patterns. Some compounds require

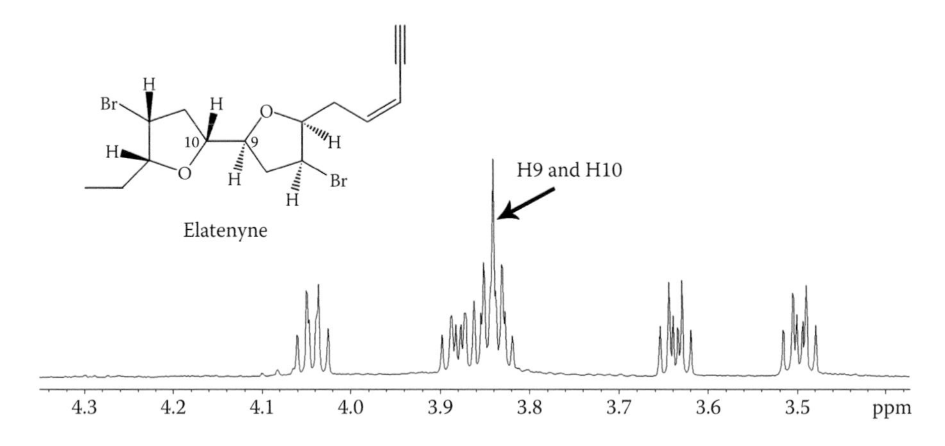


**FIGURE 7.9** Expansion of the $^1$H NMR spectrum (500 MHz, $C_6D_6$) of elatenyne.

the splitting patterns of protons to be clearly observed so that the structures may be deduced. The splitting patterns of overlapping protons may be observed by editing the standard HSQC experiment. The limiting factor is that the overlapping protons must be coupled to different carbons that do not overlap. In this variant of the HSQC NMR experiment, two parameters are changed. First, the number of data points collected in F2 (the acquisition dimension) is increased for greater resolution in the proton dimension. The second parameter that is changed is that the $^{13}$C decoupling is turned off. This means that each proton signal will now be split by its H–C coupling constant. By extracting the 1-D traces from the 2-D HSQC, the splitting patterns of the overlapping protons may now be observed.[29,30]

This particular problem was encountered in the structure of elatenyne, which was isolated from the Australian red alga *Laurencia elata*.[14,31] The two protons at the ring junction (positions 9 and 10) were overlapped. It was vital to obtain the splitting pattern and coupling constant of these overlapped ring junction protons to propose a relative configuration for elatenyne. Figure 7.9 shows the region of the $^1$H NMR spectrum for the overlapped protons at positions 9 and 10 ($\delta_H$ 3.86). Here, no splitting/coupling information about the overlapped protons could be deduced.

Figure 7.10 shows the edited gHSQCAD spectrum and extracted 1-D traces for the overlapped protons; the splitting pattern and coupling for the overlapped protons could be clearly observed. Each proton is split by their H–C coupling, and it is not uncommon for one mirrored splitting pattern to be clearer than the other. The splitting patterns were able to be deduced as a doublet of triplets and the coupling constant was measured, confirming that the two ring junction protons were in an axial–equatorial arrangement.[31]

#### *7.1.3.2.3 CIGAR-HMBC*

In rare instances, the HMBC NMR experiment will not be suited to particular compounds. The HMBC experiment uses fixed coupling constants to detect carbons correlating up to two to three bonds away. If expected correlations are not observed, the constant time inverse-detection gradient accordion rescaled (CIGAR)-HMBC may provide a better solution. The CIGAR-HMBC experiment is less sensitive than the

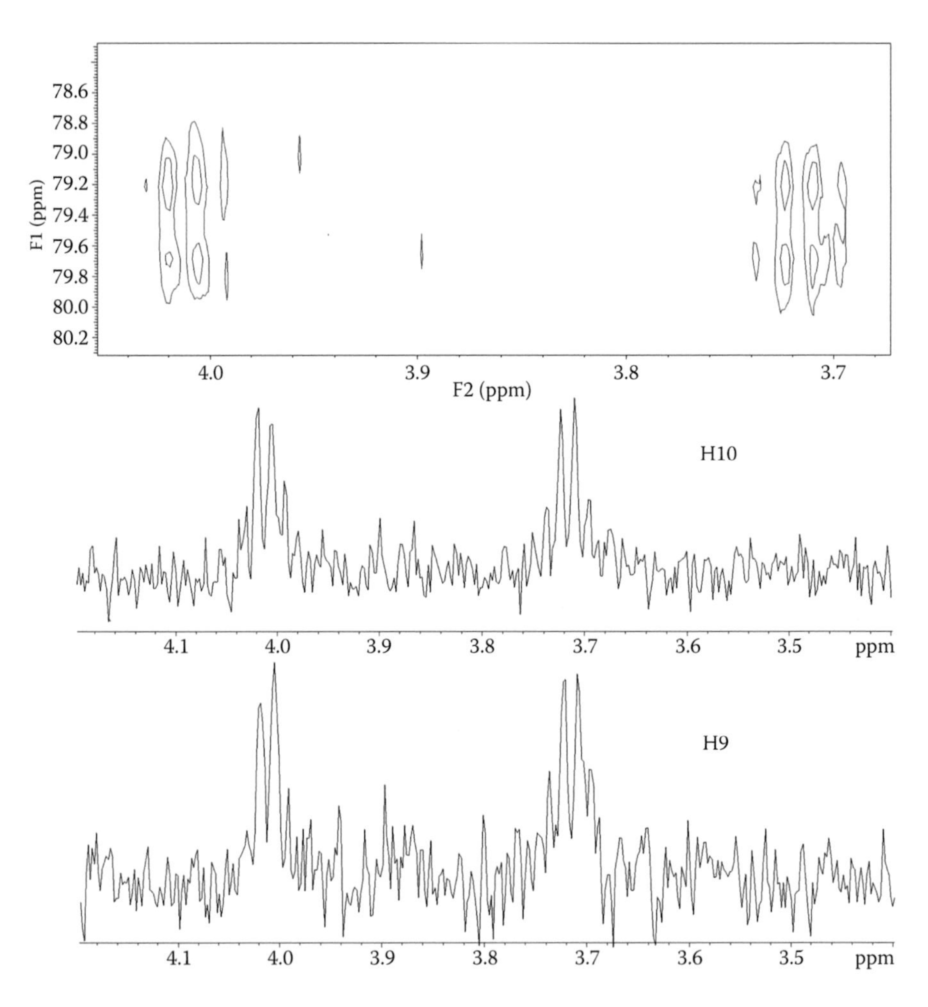


**FIGURE 7.10** Expansion of coupled HSQC NMR spectrum and extracted 1-D traces of H9 and H10 (500 MHz, $C_6D_6$) of elatenyne.

HMBC experiment; however, the CIGAR-HMBC uses variable coupling constants.[32] This experiment, therefore, provides a better opportunity to observe correlations further than three bonds away to detect situations where small coupling constants exist or instances where there might be a large range of coupling constants, as in the case of nitrogen-containing compounds.

#### *7.1.3.2.4 HSQC-NOESY*

In the previously discussed 2-D NOESY NMR experiment, one disadvantage is that, if two (or more) protons have the same chemical shift, it is difficult to distinguish which proton is, in fact, being detected. The band-selective HSQC-NOESY NMR experiment utilises both the NOESY and HSQC pulse sequences and can, therefore, be used to overcome the issues of overlapping proton chemical shifts (Chapter 3).

For example, if a proton at $\delta_H$ 3.50 shows an NOE to a proton at $\delta_H$ 6.70 with an associated carbon chemical shift at $\delta_C$ 126.7 ppm, a cross-peak between $\delta_H$ 3.50 and $\delta_C$ 126.7 ppm will be observed. If another proton also co-occurs at $\delta_H$ 6.70 but with an associated carbon chemical shift at $\delta_C$ 130.5 ppm, this distinction would be evident in a band-selective HSQC-NOESY experiment.

#### *7.1.3.2.5 Excitation-Sculptured Indirect-Detection Experiment*

The HMBC experiment provides information about protons coupled to carbons two to three bonds away. However, this experiment does not provide any information about the magnitude of these coupling constants. The excitation-sculptured indirect-detection experiment (EXIDE) is similar to the HMBC in that it shows proton-to-carbon coupling two to three bonds away but, unlike the HMBC, also displays the coupling constants.[33] This experiment displays a 2-D NMR spectrum similar to the HMBC, but each correlation appears as a doublet in the F1 dimension. By measuring the distance between each doublet and dividing by the *J* scaling factor, the coupling constants can be determined.

To illustrate the ability of the EXSIDE experiment to determine $^nJ$ C–H coupling constants, Figure 7.11 compares the results obtained in both the gHMBCAD and EXSIDE experiments for furospinosulin-1. In the gHMBCAD experiment, three distinct correlations can be observed for the proton at $\delta_H$ 7.21; however, no information about the coupling to the carbons can be deduced. The EXSIDE version still shows not only these three distinct correlations, but also the splitting of each carbon signal.

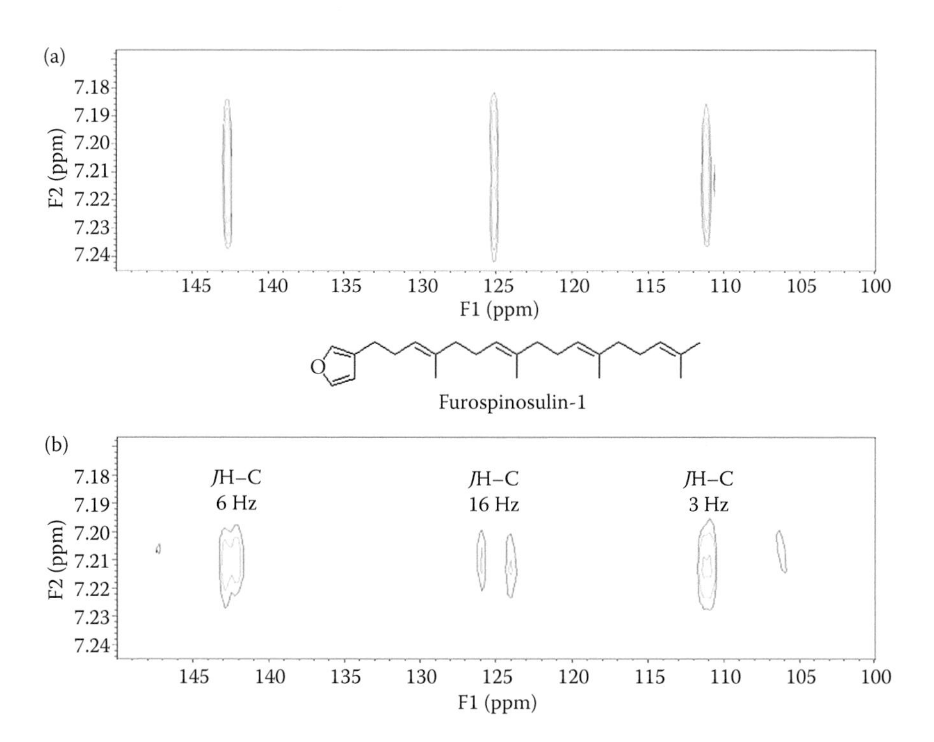


**FIGURE 7.11** Expansions of the gHMBCAD and EXSIDE NMR spectra of furospinosulin-1.

By measuring the distance between each split carbon and dividing by the *J* factor, the coupling constants of these carbons were determined to be 6, 16 and 3 Hz.

### 7.1.4 Hyphenation: HPLC-NMR

One of the more recent advancements and application of NMR spectroscopy has been the hyphenation to other separation and detection methodologies traditionally used separately in natural product isolation and characterisation. An example of such is the coupling of NMR to an HPLC system to yield the hyphenated spectroscopic technique known as 'HPLC-NMR'. This technique was developed in the 1970s;[10] however, it has only recently taken over as a robust and powerful tool for chemical profiling, dereplication, and structure elucidation. Principally, the drawback for the hyphenated spectroscopic technique included the initial high costs and the lack of sensitivity. Developments in probe design, solvent suppression, loop collection, cold probes, column trapping, capillary NMR, and microcoils have allowed it to become the powerful technique that it is today and is now finding extensive use in the field of natural products.[34]

HPLC-NMR operates in two main modes: on-flow and stop-flow.[34] On-flow mode utilises continuous $^1$H NMR data acquisition for all components as they elute from the HPLC column and, for this reason, is a far less sensitive mode of operation. The on-flow mode is, therefore, useful in providing quick general structure class information that can be used to search various databases.

The more useful, stop-flow mode (Figure 7.12), allows for any one particular component to be trapped in the NMR flow cell at a time. By trapping one component, extended acquisitions are now possible, allowing for 2-D NMR experiments to be carried out.[10,34] This increase in sensitivity is particularly important in natural

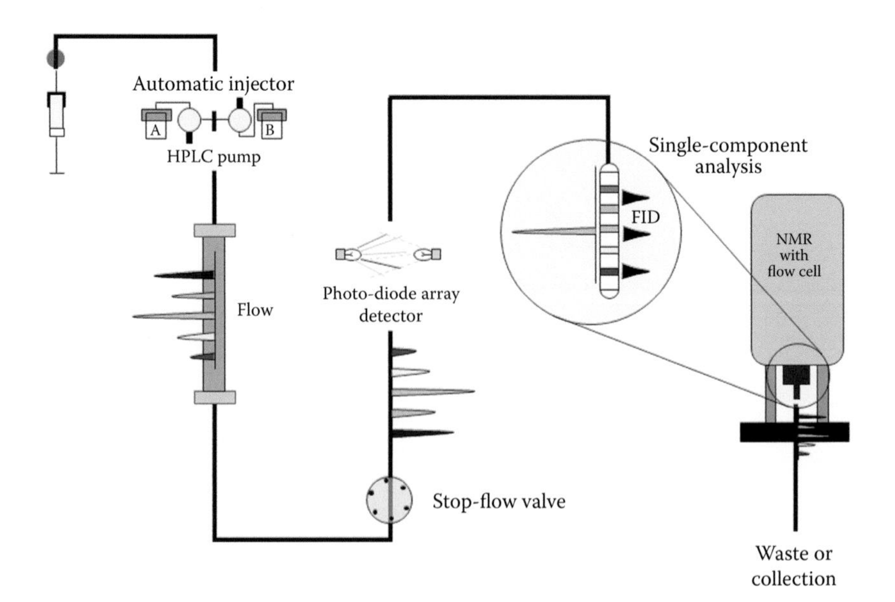


**FIGURE 7.12** Schematic of stop-flow operation of HPLC-NMR.

product identification. In general, complete 2-D NMR data can be obtained for major components in crude extracts/enriched fractions, whereas usually, only $^1$H NMR and COSY NMR data can be obtained for the minor components in the typical overnight time frame used to conduct conventional NMR experiments.

Other variants of HPLC-NMR utilise solid-phase extraction (SPE) columns to pre-concentrate analytes before NMR analysis to increase the signal-to-noise and experiment capability.[33] Loop collection allows for analytes to be separately collected and then either flushed into the flow cell one at a time or transferred into conventional NMR tubes for off-line analysis. Cold probes (or cryo probes) and microcoils have also allowed for increased sensitivity and, therefore, the NMR detection of smaller quantities of analytes.[34]

HPLC-NMR can be further extended to include MS to yield HPLC-NMR-MS. This technique has two possible set-ups, namely 'parallel' and 'series'.[34] In a parallel set-up, a splitter is placed after the column and a small amount of the analyte is directed to the mass spectrometer, with the majority directed to the NMR spectrometer. In the series configuration, the mass spectrometer is placed after the NMR spectrometer. The latter set-up can sometimes be at a disadvantage since peaks can broaden and retention times can change by the time that the analytes reach the mass spectrometer.

#### 7.1.4.1 Application to Organism Profiling and Structure Elucidation

We have conducted various studies utilising HPLC-NMR to chemically profile both marine and terrestrial natural product extracts and enriched fractions.[14,35–41] Only recently, HPLC-NMR has been utilised to chemically profile the Australian brown alga *Sargassum paradoxum*.[42] The crude dichloromethane extract of the alga was analysed, and a number of constituents were detected. The major constituent was identified using the stop-flow HPLC-NMR mode. This major constituent was present in sufficient abundance to allow for complete 2-D characterisation including gCOSY, gHSQCAD, and gHMBCAD analysis. The results obtained from the stow-flow HPLC-NMR analysis allowed the major constituent to be identified as isosargahydroquinoic acid.[41] The double bond geometry of this compound was secured by application of the 1-D NOE NMR experiment. The Wet1D (used to solvent suppress; Chapters 3 and 6), gCOSY, gHSQCAD, and gHMBCAD NMR spectra are provided in Figures 7.13 to 7.16, respectively. This represents the first instance of a HMBC-NMR spectrum being obtained in the stop-flow HPLC-NMR mode on a system equipped with a 60 µL flow cell. The ability to obtain a HMBC NMR spectrum relies on the amount of analyte present, and the limitation is the ability to dissolve the fraction in a minimal amount of solvent. The mass limitation in HPLC-NMR analyses often remains a restriction for the technique since a reasonable signal-to-noise ratio for the compound must be achieved. However, as demonstrated by this example, it is possible to obtain HMBC data for main constituents in crude extracts without the need for pre-concentration, SPE cartridges, loop storage, or cold probes.

In the gHSQCAD NMR spectrum (Figure 7.15) of isosargahydroquinoic acid obtained by stop-flow HPLC-NMR, the proton resonances between $\delta_H$ 3.00 and $\delta_H$ 4.50 ppm were attached to carbons that appeared in the negative phase (indicating methylene [$CH_2$] multiplicities) with all other carbon resonances appearing in the positive phase (indicating methine [CH] or methyl [$CH_3$] multiplicities).

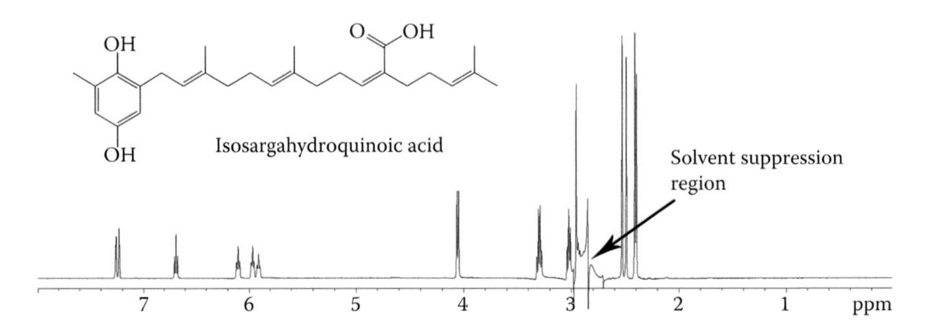


**FIGURE 7.13** Wet 1-D HPLC-NMR spectrum of isosargahydroquinoic acid (500 MHz, 75% $CH_3CN/D_2O$).

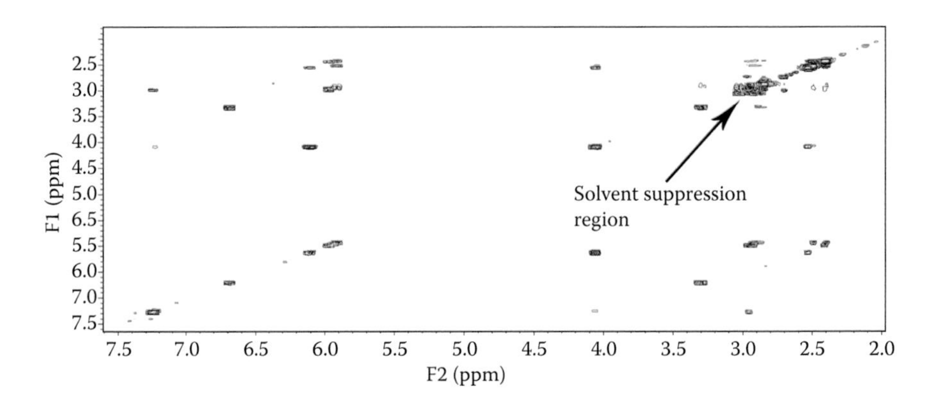


**FIGURE 7.14** gCOSY NMR spectrum of isosargahydroquinoic acid (500 MHz, 75% $CH_3CN/D_2O$).

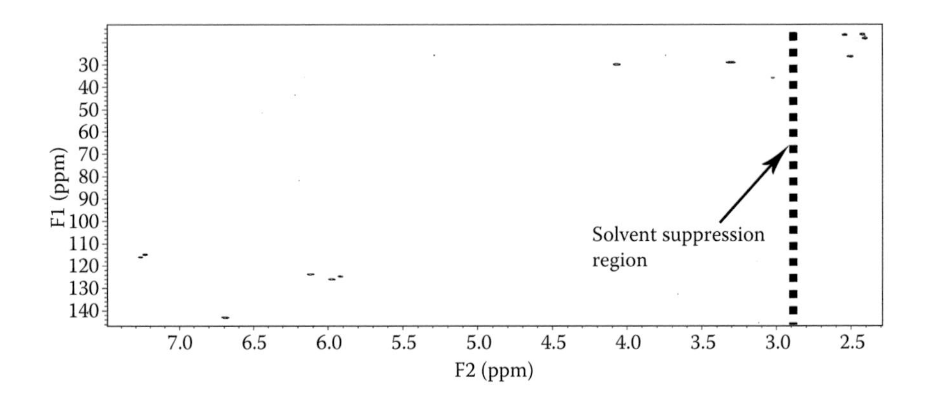


**FIGURE 7.15** gHSQCAD NMR spectrum of isosargahydroquinoic acid (500 MHz, 75% $CH_3CN/D_2O$).

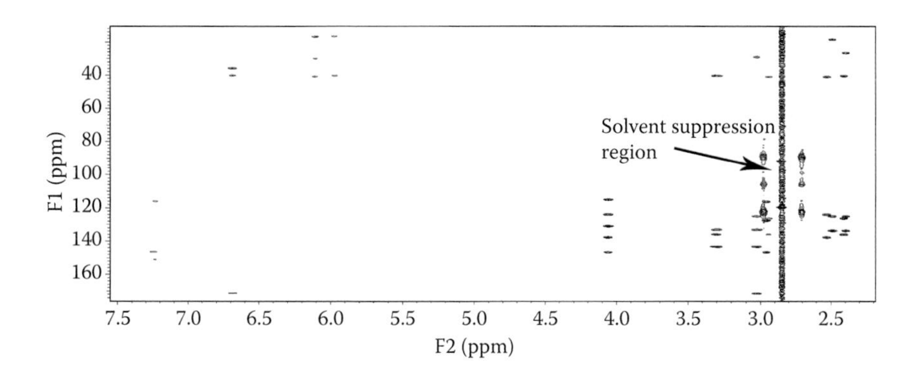


**FIGURE 7.16** gHMBCAD NMR spectrum of isosargahydroquinoic acid (500 MHz, 75% $CH_3CN/D_2O$).

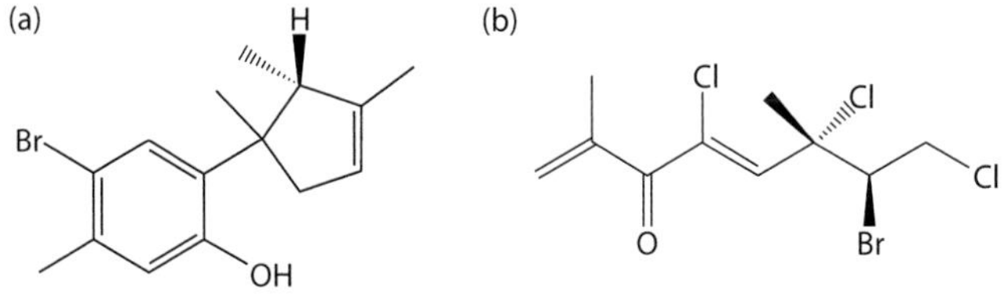

**FIGURE 7.17** Structures of two new natural products (a, Isoallolaurenterol [*Laurencia filiformis*] and b, Isoplocamenone [*Plocamium angustum*]), both of which are unstable and, hence, could only be elucidated *in situ*.

The importance of the application of HPLC-NMR was illustrated through two separate studies conducted by our research group. These included the investigation of two southern Australian red algae, namely *Laurencia filiformis*[36] and *Plocamium angustum*.[35] In both cases, new unstable natural products (Figure 7.17) were able to be identified by application of HPLC-NMR. Any attempted off-line isolation and characterisation of these two natural products could not be achieved, owing to their rapid degradation. HPLC-NMR analyses of the crude extracts of these two algae successfully established the structures *in situ*.

The unstable nature of natural products is a frequently encountered issue that is brought about by exposure to light, heat, or air oxidation, among other scenarios. The use of HPLC-NMR in such instances is invaluable in securing their structure.

### 7.1.5 Summary

The use of NMR in natural product structure elucidation has continued to evolve. From its humble beginnings in 1-D proton and carbon spectra to the advanced methods such as 2-D experiments and hyphenated spectroscopic techniques, NMR has played a large and important role in the natural product field. Continuing efforts and progress in NMR spectroscopy with regard to new and improved pulse sequences, faster and more sensitive hardware, easier access to higher field strengths, and smaller sample volumes will play an important role in natural product structure determination, allowing for the elucidation of ever-decreasing quantities. This is of vital importance in the field of natural product drug discovery where structure determination of sub-milligram quantities is now possible. The application of NMR spectroscopy in the hyphenated spectroscopic techniques such as HPLC-NMR will continue to evolve as more applications of the technique emerge, illustrating its strength and importance in chemical profiling/de-replication and in the rapid identification of natural products.

## 7.2 CARBOHYDRATES

Kristian Hollingsworth and W. Bruce Turnbull

### 7.2.1 Introduction

NMR spectroscopy is an important and useful tool in carbohydrate chemistry. It may be applied to analyse properties of monomers and small carbohydrate-based polymers with the approach to such studies falling into three basic categories: (1) identification of sugars and their connectivity in oligosaccharides, (2) analysis of their conformations in solution and (3) characterisation of their binding to proteins. Each of these facets will be considered in this chapter.

### 7.2.2 Sequence Determination by NMR

Sugars are polyhydroxy aldehydes and ketones that typically cyclise to form 5- or 6-membered rings that are known as 'furanoses' and 'pyranoses', respectively (Figure 7.18). Upon cyclisation, a new stereogenic center is formed, which is known as the 'anomeric center'; this process gives rise to α and β anomers for each sugar. While the hemiacetal forms of these sugars are in equilibrium with one another, formation of an acetal link (glycosidic link) to another monosaccharide unit gives rise to an oligosaccharide in which the stereochemistry at the anomeric center is now fixed. The structures of oligosaccharides are defined by three main criteria: the stereochemistry of the sugar moieties, the orientation of the anomeric group, and the positioning of the glycosidic linkages (Figure 7.18).[43]

In principle, most monosaccharide units could be identified from their chemical shifts and coupling constants; however, in practice, these highly complex structures contain many similar hydrogen and carbon environments, which often lead to poor chemical shift dispersion. Therefore, identification of sugar residues is often conducted in conjunction with chemical degradation and/or MS, especially when prior knowledge of biosynthetic pathways or the chemical synthesis route employed is not available.[44,45] Signal dispersion is often improved by derivatising the sugars with ester groups.[46] In contrast, derivatisation with other protecting groups commonly used in oligosaccharide synthesis, for example, benzyl ethers, may lead to even worse signal overlap as the benzylic protons also appear in the same region of the spectrum as the sugar ring protons. It is, thus, common to use TOCSY spectra to correlate all protons in a spin system (i.e. monosaccharide unit) with the anomeric signal, which will frequently come at a high-enough chemical shift (4.5–6 ppm) to

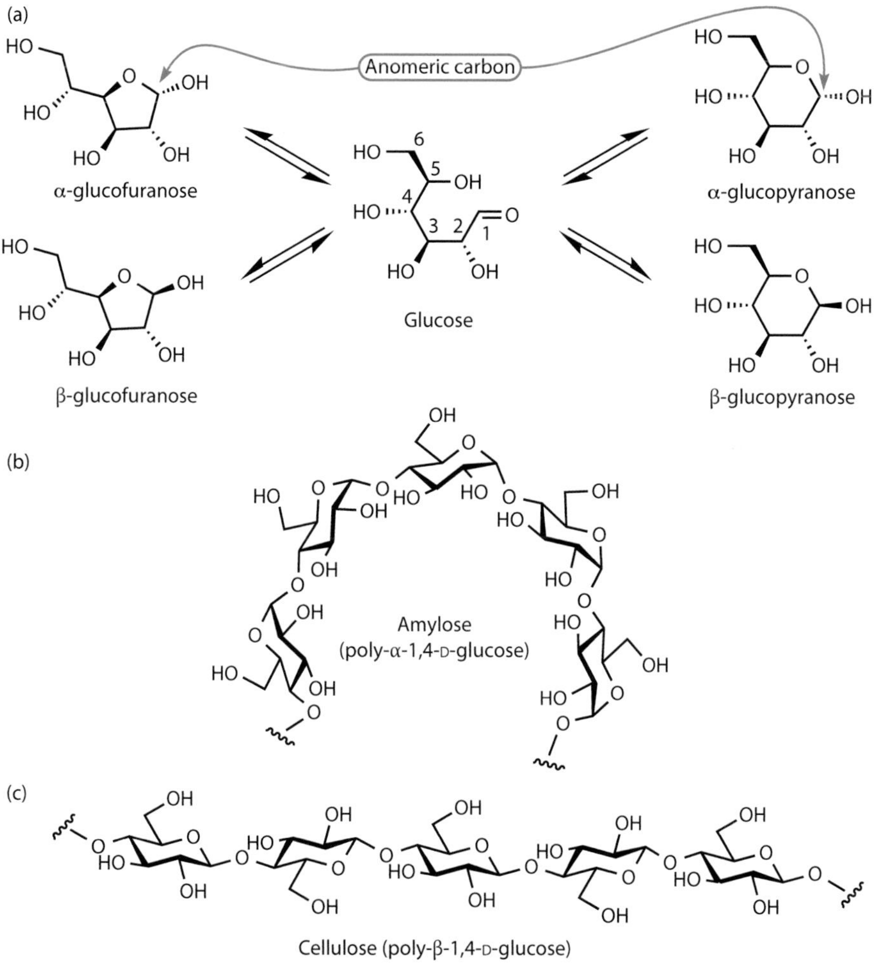

**FIGURE 7.18** (a) Equilibrium between open chain and ring forms of glucose. The atoms are numbered on the open-chain isomer; (b) glucopyranose units connected by α-1,4-linkages to form the helical polysaccharide amylose; (c) glucopyranose units connected by β-1,4-linkages to form the extended polysaccharide cellulose.

allow its separation from the rest of the sugar signals (Figure 7.19). This strategy can be more challenging for sugars in which scalar coupling constants are small, for example, in the more flexible five-membered ring furanoside sugars. In these cases, it can be helpful to use $^{13}C$ isotopic enrichment to enable the use of HCCH-TOCSY experiments, which allow the magnetisation to be shuttled through $^{1}H$–$^{13}C$ and $^{13}C$–$^{13}C$ couplings.[44]

The stereochemistry at the anomeric center can have a significant effect on the properties of an oligosaccharide; for example, the α-(1,4) configuration of glucose units in amylase gives rise to a soluble helical conformation (Figure 7.18b), whereas

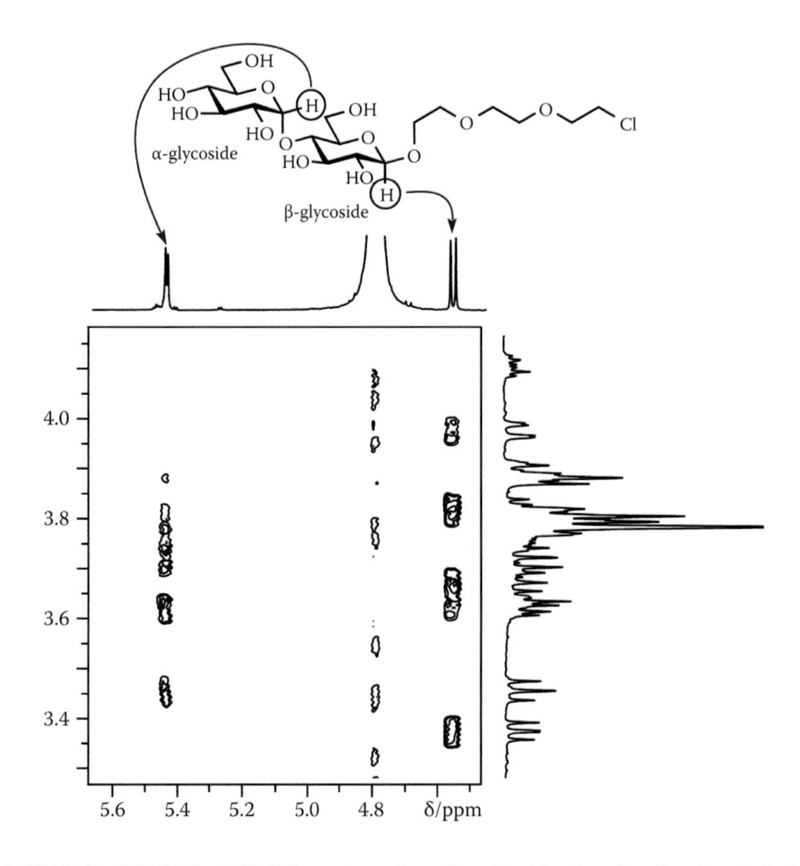


**FIGURE 7.19** 500-MHz TOCSY spectrum for a β-maltoside, showing the signals that correlate with the α- and β-anomeric protons.

the β-(1,4) linkages in cellulose gives an extended conformation (Figure 7.18c) that promotes interchain hydrogen bonding and leads to an insoluble material.[47] Therefore, the stereoselective formation of a specific anomer is important for synthetic chemists. NMR spectroscopy can allow easy determination of the anomeric configuration as 1,2-cis and 1,2-trans glycosides often have distinctive chemical shifts and coupling constants for their anomeric protons. For example, the anomeric signals for a number of α- and β-glucosides appear around 5.24 ± 0.20 ppm and 4.62 ± 0.15 ppm, respectively,[48] which conveniently fall on either side of the water signal at 300 K (Figure 7.19). However, it is more standard practice to make use of coupling constants to distinguish α- and β-configurations. For glucopyranosides, which have an equatorial oxygen substituent adjacent to the anomeric carbon, the β-anomer has a 1,2-diaxial arrangement of protons, which gives a relatively large $^3J$ coupling of 7 to 8 Hz. The corresponding α-anomer has a cis-arrangement of protons and a $^3J$ coupling of 3 to 4 Hz.[48]

The situation is less clear for other sugars such as mannopyranosides, which have an axial oxygen substituent at C-2 (Figure 7.20). Both α- and β-configurations have

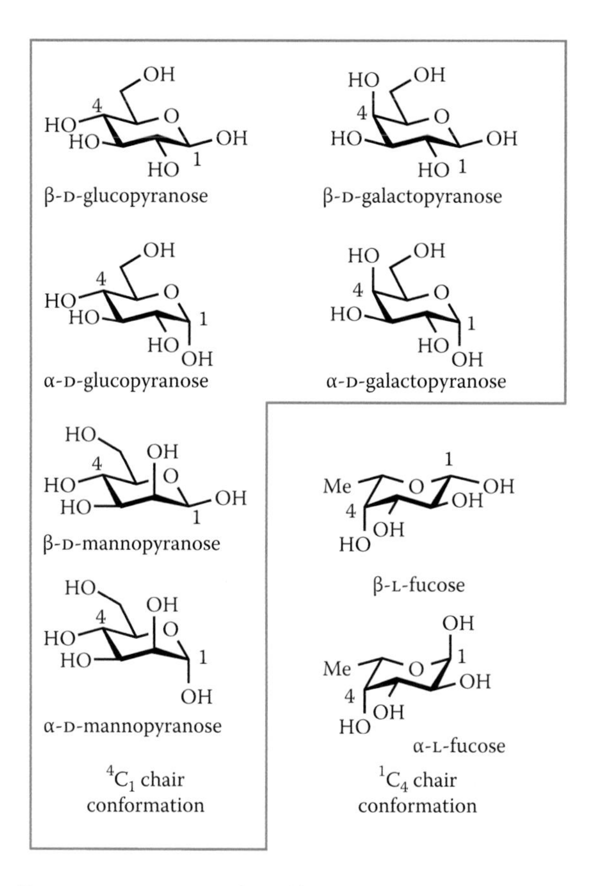


**FIGURE 7.20** Some common sugars found in mammalian oligosaccharides drawn in the most common chair conformations.

small $^3J$ coupling constants (3–4 Hz) as they are derived from trans-diequatorial or cis arrangements of protons. In such cases, the anomeric $^1H$–$^{13}C$ $^1J$ coupling constant provides a more robust diagnostic method as α- and β-glycosides have anomeric $^1J$ couplings of approximately 170 and 160 Hz, respectively.[49] Care should be taken when studying derivatives with electronegative substituents at the anomeric position as the $^1J$ couplings as both can increase; nevertheless, the 10-Hz difference between the α- and β-coupling constants is preserved.[7] This strategy is less effective for furanosides,[50] but in such cases, the anomeric $^{13}C$ chemical shift can allow 1,2-cis and 1,2-trans glycosides to be distinguished.[51]

Having identified the type of sugar residue and its anomeric configuration, the final challenge is to determine its linkage position to the neighbouring sugar in the oligosaccharide. Whereas DNA and proteins are linear polymers with well-defined linkage positions, oligosaccharides are often branched structures. HMBC experiments can allow correlation of anomeric carbon atoms with ring protons on the adjacent sugar units, the identity of which must be determined using TOCSY and COSY experiments.[44,45] The chemical shift data can also be used as an input for computer programmes such as computer-assisted spectrum evaluation of regular polysaccharides or Glynest,[52,53] which can be used to simulate NMR spectra and predict the positions of oligosaccharide linkages in an unknown sample.

## 7.2.3 Conformational Determination

### 7.2.3.1 Monosaccharide Conformation from Coupling Constants

Pyranose sugars (i.e. those with six-membered rings) usually adopt a chair conformation with ring protons in well-defined axial or equatorial positions. Therefore, with prior knowledge of the stereochemical configuration of a sugar residue, it is possible to determine its conformation by the use of the Karplus relationship for $^3J$ coupling constants. In general, trans-diaxial protons have coupling constants in the range of 7 to 9 Hz, while trans-diequatorial and cis protons have couplings of 2 to 4 Hz. For most of the common mammalian monosaccharides (glucose, galactose, mannose etc.), the ring adopts a $^4C_1$ conformation (Figure 7.20) in which carbon-4 is above the plane defined by atoms C2, C3, C5, and O5 (in which O5 is at the back of the plane) and carbon-1 is below that plane. However, the common L-configured sugar, L-fucose that is found in all blood group oligosaccharides usually adopts a $^1C_4$ conformation.

The orientation of the O6 substituent is often less clearly defined and can adopt one of the three possible conformations, which are described as gauche-gauche (*gg*), gauche-trans (*gt*), and trans-gauche (*tg*) in reference to the relationship of O6/O5 and O6/C4 (Figure 7.21). Traditionally, the ω-dihedral angle defined by O5 and O6, has been determined from the $^3J_{H5\text{–}H6proS}$ and $^3J_{H5\text{–}H6proR}$ coupling constants. However, greater accuracy can be achieved through the use of additional $^1J$ and $^2J$ C–C, C–H, and H–H coupling constants in conjunction with computer analysis programmes such as Chymesa.[54,55] If $^{13}$C couplings are to be employed, then the use of $^{13}$C-enriched carbohydrates can be advantageous to accommodate the low natural abundance of $^{13}$C atoms.[55] Often, the coupling constants can be measured directly from a 1-D NMR spectrum with sufficient accuracy, but in the case of small coupling constants, it is often necessary to simulate the spectra.[56]

The conformations of furanose sugars (i.e. those with five-membered rings) are more diverse as the five-membered rings undergo pseudorotation more easily.[57] Envelope (E) and twist (T) conformers are defined by the number of atoms in the

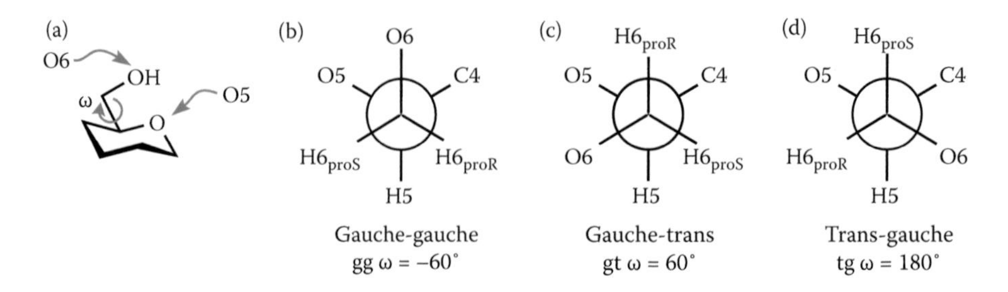


**FIGURE 7.21** Exocyclic torsion angle ω (a) can adopt three conformations (b–d). (Adapted from Stenutz, R. et al., *Journal of Organic Chemistry* 67, pp. 949–958, 2002.)

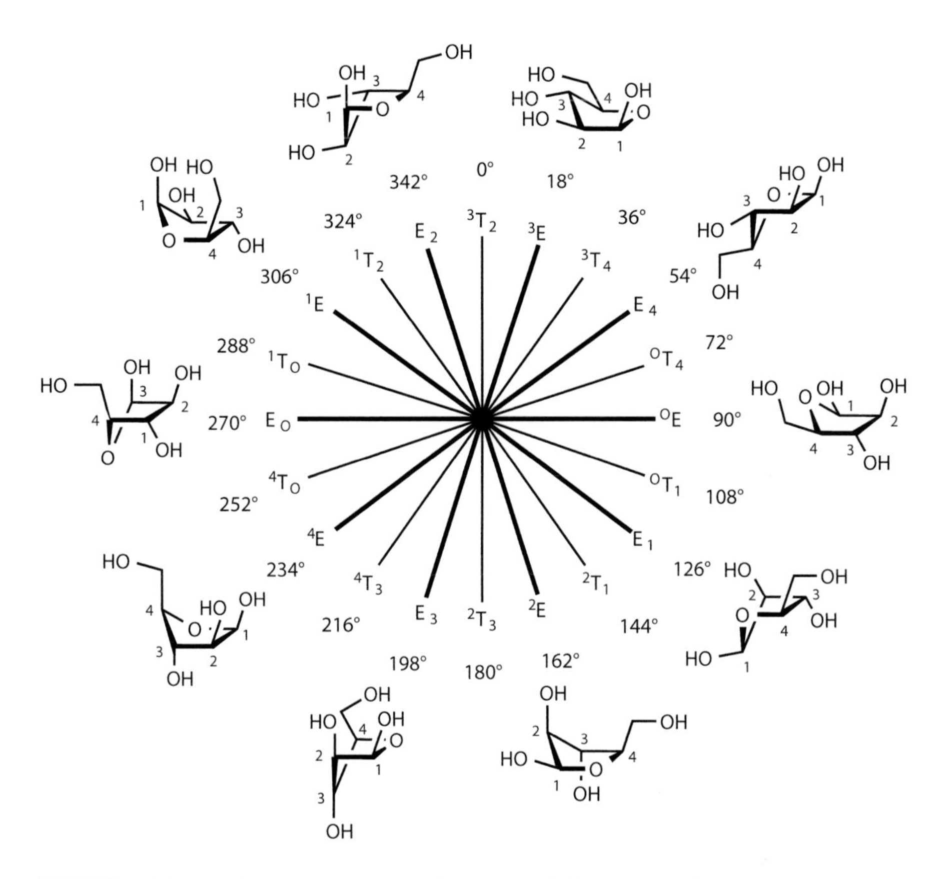


**FIGURE 7.22** Conformational wheel diagram depicting the conformations adopted by furanose rings. Only envelope conformations are shown. (Adapted from Taha, H. A. et al., *Chemical Reviews*, 113, pp. 1851–1876, 2012.)

ring that fall in the same plane: four and three atoms, respectively. The remaining atom(s) that lie above or below that plane are used to describe the conformer in a manner similar to the $^4C_1$ chair described above. For example, a $^3E$ conformation would have C3 above the plane described by atoms C1, C2, C4, and O4 (Figure 7.22), while a small displacement of C4 in this conformer to lie below the plane of atoms C1, C2, and O4 would lead to a $^3T_4$ twist conformation. An alternative description of pentose conformations is given by the Altona–Sundaralingam model which also takes into account the magnitude of displacement of atoms that do not lie in the plane.[58] In this model, the atoms that are displaced from the plane of the ring are indicated by a pseudorotational phase angle (P), and the amplitude of the ring puckering is described by a second parameter, $\phi_m$. The relationship between the E/T depictions and pseudorotational phase angle P is shown in Figure 7.22.

Individual $^3J_{HH}$ coupling constants offer limited information for furanose rings, and so a global analysis of all available $^3J_{HH}$ couplings is typically performed using the PSEUROT programme to determine the values of P and $\phi_m$ that best fit the experimental data.[59] As very few $^3J_{HH}$ coupling constants are available for a furanose ring, efforts have been made to expand the range of coupling constants used for conformational analysis. For example, Houseknecht, Lowary, and Hadad have used $^{3}J_{C1-H4}$-coupling constants to further refine the results of PSEUROT analysis of all eight possible isomers of methyl D-aldopentofuranosides.[60]

### 7.2.3.2 Conformations of Oligosaccharides

While monosaccharide conformations are dependent on the dihedral angles within the sugar ring, the conformations of oligosaccharides are defined by the dihedral angles for the acetal linkages between the monosaccharide units. In most cases, two dihedral angles $\phi$ and $\psi$) are sufficient to describe a glycosidic linkage (Figure 7.23). In NMR spectroscopy, $\phi$ and $\psi$ are defined as the angles between H1′–C1′–O–Cx and C1′–O–Cx–Hx, respectively, where x is the point of attachment on the glycosylated sugar residue. In the case of glycosidic linkages to a primary hydroxyl group (i.e. O6 for pyranosides or O5 for furanosides), an additional dihedral angle, $\omega$, is required, as defined in Section 7.2.2.1.

It is very challenging to obtain crystals of complex oligosaccharides that are suitable for X-ray diffraction studies; therefore, NMR analysis is the primary method for studying the conformations of oligosaccharides. The three NMR spectroscopy methods most commonly used for conformational studies of oligosaccharides are NOE/ROE measurements, trans-glycosidic heteronuclear couplings, and residual dipolar couplings (RDCs). Distance constraints from NOESY- or ROESY-type experiments are frequently used to identify which protons are close in space to the anomeric proton on an adjacent sugar.[61,62] Comparison of inter-residue NOEs with those within a sugar residue can provide a way of calibrating the intensity of the signals with distances separating the protons. However, frequently, there are very few such inter-residue signals that can be detected. This issue is particularly problematic for flexible linkages.

Heteronuclear $^{1}$H–$^{13}$C coupling constants can often be measured across the glycosidic linkages to obtain torsion angle information via the Karplus equation.[63] Frequently, these experiments are performed on samples that have been enriched with $^{13}$C at specific sites in the molecules. Isotopic enrichment enables experiments that are dependent on $^{13}$C atoms to be conducted at lower sample concentrations without becoming limited by the sensitivity of the experiment. Furthermore, selective $^{13}$C labelling of oligosaccharides enables experiments to be conducted on samples that have poorly dispersed NMR signals. So-called 'isotope-edited experiments' exploit an HSQC step at the start of a pulse sequence to transfer magnetisation from each proton to an attached $^{13}$C atom and back to the proton prior to the remainder of the pulse sequence.[64,65] The consequence is that signals for labelled positions are enhanced 100-fold relative to unlabelled positions, thus eliminating unimportant

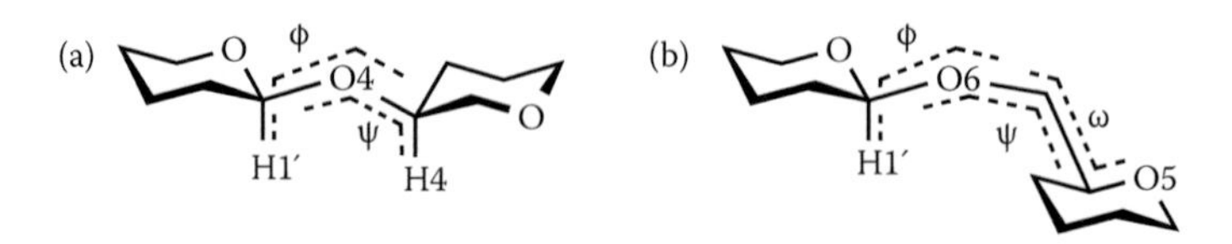


**FIGURE 7.23** Definitions of $\phi$, $\psi$ and $\omega$ torsion angles for a 1,4-glycoside (a) and a 1,6-glycoside (b).

signals from the spectra. For example, introduction of $^{13}C$ into the C6 position of one sugar in two disaccharides has allowed coupling constants between C6–C5, C6–C4, H5–C6, and H1′–C6 to be measured selectively.[61] Sometimes, a combination of experiments, for example, J-HMBC, in-phase anti-phase (IPAP)-HSQC-TOCSY-hadamard transform (HT), and $^{1}H$–$^{13}C$ HSQC-heteronuclear couplings from arbitrarily scaled shift and coupling information domain experiments with E.COSY-type cross peaks (HECADE) experiments, is used to gain sufficient coverage of the available coupling constants.[63] Analysis of dihedral angle information from such experiments is usually used in combination with NOE/ROE data as constraints for simulated annealing or molecular dynamics simulations.[66,67] It should be noted that, as glycosidic linkages are often quite flexible, it is common to find multiple conformations undergoing fast exchange on the NMR timescale (Chapter 2). Therefore, NMR-derived parameters are often population-weighted averages that need to be treated with caution when used as structural constraints in molecular dynamics simulations.

The paucity of information available to define the conformations of glycosidic linkages has been significantly supplemented with the introduction of RDC information (Chapter 3).[68,69] Dipolar couplings arise from the direct interaction between two magnetic dipoles and can be approximately 1 kHz in magnitude. The dipolar couplings are dependent not only on the distance separating two nuclei, but also on the angle of the bond vector relative to the external magnetic field. In solution, rapid tumbling of the molecules usually leads to an isotropic distribution of bond vectors, and thus, the dipolar couplings have an average value of zero. However, if the molecule under study can be partially aligned with the magnetic field, then non-zero RDCs can be observed. These data can provide information on the relative orientation of bond vectors which, in turn, can be used to determine the conformation of the molecule. Residual dipolar coupling information is very complementary to the NOE and scalar couplings described above. Whereas NOEs and inter-residue heteronuclear couplings provide short-range information for the conformational preference of a glycosidic bond, RDCs can provide long-range information on the relative orientations of two sugar residues that may be distant to one another in a large oligosaccharide. RDCs can also be used as a source of additional information for determining the conformation of flexible monosaccharides such as furanose residues.[70]

The key challenge is partial alignment of the oligosaccharide with respect to the applied magnetic field. This is usually achieved by recording the NMR spectra in the presence and absence of a liquid crystalline material that aligns itself spontaneously with the magnetic field. Common examples of alignment media include bacteriophage Pf1[71] and ‘bicelles’, for example, discrete fragments of lipid bilayers comprising mixtures of phosphatidylcholine with either long- and short-chain fatty acids.[72] An alternative alignment strategy is to append a paramagnetic metal ion such as a lanthanide to the carbohydrate.[73] This latter approach has an additional advantage of enabling other complementary experiments such as paramagnetic relaxation enhancement (PRE) and pseudocontact shifts, which can provide information on interactions of the carbohydrate with proteins. In Section 3.4.2.1.1, the choice of alignment media is discussed in more detail.

The aim of the alignment procedure is to allow approximately 1 in 1000 molecules to be aligned with the magnetic field at any given time, which leads to a

RDC that is of similar magnitude to the scalar couplings in the molecule. Standard methods for measuring scalar couplings in 1-D or 2-D NMR spectra are used in the absence and presence of the alignment medium. The difference between the two measured values provides the RDC for that given pair of nuclei. The alignment tensor for the molecule is defined by five parameters: the three principal axes describing alignment of the molecule with respect to the magnetic field and the magnitude and rhombicity of the alignment.[69] Therefore, at least five non-parallel RDC vectors need to be measured to determine the alignment of a rigid unit (e.g. pyranosyl monosaccharide residue) of the oligosaccharide. This typically requires the measurement of both homonuclear and heteronuclear couplings.[68]

Studies on the conformation of oligosaccharides have highlighted the flexibility of many glycosidic linkages. The presence of multiple discrete conformers in solution implies that carbohydrates are inherently dynamic and that there will, thus, be a loss of conformational entropy upon binding to a protein receptor. NMR spectroscopy also offers several strategies for probing ligand dynamics in solution. For example, if alignment tensors derived from RDC experiments are substantially different for each sugar residue, it is likely that the oligosaccharide is flexible.[74] Säwén et al. used a combination of *J* couplings, T-ROE measurements, RDCs, and nuclear spin relaxation measurements to probe the structure and dynamics of a human milk oligosaccharide.[74] These data allowed the results of molecular dynamics experiments to be evaluated critically and led to improved estimates of conformer populations. Identification of which regions of an oligosaccharide are flexible and which are rigid can lead to new insights into binding selectivities and, hence, the biological roles of carbohydrates.

### 7.2.4 Protein–Carbohydrate Interactions

Cell-surface protein–carbohydrate interactions mediate a wide range of biological events from initiation of the immune response to fertilisation. Many bacteria and viruses also exploit cell surface carbohydrates for initial adhesion to cell surfaces. Therefore, a detailed understanding of protein–carbohydrate interactions is essential if we are to understand the molecular biology of the cell surface. While X-ray crystallography has played an important role in revealing the structures of protein–carbohydrate interactions, NMR spectroscopy remains the technique of choice for many researchers. The relatively low affinities that are typical for protein–sugar interactions make these systems ideal for many types of NMR experiments that are dependent on relatively rapid exchange between the bound and free states of the ligand. Furthermore, solution-based experiments allow direct comparison of bound and free ligand conformations, which is insightful for interpreting binding thermodynamics. The final section here will focus on strategies for characterising protein–carbohydrate complexes.

#### 7.2.4.1 Identification of Binding Interfaces

The environment of nuclei in or near the binding sites is invariably altered upon ligand binding. The resulting perturbations in NMR signals can be used to determine if and where binding is occurring. For example, binding of a hexasaccharide to fibroblast growth factor-1 (FGF-1) has been demonstrated using a $^{1}H$–$^{15}N$ HSQC shift perturbation experiment, which shows changes in chemical shifts for N–H groups in the protein backbone and side chains.[75] Alignment of the spectra for free and bound FGF-1 indicated large perturbations in chemical shift for 12 different residues (Figure 7.24). By comparison of these data with the sequence and crystal structure of FGF-1, it was shown that the greatest changes in chemical shift were clustered around the C-terminus, indicating that this was the binding site for the oligosaccharide (Chapter 3). Care must be taken to exclude the possibility that substantial changes in protein conformation occur upon ligand binding. Such structural changes can potentially lead to changes in protein chemical shift that are distal to the binding site.

While the $^{1}H$–$^{15}N$ HSQC experiment is invaluable for showing the region on the protein surface to which a carbohydrate binds, it does not provide information about the complementary binding surface of the carbohydrate. Saturation transfer difference (STD) NMR experiments (see Chapter 3 for a detailed description of this approach) can overcome this limitation. In STD NMR, protons in the protein are saturated, and if the ligand binds to the protein, this saturation can be passed onto the ligand. Subtraction of the resulting spectrum from a reference spectrum in the absence of saturation results in a difference spectrum in which protons that interact directly with the protein will produce the strongest STD signals.

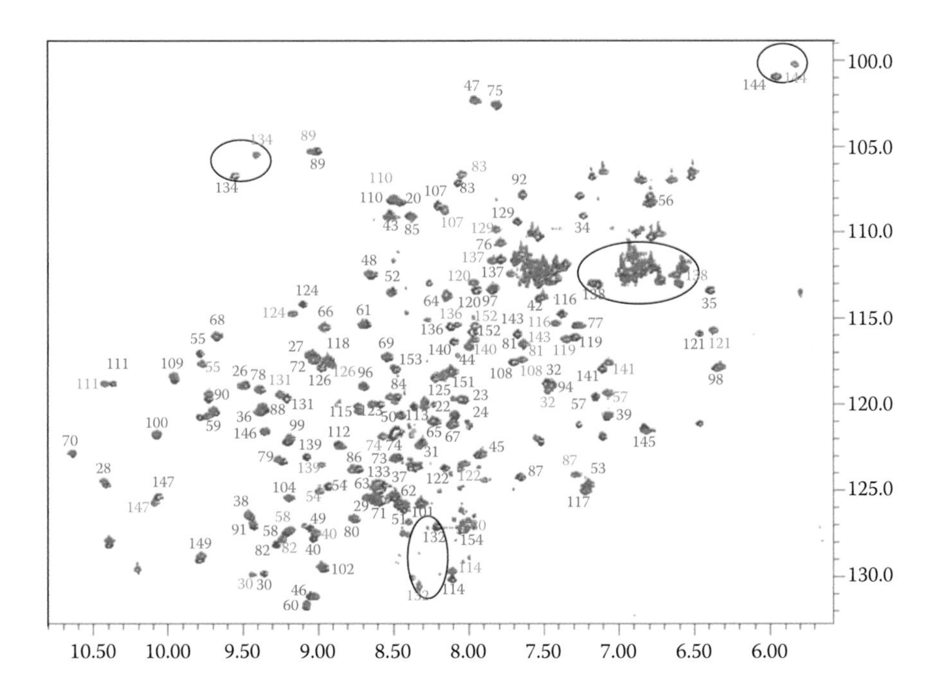


**FIGURE 7.24 (See color insert.)** The $^{15}N$–$^{1}H$ HSQC spectra of free (blue) and bound (red) FGF-1 are overlaid. Areas with the largest perturbation, indicating greatest contact with the hexasaccharide, are circled. (Reproduced from Canales-Mayordomo, A. et al., *Journal of Biomolecular NMR* 35, pp. 225–239, 2006.)

The use of STD NMR to determine carbohydrate binding has been demonstrated using *Ricinus communis* agglutinin ($RCA_{120}$), which binds terminal galactose moieties.[76] STD NMR and standard $^1H$ NMR spectra were recorded for a mixture of methyl β-galactopyranoside and $RCA_{120}$ in a 100:1 ratio. From these spectra, the difference in the intensity of the peaks in the standard and STD NMR were compared to give a value on the degrees of saturation for the protons relative to the signal for H3, which was set to 100% (Figure 7.25). H2, H3 and H4 had the highest STD intensity of 87% to 100%; H6a, H6b, and H5 had 63% to 67%; while H1 and methoxy protons had much lower values of 40% to 32%. The high degree of saturation for H2, H3 and H4 reflected their proximity to the protein and correlated with the binding specificity of $RCA_{120}$ for the non-reducing end of the galactoside.

A full understanding of protein–carbohydrate interactions requires knowledge of the role of water molecules in the binding pocket. WaterLOGSY NMR spectroscopy (Chapter 3) is analogous to STD NMR, but the magnetisation transfer is mediated by water molecules, either directly to the free ligand or via water bound to the protein surface. The result is that binding ligands give positive WaterLOGSY

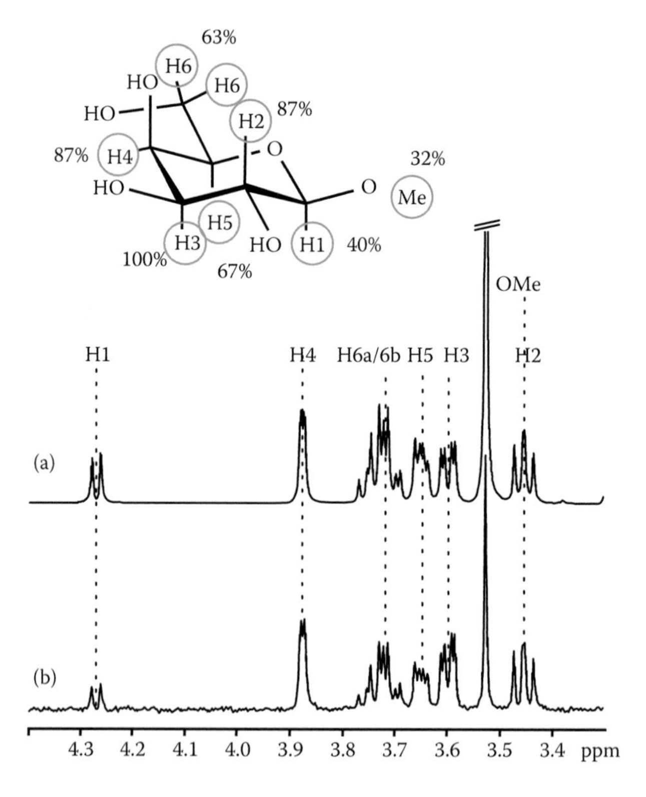


**FIGURE 7.25** (a) $^1H$ NMR of $RCA_{120}$ and methyl-β-D-galactopyranoside in a ratio of 1:100. (b) $^1H$ STD NMR of the same sample. The difference in intensity between the spectra is demonstrated by the inlay of β-GalOMe, showing the relative intensities with H3 set to 100%. (Adapted from Mayer, M. and B. Meyer, *Journal of the American Chemical Society* 123, pp. 6108–6117, 2001. With permission.)

signals, while non-binding ligands have negative signals. This technique was utilised in the evaluation of l-rhamnose-$^1$C-phosphonates as nucleotidylyltransferase inhibitors.[77] Here, a selection of analogues of α-D-glucose-1-phosphate, a substrate for nucleotidylyltransferase inhibitors, was screened for binding using the WaterLOGSY technique. Direct interactions between sugars and protein residues can be distinguished from those mediated through bound water molecules by comparison of the WaterLOGSY spectrum with an STD NMR spectrum; any signals present in the WaterLOGSY but not in the STD spectrum can be assigned to direct magnetisation from bound water.[78]

Both WaterLOGSY and STD NMR have been demonstrated to be useful in showing binding between carbohydrates and proteins, and they also excel in screening large libraries of ligands.[79,80] This method has been used to describe the binding epitope for E-selectin from a combinatorial library of lactose randomly substituted with fucosyl and sulphate residues.[81] Comparison of TOCSY and STD TOCSY spectra for a 200-compound library showed that only a small number of the components of the library were involved in binding. Having identified the spin systems for ligands that bound to E-selectin, selective 1-D TOCSY (Section 7.1.2.1) and COSY experiments could be used to identify the active ligands. The two epitopes that carried binding affinity were found to be 3-fucosyl-lactose and 2′-fucosyl-lactose, although the positions of sulfonation could not be determined using the NMR experiments alone.

STD NMR has also been used to study hSGLT1, a eukaryotic, ion-dependent co-transporter that moves glucose across the membrane.[82] In this example, it was possible to study binding interactions in whole cells with the use of high-resolution magic angle spinning NMR spectroscopy. The interaction between hSGLT1 and a glycoside inhibitor, phlorizin, was measured in human embryonic kidney (HEK) cells using 1-D STD NMR spectroscopy. The experiment was then repeated with naphthyl-β-D-galactoside, which is also a known binder for hSGLT1. Upon the addition of phlorizin to this experiment, the naphthyl-β-D-galactoside signals disappeared and were replaced with those for phlorizin, thus demonstrating that the latter inhibitor binds more strongly to hSGLT1.

There is a growing interest in the use of fluorinated sugars as chemical probes for carbohydrate-binding proteins. For example, $^{19}$F STD NMR spectroscopy can provide simplified spectra that are more straightforward to interpret.[83] Whole-cell $^{19}$F–2-D–EXSY NMR spectroscopy has also been used to determine the rate of transport of fluorinated glucose analogues across a cell membrane via a glucose transporter (Glu1).[84] The fluorinated sugars' intracellular signals are shifted downfield from their extra-cellular signals, and cross-peaks between the two in the exchange spectroscopy (EXSY) NMR spectra are observed if they move between the two locations (Chapters 2 and 3). The intensity of the cross-peak is proportional to the rate of exchange of the two groups. From this information, it was possible to assess the degree to which each fluorinated sugar was transported by the protein.

#### 7.2.4.2 Oligosaccharide Conformation When Bound

Although STD and WaterLOGSY NMR spectroscopy can provide data on which groups in a sugar are oriented toward a protein surface, they do not provide complete information on the conformation of the bound sugar. This limitation can be overcome with the use of the transfer NOE (et-NOE) (note that we have adopted the convention used in Section 3.4.1, whereby the exchange-transferred NOE is distinguished from the kinetic or transient NOE [tr-NOE]) to provide distance constraints across the glycosidic linkages. et-NOE signals can change from positive in fast-tumbling molecules to negative in slow-tumbling molecules; therefore, the signals change sign upon the ligand binding to the large slow-moving protein. et-NOE can only be used for studying ligands that have a fast dissociation rate, which makes the technique ideal for studying protein–carbohydrate interactions.

STD NMR and et-NOE experiments have been used together to determine the conformation adopted by a C-glycoside analogue of N-acetyllactosamine, which binds to galectin-1 (Gal1).[85] STD NMR demonstrated that the C-glycoside bound through the non-reducing end of the galactose residue. et-NOESY was then used to measure the distance between protons across the glycosidic linkage while bound. The results indicated that the exo-$\phi$/syn-$\psi$ conformer predominated in the bound state, which is the same as for the natural oxygen-linked substrate.

The main disadvantage of the et-NOE experiment is spin diffusion (Chapter 3), which can lead to the appearance of NOEs between two spins that are not close in space. This problem can sometimes be overcome by the use of et-ROESY. For example, et-ROESY has been used in studies of bound methyl $\alpha$-lactoside and methyl $\alpha$-allolactoside to ricin B chain.[86] et-NOESY experiment peaks were found between H1′ and H4′ as well as H1′ and H6′ for methyl $\beta$-lactoside, which were considered to be improbable given the distance between these atoms in the galactosyl residue. These cross-peaks were found to be absent in an et-ROESY experiment, which still showed strong NOEs for H1′ to H4, H3′, and H5′, as expected. As et-ROESY does not show cross-peaks for spin diffusion, this demonstrated that the cross-peaks with H4′ and H6′ were false positives in the et-NOESY experiment. The authors noted that, as et-ROESY is not as sensitive an experiment as et-NOESY, the two should be used in conjunction for the most accurate results.

### 7.2.4.3 Relative Orientations of Protein and Carbohydrate

The methods discussed so far allow (1) detection of protein–carbohydrate binding, (2) identification of which face of the ligand makes contact with which part of the protein surface and (3) the shape of the ligand when bound. With this information in hand, molecular docking procedures can sometimes be used to predict the orientation of the two binding partners in complex. However, additional distance/orientation constraints can be gleaned from et-NOEs, RDCs, and pseudo contact shifts (PCSs).

Detection of et-NOEs between the carbohydrate ligand and the protein is challenging because the experiment requires weakly bound ligands that are in fast exchange, and this often means that only a low proportion of the ligand is in the bound state. Shimizu et al. addressed this problem by using fully $^{13}C$-enriched Gb3 trisaccharide ligands for the verotoxin B sub-unit (VTB).[87] An isotope-edited HSQC-NOESY experiment allowed detection of et-NOEs from the $^{13}C$-enriched oligosaccharide to the $^{15}N$-enriched protein. The results were consistent with only a single carbohydrate binding site on each of the five VTB sub-units, which was in contrast to the crystal structure of the complex that had revealed an additional of two binding sites per VTB monomer.[88]

Subsequent RDC experiments were used to confirm that only one binding site was occupied to any significant extent in solution. Thompson et al. titrated the VTB protein into a solution of $^{13}C$-enriched Gb3 trisaccharide mixed with 1,2-dimyrisoyl-sn-glycero-3-phosphocholine (DMPC)/1,2-dihexanoyl-sn-glycero-3-phosphocholine (DHPC) bicelles.[89] By measuring the RDCs with different concentrations of VTB protein, they were able to extrapolate the RDC values to determine the RDC for 100% occupancy of the binding sites. The results were consistent with only one site being occupied with the oligosaccharide ligand under experimental conditions.

When measuring RDCs of weakly binding complexes, it can sometimes be advantageous if the protein binds tightly to the liquid crystal as the magnitude of the RDCs is then dependent on the fraction of ligand that is bound at any given time. The use of bicelles in this way has been generalised through the incorporation of nickel chelates into the bicelles to bind His tag–containing proteins.[90] There is, however, a risk that the unbound ligand could also interact directly with the bicelles if it has a hydrophobic tail, thus giving false RDC values. To avoid this problem, Zhuang et al. produced a fusion protein attaching galectin-3 to a short metal-binding peptide, which can bind to either diamagnetic $Lu^{3+}$ or paramagnetic $Dy^{3+}$ ions.[91] Binding of the fusion protein to a paramagnetic ion increased the complex's magnetic susceptibility anisotropy, which allowed the protein to be partially aligned with the magnetic field without affecting the orientation of the unbound ligand.

The use of a paramagnetic ion also allowed the measurement of PCSs to gain both distance and conformation data from the same sample.[91] PCSs lead to perturbation of the chemical shifts of various nuclei in proportion to their distance from the spin label. This effect can operate over much longer distances than standard shift perturbation assays and can be used to detect even very weakly binding ligands.

Another phenomenon that can be exploited using spin labels is PRE.[92] In PRE, the spin label enhances the relaxation of spins in proximity to it, resulting in a drop in intensity for the peak. As the degree of enhancement is dependent on distance from the unpaired electron, this technique can also be used to map distances. This effect carries up to 20 Å, a distance far greater than NOEs or STD effects. Jain et al. used a N-acetyl-lactosamine (LacNAc)-(2,2,6,6-tetramethylpiperidin-1-yl)oxy (TEMPO) disaccharide to study binding with $^{15}N$-labelled galectin-3.[92] $^{1}H$–$^{15}N$ HSQC spectra were acquired for galectin-3 in the presence and absence of the ligand. Comparison of the spectra showed a large loss in intensity for the residues in closest proximity to the spin label. The distances calculated from the PRE spectra were found to be consistent with those predicted from the X-ray crystal structure for the complex.[92]

### 7.2.5 Summary

From routine analysis of synthetic intermediates to determining solution conformations of protein–carbohydrate complexes, NMR spectroscopy is a vital analytical technique for glycoscience. The increased availability of enzymatic methods for constructing oligosaccharides will facilitate the production of isotopically enriched oligosaccharides for sophisticated NMR experiments. Furthermore, as very high field NMR spectrometers also become more widespread, it will become possible to study ever-larger oligosaccharides and proteins, which will continue to extend the application of NMR spectroscopy in this field.

## 7.3 NUCLEIC ACIDS

John A. Parkinson

### 7.3.1 Introduction

Nucleic acids, especially DNA, form the blueprint for all living organisms. Studies of their structure, behaviour, properties, and interactions with one another and with other molecules across many disciplines of science have revolutionised our understanding of the living world. The double helical structure of DNA as revealed by Watson and Crick is now common knowledge.[93] Despite this, much remains unexplored about these molecules, with discoveries being made as never before, arising through new innovations in scientific analytical methods and thinking.

For the synthetic chemist with research interests allied to nucleic acids, the forward leap that has enabled ease of access to these molecules has been made through solid-phase DNA chemical synthesis. This affords straightforward production of these sequence-defined biopolymers in quantities and with purity suitable for use in many branches of science.

In this section, the emphasis is placed on practical NMR studies of synthetic DNA. NMR studies of isotope-enriched nucleic acid samples and studies associated with RNA are not addressed here since these are much less readily accessible for everyday use. Sample handling, preparation, and the nuances and practicalities of setting up an NMR spectrometer to work with DNA are discussed. The influence of structure and behaviour on the appearance of DNA NMR data and, in the final section, methods used for data assignment and interpretation are all the areas that are addressed in this section.

### 7.3.2 Sample Selection

NMR studies of DNA begin with choices. These concern the size, length, and sequence of the molecule to be studied and its most likely assembly characteristics. This choice is governed by two main questions. First, will it be necessary to fully assign the NMR data of the DNA? Second, is there a sequence-related aspect of the DNA (such as sequence-specific recognition by an interacting molecule or a specific structural element required in the DNA) that governs the sequence to be made? In instances where some flexibility in choice of sequence may be tolerated, a major aim concerns the controlled generation of NMR data that may then be readily assigned if required. This feature is bound up in the nature of the biopolymer sequence itself.

#### 7.3.2.1 Sequence Design

Nucleic acids are biopolymers composed of a limited pallet of monomer building block 'bases'. These are attached to a sugar/phosphodiester 'backbone'. Each sugar/phosphate/base unit combination is termed a 'nucleotide'. In the case of DNA, the sugar is deoxyribose. This lacks a hydroxyl group at the C2′ position of the sugar, which is present in the ribose sugar of RNA. Unlike proteins, which are built from a general library of 20 amino acid building blocks, DNA is constructed from only 4 base units. These are the purines, adenine (A) and guanine (G), and the pyrimidines, thymine (T) and cytosine (C) (Figure 7.26).

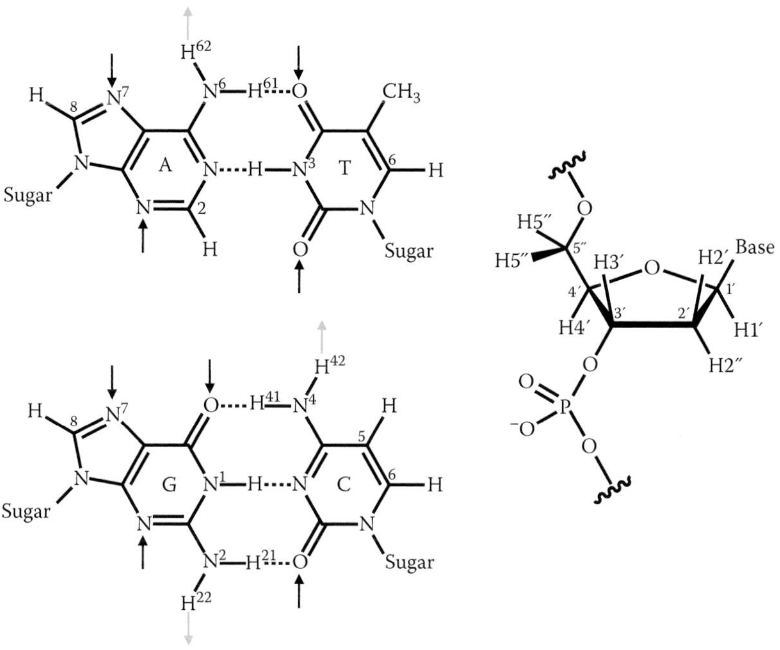

**FIGURE 7.26** Chemical structures of adenine (A), thymine (T), guanine (G) and cytosine (C) bases together with a representative deoxyribose sugar with phosphate group attached at the 3′-oxygen. Bases are represented in their standard Watson–Crick pairs A–T and G–C. Hydrogen bond acceptors are indicated by black arrows; hydrogen bond donors are represented by grey arrows. Key atoms and positions are labelled according to the scheme used for NMR signal assignment.

With a number of exceptions, the dispersion of NMR signals arising from residues of the same type within a DNA sequence is not large. This is despite sequence variations. Generally, NMR signal dispersion in DNA is only dependent upon next-door–neighbour effects. This can lead to a large degree of signal degeneracy within NMR data sets. For extended sequences of the same residue type, the NMR data easily become intractable. For this reason, it is important to be able to predict the likely pattern of NMR resonances that may emerge upon data acquisition from the DNA sequence design stage. This is particularly the case for $^1$H NMR data. It ensures, as far as possible, a high probability of success for the assignment of NMR signals when using a sequence-specific NMR data assignment protocol for DNA molecules (see in the following section). For the general case of non-isotope–enriched DNA, a strong reliance is placed on studies carried out by $^1$H NMR spectroscopy. The aromatic signal region of the 1-D $^1$H NMR spectrum and the aromatic-to-sugar H1′ and H2′/H2″ cross-peak regions of any 2-D NOESY NMR data sets become the most important features of any DNA NMR data to control, provided that this is feasible. An example of good practice in sequence choice and reflecting good quality, resolved NMR data is shown in Figure 7.27 for the DNA duplex d(CGACTAGTCTAGACG) •

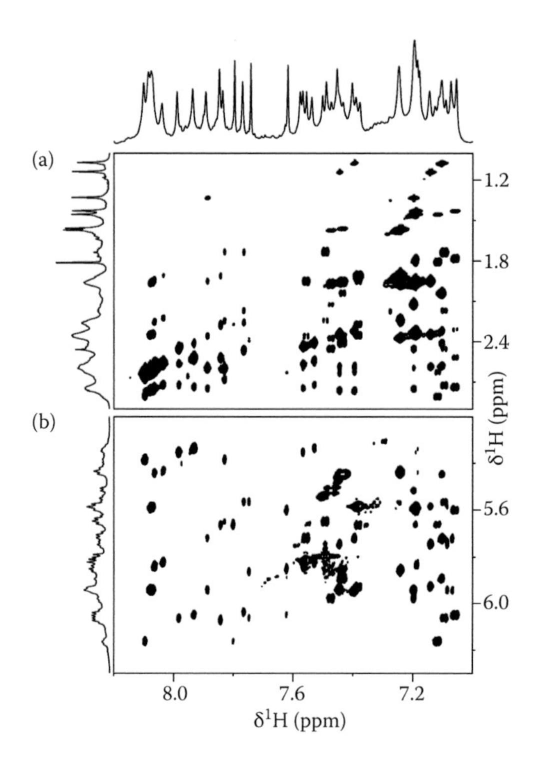


**FIGURE 7.27** Regions of the 600-MHz 2-D NOESY NMR spectrum of the non-self-complementary 15-mer DNA duplex d(CGACTAGTCTAGACG) • d(CGTCTAGACTAGTCG) acquired at 25°C with a mixing time of 150 ms: (a) aromatic-to-H2′/H2″/T-$CH_3$ cross-peaks and (b) aromatic-to-H1′/C-H5 cross-peaks. The 1-D $^1H$ NMR spectrum of the aromatic resonance region is shown as the top projection. All aromatic resonances are resolved and distinguishable in this example despite the size of the molecule.

d(CGTCTAGACTAGTCG), complemented by data presented in Table 7.1. This presents typical ranges for $^1H$ NMR chemical shifts of non-labile protons associated with standard DNA bases A, G, T, and C.

Although some overlap of the $^1H$ NMR resonance regions associated with H8, H2, and H6 protons occurs, judicious choice of DNA sequence can afford good

**TABLE 7.1**
**Typical Ranges for Proton NMR Chemical Shifts of Non-Labile Protons Associated with DNA Bases A, G, T and C**

| | Aromatic $^1H$ NMR Chemical Shift (ppm) | | | |
|---|---|---|---|---|
| **DNA Residue** | **H8** | **H2** | **H6** | **H5** |
| Adenine, A | 8.1 ± 0.1 | 7.4 ± 0.3 | – | – |
| Guanine, G | 7.7 ± 0.2 | – | – | – |
| Thymine, T | – | – | 7.2 ± 0.1 | – |
| Cytosine, C | – | – | 7.4 ± 0.2 | 5.5 ± 0.3 |

dispersion of $^1H$ NMR resonances, as shown by Figure 7.27. Such dispersion transfers to the NMR resonances arising from other protons within the DNA molecule. The dispersion that occurs for H1′, H2′ and H2″ proton resonances for the same DNA 15-mer duplex reveals this (Figure 7.27). Alternating purine–pyrimidine sequences yield the best examples of $^1H$ NMR resonance dispersion (e.g. for -ACGT-), allowing ready identification of aromatic resonances on the basis of alternating sequence-specific $^1H$ resonance frequencies. Inevitably, arranging for such a sequence is not always possible. Alternating sequences of different purines or different pyrimidines (e.g. -AGAG- or -TCTC-) serve as the next best option. Unfortunately, given the limited number of monomer residues available for DNA construction, many occasions arise when contiguous sequences comprising the same residue occur. Under these circumstances, the need for the highest available magnetic field strength becomes self-evident. This affords maximum signal dispersion, helping deliver the strongest opportunity for unambiguous signal assignment, especially for longer sequences of DNA.

#### 7.3.2.2 Length, Shape and Complementarity

The length of the DNA sequence should be considered carefully according to the nature of the work being addressed. Compliance with sequence design issues noted in Section 7.3.2.1 allows longer DNA sequences to be studied readily and in detail by NMR. This is based on the proviso that the DNA is assembled and well-structured in solution. The latter point is crucial; DNA is flexible when not associated with higher-ordered structures. NMR data provide a time average of all conformations sampled during the data acquisition period. $^1H$ NMR signal assignment of DNA is strongly reliant on the generation of high-quality 2-D NOESY NMR data. In turn, this rests upon the DNA molecule adopting relatively defined structures that have solution mobility typical of a large macromolecule. Even short, well structured, relatively rigid DNA fragments can satisfy this condition and then give rise to high-quality 2-D NOESY data in aqueous solution. In contrast, longer but non-ordered (random coil) DNA sequences give very poor responses under NOESY interrogation and do not yield easily to the classical sequence-specific assignment processes commonly used. This is mainly due to a dearth of even short-range, through-space distance information arising from such a molecule, features which normally govern successful sequential data assignment protocols. This is typically a condition associated with single-strand DNA. Consideration should, therefore, be given to the choice of sequence and the context in which the DNA is to be studied. Typically, DNA may be present in the form of unfolded single strands, folded single strands (e.g. hairpin, folded triplex or quadruplex structures) or as a structured molecule made up of separate but complementary strands (duplex, triplex, quadruplex or higher-order analogues).

When well-ordered structures are to be generated from complementary DNA strands, further choice lies in generating structures from either self-complementary or non-self-complementary strands of DNA. This choice also impacts the complexity of the NMR data that will arise from the assembled molecule. By way of example, two 8-mer DNA duplex sequences are presented in Figure 7.28.

This example shows how increased complexity in the NMR data can be ascribed simply to the non-self-complementarity of the DNA sequences used. The feature

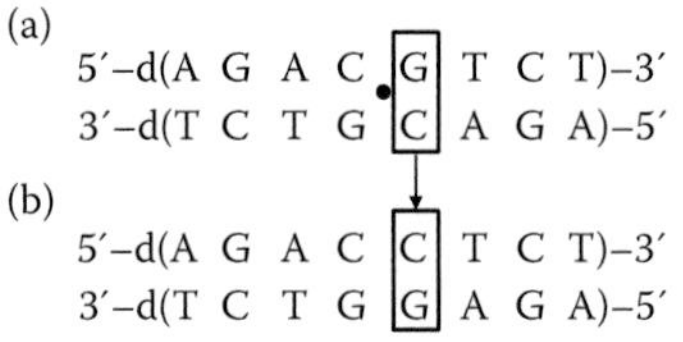

(a)
5′–d(A G A C • [G] T C T)–3′
3′–d(T C T G [C] A G A)–5′

(b)
5′–d(A G A C [C] T C T)–3′
3′–d(T C T G [G] A G A)–5′

**FIGURE 7.28** Two related duplex DNA sequences, whereby rotation of one base pair (boxed) in (a) results in (b). (a) Self-complementary duplex with rotational symmetry about "•" giving rise to 12 $^1$H NMR signals from the base unit protons alone. (b) Non-self-complementary duplex lacking rotational symmetry, giving rise to 24 $^1$H NMR signals from the base unit protons.

causes asymmetry within a structure, which leads to magnetic environments that generally yield an increased number of observable NMR responses. Contrastingly, self-complementarity gives rise to rotational symmetry resulting in simplified DNA NMR data. This simplification may be required if the NMR equipment available to the synthetic chemist is limited either in field strength or in hardware capability. It will also influence the type and ease with which NMR studies may be carried out. DNA sequence design should, therefore, be carefully considered prior to any study.

### 7.3.3 Sample Handling and Preparation for NMR Studies

Provided that the previous considerations have been thought through, the next step for attention, after either purchasing or synthesising in-house, is preparing DNA samples for NMR studies.

#### 7.3.3.1 Handling DNA

Manufactured DNA samples are usually supplied dry in pellet or powdered form. Samples should be freeze-dried prior to use. This removes any residual volatile materials that may still linger from the manufacturing and purification processes.

#### 7.3.3.2 Sample Preparation for NMR Studies

Typical DNA samples are prepared for NMR spectroscopy as aqueous solutions at neutral pH. This allows $^1$H NMR resonances arising from labile (exchangeable) protons to be observed when the sample is dissolved in $H_2O$. Sample pH may be adjusted using buffers. DNA can typically be solubilised in a solution of 100-mM NaCl, with 10- to 50-mM phosphate salts also present. Sodium azide, $NaN_3$, may also be added as a preservative in small quantities. The pH may be set using appropriate combinations of $Na_2HPO_4$ and $NaH_2PO_4$ or equivalent buffer salts if phosphate is inappropriate. If 99% $D_2O$ is the solvent of choice, deuterated salts may be used. However, these can be expensive, and a simple alternative, if the protons are labile, is to use standard protonated salts followed by freeze drying from $D_2O$ to bring about the necessary proton/deuterium exchange.

Well-structured DNA will generally give rise to a series of $^1$H NMR signals in the $\delta^1$H = 12- to 14-ppm region of the $^1$H NMR spectrum, assuming appropriate data acquisition conditions and a sample solubilised in 90% $H_2O$/10% $D_2O$. These resonances arise from labile imino protons within a structured DNA assembly when Watson–Crick base pairing predominates. In the absence of such resonances (when these are expected), it is advisable to check and readjust sample pH. Although articles concerning the study of DNA by NMR often suggest the use of 100-mM NaCl within the sample medium, in practice, this may not always be necessary at solute concentrations greater than 1 mM. At low DNA concentrations of less than 500 μM, structures are unstable in the absence of salt and tend to disassemble.

#### 7.3.3.3 Choice of Solvent

Solvent choice is largely dependent on whether $^1$H NMR resonances arising from labile protons need to be observed for the study or not. The hardware capability of the NMR spectrometer has a bearing on this choice. If the NMR spectrometer is equipped for delivery of pulsed field gradients, as most modern NMR spectrometers are, then 90% $H_2O$/10% $D_2O$ should be the initial solvent mix of choice. It allows the relevant imino proton NMR resonances to be observed, thus reporting on the integrity or otherwise of any assembled DNA structure. However, the DNA sample should also be studied in 99% $D_2O$. This has four main advantages. First, the complexity of the DNA $^1$H NMR data is reduced: signals arising from labile protons associated with imino and amino functional groups are eliminated. Second, greater digital resolution may be used in 2-D or higher-order NMR data acquisitions, the NMR frequency widths being reduced in the absence of imino proton resonances from the 12- to 14-ppm region of the $^1$H NMR spectrum. Third, $^1$H NMR signals from the DNA that resonate in the vicinity of the solvent signal (HOD = $\delta^1$H ~ 4.7 ppm) may be observed more readily without being unduly affected by solvent suppression procedures (Figure 7.29). Finally, it allows cleaner data sets to be generated, which are relatively free of artefacts when acquiring proton-detected 2-D [$^1$H,$^{31}$P] and, especially, 2-D [$^1$H,$^{13}$C] inverse correlation data using the available natural abundance $^{13}$C (Sections 7.3.4.2.3 and 7.3.4.2.4 respectively).

### 7.3.4 Data Acquisition and Processing

Nucleic acids are rich in NMR-active nuclei: $^1$H, $^{31}$P, $^{13}$C and $^{15}$N. Both $^1$H and $^{31}$P are highly abundant (99.9885% and 100%, respectively) and also possess high magnetogyric ratios, $\gamma$ (26.752 and 10.839 × $10^7$ rad s$^{-1}$ T$^{-1}$, respectively). Such properties make these nuclei ideal candidates for reporting on DNA in solution. Although $^{13}$C is a dilute spin (1.07% natural abundance), modern NMR spectrometers, especially those at higher magnetic field strengths, greater than 11.4 T ($\nu$($^1$H) = 500 MHz), are capable of generating signal responses from $^{13}$C nuclei attached to protons from DNA samples in millimolar concentrations. This is in response to proton-detected indirect-correlation methods such as [$^1$H,$^{13}$C] HSQC, especially if so-called 'sensitivity-improved methods' are adopted. $^{15}$N (0.37% natural abundance) is a more dilute spin than $^{13}$C, but under suitable conditions, natural abundance [$^1$H,$^{15}$N] correlation data may be collected, as described later in this chapter (Section 7.3.4.1.4). Thus, all four nuclei may be used to gather appropriate

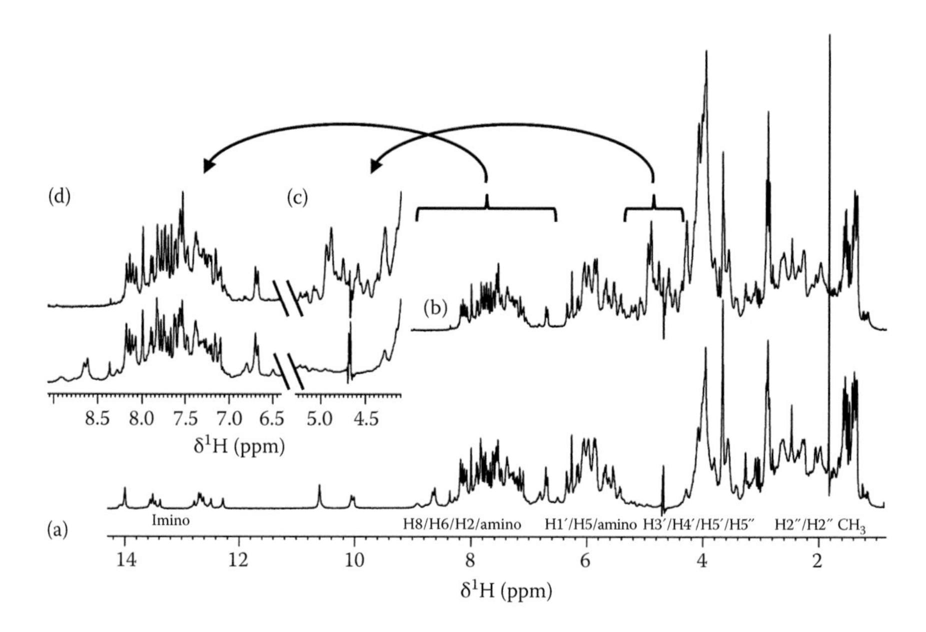


**FIGURE 7.29** 500-MHz 1-D $^{1}H$ NMR spectra in 90% $H_2O$/10% $D_2O$ (a) and 99% $D_2O$ (b) for a ligand complex of the DNA 15-mer duplex d(CGACTAGTCTAGACG) • d(CGTCTAGACTAGTCG) at 25°C. Expansions (c) show severe signal bleaching of the data in the region of $\delta^{1}H = 4.8$ ppm when using solvent suppression with the 90% $H_2O$/10% $D_2O$ solution compared with limited bleaching around the residual HOD signal when solvent-suppressing with the 99% $D_2O$ solution. Simplification occurs (d) for the $^{1}H$ NMR spectrum of the 99% $D_2O$ solution compared with the 90% $H_2O$/10% $D_2O$ solution by elimination of signals arising from amino-proton resonances.

information to study DNA in solution provided that sufficient material is available in the first instance.

Factors affecting the appearance of DNA NMR data are considered further here, including the influence of sample temperature. Various NMR methods are described for samples solubilised in both 90% $H_2O$/10% $D_2O$ and 99% $D_2O$.

### 7.3.4.1 Samples in 90% $H_2O$/10% $D_2O$

In general, it is advisable to ascribe the acquisition of 2-D $^{13}C$ and $^{31}P$ correlation data sets to samples solubilised in 99% $D_2O$ when these are proton detected (i.e. 'inverse' experiments, which depend on the observation of the $^{1}H$ NMR response). Data for 90% $H_2O$/10% $D_2O$ solutions of DNA may otherwise be severely compromised by the presence of large residual signal artefacts from the suppressed solvent signal. With this in mind, those NMR experiments more suited to this solvent mixture are considered first.

#### *7.3.4.1.1 1-D $^{1}H$ NMR Spectra*

Acquisition of 1-D $^{1}H$ NMR spectra of nucleic acids solubilised in 90% $H_2O$/10% $D_2O$ requires the excess $H_2O$ signal (from 110-M proton) to be attenuated without significantly compromising the signal response arising from the DNA solute. The recommended method for solvent suppression (Chapter 3) uses an excitation sculpting approach first reported by Hwang and Shaka in 1995 in the form of a double-pulsed field gradient spin echo (DPFGSE), otherwise referred to as 'excitation sculpting'.[94] This scheme may also be incorporated into 2-D [$^1$H,$^1$H] correlation pulse sequences for efficient solvent suppression. For nucleic acids, this method has two advantages. First, the approach is robust and efficient, suppressing the solvent signal cleanly and to a very low level. Second, the approach provides a uniform excitation profile across the full frequency window required to observe all $^1$H NMR responses from DNA protons, including, especially, the imino proton responses between $\delta^1$H = 12 and 14 ppm. $^1$H NMR signal intensity from the imino protons is, therefore, not compromised. WATERGATE[95]-based methods (Sections 3.2.2 and 6.3.1), while similar in their ability to attenuate the water signal, can suffer in this context from a non-uniform excitation profile across the frequency width, depending on the exact method of choice. If the null point of the classical WATERGATE technique is not carefully adjusted, the appearance of signals in the imino-proton resonance region may be severely compromised. Although the favoured method, the DPFSGE approach introduces echo periods that allow sufficient time for some relaxation and evolution processes to take effect. For labile protons that undergo more significant chemical exchange with the solvent, this can result in their $^1$H NMR signals being attenuated through $T_2$ relaxation and exchange effects (Chapter 2). Parameters suitable for the acquisition of typical NMR data using the DPFGSE approach for DNA samples include the use of a soft pulse with a narrow (50–100 Hz) excitation bandwidth centered on the solvent signal, which may be produced using a low-powered square-shaped pulse of typically 2-ms duration. The pulse field gradients may be sine shaped and should deliver magnetic field gradients in a typical ratio $g1{:}g2$ = 16:6 G/cm.

Pre-saturation (Chapter 3) may also be used as an alternative method for solvent suppression to allow information to be recovered in the case of protons with different chemical exchange characteristics.

###### *7.3.4.1.2 2-D $^1$H NMR Data Acquisition*

2-D NOESY and ROESY NMR data may be acquired using similar solvent suppression routines. Typical responses from ROESY are of lower intensity than those from NOESY, but each data set complements the other. ROESY provides a subset of short-range, inter-proton through-space correlations compared with NOESY data: only intra-sugar H1′–H2″ ROEs are observed, whereas both intra-sugar H1′–H2′ and H1′–H2″ NOEs occur. Acquisition of both data sets is, therefore, a useful device for distinguishing the identity of resonances arising from H2′ and H2″ protons and serves as a complement to using 2-D DQFCOSY NMR data, which carries similar information presented in a different way (Figure 7.30).

To capture $^1$H NMR resonances arising from DNA imino protons, the NMR acquisition frequency widths must be set to at least $\Delta\delta(^1$H) = 20 ppm. The carrier frequency (offset) must be centered on-resonance at the proton resonance frequency of the solvent and optimised carefully to achieve efficient and effective solvent suppression. This usually involves moving the offset frequency slightly by a few hertz

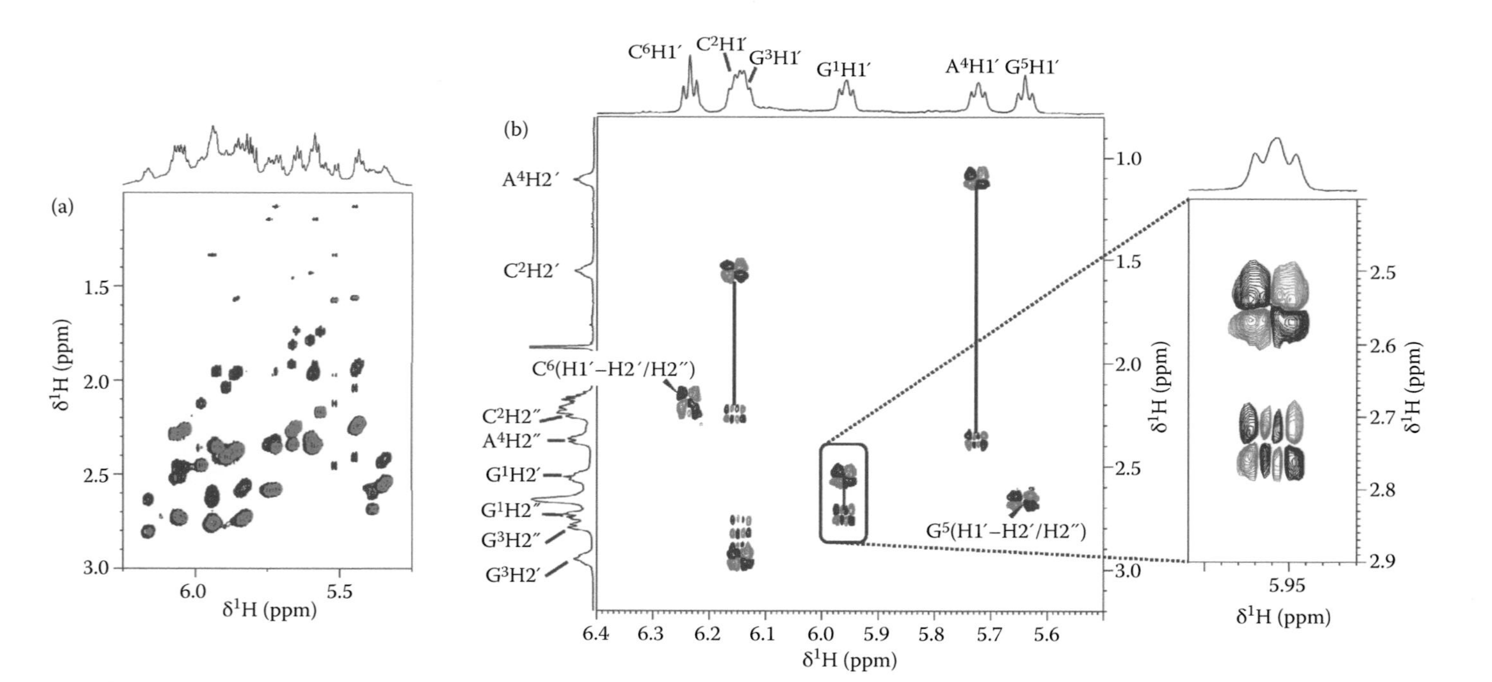


**FIGURE 7.30** Distinguishing $^{1}$H NMR resonances for H2′ from those of H2″ in DNA. (a) Example overlay of 600-MHz H1′–H2′/H2″ cross-peak region of 2-D NOESY (black contours) and ROESY (grey contours) for a 15-mer DNA duplex. ROESY only shows H1′–H2″ cross-peaks, whereas NOESY shows both H1′–H2′ and H1′–H2″ cross-peaks, allowing the resonances from H2′ and H2″ protons to be readily distinguished. (b) Alternative approach using 2-D DQFCOSY NMR data for mismatched 6-mer DNA duplex d(GCGAGC)$_2$ – correlations for H1′–H2′ differ from those for H1′–H2″ through additional passive couplings. Although the chemical shifts of H2″ protons are generally higher than those for H2′ protons in DNA, this should not be taken as a foregone conclusion, as shown for correlations G$^{3}$(H1′–H2′) and G$^{3}$(H1′–H2″) in (b). The expansion (right) shows the typical appearance of H1′–H2′ (top) and H1′–H2″ (bottom) DQFCOSY cross-peaks.

on either side of the initial set value and observing what the effect is on the solvent signal. An optimum solvent suppression frequency will be found using this approach. For 2-D data sets with extended frequency widths in both direct acquisition and indirect dimensions, digital resolution must be set high enough to allow clear resolution of cross-peaks. For short (typically, 6–8 residue) compact DNA molecules with few $^{1}$H NMR resonances, 2048 or 4096 data points are sufficient for the acquisition dimension with 256 $t_1$ increments applied in the indirect dimension. For larger, more complex DNA molecules that give rise to more densely populated $^{1}$H NMR spectra, data should be acquired at much high initial digital resolution. Typically, 8192 data points may be used in the acquisition dimension and at least 512 $t_1$ increments in the indirect dimension, assuming that a standard sampling approach is adopted. The mixing time, $t_{mix}$, for NOESY may be set between 150 and 350 ms. Shorter values of $t_{mix}$ may be used to reduce the effects of spin diffusion (Chapter 3), while longer mixing times allow further evolution of weaker NOEs. For ROESY, the mixing time should be set within the range $t_{mix}$ = 120 to 150 ms to prevent excessive signal attenuation through $T_2$ relaxation processes. The recycle delay, RD, may be set to values in the range RD = 2.0 to 7.0 s. Choice of RD values at the high end of this range arises through the fact that H2 protons of adenine display longer $T_1$ and $T_2$ relaxation times than other protons within the DNA, owing to their relative isolation within the spin relaxation network of DNA structures. $^{1}$H NMR spectra of DNA acquired under the recommended frequency width conditions appear asymmetric and unbalanced when the data acquisition is centered at the solvent frequency. The high chemical shift region of the spectrum is populated by NMR responses from imino proton resonances, whereas the equivalent low chemical shift region is devoid of resonances. This is unavoidable if efficient solvent suppression is to be achieved. It results in large regions of unoccupied frequency space in 2-D NMR data sets. Post-acquisition data handling procedures built into NMR data processing software packages can be used to remove these unoccupied regions of data space to reduce the overall size of data that must subsequently be handled.

#### *7.3.4.1.3 1-D $^{31}$P NMR Data Acquisition*

Examples of $^{31}$P–{$^{1}$H} NMR data from different DNA molecules and complexes are shown in Figure 7.31, together with representative types of conformations that occur in the phosphodiester backbone of DNA.

Figure 7.31 indicates the range of $^{31}$P NMR chemical shifts that can arise for DNA samples. In the absence of any assembled structure, the $^{31}$P chemical shift range is relatively narrow (Figure 7.31a (1)). Chemical modification in the form of phosphorothiolates or phosphorothioates (which incorporate sulphur in place of one oxygen of the phosphate group) introduces additional signal dispersion, as shown (Figure 7.31a (2)) for the phosphorothioate-modified single strand d(TATAC*CATAT), where * indicates the site of backbone modification. It is worth noting in this instance that two equally sized signals occur at significant offsets ($\delta^{31}$P = 55.10 and 55.83 ppm) compared with the main phosphorous resonance region from the remaining DNA phosphodiester backbone. The significant offset arises through the influence of sulphur directly bound to phosphorus. The fact that there are two signals at high chemical shifts arises through equivalent quantities of both R-phosphorothioate

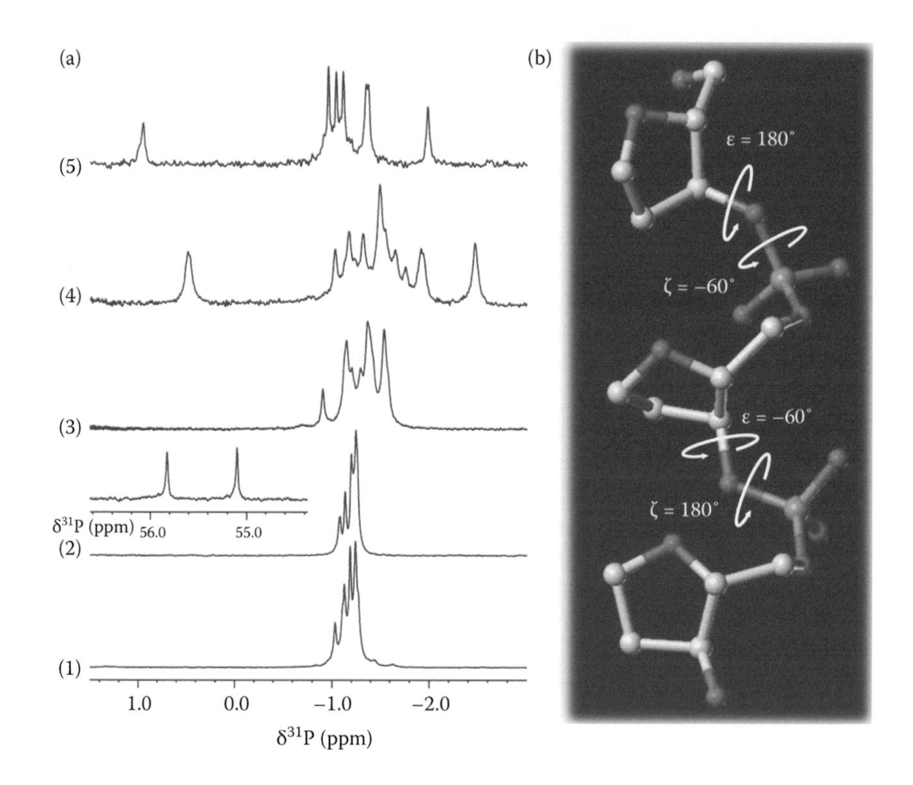


**FIGURE 7.31** (a) Representative $^{31}P$–{$^{1}H$} NMR spectra of a selection of different DNA samples: (1) single-strand d(TATACCATAT); (2) as for (1), but with phosphorothioate in place of phosphodiester substituted between the central CC residues; (3) self-complementary 16-mer DNA duplex $d(CGACTAGTACTAGTCG)_2$; (4) as for (3), but in the presence of a small-molecule binding ligand; and (5) self-complementary 8-mer DNA duplex $d(GGCGAGCC)_2$ containing the tandem-sheared G–A mismatch at the center of the structure. (b) Representation of $B_I$ (top) and $B_{II}$ (bottom) conformations in the phosphodiester backbone of DNA.

and S-phosphorothioate DNA in solution, yielding different magnetic environments. Duplex DNA generally gives rise to broader $^{31}P$ NMR resonances but with slightly greater signal dispersion compared with single-strand DNA (Figure 7.31a (3)). The two predominant conformations that may commonly be adopted by the DNA phosphodiester backbone are also represented (Figure 7.31b). The $B_I$ conformation is more common for the B-DNA duplex for which the $^{31}P$ NMR chemical shift is typically close to $\delta^{31}P = -1.0$ ppm. $^{31}P$ NMR data become more dispersed when conformational differences are either induced as a result of some change to the DNA (caused, for instance, by interactions with other molecules; Figure 7.31a (4)) or inherent due to some structural anomaly, as is the case for the tandem-sheared G–A mismatch (Figure 7.31a (5)). Adoption of the $B_{II}$ backbone conformation (Figure 7.31b) results in $^{31}P$ chemical shift differences of +1 to +2 ppm compared with values for the $B_I$ backbone conformation, thereby creating increased $^{31}P$ signal dispersion. $^{31}P$ NMR data may, therefore, be mined for information concerning DNA phosphodiester backbone conformation.

Typical $^{31}P$ NMR spectra are acquired with $^{1}H$ decoupling to remove the effects of coupling between $^{31}P$ and neighbouring H3′ and H5′/H5″ partners that exist via $^{3}J_{PH}$ coupling. Data may be acquired with relatively rapid pulsing. Typically, over a narrow frequency width of not more than $\Delta\delta = 8$ ppm centered at a frequency offset of −1.2 ppm (relative to the $^{31}P$ NMR signal from external 85% phosphoric acid), data may be acquired into 1024 data points for an acquisition time of approximately 260 ms and RD = 0.5 s. The proton decoupling offset frequency can be set to $\delta^{1}H$ = 4.0 ppm, which is central to the region in which protons resonate that are coupled to phosphorus nuclei. Using a suitable probehead, good-quality $^{31}P$–{$^{1}H$} NMR data may be acquired with typically 256 to 512 transients. Using a 9.4-T [$\nu(^{1}H)$ = 400 MHz] NMR spectrometer, this yields $^{31}P$ NMR data in approximately 3 to 5 min for a sample of DNA in the millimolar concentration range. Typical $^{3}J_{PH}$ coupling constants lie in the range 5 to 10 Hz, requiring relatively low-power decoupling.

##### *7.3.4.1.4 2-D [$^{1}H$,$^{15}N$] HSQC NMR Acquisition*

In the context of considering heteronuclear NMR data, it is worth noting the feasibility of acquiring data to correlate $^{1}H$ and $^{15}N$ nuclei in DNA samples. $^{15}N$ is naturally abundant as a dilute spin (0.37% of all nitrogen) (Chapter 5). Signal detection is, therefore, challenging when only millimolar concentrations of non-isotope–enriched DNA are considered. Nevertheless, provided that sufficient instrument time is available and larger quantities of DNA in solution, indirect detection of correlations between $^{1}H$ and $^{15}N$ nuclei is feasible via [$^{1}H$,$^{15}N$] HSQC to achieve the sensitivity increase arising from polarisation transfer between $^{1}H$ and $^{15}N$ nuclei (Chapter 5). Since labile (imino and amino) protons are involved, data acquisition must occur from a sample dissolved in 90% $H_2O$/10% $D_2O$. Provided that the labile proton resonances are observed in the 1-D $^{1}H$ NMR spectrum (slow chemical exchange on the NMR timescale; Chapter 2), 2-D [$^{1}H$,$^{15}N$] HSQC NMR data may be acquired. The $^{15}N$ offset frequency should be centered at or near to $\delta^{15}N$ = 125 ppm, and the $^{15}N$ frequency width should be set to 250 ppm. This allows capture of $^{15}N$ resonances arising from nitrogen within amino groups ($\delta^{15}N$ typically ~95 ppm) and simultaneously within imino groups ($\delta^{15}N$ typically ~145 ppm). The critical pulse sequence delay $1/(4J)$ may be set for $^{1}J_{NH}$ = 75 to 90 Hz. Since signal-to-noise is the main challenge in such data acquisitions at natural abundance, resolution in the indirect dimension may be compromised in favour of more transients per $t_1$ increment to directly improve the signal-to-noise ratio. A classic example of a 2-D [$^{1}H$,$^{15}N$] HSQC NMR data set acquired at $^{15}N$ natural abundance is shown in Figure 7.32, with acceptable signal-to-noise ratio.

#### 7.3.4.2 Samples in 99% $D_2O$

With care in recovery and handling, DNA samples may repeatedly be freeze-dried to allow solvent exchange to take place as required without any sample degradation. The advantages of solubilising a nucleic acid in $D_2O$ have already been mentioned. In general, NMR data acquisition for such samples are somewhat less complex and, in some respects, provide for greater scope in data quality and type. High-quality results may be largely devoid of artefacts that are often caused by the presence of residual solvent signals.

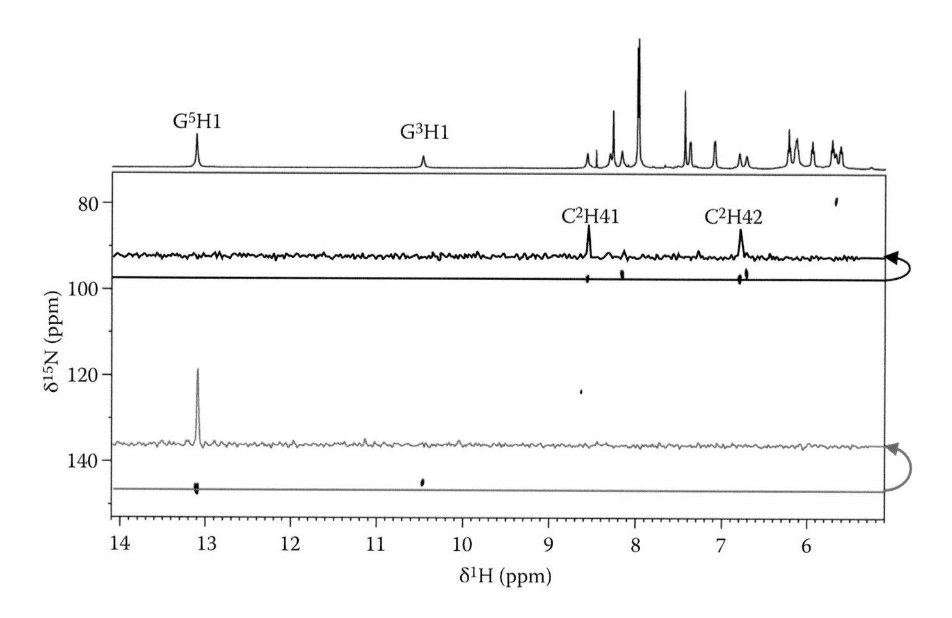


**FIGURE 7.32** 600-MHz 2-D [$^1$H,$^{15}$N] HSQC NMR spectrum of the 6-mer tandem-sheared G–A mismatch duplex DNA d(GC$^2$G$^3$AGC)$_2$ (2 mM in 90% $H_2O$/10% $D_2O$; 100-mM potassium phosphate buffer; pH = 7.4, 298 K). 360 transients, 100 $t_1$ increments using a sensitivity-improved gradient coherence selection pulse sequence.[96–98] Acquisition time, 16 h. $\omega_2$ slices are shown taken through the cross-peak positions at $\delta^{15}$N = 146.3 (grey) and 97.5 (black) ppm for which the signal-to-noise ratios are calculated as 29.45:1 and 11.63:1, respectively. The 1-D $^1$H NMR spectrum (top) acquired using the DPFGSE solvent suppression scheme[94] is shown, and assignments are indicated according to the nucleic acid numbering scheme. G$^5$ refers to the complement to C$^2$ as the Watson–Crick base-pair C$^2$ • G$^5$.

#### *7.3.4.2.1 1-D $^1$H NMR Spectra*

Loss of resonances arising from imino and amino protons simplifies, to some extent, the $^1$H NMR spectrum between 8.0 and 6.5 ppm and between 6.2 and 5.0 ppm (Figure 7.29). A further advantage is that solvent suppression may be much less severe as described previously. Although this is of less value in 1-D $^1$H NMR data acquisition, it has great value in 2-D [$^1$H,$^1$H], [$^1$H,$^{31}$P] and [$^1$H,$^{13}$C] correlation data acquisition, allowing those resonances that would otherwise be attenuated by solvent suppression to assume their full intensity. Comparisons of 1-D $^1$H and 2-D NOESY NMR data acquired with DNA in 90% $H_2O$/10% $D_2O$ and in 99% $D_2O$ are shown in Figure 7.33.

#### *7.3.4.2.2 2-D $^1$H NMR Spectra*

NOESY and ROESY NMR data sets are complemented by TOCSY and DQFCOSY, each of which further reduces data complexity, when working in $D_2O$ solution. In the former case, data are reduced to include only resonances from isolated spin systems. These include the network of spins within thymine bases (H6/$CH_3$), those within cytosine bases (H6/H5), and those associated with the sugar rings (H1′, H2′,

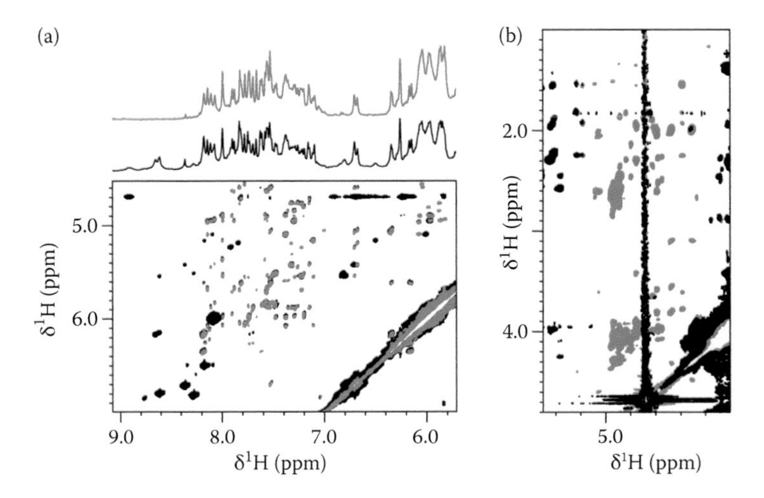


**FIGURE 7.33 (See color insert.)** Overlay of 500-MHz 2-D NOESY NMR data for a ligand complex of 15-mer duplex DNA d(CGACTAGTCTAGACG) • d(CGTCTAGACTAGTCG) at 25°C in 99% $D_2O$ (red) and 90% $H_2O$/10% $D_2O$ (black). (a) Reduction in the number of NOEs arising from the absence of amino-proton resonances in the 99% $D_2O$ solution. (b) Bleaching out of cross-peaks in the vicinity of the solvent resonance (90% $H_2O$/10% $D_2O$, black) compared with the presence of cross-peaks in the same region for the same DNA complex sample dissolved in 99% $D_2O$.

H2″, H3′, H4′, H5′, H5″). In the case of DQFCOSY spectra (e.g. Figure 7.30), data are reduced further to only show cross-peaks arising from near neighbours, largely, but not exclusively, arising from $^3J_{HH}$ couplings. Cross-peaks arising from $^4J_{HH}$ couplings are observed between methyl and H6 protons of thymine bases. Typically, TOCSY data may be acquired with a spin-lock time of 70 ms. As with all spin-locking sequences, sample heating is a consideration, and this is particularly important given the strong temperature dependence of the $^1$H (in water) resonance frequency. This can be handled by using a significant dummy acquisition period (typically, 5 min).

### *7.3.4.2.3 2-D [$^1$H,$^{31}$P] HSQC NMR Acquisition*

Correlation data acquired for nucleic acids between $^{31}$P and $^1$H nuclei enable through-bond relationships to steer the signal assignment process for both $^{31}$P NMR resonances and resonances arising from H3′, H4′ and H5′/H5″ protons. Signals from the latter crowd the chemical shift region of the $^1$H NMR spectrum generally between $\delta^1$H = 5.2 and 3.7 ppm, although this frequency window may need to be widened if the relevant protons experience substantially greater shielding or de-shielding than is normally the case (Figure 7.34b). Such crowding often reduces the utility of 2-D [$^1$H,$^1$H] correlation data when used alone in determining $^1$H NMR signal assignments for DNA. However, when supplemented with data from 2-D [$^1$H,$^{31}$P] HSQC or similar correlation experiments, the benefits to the assignment process become clear.

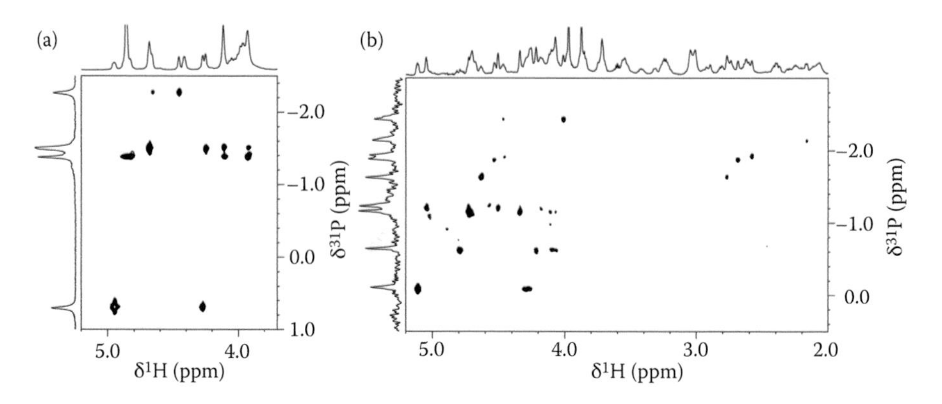


**FIGURE 7.34** Examples of 2-D [$^1$H,$^{31}$P] HSQC NMR data sets arising from DNA: (a) Data from tandem-sheared G–A mismatch duplex d(ACGAGT)$_2$, and (b) data from a ligand complex of the 10-mer DNA duplex d(CGACTAGTCG)$_2$. Note that, in both cases, the chemical shift dispersion in the $^{31}$P frequency dimension brought about through alterations to the DNA backbone conformation from the standard B$_I$ conformation. It is also notable in (b) that cross-peaks appear in the $^1$H chemical shift region between δ$^1$H = 2.0 and 3.0 ppm. These signals arise from H4′ protons undergoing magnetic shielding, and in this instance, the cross-peaks exist via $^5J_{PH}$ $^{31}$P–H4′ couplings.

Given that correlations between $^{31}$P and $^1$H nuclei are generally specific to protons resonating between δ$^1$H = 5.2 and 3.7 ppm, it is possible to reduce the frequency width of the $^1$H NMR data acquisition window when acquiring such data. This is because responses only arise as a result of signal filtering through the $^{31}$P NMR nucleus. A smaller frequency width is generally associated with the $^{31}$P NMR spectrum except in cases where the phosphodiester backbone is chemically modified in some way, as described previously. 2-D [$^1$H,$^{31}$P] HSQC and related NMR data may, therefore, generally be acquired using a small data set of typically 512 data point in the acquisition dimension and 64 $t_1$ increments in the indirectly detected dimension. HSQC requires the choice of delay, $1/(4J)$, to optimise the signal response. The value $J$, in this instance, may be set to between 5 and 10 Hz, resulting in reasonably large inter-pulse delays. It is often advisable to acquire several data sets with different values for the delay period $1/(4J)$, which allows the optimum conditions for data acquisition to be set. Data arising from such acquisition conditions for a DNA complex and for a short tandem-sheared G–A mismatch duplex are shown in Figure 7.34. As indicated, filtering of $^1$H NMR resonances as a function of $^{31}$P chemical shift allows resonance dispersion to be achieved. In the best cases, this allows the link to be made between resonances associated with adjacent sugar rings.

#### *7.3.4.2.4 2-D [$^1$H,$^{13}$C] HSQC NMR Acquisition*

In a similar way to that described for [$^1$H,$^{31}$P] correlations, 2-D [$^1$H,$^{13}$C] HSQC for nucleic acids is feasible provided that sufficient material is present in the sample solution. Such data acquisition can typically be attempted for DNA samples at concentrations above 2 mM using a sufficiently sensitive probehead (typical $^1$H signal to noise equivalent to 1400:1 or above on a standard 0.1% ethylbenzene in $CDCl_3$

sample). Under these conditions, data acquisition still requires many (e.g. typically, 64 or more) transients per $t_1$ increment to generate sufficient signal-to-noise for the $^1$H–$^{13}$C cross-peaks to become visible within the Fourier-transformed data sets. However, the advantage of such data acquisition lies in the $^{13}$C frequency separation that is generated for $^1$H NMR resonances. This enables clear identification of clusters of proton types and specific correlations associated with $^1J_{CH}$ couplings. Such an approach is especially helpful when large changes in proton chemical shifts occur as a result of alterations to a DNA molecule, either through conformational changes or via the effects of interactions with other molecules. Whereas $^1$H chemical shifts may change dramatically, $^{13}$C chemical shifts of the carbons to which these protons are attached remain largely invariant. Such an effect can, therefore, be used to best advantage when attempting to identify specific signals. A typical 2-D [$^1$H,$^{13}$C] HSQC NMR data set for DNA (Figure 7.35) shows how the signals cluster, thereby allowing

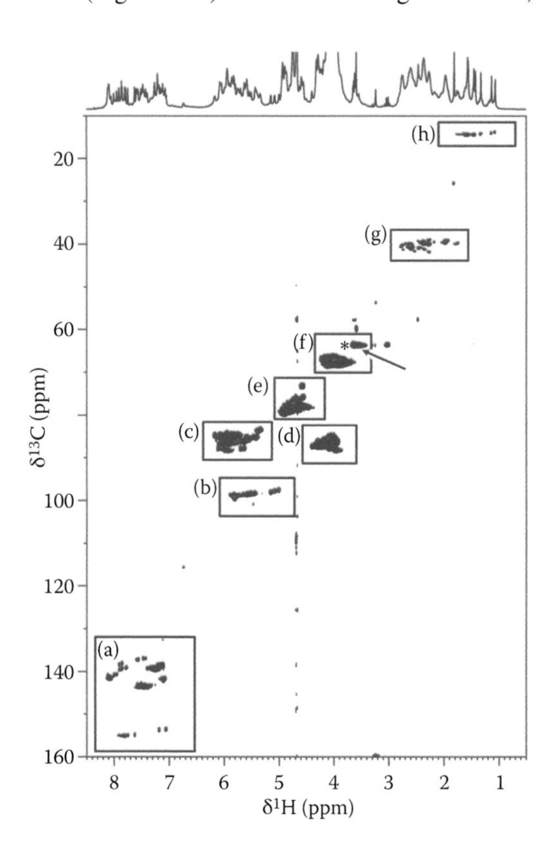


**FIGURE 7.35** 600-MHz 2-D [$^1$H,$^{13}$C] HSQC NMR spectrum acquired on 15-mer DNA duplex d(CGACTAGTCTAGACG) • d(CGTCTAGACTAGTCG), with different regions of data highlighted for different types of one-bond proton-carbon correlations: (a) aromatic H8–C8/H6–C6/H2–C2, (b) cytosine H5–C5, (c) H1′–C1′, (d) H4′–C4′, (e) H3′–C3′, (f) H5′/H5″–C5′, (g) H2′/H2″–C2′ and (h) thymine methyl H/C correlations. The arrow highlights the cross-peak (labelled '*') that is associated with the 5′-terminal H5′/H5″–C5′ cross-peak, which can be key to making a successful data assignment.

identification and scope for signal assignment. The figure shows labelled those regions of data into which particular clusters of resonances fall. For 2-D [$^{1}$H,$^{13}$C] HSQC spectra, the key delay, 1/(4$J$), may be set to a value of $^{1}J_{CH}$ = 140 to 150 Hz. As shown (Figure 7.35), correlations within the $^{13}$C HSQC are spread roughly along a diagonal from top right to bottom left, a common feature of most [$^{1}$H,$^{13}$C] correlation data, being a function of the increase in de-shielding experienced by the respective protons and carbons to which they are directly attached. One way of increasing digital resolution in the $^{13}$C indirect detection dimension is to purposely fold the data along $\omega_1$ ($F_1$) and acquire two data sets instead of one. Thus, the center of the $^{13}$C indirect detection dimension may be set to $\delta^{13}$C = 110 ppm. This is combined with a narrow $\omega_1$ frequency width equivalent to $\Delta\delta^{13}$C = 100 ppm. The result is to fold the regions labelled (g) and (h) in Figure 7.35 into the bottom right-hand corner of the data set. Such folding will not interfere with the unfolded regions (a) to (f) and will also allow greater digital resolution along $\omega_1$. Similarly, the center of the $^{13}$C indirect detection dimension may be set to $\delta^{13}$C = 80 ppm with the $\omega_1$ frequency width set to $\Delta\delta^{13}$C = 100 ppm, thereby folding the region marked (a) in Figure 7.35 into the top left-hand corner of the data. The additional benefit that this introduces is an ability to more efficiently decouple the $^{13}$C frequency range, especially at higher magnetic fields. The $^{13}$C decoupler power during free induction decay (FID) collection can, therefore, be reduced. The compromise is that the folded signals may not be decoupled, owing to the narrower decoupling field applied as a result of the reduction in relative decoupler power, hence the recommendation and benefit of acquiring two data sets with overlapping frequency windows in the $^{13}$C dimension.

#### 7.3.4.3 2-D NMR Data Processing

While there are no special rules associated with the processing of NMR data from DNA, it is worth reiterating certain aspects of good practice in relation to post-acquisition handling of such data to achieve best results. Typically, data should be acquired in phase-sensitive mode with pure absorption line shape to ensure optimum cross-peak shape and resolution. Apodisation to improve signal to noise or resolution, followed by zero filling (to improve digital resolution), should be conducted as normal. Base-plane correction, the removal of $t_1$ ridge artefacts, and the application of a mild frequency filter in the acquisition dimension at the solvent resonance frequency are all required to ensure optimum quality data. Digital enhancement, including linear prediction to extend data resolution in the indirect dimension, may also be helpful when considering the use to which such data will be put, especially if data are significantly crowded as for longer DNA sequences. A further approach to reducing the size of large data sets is to strip-transform only those regions of data that carry NMR responses. This helps eliminate large regions of empty frequency space. A key aspect in all data processing is consistent referencing across all data sets. Ideally, this is best achieved by inclusion of an internal standard within the sample. Classically, TSP-$d_4$ (the sodium salt of 3-trimethylsilyl-propionic acid-$d_4$ in which only the methyl protons bear $^{1}$H nuclei) is used for such referencing. Once correctly referenced, all data sets acquired during the same data acquisition run will be frequency-registered with respect to one another. Provided that full alignment of all data sets is accurately made, coordinated displays can be used for data examination and signal assignment.

Such processes are facilitated by readily available software packages such as Sparky,[99] which are designed to assist with data visualisation and assignment.

#### 7.3.4.4 Temperature Considerations

The influence of temperature on the appearance of DNA NMR data should be carefully considered when designing a study. Nucleic acids, in general, possess all the thermodynamic drive to propel them spontaneously toward any number of assembled states, mainly being a function of DNA sequence complementarity. Provided that appropriate solution conditions persist, nucleic acids will invariably assemble themselves into low-energy, stable states. The nature of these is governed by hydrogen bond, π–π stacking and hydrophobic interactions combined with stabilisation through solvent hydration and counter-ion association. Base-pair complementarity lies at the heart of most DNA assembly. Such stable states form one side of an equilibrium process, the other side of which is represented by a non-assembled molecular system.

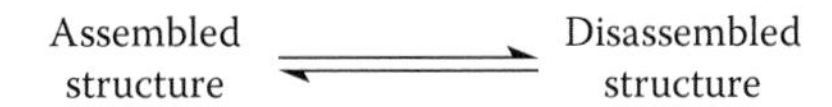


The left-hand side of this equilibrium is likely to be represented by the lowest energy structure. In solution, an ensemble average describes this by means of relatively tight sampling of conformational space. The right-hand side of the equilibrium, on the other hand, may be represented by an unfolded ensemble average structure describing very wide sampling of conformational space. The effects of introducing heat to shift the position of this equilibrium are readily observed by reference to either 1-D $^1H$ or $^{31}P$–{$^1H$} NMR data as a function of temperature. In this context, it is important to reiterate the use of a reference standard whose chemical shift is invariant with temperature if the true effects of temperature on the appearance of the solute DNA data are to be shown correctly. Tetramethyl-ammonium chloride ($\delta^1H$ = 3.178 ppm) is one such reference standard that is suitable for $^1H$ NMR spectroscopy. Two features of DNA NMR data generally change as a function of temperature: chemical shift and resonance line shape. The exact appearance of the NMR data will depend strongly on the equilibrium condition. Two examples are shown that represent different positions on the equilibrium continuum (Figure 7.36).

The condition associated with fast exchange on the NMR (chemical shift) timescale (Chapter 2) is shown in Figure 7.36a for temperatures above 323 K and in Figure 7.36b for temperatures above 298 K. Under this condition, resonances sharpen and move gradually to higher chemical shift with each temperature increase. In general, $^1H$ NMR chemical shifts of DNA samples tend toward different limiting values at high and low temperatures. These represent the limiting chemical shift values for completely disassembled and completely assembled materials, respectively. Between these two extremes, the chemical shift of each signal is a population-weighted average that describes the position of the equilibrium at a given temperature. The gradual drift in chemical shift as a function of temperature in the fast exchange limit suggests comparatively weak interactions. This type of condition is indicative of an equilibrium position tending toward total disassembly, that is, single-strand DNA. In the case of d(GCAGGC) (Figure 7.36b), evidence of a shift in the equilibrium position

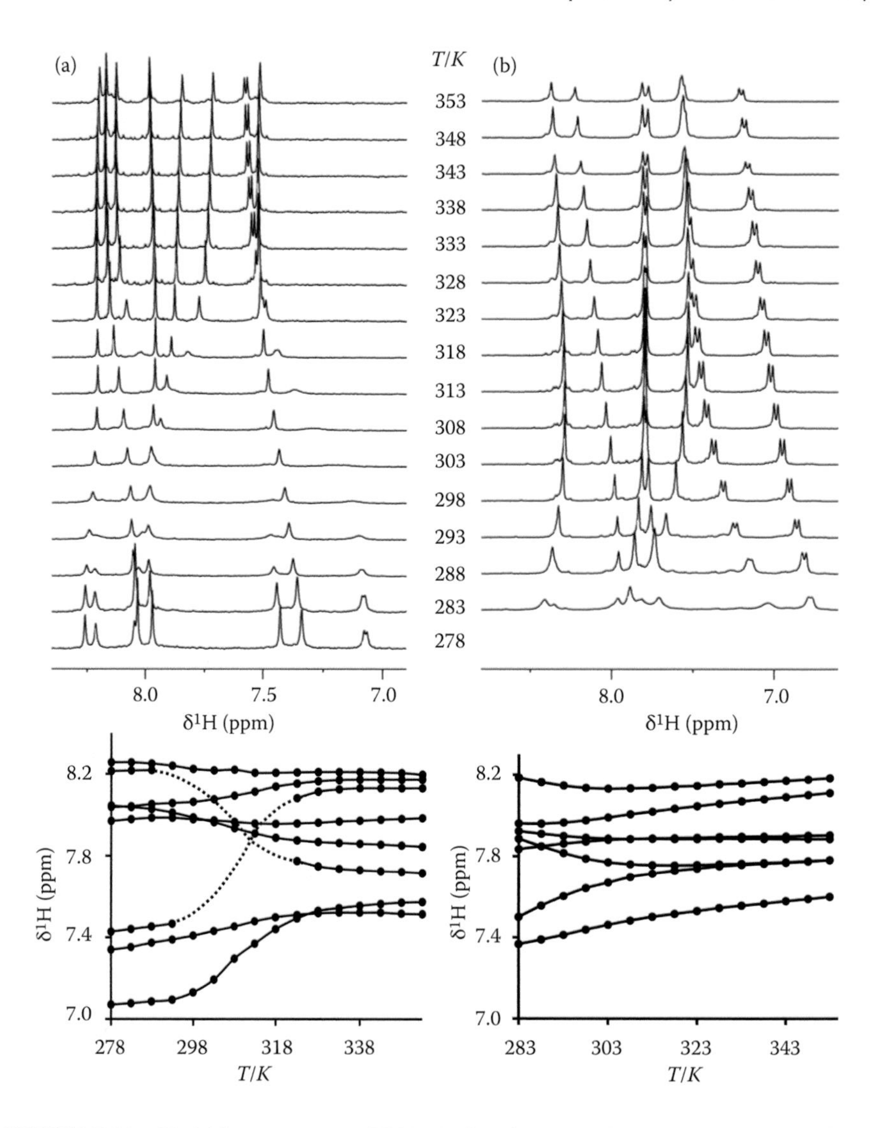


**FIGURE 7.36** Variable temperature $^{1}H$ NMR data for aromatic resonance regions of two 6-mer oligonucleotides: (a) d(ACGAGT) and (b) d(GCAGGC). NMR spectra are shown stacked as a function of temperature, with data reproduced below each stacked plot in the form of chemical shifts plotted as a function of sample temperature.

toward an assembled structure only becomes apparent at 283 K as the sample begins to enter the intermediate exchange condition. The implication for this molecule is that it does not spontaneously form a stable assembled structure at room temperature and requires cooling to induce any form of assembly. In contrast, at this temperature, the data for d(ACGAGT) is relatively sharp and well defined, but on heating, resonance broadening and significant changes in chemical shift become evident (Figure 7.36a). At low temperatures, this DNA shows strong evidence of assembly consistent with the slow exchange limit (Chapter 2) and a mismatched duplex DNA molecule in line with expectations for a tandem-sheared G–A structure. With heating, the duplex begins to melt, the data taking on intermediate exchange characteristics at the midpoint of the melting process. At low temperature, the equilibrium entirely favours a fully assembled structure. Significant chemical shift changes take place with increasing temperature, which are accompanied by some resonance broadening. This becomes dramatic to the point that some signals merge into the baseline and effectively disappear from the NMR spectrum at the melting temperature. Further increase in temperature results in new signals emerging from the baseline and in the whole spectrum sharpening up. At the highest temperature, the NMR spectrum represents the completely disassembled DNA. Further examples of slow and intermediate exchange conditions are represented in the following section in the context of small molecule complexes of DNA (Section 7.3.5), and in Chapter 2.

Although chemical shifts generally tend to move to higher values as DNA structures disassemble, exceptions do occur as noted for the tandem-sheared G–A mismatch (Figure 7.36a). The more unusual sheared structure places the relevant protons in substantially different environments compared with standard duplex DNA. The G–A mismatch base pair is shown represented in Figure 7.37. Upon melting, the NMR chemical shifts assume their typical random coil values. Magnetic shielding arises as a result of $\pi$–$\pi$ stacking interactions, and as unfolding takes place, the shielding effects arising from stacked, adjacent, aromatic rings are lost. G–A mismatches are interesting species with sometimes exceptional stability, and more may be read about these in the context of different studies including DNA,[100] RNA[101] and DNA complexes with small molecules.[102,103]

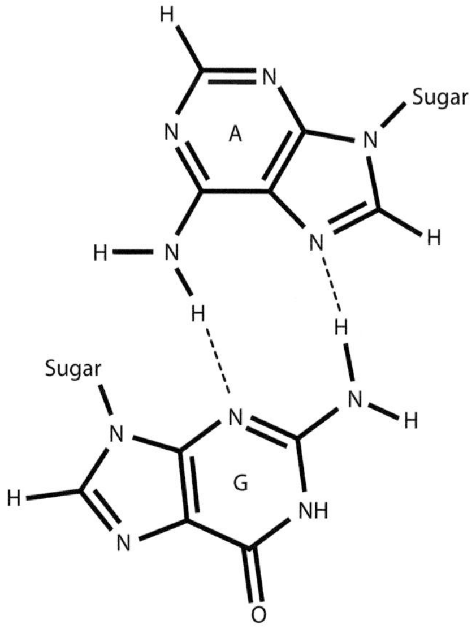

**FIGURE 7.37** Representation of the sheared base-pairing arrangement within a tandem-sheared G–A base-pair mismatch.

### 7.3.5 Assessment of Ligand Binding Using 1-D NMR

A benefit of using NMR spectroscopy to study DNA molecules lies in its ability to reveal binding processes between DNA and other molecules (Chapter 3). Changes in the appearance of the DNA NMR spectrum generally report on the nature of any interaction/complex forming processes. These will likely conform to the general equilibrium scheme shown in the following:

Disrupted DNA ⇌ Assembled structure + Ligand ⇌ Ligand/DNA complex

As with the DNA assembly/disassembly, the appearance of NMR data representing a position along the equilibrium continuum in the presence of a ligand will depend on the rate of exchange between the ligand and the DNA. This, in turn, is dictated by the binding constant, $K_a$. In the extreme case of ligand attachment via covalent binding to DNA, the 'equilibrium' lies at the extreme right-hand side of this scheme. For ligands bound non-covalently but tightly to their DNA target, NMR data will conform to the slow exchange limit (Chapter 2). Weaker binding will result either in line broadening effects, representing intermediate exchange, or incremental chemical shift changes, representing the fast exchange limit as DNA is titrated or reacted with ligand (cf. Chapter 3 and Section 7.2 for analogous observations for protein-based systems). In the worst-case scenario, the presence of ligand could disrupt the DNA structure altogether. In the absence of data assignment knowledge, recorded NMR data could simply be used for screening ligand binding. Changes in the appearance of data represent evidence for binding, which, in turn, can be further investigated through detailed NMR data analysis if required.

Non-covalent binding studies between ligands and the DNA minor groove are multifaceted and widely represented throughout the scientific literature. The DNA minor groove serves as one target for small-molecule ligand design and recognition. It accepts positively charged ligands that are planar and curved and that possess a suitable array of potential hydrogen-bond partners. If shape, charge, and hydrogen bond capacity are satisfied, ligand binding may be favoured. Screening for binding by NMR is achieved most readily by reference to changes in the appearance of the imino-proton resonance region of the $^1$H NMR spectrum upon addition of ligand to DNA solution. This is represented for tight (but non-covalent) binding in the left-hand portion of Figure 7.38. Here, a 50:50 mixture (Figure 7.38a (2)) of ligand-bound and ligand-free DNA strongly suggests ligand binding to DNA in an anti-parallel, face-to-face fashion – the data report that structural symmetry is retained. If asymmetry was imposed by the ligand, which is asymmetric, twice as many imino-proton resonances would be observed for the complex. In this mode, chemical exchange is slow on the NMR timescale (Chapter 2). Rotational symmetry is assessed by comparison of the number of expected imino-proton NMR resonances and the number of resonances observed, as described previously. In contrast, weak, non-specific association is indicated by the data shown centered in Figure 7.38. Weak, non-specific association is suggested by the mildly broadened resonances detected upon addition of ligand to the solution. In this instance, fast exchange is implied since no change in

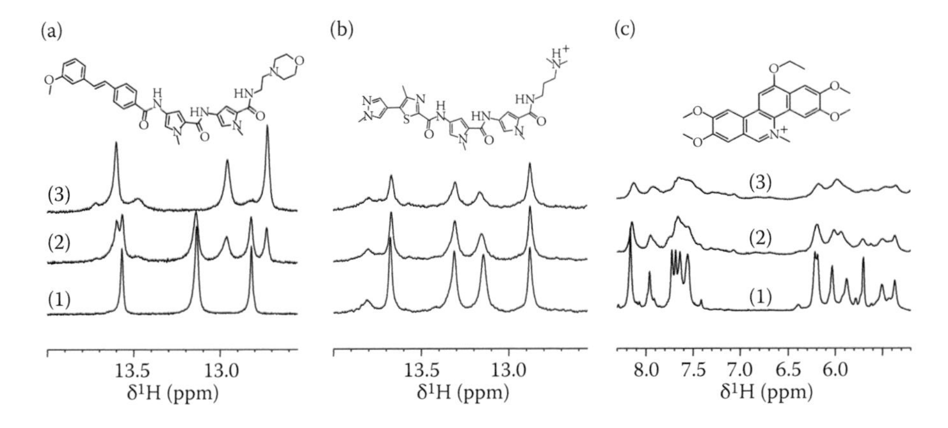


**FIGURE 7.38** Screening of non-covalent binding by ligands to DNA: (1) no ligand, (2) 1:1 (0.5:1) ligand:DNA and (3) 2:1 (1:1) ligand:DNA (figures in parenthesis relate to data at (c)). (a) Tight sequence-specific binding of ligand to the DNA minor groove of duplex $d(CGATATATCG)_2$ in 90% $H_2O$/10% $D_2O$. (b) Weak electrostatic minor groove/phosphodiester association between ligand and DNA duplex $d(GCACATATGTGC)_2$ (materials supplied courtesy of F. Brucoli). (c) Screening of intercalation between ligand and DNA duplex $d(TTCGAA)_2$.

proton chemical shift is detected for any of the imino-proton resonances. Neither do new signals appear in the data. Any association under these circumstances is likely only via electrostatic interaction between the ligand and the phosphodiester backbone of DNA in a non-sequence specific manner, as evidenced by slight broadening of resonances upon ligand addition.

Intercalation, the process by which small, flat, aromatic molecules become sandwiched between DNA base pairs, is represented in Figure 7.38c. Addition of an intercalating ligand such as the one shown results in substantially less well-defined NMR spectra. This is largely the result of the complex's lifetime. The substantial line broadening in the NMR data severely degrades the quality of the results. Nevertheless, this is a very good indicator that intercalation processes are present. The resonance broadening causes issues regarding clear definition of the nature of any complex formed, and it is often difficult to generate high-quality NOE data under such circumstances. Overall, the use of either 1-D $^1H$ or $^{31}P-\{^1H\}$ NMR data to screen ligand binding (Chapter 3), in this way, is very straightforward and provides an instant suggestion of the nature of any complex formed without the need to resort to more complex NMR methods if the full structural details of the binding process are not required.

### 7.3.6 Resonance Assignment

Detailed assignment of NMR data from DNA samples and their complexes with other molecules follows a logical process and requires cross-referencing between different types of NMR data to gain confidence in signal identification. NMR data from small DNA molecules are inevitably less crowded than for larger DNA structures, showing sharper signals and greater resolution. The assignment principles remain the same, however. To illustrate this process, the mismatch DNA duplex d(GCGAGC)$_2$ system is employed, the data for which show good dispersion and allow the principles of data assignment to be demonstrated readily. The 1-D $^1$H NMR spectrum of the sample in both 99% $D_2O$ and 90% $H_2O$/10% $D_2O$ in the chemical shift range $\delta^1$H = 0.5 to 9 ppm is shown (Figure 7.39).

Starting with the 1-D $^1$H NMR spectrum in 99% $D_2O$, the signals may be identified according to shape, chemical shift, and integral. The characteristic appearance of particular signals should be noted. Cytosine H6 and H5 resonances appear at their respective chemical shifts as doublets with typical $^3J_{HH}$ = 7.4 Hz. H1′ resonances appear as either triplets or double doublets (often broadened) through $^3J_{HH}$ coupling to both H2′ and H2″ protons. H8 and H2 resonances appear as singlets at chemical shifts characteristic of the bases to which they are associated (Table 7.1). Amino $^1$H NMR resonances are generally only observed for cytosine H41 (hydrogen bonded through Watson–Crick pairing; typically, $\delta^1$H > 8.0 ppm) and H42 protons (typically, $\delta^1$H ~ 6.7 ppm). Amino-proton resonances from other bases (adenine H61/H62, guanine H21/H22) are generally not observed due to intermediate exchange processes associated with these groups, causing the respective resonances to broaden into the baseline and beyond detection. The example of d(G$^1$C$^2$G$^3$A$^4$G$^5$C$^6$)$_2$ is an exception to this rule. Its data show responses arising from G$^3$H22 ($\delta^1$H = 9.45 ppm), A$^4$H61 ($\delta^1$H = 8.27 ppm), and A$^4$H62 ($\delta^1$H = 5.65 ppm).[100] Signal counting, inspection and

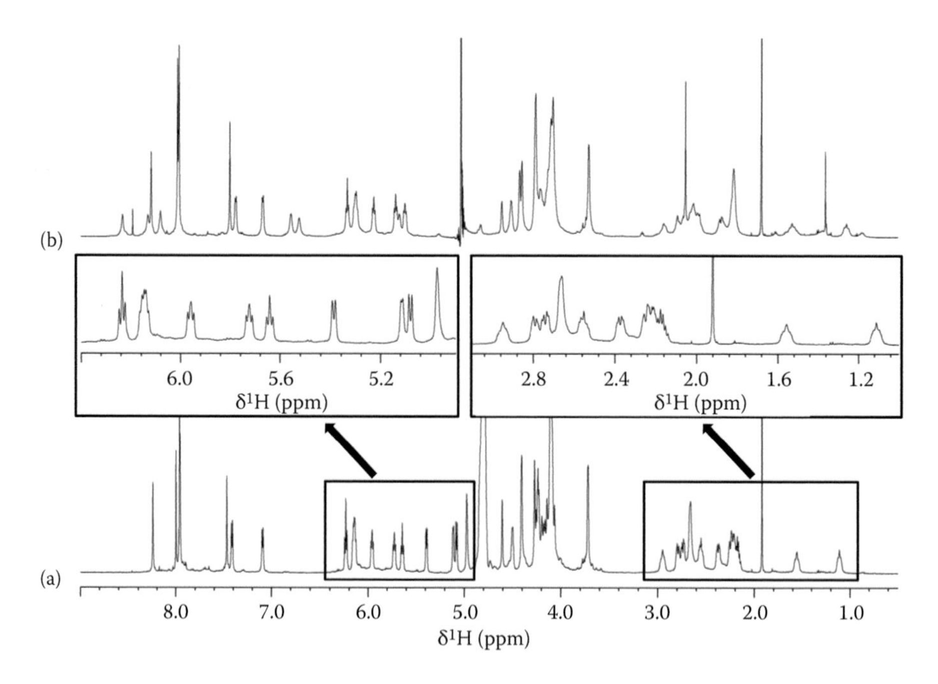


**FIGURE 7.39** 600-MHz 1-D $^1$H NMR spectrum of 2-mM d(GCGAGC)$_2$ in 99% $D_2O$ (a) and 90% $H_2O$/10% $D_2O$ (b), showing the chemical shift region between $\delta^1$H = 0.5 and 9.0 ppm. Expansions are shown for the H1′/C$^n$H5 and H2′/H2″ resonance regions. Additional signals in (b) between $\delta^1$H = 6.5 and 8.5 ppm arise from labile amino-proton resonances.

integration of the data shown in Figure 7.39 suggest five singlets and two doublets in the aromatic region, consistent with seven expected aromatic signals for the requisite DNA sequence. Six multiplets between $\delta^1H = 5.6$ and 6.3 ppm are consistent with the expected number of signals arising from H1′ resonances for six different sugar rings in the sequence.

The next step is to examine the 2-D [$^1H$,$^1H$] NOESY NMR spectrum of the DNA sample solubilised in 99% $D_2O$. Cross-peak regions are shown labelled in the full data set (Figure 7.40).

The key region from which a starting point and a subsequent sequential assignment may be made is labelled (e) in Figure 7.40, which is shown expanded in Figure 7.41.

The signal assignment process follows a sequential logic and provides a unique solution for a stable assembled structure. In the first instance, however, this process is non-linear and remains so until assurance is gained of the first cross-peak assignment. Patterns of intra-residue and inter-residue proton–proton NOEs are, eventually, identified in a stepwise fashion, with a signal assignment pathway being traced along the entire DNA sequence from 5′ to 3′ ends. The link is made between

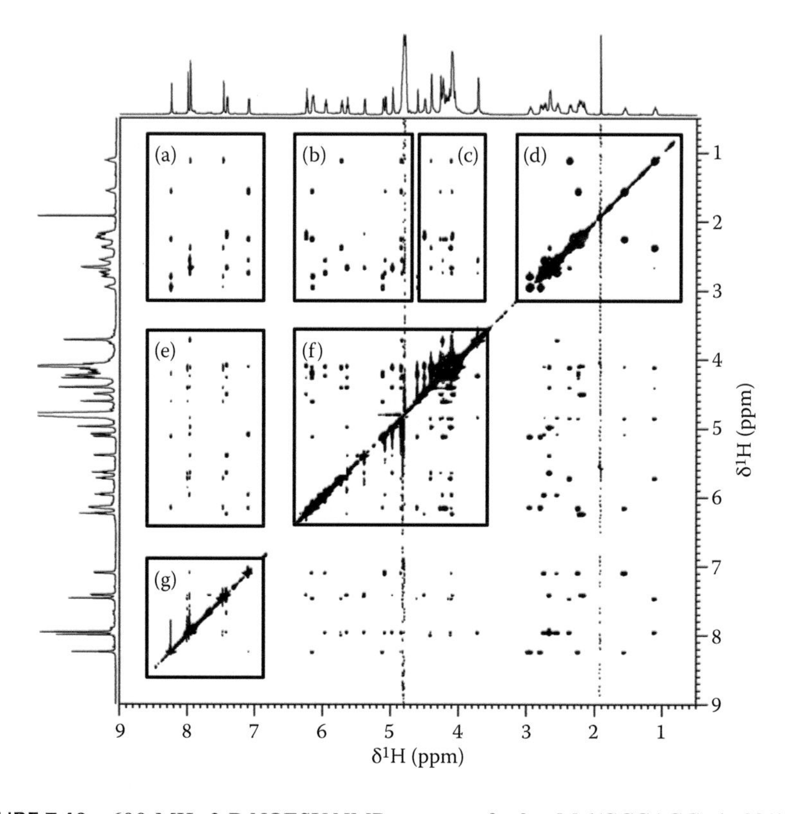


**FIGURE 7.40** 600-MHz 2-D NOESY NMR spectrum for 2-mM d(GCGAGC)$_2$ in 99% $D_2O$ with 100-mM phosphate buffer at pH = 7.4. Cross-peak regions highlighted are aromatic-H2′/H2″/$CH_3$ (a), H1′/H5/H3′–H2′/H2″ (b), H4′/H5′/H5″–H2′/H2″/$CH_3$ (c), H2″–H2′/$CH_3$ (d), aromatic-H1′/H5/H3′/H4′/H5′/H5″ (e), sugar protons and H5 interactions excluding H2′/H2″ (f) and aromatic–aromatic (g).

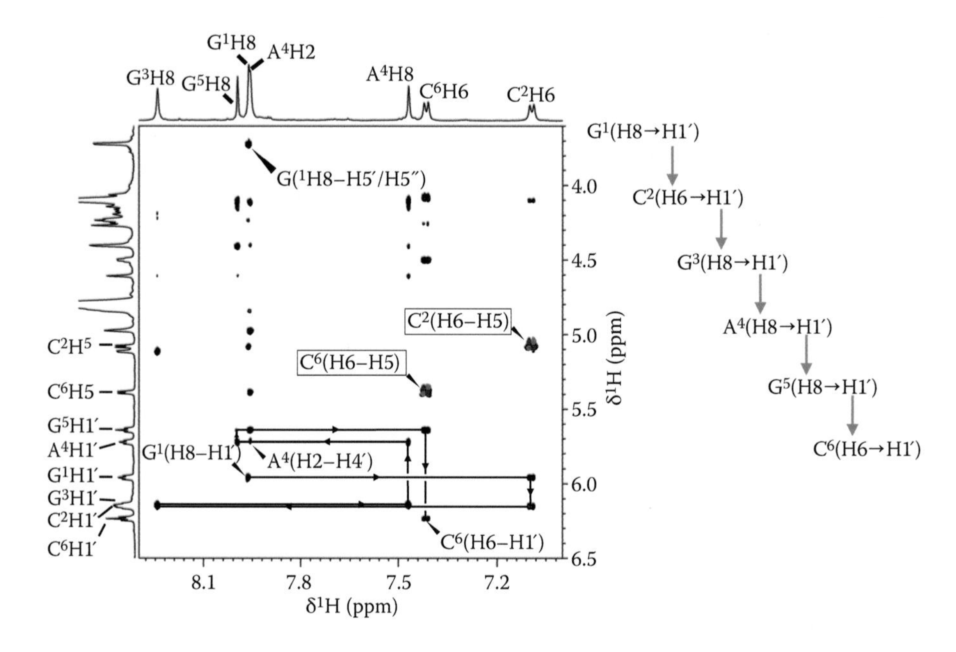


**FIGURE 7.41** Expansion of region (e) of the 2-D [¹H,¹H] NOESY NMR spectrum of the mismatched DNA duplex d(G¹C²G³A⁴G⁵C⁶)$_2$ from Figure 7.40. Bases are numbered according to ascending sequence from the 5′-terminus. Intra-residue NOE cross-peaks are labelled as BASE$^{\text{Position}}$($\omega_2$ chemical shift assignment–$\omega_1$ chemical shift assignment). NOESY cross-peaks coincident with COSY cross-peaks are shown overlaid (with DQFCOSY cross-peaks in boxed) for cytosine H6–H5 correlations.

pairs of protons by means of NOE correlations. Any one of a number of possible proton–proton correlation pathways may be traced along the DNA sequence. In the example, the pathway traced for d(G¹C²G³A⁴G⁵C⁶) in Figure 7.41 is between aromatic and H1′ proton resonances. Similar NOE pathways may be traced between aromatic and H2′/H2″ protons or within the aromatic proton resonance group only. In this example, the set of NOEs that need to be identified uniquely for the sequential assignment via aromatic and H1′ resonances are shown diagrammatically to the right of Figure 7.41.

Identification of the first resonance in the sequence starts at the 5′-terminal residue. A clearly identifiable cross-peak is observed between the 5′-terminal aromatic and the H5′/H5″ protons of the sugar ring to which the 5′-base is attached. Invariably, for synthetic DNA, the 5′-terminal sugar does not possess a phosphate group attached to the 5′ oxygen. The chemical shift of the associated H5′ and H5″ proton resonances always shows a signature ¹H NMR resonance at or close to δ¹H = 3.72 ppm. This is clearly resolved from the remaining phosphodiester-associated H5′/H5″ resonances, which occur at higher chemical shifts. In the example (Figure 7.41), the resolved signal at δ¹H = 3.72 ppm correlates with a resonance at δ¹H = 7.96 ppm via the NOE cross-peak at chemical shift coordinates δ¹H($\omega_2$, $\omega_1$) = 7.96, 3.72 ppm. This correlation can only arise between the aromatic proton of the H5′-terminal residue ($G^1$H8) and $G^1$H5′/H5″. The NOE cross-peak is, thus, labelled $G^1$(H8–H5′/H5″). Note here that H5′ and H5″ protons resonate at the same frequency. This is often, but not always, the case for H5′ and H5″ resonances.

The resonance assigned to $G^1$H8 is coincident with a further aromatic proton resonance (subsequently identified as $A^4$H2). Such coincidence is not uncommon in NMR studies of DNA. Ambiguity in resonance identification at this stage must be resolved, or else, it will lead to mistaken signal assignment and ultimately incorrect structure identification. To overcome this matter and to resolve potential assignment conflicts, 2-D NOESY data sets must be acquired on the same sample but at different temperatures within a ±10°C range. For structures that are firmly assembled at room temperatures, this represents a safe range within which data may alter but the structure is unlikely to melt. Temperature variation induces relative chemical shift changes among the resonances. When such data sets are compared, cross-peak relationships can generally be resolved and conflicts can be identified.

Having identified the aromatic resonance associated with the 5′-base, the next stage is to identify the H1′ resonance of the 5′-terminal sugar. Chemical shifts for DNA H1′ protons lie in the range $\delta^1$H = 5.5 to 6.3 ppm. The example shown (Figure 7.41) reveals a number of correlations that could be attributed to the through-space relationship $G^1$(H8–H1′), but only one cross-peak can be ascribed this identity. How can the correct cross-peak assignment be made? It is at this early point in the process of data assignment that the pathway must widen to encompass other data, before firm cross-peak assignment may be made that ascribes resonances confidently to both $G^1$H8 and $G^1$H1′. While examining the data for an NOE cross-peak, which satisfies the relationship $G^1$(H8-H1′), the corresponding inter-residue NOE $n$H1′–($n$ + 1) (aromatic) must also be identified. Here, $n$ and $n + 1$ represent residues that share a 5′/3′ relationship to one another, respectively (represented by $G^1$H1′–$C^2$H6 in this case). At this point, the identification of a doublet resonance corresponding to cytosine H6 is required. Two possibilities exist in the example since there are two cytosine residues in the sequence.

Resolution of the issue of which resonance is associated with which residue is achieved by considering the identity of neighbouring residues and the likely pattern of NOEs that will be observed. Consider cytosine at position 2 in the sequence, $C^2$. $C^2$H6 would correlate with $G^1$H1′ and $C^2$H1′ according to the pathway scheme of Figure 7.41. $C^2$H1′, in turn, would correlate with $G^3$H8. In other words, the pattern of NOE correlations associated with cytosine residue $C^2$ would be consistent with having both 5′- and 3′-nucleotide neighbours. $C^6$H6 would also give rise to two H1′ correlations, namely $G^5$H1′ and $C^6$H1′. However, in contrast to cytosine residue $C^2$, $C^6$ does not possess a 3′-neighbouring residue. It lies at the 3′-terminus of the DNA sequence. The assignment 'walk' would, therefore, also terminate at the cross-peak representing $C^6$(H6–H1′). Such a cross-peak may normally be identified by its relative isolation – no other cross-peaks lie at the same $\omega_1$ chemical shift. By elimination, therefore, the doublet resonance in the example at $\delta^1$H = 7.09 ppm may, therefore, be assigned to $C^2$H6. Several NOEs lie along the $\omega_1$ coordinate at $\omega_2$ = $C^2$H6. Two of these correlate $C^2$H6 to H1′ resonances in the region $\delta^1$H = 5.5 to 6.3 ppm. One of these correlations lies at the same $\omega_1$ chemical shift as a cross-peak located at $\omega_2$ = $G^1$H8. This cross-peak is then clearly identified corresponding to $G^1$(H8–H1′).

With this assignment confirmed, an assignment 'walk' can proceed in a stepwise manner. Sequential identification of intra- followed by inter-residue NOEs is made as shown by the trace linking cross-peaks in Figure 7.41. The key to this process is to ensure that, at every step, the identity of a cross-peak is confirmed by considering the context of surrounding cross-peaks. Identification of other features in the data aids this process. For instance, overlaying data allow some cross-peaks to be eliminated from consideration. In the example, this is shown for those cross-peaks between cytosine H6 and H5 protons. These give rise not only to NOEs in NOESY data, but also to correlations via $^3J_{HH}$ in COSY data. Other examples of this type may be identified by comparing different data. From this point forward, once an NOE cross-peak assignment pathway through the DNA sequence is identified, the assignment process can be broadened to encompass assignments of other proton resonances arising from the molecule. Subsequently, these may be further used to assign signals arising within other data sets (e.g. 90% $H_2O$/10% $D_2O$ data) and from other nuclei according to the type of data previously described for $^{31}$P, $^{13}$C and $^{15}$N.

Coordinated use of data sets to assign NMR data arising from DNA allows the assignment process to be resolved relatively efficiently in instances where the DNA structure allows this process to occur. The use of a variety of different types of data allows spin networks and spatial proximity between atoms to be firmly established. Torsion angle information becomes an extension to this that may be drawn from some of these data sets with implications regarding conformation and structure refinement. These points are beyond the scope of the current section. The process described by example is suitable when right-handed duplex DNA forms and for related triplex structures. Such assignment pathways are not wholly appropriate for instances when bases adopt syn- rather than anti-configurations about the C1′–N glycosidic bond or for looped-out elements of structures. In these cases, alternative data assignment strategies must be brought into play, which are not discussed further in this context.

### 7.3.7 Summary

For those interested in DNA or who may be interested in understanding how DNA may react or interact with other molecules or materials, the ability to study and use this substance as a tool is readily apparent. NMR spectroscopy is one of the, overriding analytical techniques that is used by a synthetic chemist to report on materials being manufactured in the laboratory or being examined for their behaviour and characteristics. The properties of DNA molecules, whether modified synthetically or not, are fascinating and unique. In this section, the manner by which DNA may be used and analysed using the powerful analytical toolbox that is NMR spectroscopy has been described, not only in terms of experimental approaches and data analysis, but also in terms of the fundamental design of the molecule being used for investigations. Careful forethought at the design stages goes a long way toward easing the process of understanding DNA NMR data and helps avoid pitfalls that can so easily be encountered through signal degeneracy and data complexity.

## REFERENCES

1. Newman, D. J., G. M. Cragg and K. M. Snader. 2003. Natural products as sources of new drugs over the period 1981–2002. *Journal of Natural Products* 66, pp. 1022–1037.
2. Newman, D. J. and G. Cragg. 2012. Natural products as sources of new drugs over 30 years from 1981 to 2010. *Journal of Natural Products* 75, pp. 311–335.
3. Wright, G. D. 2012. Antibiotics: A new hope. *Chemistry and Biology* 19, pp. 3–10.
4. Colwell, R. R. 2002. Fulfilling the promise of biotechnology. *Biotechnology Advances* 20, pp. 215–228.
5. Hill, R. T. and W. Fenical. 2010. Pharmaceuticals from marine natural products: Surge or ebb? *Current Opinion in Biotechnology* 21, pp. 777–779.
6. Hughes, C. C. and W. Fenical. 2010. Antibacterials from the sea. *Chemistry: A European Journal* 16, pp. 12,512–12,525.
7. Rinehart, K. and T. G. Holt. 1987. *Purification and Characterisation of Ecteinascidins 729, 743, 745, 759a, 759b, and 770 Having Antibacterial and Anti-Tumour Properties.* Illinois: University of Illinois, pp. 1–32.
8. Sigel, M. M., L. L. Wellham, W. Lichter, L. E. Dudeck, J. Gargus and A. H. Lucas. 1969. In: Food-Drugs from the Sea Proceedings. Younghen H. W. Jr, ed. Washington, DC: Marine Technology Society, pp. 281–294.
9. Wani, M. C., H. L. Taylor, M. E. Wall, P. Coggon and A. T. McPhail. 1971. Plant anti-tumour agents: VI. Isolation and structure of taxol, a novel anti-leukemic and anti-tumour agent from *Taxus brevifolia. Journal of the American Chemical Society* 93, pp. 2325–2327.
10. Urban, S. and F. Separovic. 2005. Developments in hyphenated spectroscopic methods in natural product profiling. *Frontiers in Drug Design and Discovery* 1, pp. 113–166.
11. Claridge, T. D. W. 1999. *High-Resolution NMR Techniques in Organic Chemistry.* Oxford, United Kingdom: Elsevier Science, Ltd, pp. 139–146, 227–312 and 328–334.
12. Webb, G. A. 2008. *Modern Magnetic Resonance: Applications in Chemistry, Biological, and Marine Sciences.* Dordrecht, The Netherlands: Springer, pp. 409–412.
13. Keeler, J. 2005. *Understanding NMR Spectroscopy.* Chichester, United Kingdom: John Wiley & Sons, pp. 194–204 and 279–291.
14. Dias, D. A. and S. Urban. 2011. Phytochemical studies of the Southern Australian marine alga *Laurencia elata. Phytochemistry* 72, pp. 2081–2089.
15. Morris, G. A. and J. W. Emsley. 2010. *Multi-dimensional NMR Methods for the Solution State.* West Sussex, England: John Wiley & Sons, Ltd, pp. 205–219 and 233–258.
16. Cimino, G., S. de Stefano and L. Minale. 1972. Polyprenyl derivatives from the sponge *Ircinia spinosula. Tetrahedron* 28, pp. 1315–1324.
17. Brkljaca, R. and S. Urban. Phytochemical investigation of the sea sponge *Dactylospongia* sp. (unpublished results).
18. Jacobsen, N. E. 2007. *NMR Spectroscopy Explained.* New Jersey: John Wiley & Sons, Inc, pp. 220–222.
19. Becker, E. D. 2000. *High-Resolution NMR: Theory and Chemical Applications.* San Diego, California: Academic Press, pp. 330–334.
20. Cooke, R. G. and I. J. Rainbow. 1977. Coloring matters of Australian plants: XIX. Haemocorin: Unequivocal synthesis of the aglycone an some derivatives. *Australian Journal of Chemistry* 30, pp. 2241–2247.
21. Morrison, G. A. and B. Laundon. 1971. Naturally occuring compounds related to phenalenone: Part II. The synthesis of haemocorin aglycone. *Journal of the Chemical Society* 9, pp. 1694–1704.
22. Brkljaca, R. and S. Urban. Phytochemical investigation of the Australian plant *Haemodorum* sp. (unpublished results).
23. Friebolin, H. 2005. *Basic One- and Two-Dimensional NMR Spectroscopy.* Weinheim, Germany: Wiley-VCH, pp. 276–280 and 295–296.

24. Hadden, C. E. 2005. Adiabatic pulses in $^{1}H$–$^{15}N$ direct and long-range heteronuclear correlations. *Magnetic Resonance in Chemistry* 43, pp. 330–333.
25. Hurd, R. E. 1990. Gradient-enhanced spectroscopy. *Journal of Magnetic Resonance* 87, pp. 422–428.
26. Ruiz-Cabello, J., G. W. Vuister, C. T. W. Moonen, P. van Gelderen, J. Cohen and P. C. M. van Zijl. 1992. Gradient-enhanced heteronuclear correlation spectroscopy: Theory and experimental aspects. *Journal of Magnetic Resonance* 100, pp. 282–302.
27. Bax, A. and M. F. Summers. 1986. $^{1}H$ and $^{13}C$ sssignments from sensitivity-enhanced detection of heteronuclear multiple-bond connectivity by 2-D multiple quantum NMR. *Journal of the American Chemical Society* 108, pp. 2093–2094.
28. Urban, S., R. Brkljaca, J. M. White and M. A. Timmers. 2013. Phenylphenalenones and oxabenzochrysenones from the Australian plant *Haemodorum simulans*. *Phytochemistry* 95, pp. 351–359; 96, p. 465.
29. Mucci, A., F. Parenti and L. Schenetti. 2002. On the recovery of $^{3}J_{H,H}$ and the reduction of molecular symmetry by simple NMR inverse detection experiments. *European Journal of Organic Chemistry* 5, pp. 938–940.
30. Ryan, J. M. 2008. *Novel Secondary Metabolites from New Zealand Marine Sponges*. Wellington, New Zealand: Victoria University of Wellington, pp. 67–69.
31. Brkljača, R. and S. Urban. 2013. Relative configuration of the marine natural product elatenyne using NMR spectroscopic and chemical derivatisation methodoligies. *Natural Product Communications* 8, pp. 729–732.
32. Hadden, C. E., G. E. Martin and V. V. Krishnamurthy. 2000. Constant time inverse-detection gradient accordion rescaled heteronuclear multiple-bond correlation spectroscopy: CIGAR-HMBC. *Magnetic Resonance in Chemistry* 38, pp. 143–147.
33. Krishnamurthy, V. V. 1996. Excitation-sculptured indirect-detection experiment (EXSIDE) for long-range CH coupling-constant Measurement. *Journal of Magnetic Resonance Series A* 121, pp. 33–41.
34. Brkljača, R. and S. Urban. 2011. Recent advancements in HPLC-NMR and applications for natural product profiling and identification. *Journal of Liquid Chromatography and Related Technologies* 34, pp. 1063–1076.
35. Timmers, M. A., D. A. Dias and S. Urban. 2012. Application of HPLC-NMR in the identification of plocamenone and isoplocamenone from the marine red alga *Plocamium angustum*. *Marine Drugs* 10, pp. 2089–2102.
36. Dias, D., J. M. White and S. Urban. 2009. *Laurencia filiformis*: Phytochemical profiling by conventional and HPLC-NMR approaches. *Natural Products Communication* 4, pp. 157–172.
37. Timmers, M. A., D. A. Dias and S. Urban. 2013. HPLC-NMR chemical profiling of the Australian carnivorous plant *Drosera erythrohiza* subspecies magna. *Natural Products Journal* 3, pp. 35–41.
38. Dias, D. A. and S. Urban. 2009. Application of HPLC-NMR for the rapid chemical profiling of a Southern Australian sponge *Dactylospongia* sp. *Journal of Separation Science* 32, pp. 542–548.
39. Dias, D. A. and S. Urban. 2008. Phytochemical analysis of the Souhtern Australian marine alga *Plocamium mertensii* using HPLC-NMR. *Phytochemical Analysis* 19, pp. 453–470.
40. Urban, S. and M. Timmers. 2013. HPLC-NMR chemical profiling and dereplication studies of the marine brown alga *Cystophora torulosa*. *Natural Products Communication* 8, pp. 715–719.
41. Timmers, M. and S. Urban. 2011. Online (HPLC-NMR) and offline phytochemical profiling of the Australian plant *Lasiopetalum macrophyllum*. *Natural Products Communication* 6, pp. 1605–1616.
42. Brkljaca, R. and S. Urban. Phytochemical profiling and biological activity studies of the Southern Australian marine alga *Sargassum paradoxum* (unpublished results).

43. Stoddart, J. F. 1971. *Stereochemistry of Carbohydrates*. New York: Wiley-Interscience, pp. 109–118.
44. Treumann, A., F. Xidong, L. McDonnell, P. J. Derrick, A. E. Ashcroft, D. Chatterjee and S. W. Homans. 2002. 5-methylthiopentose: A new substituent on lipoarabinomannan in *Mycobacterium tuberculosis*. *Journal of Molecular Biology* 316, pp. 89–100.
45. de Castro, C., A. Molinaro, F. Piacente, J. R. Gurnon, L. Sturiale, A. Palmigiano, R. Lanzetta, M. Parrilli, D. Garozzo, M. G. Tonetti and J. L. van Etten. 2013. Structure of N-linked oligosaccharides attached to chlorovirus PBCV-1 major capsid protein reveals unusual class of complex N-glycans. *Proceedings of the National Academy of Science U.S.A.* 110, pp. 13,956–13,960.
46. Turnbull, W. B., S. A. Kalovidouris and J. F. Stoddart. 2002. Large oligosaccharide-based glycodendrimers. *Chemistry: A European Journal* 8, pp. 2988–3000.
47. Collins, P. M. and R. J. Ferrier. 1995. *Monosaccharides: Their Chemistry and Their Roles in Natural Products*. Chichester, United Kingdom: John Wiley & Sons, pp. 488–494.
48. Roslund, M. U., P. Tähtinen, M. Niemitz and R. Sjöholm. 2008. Complete assignments of the $^{1}$H and $^{13}$C chemical shifts and $J_{H,H}$ coupling constants in NMR spectra of D-glucopyranose and all D-glucopyranosyl-D-glucopyranosides. *Carbohydrate Research* 343, pp. 101–112.
49. Bock, K. and C. Pedersen. 1974. A study of $^{13}$CH coupling constants in hexopyranoses. *Journal of the Chemical Society, Perkin Transactions* 2, pp. 293–297.
50. Mizutani, K., R. Kasai, M. Nakamura, O. Tanaka and H. Matsuura. 1989. NMR spectral study of α- and β-L-arabinofuranosides. *Carbohydrate Research* 185, pp. 27–38.
51. Turnbull, W. B., K. H. Shimizu, D. Chatterjee, S. W. Homans and A. Treumann. 2004. Identification of the 5-methylthiopentosyl substituent in *Mycobacterium tuberculosis* lipoarabinomannan. *Angewandte Chemie International Edition* 43, pp. 3918–3922.
52. Jansson, P.-E., R. Stenutz and G. Widmalm. 2006. Sequence determination of oligosaccharides and regular polysaccharides using NMR spectroscopy and a novel Web-based version of the computer program CASPER. *Carbohydrate Research* 341, pp. 1003–1010.
53. Loß, A., R. Stenutz, E. Schwarzer and C.-W. von der Lieth. 2006. GlyNest and CASPER: Two independent approaches to estimate $^{1}$H and $^{13}$C NMR shifts of glycans available through a common Web interface. *Nucleic Acids Research* 34, pp. W733–W737.
54. Stenutz, R., I. Carmichael, G. Widmalm and A. S. Serianni. 2002. Hydroxymethyl group conformation in saccharides: Structural dependencies of $^{2}J_{HH}$, $^{3}J_{HH}$, and $^{1}J_{CH}$ spin–spin coupling constants. *Journal of Organic Chemistry* 67, pp. 949–958.
55. Thibaudeau, C., R. Stenutz, B. Hertz, T. Klepach, S. Zhao, Q. Wu, I. Carmichael and A. Serianni. 2004. Correlated C–C and C–O bond conformations in saccharide hydroxymethyl groups: Parametrisation and application of redundant $^{1}$H–$^{1}$H, $^{13}$C–$^{1}$H, and $^{13}$C–$^{13}$C NMR *J* couplings. *Journal of the American Chemical Society* 126, pp. 15,668–15,685.
56. Toukach, F. V. and V. P. Ananikov. 2013. Recent advances in computational predictions of NMR parameters for the structure elucidation of carbohydrates: Methods and limitations. *Chemical Society Reviews* 42, pp. 8376–8415.
57. Taha, H. A., M. R. Richards and T. L. Lowary. 2012. Conformational analysis of furanoside-containing mono- and oligosaccharides. *Chemical Reviews* 113, pp. 1851–1876.
58. Altona, C. and M. Sundaralingam. 1972. Conformational analysis of the sugar ring in nucleosides and nucleotides: New description using the concept of pseudorotation. *Journal of the American Chemical Society* 94, pp. 8205–8212.
59. de Leeuw, F. A. A. M. and C. Altona. 1983. Computer-assisted pseudorotation analysis of five-membered rings by means of proton spin–spin coupling constants: Program PSEUROT. *Journal of Computational Chemistry* 4, pp. 428–437.
60. Houseknecht, J. B., T. L. Lowary and C. M. Hadad. 2002. Improved Karplus equations for $^{3}J_{C1,H4}$ in aldopentofuranosides: Application to the conformational preferences of the methyl aldopentofuranosides. *Journal of Physics and Chemistry A* 107, pp. 372–378.

61. Olsson, U., E. Säwén, R. Stenutz and G. Widmalm. 2009. Conformational flexibility and dynamics of two (1 → 6)-linked disaccharides related to an oligosaccharide epitope expressed on malignant tumour cells. *Chemistry: A European Journal* 15, pp. 8886–8894.
62. Säwén, E., T. Massad, C. Landersjo, P. Damberg and G. Widmalm. 2010. Population distribution of flexible molecules from maximum entropy analysis using different priors as background information: Application to the φ,ψ-conformational space of the α-(1 → 2)-linked mannose disaccharide present in N- and O-linked glycoproteins. *Organic and Biomolecular Chemistry* 8, pp. 3684–3695.
63. Säwén, E., M. U. Roslund, I. Cumpstey and G. Widmalm. 2010. Synthesis and conformational analysis of carbasugar bioisosteres of α-L-iduronic acid and its methyl glycoside. *Carbohydrate Research* 345, pp. 984–993.
64. Milton, M. J., R. Harris, M. A. Probert, R. A. Field and S. W. Homans. 1998. New conformational constraints in isotopically $^{13}$C-enriched oligosaccharides. *Glycobiology* 8, pp. 147–153.
65. Wu, J. and A. S. Serianni. 1992. Isotope-edited 1-D and 2-D NMR: Spectroscopy of $^{13}$C-substituted carbohydrates. *Carbohydrate Research* 226, pp. 209–218.
66. Sayers, E. W. and J. H. Prestegard. 2000. Solution conformations of a trimannoside from nuclear magnetic resonance and molecular dynamics simulations. *Biophysical Journal* 79, pp. 3313–3329.
67. Rutherford, T. J. and S. W. Homans. 1994. Restrained versus free dynamics simulations of oligosaccharides: Application to solution dynamics of biantennary and bisected biantennary N-linked glycans. *Biochemistry* 33, pp. 9606–9614.
68. Kiddle, G. R. and S. W. Homans. 1998. Residual dipolar couplings as new conformational restraints in isotopically $^{13}$C-enriched oligosaccharides. *FEBS Letters* 436, pp. 128–130.
69. Canales, A., J. Jiménez-Barbero and M. Martín-Pastor. 2012. Review: Use of residual dipolar couplings to determine the structure of carbohydrates. *Magnetic Resonance in Chemistry* 50, pp. S80–S85.
70. Freedberg, D. I. 2002. An alternative method for pucker determination in carbohydrates from residual sipolar couplings: A solution NMR study of the fructofuranosyl ring of sucrose. *Journal of the American Chemical Society* 124, pp. 2358–2362.
71. Almond, A. and J. Duus. 2001. Quantitative conformational analysis of the core region of N-glycans using residual dipolar couplings, aqueous molecular dynamics, and steric alignment. *Journal of Biomolecular NMR* 20, pp. 351–363.
72. Losonczi, J. and J. Prestegard. 1998. Improved dilute bicelle solutions for high-resolution NMR of biological macromolecules. *Journal of Biomolecular NMR* 12, pp. 447–451.
73. Mallagaray, A., A. Canales, G. Dominguez, J. Jimenez-Barbero, and J. Perez-Castells. 2011. A rigid lanthanide binding tag for NMR structural analysis of carbohydrates. *Chemical Communications* 47, pp. 7179–7181.
74. Saŵeń, E., B. Stevensson, J. Os̈tervall, A. Maliniak and G. Widmalm. 2011. Molecular conformations in the pentasaccharide LNF-1 derived from NMR spectroscopy and molecular dynamics simulations. *Journal of Physics and Chemistry B* 115, pp. 7109–7121.
75. Canales-Mayordomo, A., R. Fayos, J. Angulo, R. Ojeda, M. Martín-Pastor, P. M. Nieto, M. Martín-Lomas, R. Lozano, G. Giménez-Gallego and J. Jiménez-Barbero. 2006. Backbone dynamics of a biologically active human FGF-1 monomer, complexed to a hexasaccharide heparin analogue, by $^{15}$N NMR relaxation methods. *Journal of Biomolecular NMR* 35, pp. 225–239.
76. Mayer, M. and B. Meyer. 2001. Group epitope mapping by saturation transfer difference NMR to identify segments of a ligand in direct contact with a protein receptor. *Journal of the American Chemical Society* 123, pp. 6108–6117.
77. Loranger, M. W., S. M. Forget, N. E. McCormick, R. T. Syvitski and D. L. Jakeman. 2013. Synthesis and evaluation of l-rhamnose $^{1}$C-phosphonates as nucleotidylyltransferase inhibitors. *Journal of Organic Chemistry* 78, pp. 9822–9833.

78. Szczepina, M. G., D. W. Bleile, J. Müllegger, A. R. Lewis and B. M. Pinto. 2011. WaterLOGSY NMR experiments in conjunction with molecular dynamics simulations identify immobilised water molecules that bridge peptide mimic MDWNMHAA to anti-carbohydrate antibody SYA/J6. *Chemistry: A European Journal* 17, pp. 11,438–11,445.
79. Dalvit, C., P. Pevarello, M. Tatò, M. Veronesi, A. Vulpetti and M. Sundström. 2000. Identification of compounds with binding affinity to proteins via Magnetisation transfer from bulk water. *Journal of Biomolecular NMR* 18, pp. 65–68.
80. Mayer, M. and B. Meyer. 1999. Characterisation of ligand binding by saturation transfer difference NMR spectroscopy. *Angewandte Chemie International Edition* 38, pp. 1784–1788.
81. Meyer, B., J. Klein, M. Mayer, R. Meinecke, H. Möller, A. Neffe, O. Schuster, J. Wülfken, Y. Ding, O. Knaie, J. Labbe, M. M. Palcic, O. Hindsgaul, B. Wagner and B. Ernst. 2004. Saturation transfer difference NMR spectroscopy for identifying ligand epitopes and binding specificities. In *Leucocyte Trafficking*. Heidelberg, Germany: Springer Berlin, pp. 149–167.
82. Airoldi, C., S. Giovannardi, B. la Ferla, J. Jiménez-Barbero and F. Nicotra. 2011. Saturation transfer difference NMR experiments of membrane proteins in living cells under HR-MAS conditions: The interaction of the SGLT1 co-transporter with its ligands. *Chemistry: A European Journal* 17, pp. 13,395–13,399.
83. Diercks, T., J. P. Ribeiro, F. J. Cañada, S. André, J. Jiménez-Barbero and H.-J. Gabius. 2009. Fluorinated carbohydrates as lectin ligands: Versatile sensors in $^{19}$F-detected saturation transfer difference NMR spectroscopy. *Chemistry: A European Journal* 15, pp. 5666–5668.
84. Bresciani, S., T. Lebl, A. M. Z. Slawin and D. O'Hagan. 2010. Fluorosugars: Synthesis of the 2,3,4-trideoxy-2,3,4-trifluoro hexose analogues of D-glucose and D-altrose and assessment of their erythrocyte trans-membrane transport. *Chemical Communications* 46, pp. 5434–5436.
85. García-Aparicio, V., M. Sollogoub, Y. Blériot, V. Colliou, S. André, J. L. Asensio, F. J. Cañada, H.-J. Gabius, P. Sinaÿ and J. Jiménez-Barbero. 2007. The conformation of the C-glycosyl analogue of N-acetyllactosamine in the free state and bound to a toxic plant agglutinin and human adhesion/growth-regulatory galectin-1. *Carbohydrate Research* 342, pp. 1918–1928.
86. Asensio, J. L., F. J. Cañada and J. Jiménez-Barbero. 1995. Studies of the bound conformations of methyl α-lactoside and methyl β-allolactoside to ricin B chain using transferred NOE experiments in the laboratory and rotating frames, assisted by molecular mechanics and dynamics calculations. *European Journal of Biochemistry* 233, pp. 618–630.
87. Shimizu, H., R. A. Field, S. W. Homans and A. Donohue-Rolfe. 1998. Solution structure of the complex between the B-sub-unit homopentamer of verotoxin VT-1 from *Escherichia coli* and the trisaccharide moiety of globotriaosylceramide. *Biochemistry* 37, pp. 11,078–11,082.
88. Ling, H., A. Boodhoo, B. Hazes, M. D. Cummings, G. D. Armstrong, J. L. Brunton and R. J. Read. 1998. Structure of the shiga-like toxin I B-pentamer complexed with an analogue of its receptor $Gb_3$. *Biochemistry* 37, pp. 1777–1788.
89. Thompson, G. S., H. Shimizu, S. W. Homans and A. Donohue-Rolfe. 2000. Localisation of the binding site for the oligosaccharide moiety of $Gb_3$ on verotoxin 1 using NMR residual dipolar coupling measurements. *Biochemistry* 39, pp. 13,153–13,156.
90. Seidel, R. D., T. Zhuang and J. H. Prestegard. 2007. Bound-state residual dipolar couplings for rapidly exchanging ligands of His-tagged proteins. *Journal of the American Chemical Society* 129, pp. 4834–4839.
91. Zhuang, T., H.-S. Lee, B. Imperiali and J. H. Prestegard. 2008. Structure determination of a galectin-3–carbohydrate complex using paramagnetism-based NMR constraints. *Protein Science* 17, pp. 1220–1231.

92. Jain, N. U., A. Venot, K. Umemoto, H. Leffler and J. H. Prestegard. 2001. Distance mapping of protein-binding sites using spin-Labelled oligosaccharide ligands. *Protein Science* 10, pp. 2393–2400.
93. Watson, J. D. and F. H. C. Crick. 1953. Molecular structure of nucleic acids: A structure for deoxyribose nucleic acid. *Nature* 171, pp. 737–738.
94. Hwang, T. L. and J. A. Shaka. 1995. Water suppression that works: Excitation sculpting using arbitrary waveforms and pulsed field gradients. *Journal of Magnetic Resonance Series A* 112, pp. 275–279.
95. Piotto, M., V. Saudek and V. Sklenár. 1992. Gradient-tailored excitation for single-quantum NMR spectroscopy of aqueous solutions. *Journal of Biomolecular NMR* 2, pp. 661–665.
96. Palmer, A. G., J. Cavanagh, P. E. Wright and M. Rance. 1991. Sensitivity improvement in proton-detected heteronuclear correlation experiments. *Journal of Magnetic Resonance* 93, pp. 151–170.
97. Kay, L. E., P. Keifer and T. Saarinen. 1992. Pure absorption gradient–enhanced heteronuclear single-quantum correlation spectroscopy with improved sensitivity. *Journal of the American Chemical Society* 114, pp. 10,663–10,665.
98. Schleucher, J., M. Schwendinger, M. Sattler, P. Schmidt, O. Schedletzky, S. J. Glaser, O. W. Sørensen and C. Griesinger, C. 1994. A general enhancement scheme in heteronuclear multi-dimensional NMR employing pulsed field gradients. *Journal of Biomolecular NMR* 4, pp. 301–306.
99. Goddard, T. D. and D. G. Kneller. Sparky 3. Available at http://www.cgl.ucsf.home.sparky. Accessed on October 30, 2013.
100. Evstigneev, M. P., J. A. Parkinson, A. O. Lantushenko, V. V. Kostjukov and V. I. Pahomov. 2010. Hexamer oligonucleotide topology and assembly under solution phase NMR and theoretical modelling scrutiny. *Biopolymers* 93, pp. 1023–1038.
101. Heus, H. A., S. S. Wijmenga, H. Hoppe and C. W. Hilbers. 1997. The detailed structure of tandem G–A mismatched base-pair motifs in RNA duplexes is context dependent. *Journal of Molecular Biology* 27, pp. 147–158.
102. Harvie, S., O. Wilson and J. A. Parkinson. 2011. Insights into DNA platination within unusual structural settings. *International Journal of Inorganic Chemistry*. Article ID 319757. Available at http://www.hindawi.com/journals/ijic/2011/319757/. Accessed on October 30, 2013.
103. Chen, H. L., P. Yang, C. X. Yuan and X. H. Pu. 2005. Study on the binding of base-mismatched oligonucleotide d(GCGAGC)$_2$ by cobalt(III) complexes. *European Journal of Inorganic Chemistry* 15, pp. 3141–3148.

## FURTHER READINGS

Cowburn, D. 1996. Nucleic acids: Chemical shifts. In Grant, D. M. and R. K. Harris, eds. *Encyclopedia of NMR*. Chichester: John Wiley & Sons Inc., pp. 3336–3339.

Gorenstein, D. G. 1996. Nucleic acids: phosphorus-31 NMR. In Grant, D. M. and R. K. Harris, eds. *Encyclopedia of NMR*. Chichester, United Kingdom: John Wiley & Sons, Inc., pp. 3340–3344.

Hilbers, C. W. and S. S. Wijmenga. 1996. Nucleic acids: Spectra, structures, and dynamics. In Grant, D. M. and R. K. Harris, eds. *Encyclopedia of NMR*. Chichester, United Kingdom: John Wiley & Sons, Inc., pp. 3346–3359.

James, T. L. 1996. Nucleic acid structures in solution: Sequence dependence. In Grant, D. M. and R. K. Harris, eds. *Encyclopedia of NMR*. Chichester, United Kingdom: John Wiley & Sons, Inc., pp. 3320–3332.

Schweizer, M. P. 1996. Nucleic acids: Base stacking and base pairing interactions. In Grant, D. M. and R. K. Harris, eds. *Encyclopedia of NMR*. Chichester, United Kingdom: John Wiley & Sons, Inc., pp. 3332–3335.

# Index

Page numbers followed by f and t indicate figures and tables, respectively.


## A

Acetoacetate, 254, 255f
Acetone, 28f, 201, 242f, 255f
Acrylamide, 107
Activation energy, 26, 27f, 35
Adenine, 296, 297f, 298t, 305, 318
Adenosine, 213, 213f
Alanine, 234f, 245f, 255f
Albumin, 230, 239, 239f
Alga, crude dichloromethane extract of, 279
Alignment of molecules, 107
Allantoin, 255f
Altona–Sundaralingam model, 287
Amide proton, 23, 33, 34f
Amino acid, 91, 92f, 93, 95f, 111
Amino-proton resonances, 302f, 309f, 318, 318f
α-naphthylisothiocyanate (ANIT), 256, 258f, 259, 260f
Anhydrous molecule, 149t
Anisotropic diffusion, 128, 129f
Anisotropic magnetic susceptibility, 6, 108
Anomalous diffusion, 128
Anomeric signals, 282, 284
Anti-tumour agents, structure of, 265f
Apodisation, 312
Apparent diffusion coefficient, 128, 129f, 143
APT. *See* Attached proton test (APT)
Arginine, 111
Arginine-glycine-aspartic acid, 91, 104
Aromatic protons, 50f, 51f
Aromatic resonance, 298f, 299, 314f, 321
Aroma WaterLOGSY, 100
Arrhenius analysis, 36
Associating systems, 136–139. *See also* Diffusive averaging effects
Attached proton test (APT), 269, 270f
Attenuation, 133
Avogadro's number, 146

## B

Band-selective optimised flip-angle short-transient, 113
Band-shape analysis, 198
Bare nucleus, 5
Batch processing, 227–228, 252–254. *See also* Pattern recognition
  batch level data analysis in, 253–254
  observation level data analysis in, 253
Benzene, 7f, 22f, 23
Bicelles, 108, 109, 110
Bilinear rotation decoupling (BIRD), 194
Bimolecular association exchange, 70f
Bimolecular reaction, 24
Binding equilibria, 142–144, 143f. *See also* Diffusive averaging effects
Binding interfaces, identification of, 290–293, 291f, 292f. *See also* Protein-carbohydrate interactions
Binning, 247f, 248
Bioactive natural products, 266f
Bioassay-guided isolation, 266
Biofluids, 226, 230, 244, 246f
Biological fluids, 233, 235t
Biopolymers, 296
Bipolar gradient pulse, 159
Bloch equations. *See also* Dynamic nuclear magnetic resonance
  in frequency domain, 37–39
  in time domain, 39–44
  with unequal populations, 45–46
Blood plasma, 229t, 230, 235t, 237f, 246, 254
Boltzmann constant, 35, 130
Boltzmann distribution, 2
Boltzmann population, 24
Bovine fetal serum (BFS), 93
Bovine neurophysin, 102
Bovine serum albumin (BSA) system, 68, 79f
Bristol-Myers Squibb (BMS), 265
Bromines, 20
Brownian motion, 127f
Bruker convention, 75
Bruker spectrometer, 186, 188f
Bulk magnetisation, 2, 3, 65

## C

Carbohydrates. *See also* Nuclear magnetic resonance (NMR), applications
  conformational determination

monosaccharide conformation from coupling constants, 286–288, 286f, 287f
oligosaccharides, conformations of, 288–290, 288f
protein-carbohydrate interactions
binding interfaces, identification of, 290–293, 291f, 292f
oligosaccharide conformation when bound, 293–294
relative orientation of, 294–295
sequence determination by NMR, 282–286, 283f, 284f, 285f
Carbon–carbon double bond, 20
Carbonyl bond, 7f
Cardiolipin, 246, 246f
Carr-Purcell-Meiboom-Gill (CPMG) pulse sequence, 30, 31, 34f, 72, 73, 97, 235, 236f
Case studies, NMR and complex mixtures
application to plasma, 254, 255f
application to saliva, 255–256, 256f
batch analysis, 256–259, 257f, 258f, 260f
Cellulose, 283f, 284
Central excitation band, 80
Chemical exchange, 18, 231. *See also* Exchange
in methyl 3-dimethylamino-13 2-cyanocrotonate (MDACC), 19f
saturation transfer, 29
Chemical shift, 5–8, 6t, 7f, 51f, 78t, 81f, 185–186, 241, 321. *See also* Multi-nuclear magnetic resonance spectroscopy
perturbations
binding affinities from, 115–118, 116f, 117f
in drug discovery, 110–114, 112f
quantifying, 114–115
Chemical shift anisotropy (CSA), 26, 101
Chemical shift mapping, 114
Chiral derivatising agents, 211
Chloroform–methanol, 228
Cholesterol, 245f
Choline, 245f
CIGAR-HMBC experiment. *See* Constant time inverse-detection gradient accordion rescaled (CIGAR)-HMBC experiment
Cisplatin, 211
Citrate, 255f
Close-to-water resonances, 160
Cloud ligand, 70, 70f
C-methyl signals, 34f
Coalescence, 38, 39, 41
Cobalamin biosynthesis, 112f
Cold probes, 279
Complete relaxation and conformational exchange matrix analysis (CORCEMA), 86
Concentration diffusion, 126
Constant sum (CS), 248
Constant-time experiment, 31
Constant time inverse-detection gradient accordion rescaled (CIGAR)-HMBC experiment, 275–276
Continuous wave (CW), 28
Contribution plots, 253, 259
Convection, 162–163. *See also* Nuclear magnetic resonance (NMR)
Coordination number, 147
Correlation spectroscopy (COSY), 268, 270–271
Correlation through long-range coupling (COLOC), 273
COSY. *See* Correlation spectroscopy (COSY)
Coupled heteronuclear single quantum coherence (HSQC), 274–275, 275f, 276f. *See also* Heteronuclear single quantum coherence (HSQC)
Coupled spin system, 46
Coupling constants, 198
Creatinine, 234f, 237f, 255f
Crude dichloromethane extract of alga, 279
Crude extract, 264, 265, 266f, 267, 279, 281
Cryoprobes, 244
Cyclisation, 282
Cycloelatanene, 268, 268f
Cyclohexane, 20, 35
Cytosine, 296, 297f, 298t, 308, 311f, 318, 320f, 321, 322


## D


Dark-state exchange saturation transfer, 29
Data acquisition and processing, 301–315, 302f. *See also* Nucleic acids
2-D NMR data processing, 312–313
samples in 99% D2O
2-D HSQC NMR acquisition, 309–312
1-D NMR spectra, 308
2-D NMR spectra, 308–309, 309f
samples in 90% H2O/10% D2O
2-D H NMR data acquisition, 303–305
1-D H NMR spectra, 302–303
2-D HSQC NMR acquisition, 307
1-D P NMR data acquisition, 305–307, 306f
temperature considerations, 313–315, 314f, 315f
Data normalisation, NMR and complex mixtures, 248
Data reduction, NMR and complex mixtures, 247–248
Deconvolution, 89, 106, 110
Decoupler pulse, 187, 192
Delays alternating with nutations for tailored excitation (DANTE), 197

Density matrix, defined, 46
Density matrix treatment, 46–47. *See also* Dynamic nuclear magnetic resonance
complex systems, 48–49
coupled spin systems, 49–51, 50f, 51f
NOESY/EXSY 2-D experiment, 55–57
spin-lattice relaxation experiments, 53–55
*z* magnetisations, 51–52
Deoxyribose, 296, 297f
DEPT. *See* Distortionless enhancement by polarisation transfer (DEPT)
Dereplication, 266f, 267, 278
Deuterium, 228, 229t
DHPC (1,2-dihexanyl-sn-glycero-3-phosphocholine), 108
Diamagnetic shift, 5
Diamagnetic term, 185
Dichloromethane extract of alga, crude, 279
Difference spectroscopy, 67
Diffusion
anisotropic, 128, 129f
diffusive averaging effects
association and polydispersity, 136–139, 137f, 138f
binding equilibria, 142–144, 143f
obstructions, 139–142, 140f
and ion/molecule radius in solution
hydration and solvation, 145–155, 148t–151t
Stokes–Einstein–Sutherland equation, 144–145
isotropic, 128, 129f
NMR in measuring
effect on gradients, 136
relaxation/gradients/echoes, 129–136, 130t, 131f, 132f, 135f, 136f
NMR measurement interference, 155–163
convection, 162–163
Eddy currents, 156–158, 156f, 157f, 158f
field inhomogeneities, 158–159, 159f
solvent suppression, 160–162, 160f, 161f, 162f
overview, 125–129, 126f, 127f, 129f
restricted, 127, 127f
unrestricted, 128
Diffusion coefficient, 128, 238
Diffusion-edited spectrum of blood plasma, 239, 239f
Diffusion editing, 238–239, 239f. *See also* Nuclear magnetic resonance (NMR) and complex mixtures
Diffusion ordered spectroscopy (DOSY), 129, 201
Diffusion tensor, 129, 135
Diffusive averaging effects. *See also* Diffusion
association and polydispersity, 136–139, 137f, 138f
binding equilibria, 142–144, 143f
obstructions, 139–142, 140f
Digital enhancement, 312
Diglyceride, 245f
1,2-dihexanoyl-sn-glycero-3-phosphocholine (DHPC), 295
3-dimethylamino-7-methyl-1,2,4-benzotriazine, 16
4,4-dimethyl-4-silapentane-1-ammonium trifluoroacetate (DSA), 230
2,2-dimethyl-2-silapentane-5-sulfonate sodium salt (DSS), 230
1,2-dimyrisoylsn-glycero-3-phosphocholine (DMPC), 295
1,6-dipivaloyl-3,4,7,8-tetramethyl-2,5-dithioglycoluril, 32, 33f
Dipolar coupling, 105, 289
Dipole-dipole process, 12
Dipole-dipole relaxation, 27, 55
Dipyrazolylmethane complexes, 213
Direct detection of inorganic nuclei, 189–190. *See also* Multi-nuclear magnetic resonance spectroscopy
Disaccharide, 100
Dissociation constant, 71, 72, 117f
Dissociation equilibrium constant, 143
Dissociation event in STD NMR, 71–73
Distortionless enhancement by polarisation transfer (DEPT), 189, 269, 270f
DMPC (1,2-dimyristoyl-sn-glycero-3-phosphocholine), 108
DNA construction, 296, 297f, 298t, 299
Dole–Jones viscosity, 153
DOSY. *See* Diffusion ordered spectroscopy (DOSY)
Double bond, 18
Double-pulsed field gradient spin echo (DPFGSE), 303
Double-quantum– filtered COSY (DQFCOSY), 271
Double refocussed spin echo, 157, 158f
Double spin echo method, 242
Drug discovery, chemical shift perturbations in, 110–114, 112f
Duplex DNA, 306
Dynamic nuclear magnetic resonance
Bloch equations
in frequency domain, 37–39
in time domain, 39–44
with unequal populations, 45–46
density matrix treatment, 46–57, 50f, 51f
exchange regimes, 18–21, 19f, 21f
fast exchange, 31–35, 32f, 33f, 34f
slow exchange, 27–31, 28f, 30f, 31f
unequal/equal populations, 21–26, 22f, 23f, 24f, 25f, 26f, 27f
kinetics, 35–37, 36f
NMR timescales, 15–18, 16f, 17f

## E


Ecteinascidin 743, 264, 265, 265f
Eddy currents, 156–158, 156f, 157f, 158f. *See also* Nuclear magnetic resonance (NMR)
Edge-bridging carbonyls, 195
Eigenvectors, 40, 42, 47, 52
Elatenyne, 275, 276f
Electric field gradients, 214, 216
Electron distribution, 5, 6
Energy, 9f, 10, 21f
  of activation, 26, 27f, 35
Enhanced protein hydration observed through gradient spectroscopy (ePHOGSY), 97
Enthalpy, 36, 36f
  of activation, 36f
Entropy, 19f, 20, 21, 36, 36f
E-selectin, 293
Ethylenediaminetetraacetic acid (EDTA), 93
et-NOE. *See* Exchange tr-NOE
Exchange matrix, 40, 45
Exchange (chemical) regimes, 18–21, 19f, 21f
  fast exchange, 31–35, 32f, 33f, 34f
  slow exchange, 27–31, 28f
    selective inversion relaxation experiments, 28–31, 30f, 31f
  unequal/equal populations, 21–26, 22f, 23f, 24f, 25f, 26f, 27f
Exchange spectroscopy (EXSY), 199, 200f, 211f, 293
Exchange transferred nuclear overhauser effect (tr-NOE), 102–105, 103f
Exchange tr-NOE (et-NOE), 102–104, 103f
  illustrations of, 104–105
Excitation sculpting, 75, 303
Excitation-sculptured indirect-detection experiment (EXIDE), 277–278, 277f
Exocyclic torsion angle, 286f
Exposed amides with TROSY (SEA-TROSY), 114
Eyring plot, 36, 36f


## F


Face-bridging carbonyl group, 194
Fast exchange, 20, 31–35, 32f, 33f, 34f, 45, 115, 116. *See also* Exchange (chemical) regimes
Fast relaxation, 185
Fatty acid, 289
Fatty acid acyl chains, 239
Fatty acids for energy, 254
Fatty acid signals, 254
Fatty acyl groups, 255f
Ferry equation, 155
Fibroblast growth factor (FGF), 291
Field gradient pulses, 192
Field inhomogeneities, 158–159, 159f. *See also* Nuclear magnetic resonance (NMR)
Filamentous Pf1 phage, 107–108
Flip-angle adjustable 1-D NOESY (FLIPSY), 231
Fluorine, 27, 101
Folch method, 228
Foot-and-mouth disease virus (FMDV), 93
Formate, 255f
Fourier transformation, 43, 44
Fourier-transformed 1-D (pulse-and-acquire) NMR spectrum, 243
Fourier transform spectrometer, 37
Fragment-based drug discovery (FBDD), 64
Free diffusion, 126, 128
Free energy, 35, 36
Free induction decays (FID), 3, 4f, 67, 189, 243, 312
Free ligands, 72, 74, 75, 90, 101, 102, 103f, 104
Free peptide, 105
Free protein, 115
Free water, 99
Frequency domain, Bloch equations in, 37–39
Frictional coefficient, 144
Furanose, 282, 287f, 289
Furanose sugars, 286
Furospinosulin-1, 268, 278f


## G


Gaussian pulse, 80
Gaussian pulse length, 82
GCOSY NMR spectrum, 272f, 280f
Gibbs free energy, 144
Glucopyranose units, 283f
Glucose, 148t, 245f
  transporter, 293
Glutamate, 245, 245f
Glutathione, 213
Glycerophosphorylcholine (GPC), 245
Glycoprotein N-acetyl, 234f, 237f, 255f
Glycopyranosides, 93
Gradient-accelerated HOESY spectroscopy, 200
Gradient-based methods, 231, 232t, 233
Gradient-based relaxation-edited spectroscopy (G-RESY), 243
Gradient heteronuclear multiple-bond correlation adiabatic (gHMBCAD) NMR spectra, 274, 281f
Gradient heteronuclear single quantum coherence adiabatic (gHSQCAD) NMR spectra, 274
Gradient or diffusion weighting factor, 133
Gradient pulses, 133, 156, 156f
Ground-state X-ray, 22f
Group epitope mapping (GEM), 78

Group epitope mapping considering relaxation of ligand (GEM-CRL), 86, 87
Guanine, 296, 297f
Guanosine, 213

## H

Hahn spin echo, 131f
Hamiltonian, 10, 39, 46
Hard pulses, 80
Heavy atom, 9
Hermitian matrix, 40, 52
Heteronuclear 2-D NMR approaches. *See also* 2-D NMR experiments
 CIGAR-HMBC experiment, 275–276
 coupled HSQC, 274–275, 275f, 276f
 excitation-sculptured indirect-detection experiment (EXIDE), 277–278, 277f
 HSQC and HMBC, 271, 273–274, 274f
 HSQC-NOESY, 276–277
Heteronuclear multiple bond correlation (HMBC) spectroscopy, 190, 196
Heteronuclear multiple-quantum coherence (HMQC), 186, 191–196, 271, 273–274, 274f
 gradient *versus* phase cycle selection, 192–194
 spin systems involving coupling to multiple insensitive spins, 194–195, 195f, 196f
Heteronuclear NMR spectroscopy, 178t, 179t
Heteronuclear scalar coupling, 179
Heteronuclear shift correlation (HETCOR), 273
Heteronuclear single quantum coherence (HSQC), 89, 94f, 106f, 112, 186, 271, 273–274, 274f, 307, 309–312. *See also* Heteronuclear multiple-quantum coherence (HMQC)
 coupled, 274–275, 275f, 276f
Heteronuclear single quantum coherence (HSQC)-Nuclear overhauser effect spectroscopy (NOESY), 276–277
Heteronuclear single-quarter coherence (HSCQ), 22
Heteronuclides, 189
H frequency, 186, 187, 216
High-molecular-weight molecules, 238–239, 239f
High-performance liquid chromatography (HPLC), 266, 278–281, 278f
Histidine, 111, 234f, 255f
Histo-blood group antigens (HBGA), 96
HMBC spectroscopy. *See* Heteronuclear multiple bond correlation spectroscopy
H-NMR spectroscopy, 227, 232t, 245f
Hoffman–Forsen experiment, 28
Hofmeister effect, 146
Homonuclear 2-D NMR approaches. *See also* 2-D NMR experiments
 correlation spectroscopy, 270–271
 NOESY and ROESY, 271, 273f
Homonuclear scalar coupling, 233
Homotrimer protein, 109
HPLC. *See* High-performance liquid chromatography (HPLC)
HSQC. *See* Heteronuclear single quantum coherence (HSQC)
Hubbard–Onsager model, 152
Human blood plasma, 234f, 237f
Human embryonic kidney (HEK), 293
Human follicular fluid, 241f
Human serum albumin (HSA), 88
Hydration. *See also* Ion/molecule radius in solution, diffusion and
 of ions, 146–154, 148t–151t
 of non-electrolytes, 146
 number, 147, 150t, 154
 shells, 147
 and solvation, 145–155
Hydrogen atoms in ibuprofen, 11f
Hydrogen bond, 8, 112, 153
Hydrogenobyrinic acid, 112f
(hydroxyethyl) piperazine-1-ethanesulfonic acid (HEPES), 242
Hyperpolarisation, 244
Hyphenation, HPLC-NMR, 278–281, 278f
 application to organism profiling/structure elucidation, 279–281, 280f, 281f

## I

Imino-proton resonances, 317
Indirect detection of inorganic nuclei, 190–196, 191t. *See also* Multi-nuclear magnetic resonance spectroscopy
 HMBC spectroscopy, 196
 HMQC/HSQC approaches, 191–196
  gradient *versus* phase cycle selection, 192–194
  spin systems involving coupling to multiple insensitive spins, 194–195, 195f, 196f
 instrumental considerations, 191
INEPT. *See* Insensitive nucleus enhancement by polarization transfer (INEPT)
Inorganic nuclei, 178
 direct detection of, 189–190
 indirect detection of. *See* Indirect detection of inorganic nuclei
Inorganic nuclides
 measurements of. *See also* Multi-nuclear magnetic resonance spectroscopy
  $^{15}N$ NMR, applications of, 203–207, 205t, 206f, 207f
  $^{195}Pt$ NMR, applications of, 207–214, 208f, 209f, 210t, 211f, 212f, 213f

NMR spectrum of, acquisition of. *See* Nuclear magnetic resonance (NMR) spectrum of inorganic nuclide, acquisition of
Inorganic phosphate, 246f
In-phase anti-phase (IPAP)-HSQC-TOCSY-hadamard transform (HT), 289
Insensitive nucleus enhancement by polarization transfer (INEPT), 186, 189–190, 204
Insensitive spin satellites, selective observation and, 197, 197f. *See also* Multi-nuclear magnetic resonance spectroscopy
Insulin-treated diabetic patients, 254
Intercalation, 317
Interdiffusion, 126
Intermediate exchange, 18, 27
Intermolecular exchange of tellurium, 25f
Intermolecular reactions, 18
International Union of Pure and Applied Chemistry (IUPAC) scale, 187
Inter-pulse delay, 11f
Intramolecular exchange, 18
Inversion–recovery experiment, 86
Ion/molecule radius in solution, diffusion and. *See also* Diffusion
  hydrated radius by experiment, 154–155
  hydration
    of ions, 146–154, 148t–151t
    of non-electrolytes, 146
    and solvation, 145–155
  Stokes-Einstein-Sutherland equation, 144–145
Ions, hydration of, 146–154, 148t–151t. *See also* Hydration
Irradiation, 67
Isoallolaurenterol, 281f
Isoleucine, 234f
Isomerisation process, 209
Isomers, 212
Isoplocamenone, 281f
Isosargahydroquinoic acid, 280f, 281f
Isotope-edited experiments, 288
Isotopic enrichment, 288
Isotropic diffusion, 128, 129f
Iterative STOCSY (I-STOCSY), 250t

## J

J coupling constants, 285, 286
J-resolved (JRES) experiment, 236
J-resolved NMR spectroscopy, 241f
JRES spectroscopy, 239–242, 240f, 241f, 242f

## K

Karplus equation, 288
Karplus relationship, 286
Ketogenesis, 254
Ketones, 282
Ketone signals, 254
Kinetic matrix, 48
Kinetics, 35–37, 36f. *See also* Dynamic nuclear magnetic resonance
Kubo–Sack matrix, 47

## L

Lactate, 255f
Lammelle, 108
Lanthanides, 108
Larmor (or resonance) frequency, 3, 12, 136f, 204
Latex rubber, 17
Leucine, 234f
Levenberg–Marquardt non-linear least squares algorithm, 134
Ligand binding, 103f, 111
Ligand binding studies, NMR in. *See* Nuclear magnetic resonance (NMR) in ligand binding studies
Ligand binding using 1-D NMR, assessment of, 316–317, 317f. *See also* Nucleic acids
Ligand–biomolecule interactions, 161
Ligand excess, 76
Ligand-observe screening, 100–101
Liouville–von Neumann equation, 46, 55
Liouvillian matrix, 39, 45
Lipid extraction techniques, 228
Lipidomics, 227
Liquid crystal, 295
Lithium battery, 17, 27
Loadings plot, 249, 251f
Longitudinal Eddy current delay (LED), 157, 157f
Lorentzian line, 41
Low-density lipoprotein (LDL), 227, 234f, 237f
Lower-affinity ligands, 116
Lozenge ligand, 70f
L-tryptophan (L-Trp), 65, 69f, 78t
Lyotropic liquid crystals (LLC), 154
Lysophosphatidylcholine (LPC), 246f
Lysozyme, 137f, 138f

## M

Macromolecules, 68, 135, 140, 143f
Magic angle spinning (MAS), 228, 243, 244
Magnetic anisotropy, 7
Magnetic gradient, 136f
  coil, 136
Magnetic moments, 106f
Magnetic resonance imaging (MRI), 126, 130t
Magnetic shielding, 315
Magnetic susceptibility, 7, 8
Magnetisation, 11, 132

longitudinal placement of, 156
sinks, 114
vector, 66f
Magnetogyric ratio, 2
Magnets, nuclear, 8
Mammalian oligosaccharides, 285f, 286f
Mannopyranosides, 284
Mannose-binding protein (MBP), 109
Marine and terrestrial natural product (MATNAP), 268
Mass spectrometry (MS), 227
Materials science, 1
Mean-squared displacement (MSD), 126, 128
Melatonin, 105
MEshcher–GArwood (MEGA), 231, 232t
Metabonomics, 226–227, 226f
Metal–metal bond, 211
Methine/methyl (CH/CH3) multiplicities, 273
Methylated derivatives, 207f
Methyl-β-D-galactopyranoside, 292f
Methyl 3-dimethylamino-13 2-cyanocrotonate (MDACC), 19f
Methyl protons, 68, 69f, 81
Millimolar-to-micromolar binding events, 72
Molecular-weight, 233, 234f
Monosaccharide conformations, 288
from coupling constants, 286–288, 286f, 287f
Multi-nuclear magnetic resonance spectroscopy
acquisition of NMR spectrum of inorganic nuclide
referencing, 188
running 'off lock,' 189
solvent and temperature, 189
spectrometer, 186–187
test sample, 187–188, 188f
chemical shift, 185–186
choice of experiment
direct detection, INEPT, 189–190
indirect detection, 190–196, 191t
selective observation and insensitive spin satellites, 197, 197f
dynamic systems
exchange spectroscopy (EXSY), 199, 200f
saturation transfer exchange spectroscopy, 198–199, 199f
simulation of VT spectra, 198
gradient-accelerated HOESY spectroscopy, 200
heteronuclear NMR spectroscopy, 178t, 179t
inorganic nuclides, measurements of
$^{15}$N NMR, applications of, 203–207, 205t, 206f, 207f
$^{195}$Pt NMR, applications of, 207–214, 208f, 209f, 210t, 211f, 212f, 213f
overview, 178
para-hydrogen-enhanced spectroscopy, 203
properties of selected spin-1/2 nuclei, 180t–181t
pulsed field gradient spin echo (PFGSE) diffusion spectroscopy, 201–203, 202f
quadrupolar nuclei, 203
measurement of, 214–218, 215f, 217f, 217t, 218f, 218t
spin properties of selected quadrupolar nuclei, 182t–184t
Multivariate statistical analyses, 249–252, 250t–251t, 251f. *See also* Pattern recognition
scores and loadings plots, interpretation of, 249, 251f, 252
Mutual exchange, 19, 21

## N

N-acetyl-glucosaminyltransferase V (GnTV), 109
Nanomax-50 membrane, 155
Naproxen, 88
inhibitor, 88f
Natural products. *See also* Nuclear magnetic resonance (NMR), applications
1-D NMR experiments
DEPT and APT, 269, 270f
1-D NOE and 1-D TOCSY experiments, 267–268, 268f, 269f
2-D NMR experiments
heteronuclear approaches, 271–278, 272f, 273f, 274f, 275f, 276f, 277f
homonuclear approaches, 270–271
hyphenation, HPLC-NMR, 278–281, 278f
application to organism profiling/structure elucidation, 279–281, 280f, 281f
overview, 264–267, 265f, 266f
structure elucidations, 267
Navier–Stokes equations, 142
N chemical shifts, 205t
Nickel, 216, 218t
chelates, 295
$^{61}$Ni NMR as quadrupolar nuclide. *See also* Quadrupolar nuclei, measurement of
chemical shifts range and coupling constants, 216
properties of $^{61}$Ni isotope, 216
use of $^{61}$Ni NMR, 216–218, 217f, 217t, 218f, 218t
Nitrobenzotriazole, 207f
Nitrogen, 111, 112
Nitrogen compounds, 215f
Nitromethane, 215f
N-methyl signals, 34f
NMR. *See* Nuclear magnetic resonance (NMR)
N,N-dimethyl-4-nitrosoaniline, 49, 50f, 51f
$^{15}$N NMR, applications of, 203–207, 205t, 206f, 207f. *See also* Inorganic nuclides, measurements of
NOE. *See* Nuclear overhauser effect (NOE)

NOESY. *See* Nuclear overhauser effect spectroscopy (NOESY); Nuclear overhauser spectroscopy
Non-covalent binding studies, 316, 317f
Non-electrolytes, hydration of, 146. *See also* Hydration
Non-linear least-squares software, 29
Non-overlapping resonances, 137
Non-selective inversion pulse, 53
Non-selective inversion recovery, 29
Normalisation, 248
Nuclear de-shielding, 5
Nuclear magnetic moments, 10
Nuclear magnetic resonance (NMR), 125–126. *See also* Diffusion
- diffusometry, 129
- dynamic. *See* Dynamic nuclear magnetic resonance
- measurement interference, 155–163
  - convection, 162–163
  - Eddy currents, 156–158, 156f, 157f, 158f
  - field inhomogeneities, 158–159, 159f
  - solvent suppression, 160–162, 160f, 161f, 162f
- in measuring diffusion
  - effect on gradients, 136
  - relaxation/gradients/echoes, 129–136, 130t, 131f, 132f, 135f, 136f
- sequence determination by, 282–286, 283f, 284f, 285f
- techniques, 1
- timescales, 15–18, 16f, 17f

Nuclear magnetic resonance (NMR), applications
- carbohydrates
  - conformational determination, 286–290, 286f, 287f, 288f
  - protein–carbohydrate interactions, 290–295, 291f, 292f
  - sequence determination by NMR, 282–286, 283f, 284f, 285f
- natural products
  - 1-D NMR experiments, 267–269, 268f, 269f, 270f
  - 2-D NMR experiments, 270–278, 272f, 273f, 274f, 275f, 276f, 277f
  - hyphenation, HPLC-NMR, 278–281, 278f, 280f, 281f
  - overview, 264–267, 265f, 266f
- nucleic acids
  - data acquisition and processing, 301–315, 302f, 304f, 306f, 308f, 309f, 310f, 311f, 314f, 315f
  - ligand binding using 1-D NMR, assessment of, 316–317, 317f
  - resonance assignment, 317–322, 318f, 319f, 320f
  - sample handling and preparation for NMR studies, 300–301
  - sample selection, 296–300, 297f, 298t

Nuclear magnetic resonance (NMR) and complex mixtures
- batch processing, 227–228
- case studies
  - application to plasma, 254, 255f
  - application to saliva, 255–256, 256f
  - batch analysis, 256–259, 257f, 258f, 260f
- data treatment for statistical analysis
  - data normalisation, 248
  - data reduction, 247–248
  - scaling and mean centering, 249
  - spectral processing, 246–247, 247f
- diffusion editing, 238–239, 239f
- JRES spectroscopy, 239–242, 240f, 241f, 242f
- lipidomics, 227
- metabonomics, 226–227, 226f
- pattern recognition
  - data analysis in batch processing, 252–254
  - multivariate statistical analyses, 249–252, 250t–251t, 251f
- sample preparation, 228–230, 229t
  - concentration references, 230
- solid-state studies, approaches to, 243–244
- $T_2$ relaxation editing, 235–238, 236f, 237f, 238f
- water attenuation, 230–234, 231f, 232t–233t, 234f, 235t
- X-nucleus detection, 244–246, 245f, 246f

Nuclear magnetic resonance (NMR) in ligand binding studies
- chemical shift, quantifying, 114–115
- chemical shift perturbations
  - binding affinities from, 115–118, 116f, 117f
  - in drug discovery, 110–114, 112f
- protein-based observation of ligand binding, 110–118
- saturation in NMR
  - NOE difference experiment, 65, 67
  - saturation, defined, 64–65, 66f
  - in solvent suppression, 65, 67f
- STD NMR
  - basic analysis, 76–80, 77f, 78t, 79f
  - 2-D, 88–90, 90f
  - dissociation event in, 71–73
  - experimental approach, 73–76, 73f, 74f
  - and ligand binding competition, 87–88, 88f
  - ligand-observe screening, 100–101
  - and peptide-based ligands, 91–96, 92f, 94f, 95f
  - saturation times in, 83–87, 84f, 85f, 86t

selective excitation pulse for protein saturation in, 80–83, 81f, 82f
STD and protein, 68–71, 69f, 70f
and very large proteins, 96–97
WaterLOGSY (transfer methods involving water), 97–100, 98f
tr-NOE and transferred RDC, 101–102
exchange tr-NOE, 102–105, 103f
transferred RDC, 105–110, 106f
Nuclear magnetic resonance (NMR) spectra
1-D, 308
2-D, 308–309, 309f
Nuclear magnetic resonance spectrum of inorganic nuclide, acquisition of. *See also* Multi-nuclear magnetic resonance spectroscopy
referencing, 188
running 'off lock,' 189
solvent and temperature, 189
spectrometer, 186–187
test sample, 187–188, 188f
Nuclear magnetisation, 2–5
perturbation of, 3–5
radio-frequency (RF) pulses, 3
vector representation of, 4f
Nuclear magnets, 8
Nuclear overhauser effect (NOE), 12, 99, 102, 103f, 203, 204, 288
saturation in NMR and, 65, 67
transient, 294
Nuclear overhauser effect spectroscopy (NOESY), 230, 234, 234f, 271
Nuclear overhauser effect spectroscopy (NOESY)/exchange spectroscopy (EXSY) 2-D experiment, 55–57. *See also* Density matrix treatment
Nuclear shielding, 5
Nuclear spin magnetic moments, 132f
Nuclear spin relaxation, 10–12, 235
longitudinal relaxation time constant, 11–12, 11f
Nucleic acids. *See also* Nuclear magnetic resonance (NMR), applications
data acquisition and processing, 301–315, 302f
2-D NMR data processing, 312–313
samples in 99% D2O, 307–312, 308f, 309f, 310f, 311f
samples in 90% H2O/10% D2O, 302–307, 304f, 306f
temperature considerations, 313–315, 314f, 315f
ligand binding using 1-D NMR, assessment of, 316–317, 317f
resonance assignment, 317–322, 318f, 319f, 320f
sample handling and preparation for NMR studies
handling DNA, 300
sample preparation for NMR studies, 300–301
solvent choice, 301
sample selection
length/shape/complementarity, 299–300, 300f
sequence design, 296–299, 297f, 298f, 298t
N-unsubstituted benzotriazoles, 207

## O

Observation-level batch processing, 253, 258f
Obstruction factor, diffusion, 139–142, 140f
Octahedral aluminum complex, 214
Octanoic acid, 96
Off-resonance experiment, 67
Off-resonance NMR, 67
Oleuropein, 105
Oligomers, 139
Oligosaccharide conformation, 288–290, 288f, 293–294
Oligosaccharides, structure of, 282
One-bond proton-carbon correlations, 311f
1-D H NMR spectra, 302–303
1-D NMR experiments. *See also* Natural products
DEPT and APT, 269, 270f
1-D NOE and 1-D TOCSY experiments, 267–268, 268f, 269f
1-D NMR spectra, 308
1-D NOE and 1-D TOCSY experiments, 267–268, 268f, 269f
1-D P NMR data acquisition, 305–307
1-D TOCSY experiments, 1-D NOE and, 267–268, 268f, 269f
On-resonance data acquisitions, 67
Organometallic compound, 30f
Organometallic molecules, 18
Orthogonal projections to latent structures (OPLS), 250t
Orthogonal projections to latent structures-discriminant analysis (OPLS-DA), 250t
Overhauser effect, 29

## P

Para-hydrogen-enhanced spectroscopy, 203. *See also* Multi-nuclear magnetic resonance spectroscopy
Paramagnetic ion, 295
Paramagnetic relaxation enhancement (PRE), 289
Paramagnetic shift, 5
Pareto scaling, 249

Partial least squares (PLS), 249, 249f, 250t, 254
Partial least squares-discriminant analysis (PLS-DA), 255, 256, 257f
Passive coupling, 192
Pattern recognition. *See also* Nuclear magnetic resonance (NMR) and complex mixtures
  data analysis in batch processing, 252–254
  multivariate statistical analyses, 249–252, 250t–251t, 251f
Pentamethylcyclopentadiene, 29
Peptide-based ligands, 91–96, 92f, 94f, 95f. *See also* Saturation transfer difference nuclear magnetic resonance (STD NMR)
Perfect rectangular pulses, 134
Phase cycling, 192, 194, 197f
Phenylalanine, 237f
Phlorizin, 293
Phosphatidylcholine (PC), 245f, 246, 246f, 289
Phosphatidylethanolamine (PE), 245f, 246f
Phosphatidylinositol (PI), 245f, 246f
Phosphatidylserine, 246f
Phosphine oxides, 189
Phospholipids, 227, 244
Phosphorothioates, 305, 306f
Phosphorus-carbon-phosphorus (PCP)-type ligands, 24, 25f
Planck's constant, 2
Plastic crystals, 17
Platinum–bipyrimidine complex, 209
Platinum carboxamide complexes, 209
Platinum compounds, 208
Platinum–nitrogen complexes, 213
Platinum NMR spectra, 208
P-NMR spectroscopy, 227
Polarisation transfer, 307
  agent, 244
  methods, 203
Polyacrylamide gels, 107, 108, 109
Polydispersity, 136–139. *See also* Diffusive averaging effects
Polyethylene glycol ether (PEG) bilayers, 108
Polyhydroxy aldehydes, 282
Polymers, 17
Polymetallic compounds, 195
Potential energy (PE), 20
Pre-saturation, 65, 67f
Primary drug screening, 113
Principal components analysis (PCA), 249, 251f, 256f
Proteasome, 29
Protein-based observation of ligand binding, 110–118. *See also* Chemical shift perturbations; Nuclear magnetic resonance (NMR) in ligand binding studies
Protein-carbohydrate interactions. *See also* Carbohydrates
  binding interfaces, identification of, 290–293, 291f, 292f
  oligosaccharide conformation when bound, 293–294
  relative orientation of, 294–295
Protein data bank (PDB), 109
Protein–ligand complex, 71
Protein-observe methods, 110
Protein–peptide interactions, 93
Proton, 5
Protonation, 112
Proton-detected indirect-correlation methods, 301
Proton spectrum, 16f, 20
Pseudo contact shifts (PCS), 294, 295
*Pseudomonais aeruginosa*, 107
PSEUROT analysis, 287, 288
$^{195}$Pt NMR, applications of, 207–214, 208f, 209f, 210t, 211f, 212f, 213f. *See also* Inorganic nuclides, measurements of
Pulse-and-acquire sequence, 230, 231f
Pulse data acquisition, 3
Pulsed field gradients (PFG), 1
Pulsed field gradient spin echo (PFGSE) diffusion spectroscopy, 201–203, 202f
Pulsed gradient spin echo (PGSE), 126
Pulsed gradient stimulated echo (PGSTE), 131, 134, 135f, 157
Pulse sequences, 162
Purines, 296
Push–pull ethylenes, 19
Pyranoses, 282
Pyranose sugars, 286
Pyrimidines, 296
Pythagoras' theorem, 114


## Q


QSTD (semi-quantitative saturation transfer difference), 86
Quadrupolar nuclei, 203
Quadrupolar nuclei, measurement of. *See also* Multi-nuclear magnetic resonance spectroscopy
  $^{61}$Ni NMR as quadrupolar nuclide
    chemical shifts range and coupling constants, 216
    properties of $^{61}$Ni isotope, 216
    use of $^{61}$Ni NMR, 216–218, 217f, 217t, 218f, 218t


## R


Rabbit haemorrhagic disease virus (RHDV), 96
Radio frequency (RF)

inhomogeneity, 131
pulses, 3–5, 4f, 230
RDC. *See* Residual dipolar couplings (RDC)
Redfield-type matrix, 47
Referencing of heteronuclear NMR spectrum, 188
Relaxation delay (RD), 3, 4f, 75, 240
Relaxation measurements, 129
Residual dipolar couplings (RDC), 288, 289. *See also* Nuclear magnetic resonance (NMR) in ligand binding studies
basics of, 105–108, 106f
alignment media, 107–108
bicelles, 108
filamentous Pf1 phage, 107–108
lammelle, 108
lanthanides, 108
polyacrylamide gels, 107
hints and tips for obtaining, 108–109
illustrations of RDC use, 109–110
transferred, 105–110, 106f
Residual HDO resonance, 77f
Resonance assignment, 317–322, 318f, 319f, 320f. *See also* Nucleic acids
Resonance frequency, 189
Resonance tracking, 116
Response matrix, 253
Restricted diffusion, defined, 127, 127f
RGD motif, 91
*Ricinus communis* agglutinin (RCA), 78, 292, 292f
Ring current shift (RCS), 7, 7f
ROESY. *See* Rotating frame Overhauser effect spectroscopy (ROESY)
Root mean-squared displacement (RMSD), 128
Rotating frame Overhauser effect spectroscopy (ROESY), 271
Rotational symmetry, 316

## S

SALMON (modified WaterLOGSY approach), 100
Sample(s)
in 99% D2O. *See* Data acquisition and processing
in 90% H2O/10% D2O. *See* Data acquisition and processing
Sample handling and preparation for NMR studies. *See also* Nucleic acids
handling DNA, 300
sample preparation for NMR studies, 300–301
solvent choice, 301
Sample preparation, 228–230, 229t
concentration references, 230
Saturation
defined, 64–65, 66f
in solvent suppression, 65, 67f
transfer, 78, 101
Saturation in NMR
NOE difference experiment, 65, 67
saturation, defined, 64–65, 66f
in solvent suppression, 65, 67f
Saturation period, defined, 75
Saturation times in STD NMR, 83–87, 84f, 85f, 86t
Saturation transfer difference (STD), 291
amplification factors, 78t, 95f
and protein, 68–71, 69f, 70f
Saturation transfer difference nuclear magnetic resonance (STD NMR)
basic analysis, 76–80, 77f, 78t, 79f
2-D, 88–90, 90f
dissociation event in, 71–73
experimental approach, 73–76, 73f, 74f
and ligand binding competition, 87–88, 88f
ligand-observe screening, 100–101
and peptide-based ligands, 91–96, 92f, 94f, 95f
saturation times in, 83–87, 84f, 85f, 86t
selective excitation pulse for protein saturation in, 80–83, 81f, 82f
STD and protein, 68–71, 69f, 70f
and very large proteins, 96–97
WaterLOGSY (transfer methods involving water), 97–100, 98f
Saturation transfer double-difference (STDD), 96
Saturation transfer exchange spectroscopy, 198–199, 199f
Scalar couplings, 9
Scaling and mean centering, 249
Scores plot, 249, 251f
Secondary chemical shift, 111
Selective excitation pulse for protein saturation in STD NMR, 80–83, 81f, 82f
Selective inversion, 54
relaxation experiments, 28–31, 30f, 31f
Self-diffusion, 126f
categories, 127
coefficients, 141
Serum albumin, 100
Shaped pulses, 80
Shells, hydration, 147. *See also* Hydration
Signal attenuation, 133
Signal detection, 307
Signal dispersion, 282
Signal overlap, reducing, 239–242, 240f, 241f, 242f
Signal-to-noise ratios, 307, 308f
Single bond, 18
Slow exchange, 27–31, 28f, 115. *See also* Exchange (chemical) regimes
selective inversion relaxation experiments, 28–31, 30f, 31f

Sodium azide, 300
Soft pulses, 80
Solid-phase extraction (SPE), 279
Solid-state studies, approaches to, 243–244
Solution-state NMR, 244
Solvation, hydration and, 145–155. *See also* Hydration
Solvent and temperature, 189
Solvent choice, 301
Solvent-optimised double gradient spectroscopy (SOGGY), 231, 232t
Solvent suppression, 160–162, 160f, 161f, 162f, 233. *See also* Nuclear magnetic resonance (NMR)
  saturation in, 65, 67f
Spectral processing, 246–247, 247f
Spectrometer, 186–187
Spherocylinder, 141
Sphingomyelin (SPH), 245f, 246f
Spin diffusion, 68, 104
Spin echo, defined, 236
Spin echo pulse sequence, 235, 236f, 238, 240
Spin-lattice relaxation, 28, 29
  experiments, 53–55. *See also* Density matrix treatment
  time, 199
Spin lock, 74, 76
Spin properties of selected quadrupolar nuclei, 182t–184t
Spin relaxation, 16
Spin-spin coupling, 8–10
  'AX' spin system, 9–10, 9f
Spin–spin relaxation, 12
  time, 138
Spin systems involving coupling to multiple insensitive spins, 194–195, 195f, 196f
Statistical heterospectroscopy (SHY), 251t
Statistical total correlation spectroscopy (STOCSY), 250t
Statistical total correlation spectroscopy editing (E-STOCSY), 250t
Statistical total correlation spectroscopy scaling ($STOCSY^S$), 251t
STD NMR. *See* Saturation transfer difference nuclear magnetic resonance (STD NMR)
Stereochemistry, 207, 283
Steric crowding forces, 32
STOCSY. *See* Statistical total correlation spectroscopy (STOCSY)
STOCSY editing (E-STOCSY), 250t
STOCSY scaling ($STOCSY^S$), 251t
Stokes–Einstein equation, 201
Stokes-Einstein-Sutherland equation, 137, 144–145. *See also* Ion/molecule radius in solution, diffusion and
Stokes radii, 150t, 152
Stop-flow operation of HPLC-NMR, 278f
Sucrose, 99, 148t
Sugars, 282
  furanose, 286
  pyranose, 286
Supervised multivariate statistical analyses, 250t
Surface relaxivity, 128

## T

Tautomerism, 205
Taxol, 265f, 274
Tellurium
  intermolecular exchange of, 25
  transfer, 26f, 27f
Temperature gradients, 201
Tetraalkyl ammonium ions, 152, 155
Tetrahydrofuran (THF), 216, 217f, 218f
Tetramethyl-ammonium chloride, 313
Tetramethyl ammonium ion, 152
Tetramethylsilane (TMS), 187
Thermal jacket, 201
Thermodynamics, 35
Thioflavin T, 105
3-hydroxybutyrate (3HB), 237f, 245f, 255f
Thymine, 296
Time-dependent Eddy current, 158
Time domain, Bloch equations, 39–44
Timescale, NMR, 15–18
Tissue compartmentalisation, 243
T4 lysozyme (T4L), 22, 22f, 23f
Tokuyama's obstruction model for monomeric lysozyme, 138f
Toluene, 25f
Torrey–Bloch equations, 133
Trabectedin, 265
Trans-1,4-dibromodicyanocyclohexane, 21f, 22
Transfer methods involving water, 97–100
Transfer NOE (et-NOE), 294
Transferred nuclear overhauser effect (tr-NOE). *See also* Nuclear magnetic resonance (NMR) in ligand binding studies
  exchange, 102–105, 103f
  and transferred RDC, 101–102
Transferred residual dipolar coupling, 105–110, 106f
Transient NOE, 294. *See also* Nuclear overhauser effect (NOE)
Transmitter pulse measurement, 188f
Transport diffusion, 126
Transverse relaxation time, 12
$T_2$ relaxation editing, 235–238, 236f, 237f, 238f. *See also* Nuclear magnetic resonance (NMR) and complex mixtures
Tricarbonylnickel(0) phosphine complexes, 216, 218t
Trifluoroethanol (TFE), 65, 91, 95f, 96

Triglyceride (TG), 245f
Trimethylamine-N-oxide (TMAO), 245f, 256, 257f
Trimethylphosphate, 215f
ligands, 214
3-(trimethylsilyl) propionic (TSP), 230, 237f
Trim pulses, 75
Twice-refocussed spin echo (TRSE), 157, 158f
2-D correlation spectroscopy, 190
2-D H NMR data acquisition, 303–305
2-D HSQC NMR acquisition, 307, 309–312
2-D NMR experiments. *See also* Natural products
heteronuclear approaches
CIGAR-HMBC experiment, 275–276
coupled HSQC, 274–275, 275f, 276f
excitation-sculptured indirect-detection experiment (EXIDE), 277–278, 277f
HSQC and HMBC, 271, 273–274, 274f
HSQC-NOESY, 276–277
homonuclear approaches
correlation spectroscopy, 270–271
NOESY and ROESY, 271, 273f
2-D NMR spectra, 308–309, 309f
2-D saturation transfer difference nuclear magnetic resonance (STD NMR), 88–90, 90f
Tyrosine, 255f

## U

Ultraviolet-visible (UV-VIS) spectroscopy, 266
Unequal/equal populations, 21–26, 22f, 23f, 24f, 25f, 26f, 27f, 32f. *See also* Exchange (chemical) regimes
Unrestricted diffusion, defined, 128
Unsaturated fatty acid (UFA), 245f
Unsupervised multivariate statistical analyses, 250t

## V

Valine (Val), 245f, 255f
Van der Waals effects, 7
Variable temperature (VT) spectra, simulation of, 198
Vector representation of RF pulse, 4f
Verotoxin B sub-unit (VTB), 294
Very low–density lipoprotein (VLDL), 227, 237f, 239f
Virus-like particles (VLP), 96

## W

Water attenuation, 230–234, 231f, 232t–233t, 234f, 235t. *See also* Nuclear magnetic resonance (NMR) and complex mixtures
WATERGATE solvent signal suppression unit, 161f
WaterLOGSY (transfer methods involving water), 97–100, 98f, 292
Water suppression by gradient-tailored excitation (WATERGATE), 231, 232t
Watson-Crick pairing, 301, 318
Weight-averaged diffusion coefficient, 139
Wilkinson's catalyst, 193

## X

Xenobiotic-induced hepatotoxic lesions, 258f
X-nucleus detection, 244–246, 245f, 246f

## Z

*Z* magnetisations, 30f, 51–52. *See also* Density matrix treatment